Advanced Markov Chain Monte Carlo Methods

Wiley Series in Computational Statistics

Wiley Series in Computational Statistics is comprised of practical guides and cutting edge research books on new developments in computational statistics. It features quality authors with a strong applications focus. The texts in the series provide detailed coverage of statistical concepts, methods and case studies in areas at the interface of statistics, computing, and numerics.

With sound motivation and a wealth of practical examples, the books show in concrete terms how to select and to use appropriate ranges of statistical computing techniques in particular fields of study. Readers are assumed to have a basic understanding of introductory terminology.

The series concentrates on applications of computational methods in statistics to fields of bioinformatics, genomics, epidemiology, business, engineering, finance and applied statistics.

Titles in the Series

Billard and Diday - Symbolic Data Analysis: Conceptual Statistics and Data Mining
Bolstad - Understanding Computational Bayesian Statistics
Borgelt, Steinbrecher and Kruse - Graphical Models, 2e
Dunne - A Statistical Approach to Neutral Networks for Pattern Recognition
Liang, Liu and Carroll - Advanced Marcov Chain Monte Carlo Methods
Ntzoufras - Bayesian Modeling Using WinBUGS

Advanced Markov Chain Monte Carlo Methods

Learning from Past Samples

Faming Liang
Department of Statistics, Texas A&M University

Chuanhai Liu
Department of Statistics, Purdue University

Raymond J. Carroll
Department of Statistics, Texas A&M University

A John Wiley and Sons, Ltd., Publication

This edition first published 2010

Registered office
John Wiley & Sons Ltd, The Atrium, Southern Gate, Chichester, West Sussex, PO19 8SQ, United Kingdom

For details of our global editorial offices, for customer services and for information about how to apply for permission to reuse the copyright material in this book please see our website at www.wiley.com.

Library of Congress Cataloging-in-Publication Data
Liang, F. (Faming), 1970-
Advanced Markov Chain Monte Carlo methods : learning from past samples / Faming Liang, Chuanhai Liu, Raymond J. Carroll.
p. cm.
Includes bibliographical references and index.
ISBN 978-0-470-74826-8 (cloth)
1. Monte Carlo method. 2. Markov processes. I. Liu, Chuanhai, 1959- II. Carroll, Raymond J. III. Title.
QA298.L53 2010
518′.282 – dc22

2010013148

A catalogue record for this book is available from the British Library.

ISBN 978-0-470-74826-8

Typeset in 10/12 cmr10 by Laserwords Private Limited, Chennai, India

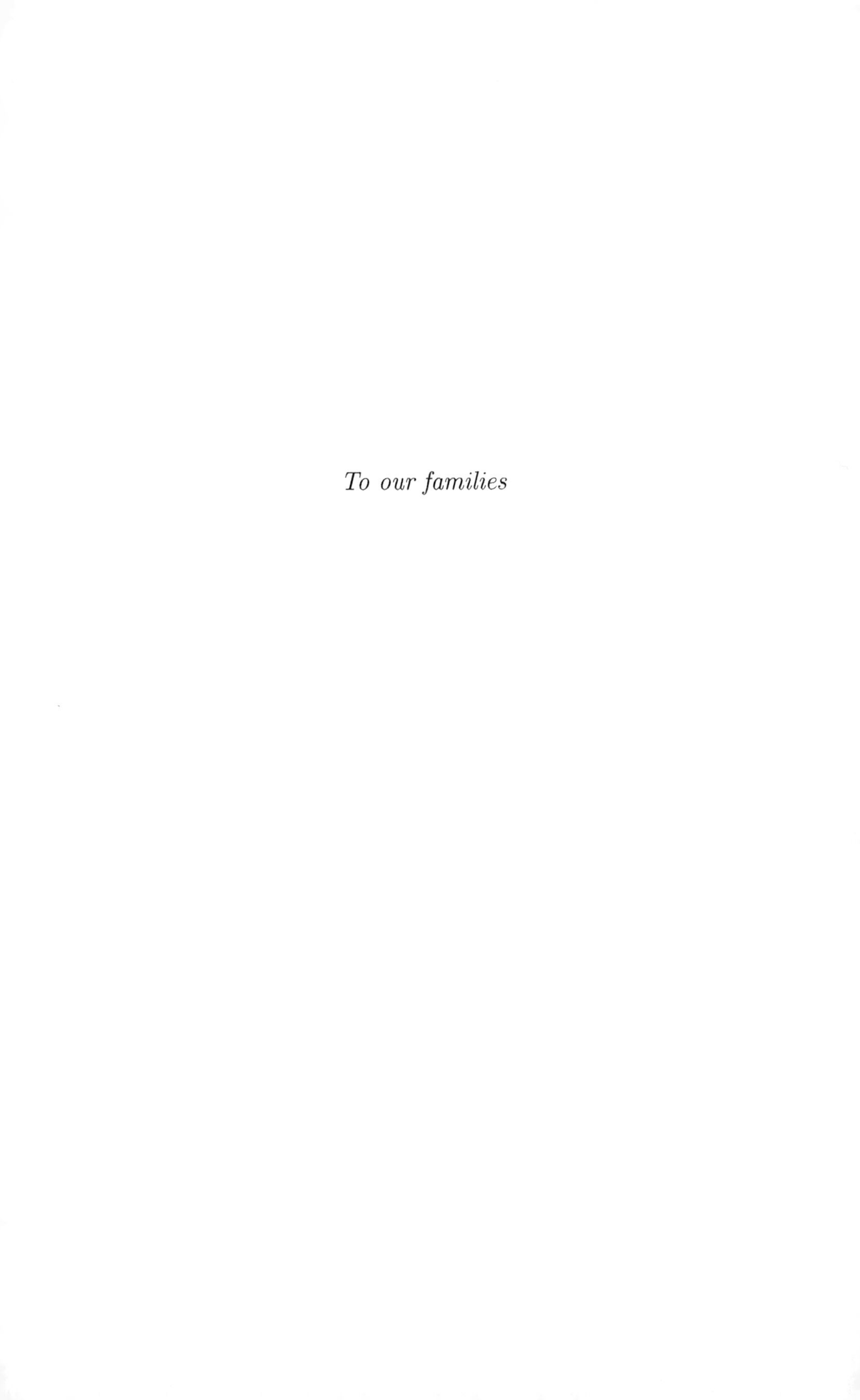

To our families

Contents

Preface

The Markov Chain Monte Carlo (MCMC) method is rooted in the work of physicists such as Metropolis and von Neumann during the period 1945–55 when they employed modern electronic computers for the simulation of some probabilistic problems in atomic bomb designs. After five decades of continual development, it has become the dominant methodology in the solution of many classes of computational problems of central importance to science and technology.

Suppose that one is interested in simulating from a distribution with the density/mass function given by $f(x) \propto \exp\{-H(x)/t\}$, $x \in \mathcal{X}$, where $H(x)$ is called the energy function, and t is called the temperature. The Metropolis algorithm (Metropolis *et al.*, 1953) is perhaps the first sampling algorithm for iterative simulations. It has an extremely simple form. Starting with any point $x_0 \in \mathcal{X}$, it proceeds by iterating between the following two steps, what we call the proposal-and-decision steps:

1. (*Proposal*) Propose a random 'unbiased perturbation' of the current state x_t generated from a symmetric proposal distribution $T(x_t, y)$, i.e., $T(x_t, y) = T(y, x_t)$.

2. (*Decision*) Calculate the energy difference $\Delta H = H(y) - H(x_t)$. Set $x_{t+1} = y$ with probability $\min\{1, \exp(-\Delta H/t)\}$, and set $x_{t+1} = x_t$ with the remaining probability.

This algorithm was later generalized by Hastings (1970) to allow asymmetric proposal distributions to be used in generating the new state y. The generalized algorithm is usually called the Metropolis-Hastings algorithm. A fundamental feature of the Metropolis-Hastings update is its localness, the new state being generated in a neighborhood of the current state. This feature allows one to break a complex task into a series of manageable pieces. On the other hand, however, it tends to suffer from the local-trap problem when the energy function has multiple local minima separated by high energy barriers. In this situation, the Markov chain will be indefinitely trapped in local energy minima. Consequently, the simulation process may fail to sample from the relevant parts of the sample space, and the quantities of interest cannot be estimated with satisfactory accuracies. Many applications of the MCMC method, such as protein folding, combinatorial optimization, and spin-glasses, can be dramatically enhanced by sampling algorithms which allow the process to avoid being trapped in local energy minima.

Developing MCMC sampling algorithms that are immune to the local-trap problem has long been considered as one of the most important topics in MCMC research. During the past two decades, various advanced MCMC algorithms which address this problem have been developed. These include: the Swendsen-Wang algorithm (1987); parallel tempering (Geyer, 1991; Hukushima and Nemoto, 1996), multicanonical Monte Carlo (Berg and Neuhaus, 1991, 1992); simulated tempering (Marinari and Parisi, 1992; Geyer and Thompson, 1995); dynamic weighting (Wong and Liang, 1997; Liu *et al.*, 2001; Liang, 2002b); slice sampler (Higdon, 1998; Edwards and Sokal, 1988); evolutionary Monte Carlo (Liang and Wong, 2000, 2001b), adaptive Metropolis algorithm (Haario *et al.*, 2001); the Wang-Landau algorithm (Wang and Landau, 2001; Liang, 2005b); equi-energy sampler (Kou *et al.*, 2006); sample Metropolis-Hastings algorithm (Lewandowski and Liu, 2008); and stochastic approximation Monte Carlo (Liang *et al.*, 2007; Liang, 2009b), among others.

In addition to the local-trap problem, the Metropolis-Hastings algorithm also suffers from the inability in sampling from distributions with the mass/density function involving intractable integrals. Let $f(x) \propto c(x)\psi(x)$, where $c(x)$ denotes an intractable integral. Clearly, the Metropolis-Hastings algorithm cannot be applied to simulate from $f(x)$, as the acceptance probability would involve the intractable ratio $c(y)/c(x)$, where y denotes the candidate sample. To overcome this difficulty, advanced MCMC algorithms have also been proposed in recent literature. These include the Møller algorithm (Møller *et al.*, 2006), the exchange algorithm (Murray *et al.*, 2006), the double Metropolis-Hastings algorithm (Liang, 2009c; Jin and Liang, 2009), the Monte Carlo dynamically weighted importance sampling algorithm (Liang and Cheon, 2009), and the Monte Carlo Metropolis-Hastings sampler (Liang and Jin, 2010), among others.

One common key idea behind these advanced MCMC algorithms is *learning from past samples*. For example, stochastic approximation Monte Carlo (Liang *et al.*, 2007) modifies its invariant distribution from iteration to iteration based on its past samples in such a way that each region of the sample space can be drawn from with a desired frequency, and thus the local-trap problem can be avoided essentially. The adaptive Metropolis algorithm modifies its proposal distribution from iteration to iteration based on its past samples such that an "optimal" proposal distribution can be achieved dynamically. In the dynamic weighting algorithm, the importance weight carries the information of past samples, which helps the system move across steep energy barriers even in the presence of multiple local energy minima. In parallel tempering and evolutionary Monte Carlo, the state of the Markov chain is extended to a population of independent samples, for which, at each iteration, each sample can be updated based on entire samples of the current population. Hence, parallel tempering and evolutionary Monte Carlo can also be viewed as algorithms for learning from past samples, although they can only learn within a fixed horizon.

Meanwhile, many advanced techniques have been developed in the literature to accelerate the convergence of the Metropolis-Hastings algorithm and the Gibbs sampler; the latter can be viewed as a special form of the Metropolis-Hastings algorithm, with each component of the state vector being updated via a conditional sampling step. Such techniques include: blocking and collapsing (Liu *et al.*, 1994); reparameterization (Gelfand *et al.*, 1995); parameter expansion (Meng and van Dyk, 1999; Liu and Wu, 1999); multiple-try (Liu *et al.*, 2000); and alternating subspace-spanning resampling (Liu, 2003), among others.

The aim of this book is to provide a unified and up-to-date treatment of advanced MCMC algorithms and their variants. According to their main features, we group these advanced MCMC algorithms into several categories. The Gibbs sampler and acceleration methods, the Metropolis-Hastings algorithm and extensions, auxiliary variable-based MCMC algorithms, population-based MCMC algorithms, dynamic weighting, stochastic approximation Monte Carlo, and MCMC algorithms with adaptive proposals are described in Chapters 2–8. Chapter 1 is dedicated to brief descriptions of Bayesian inference, random number generation, and basic MCMC theory. Importance sampling, which represents another important area of Monte Carlo other than MCMC, is not fully addressed in this book. Those interested in this area should refer to Liu (2001) or Robert and Casella (2004).

This book is intended to serve three audiences: researchers specializing in Monte Carlo algorithms; scientists interested in using Monte Carlo methods; and graduate students in statistics, computational biology, engineering, and computer sciences who want to learn Monte Carlo methods. The prerequisites for understanding most of the material presented are minimal: a one-semester course on probability theory (Ross, 1998) and a one-semester course on statistical inference (Rice, 2007), both at undergraduate level. However, it would also be more desirable for readers to have a background in some specific scientific area such as Bayesian computation, artificial intelligence, or computational biology. This book is suitable as a textbook for one-semester courses on Monte Carlo methods, offered at the advanced Master's or Ph.D. level.

Faming Liang, Chuanhai Liu, and Raymond J. Carroll
December, 2009
`www.wiley.com/go/markov`

Acknowledgments

Faming Liang is most grateful to his PhD advisor professor, Wing Hung Wong, for his overwhelming passion for Markov Chain Monte Carlo and scientific problems, and for his constant encouragement. Liang's research was partially supported by grants from the National Science Foundation (DMS-0607755 and CMMI-0926803).

Chuanhai Liu's interest in computational statistics is due largely to the support and encouragement from his M.S. advisor professor, Yaoting Zhang and PhD advisor professor, Donald B. Rubin. In the mid-1980s, Chuanhai Liu learned from Professor Yaoting Zhang the importance of statistical computing. Over a time period of more than ten years from late 1980s, Chuanhai Liu learned from Professor Donald B. Rubin statistical thinking in developing iterative methods, such as EM and Gibbs-type algorithms.

Raymond Carroll's research was partially supported by a grant from the National Cancer Institute (CA57030).

Finally, we wish to thank our families for their constant love, understanding and support. It is to them that we dedicate this book.

F.L., C.L. and R.C.

Publisher's Acknowledgments

The publisher wishes to thank the following for permission to reproduce copyright material:

Table 3.2, Figure 3.4: Reproduced by permission of Licensee BioMed Central Ltd.

Table 4.1, Figure 4.2, Table 4.3: Reproduced by permission of Taylor & Francis.

Figure 5.1, Figure 5.5, Figure 5.6: Reproduced by permission of International Chinese Statistical Association.

Figure 5.2, Figure 5.3, Figure 6.5, Figure 7.1, Figure 7.3, Figure 7.10: Reproduced by permission of American Statistical Association.

Table 5.4, Figure 5.4: Reproduced by permission of American Physical Society.

Figure 5.7, Table 5.5, Figure 5.8: Reproduced by permission of American Institute of Physics.

Figure 5.9, Figure 7.11, Table 7.7, Figure 8.1, Figure 8.2, Figure 8.3: Reproduced by permission of Springer.

Figure 6.1, Figure 6.2, Table 7.2, Figure 7.2, Table 7.4, Figure 7.6, Table 7.5, Figure 7.7, Figure 7.8, Figure 7.9: Reproduced by permission of Elsevier.

Figure 6.3: Reproduced by permission of National Academy of Science.

Figure 7.4, Figure 7.5: Reproduced by permission of National Cancer Institute.

Every effort has been made to trace rights holders, but if any have been inadvertently overlooked the publishers would be pleased to make the necessary arrangements at the first opportunity.

Chapter 1

Bayesian Inference and Markov Chain Monte Carlo

1.1 Bayes

Bayesian inference is a probabilistic inferential method. In the last two decades, it has become more popular than ever due to affordable computing power and recent advances in Markov chain Monte Carlo (MCMC) methods for approximating high dimensional integrals.

Bayesian inference can be traced back to Thomas Bayes (1764), who derived the inverse probability of the success probability θ in a sequence of independent Bernoulli trials, where θ was taken from the uniform distribution on the unit interval $(0, 1)$ but treated as unobserved. For later reference, we describe his experiment using familiar modern terminology as follows.

■ **Example 1.1 The Bernoulli (or Binomial) Model With Known Prior**

Suppose that $\theta \sim \text{Unif}(0, 1)$, the uniform distribution over the unit interval $(0, 1)$, and that $x_1, \ldots, x_n$ is a sample from Bernoulli(θ), which has the sample space $\mathcal{X} = \{0, 1\}$ and probability mass function (pmf)

$$\Pr(X = 1|\theta) = \theta \quad \text{and} \quad \Pr(X = 0|\theta) = 1 - \theta, \tag{1.1}$$

where X denotes the Bernoulli random variable (r.v.) with $X = 1$ for *success* and $X = 0$ for *failure*. Write $N = \sum_{i=1}^{n} x_i$, the observed number of successes in the n Bernoulli trials. Then $N|\theta \sim \text{Binomial}(n, \theta)$, the Binomial distribution with parameters size n and probability of success θ.

Advanced Markov Chain Monte Carlo Methods: Learning from Past Samples
Faming Liang, Chuanhai Liu and Raymond J. Carroll

The inverse probability of θ given $x_1, \ldots, x_n$, known as the posterior distribution, is obtained from Bayes' theorem, or more rigorously in modern probability theory, the definition of conditional distribution, as the Beta distribution Beta$(1+N, 1+n-N)$ with probability density function (pdf)

$$\frac{1}{B(1+N, 1+n-N)} \theta^{(1+N)-1}(1-\theta)^{(1+n-N)-1} \qquad (0 \leq \theta \leq 1), \qquad (1.2)$$

where $B(\cdot, \cdot)$ stands for the Beta function.

1.1.1 Specification of Bayesian Models

Real world problems in statistical inference involve the unknown quantity θ and observed data X. For different views on the philosophical foundations of Bayesian approach, see Savage (1967a, b), Berger (1985), Rubin (1984), and Bernardo and Smith (1994). As far as the mathematical description of a Bayesian model is concerned, Bayesian data analysis amounts to

(i) specifying a sampling model for the observed data X, conditioned on an unknown quantity θ,

$$X \sim f(X|\theta) \qquad (X \in \mathcal{X}, \theta \in \Theta), \qquad (1.3)$$

where $f(X|\theta)$ stands for either pdf or pmf as appropriate, and

(ii) specifying a marginal distribution $\pi(\theta)$ for θ, called the prior distribution or simply the *prior* for short,

$$\theta \sim \pi(\theta) \qquad (\theta \in \Theta). \qquad (1.4)$$

Technically, data analysis for producing inferential results on assertions of interest is reduced to computing integrals with respect to the posterior distribution, or *posterior* for short,

$$\pi(\theta|X) = \frac{\pi(\theta)L(\theta|X)}{\int \pi(\theta)L(\theta|X)d\theta} \qquad (\theta \in \Theta), \qquad (1.5)$$

where $L(\theta|X) \propto f(X|\theta)$ in θ, called the likelihood of θ given X. Our focus in this book is on efficient and accurate approximations to these integrals for scientific inference. Thus, limited discussion of Bayesian inference is necessary.

1.1.2 The Jeffreys Priors and Beyond

By its nature, Bayesian inference is necessarily subjective because specification of the full Bayesian model amounts to practically summarizing available information in terms of precise probabilities. Specification of probability models is unavoidable even for frequentist methods, which requires specification of the sampling model, either parametric or non-parametric, for the observed data X. In addition to the sampling model of the observed data X for developing frequentist procedures concerning the unknown quantity θ, Bayesian inference demands a fully specified prior for θ. This is natural when prior information on θ is available and can be summarized precisely by a probability distribution. For situations where such information is neither available nor easily quantified with a precise probability distribution, especially for high dimensional problems, a commonly used method in practice is the Jeffreys method, which suggests the prior of the form

$$\pi_J(\theta) \propto |I(\theta)|^{1/2} \qquad (\theta \in \Theta), \tag{1.6}$$

where $I(\theta)$ denotes the Fisher information

$$I(\theta) = -\int \frac{\partial^2 \ln f(x|\theta)}{\partial\theta(\partial\theta)'} f(x|\theta)dx.$$

The Jeffreys priors have the appealing property that they are invariant under reparameterization. A theoretical adjustment in terms of frequency properties in the context of large samples can be found in Welch and Peers (1963). Note that prior distributions do not need to be proper as long as the posteriors are proper and produce sensible inferential results. The following Gaussian example shows that the Jeffreys prior is sensible for single parameters.

■ Example 1.2 The Gaussian $N(\mu, 1)$ Model

Suppose that a sample is considered to have taken from the Gaussian population $N(\mu, 1)$ with unit variance and unknown mean μ to be inferred. The Fisher information is obtained as

$$I(\mu) = \int_{-\infty}^{\infty} \phi(x-\mu)dx = 1,$$

where $\phi(x-\mu) = (2\pi)^{-1/2}\exp\{-\frac{1}{2}(x-\mu)^2\}$ is the pdf of $N(\mu, 1)$. It follows that the Jeffreys prior for θ is the flat prior

$$\pi_J(\mu) \propto 1 \qquad (-\infty < \mu < \infty), \tag{1.7}$$

resulting in the corresponding posterior distribution of θ given X

$$\pi_J(\mu|X) = N(X, 1). \tag{1.8}$$

Care must be taken when using the Jeffreys rule. For example, it is easy to show that applying the Jeffreys rule to the Gaussian model $N(\mu, \sigma^2)$ with both mean μ and variance σ^2 unknown leads to the prior

$$\pi_J(\mu, \sigma^2) \propto \frac{1}{\sigma^3} \qquad (-\infty < \mu < \infty; \sigma^2 > 0).$$

However, this is not the commonly used prior that has better frequency properties (for inference about μ or σ) and is given by

$$\pi(\mu, \sigma^2) \propto \frac{1}{\sigma^2} \qquad (-\infty < \mu < \infty; \sigma^2 > 0),$$

that is, μ and σ^2 are independent and the distributions for both μ and $\ln \sigma^2$ are flat. For high dimensional problems with small samples, the Jeffreys rule often becomes even less appealing. There are also different perspectives, provided by the extensive work on reference priors by José Bernardo and James Berger (see, e.g., Bernardo, 1979; Berger, 1985). For more discussion of prior specifications, see Kass and Wasserman (1996).

For practical purposes, we refer to Box and Tiao (1973) and Gelman *et al.* (2004) for discussion on specification of prior distributions. The general guidance for specification of priors when no prior information is available, as is typical in Bayesian analysis, is to find priors that lead to posteriors having good frequency properties (see, e.g., Rubin, 1984; Dawid, 1985). Materials on probabilistic inference without using difficult-to-specify priors are available but beyond the scope of Bayesian inference and therefore will not be discussed in this book. Readers interested in this fascinating area are referred to Fisher (1973), Dempster (2008), and Martin *et al.* (2009). We note that MCMC methods can be applied there as well.

1.2 Bayes Output

Bayesian analysis for scientific inference does not end with posterior derivation and computation. It is thus critical for posterior distributions to have clear interpretation. For the sake of clarity, probability used in this book has a long-run frequency interpretation in repeated experiments. Thus, standard probability theory, such as conditioning and marginalization, can be applied. Interpretation also suggests how to report Bayesian output as our assessment of assertions of interest on quantities in the specified model. In the following two subsections, we discuss two types of commonly used Bayes output, credible intervals for estimation and Bayes factors for hypothesis testing.

1.2.1 Credible Intervals and Regions

Credible intervals are simply posterior probability intervals. They are used for purposes similar to those of confidence intervals in frequentist statistics and thereby are also known as Bayesian confidence intervals. For example, the 95% left-sided Bayesian credible interval for the parameter μ in the Gaussian Example 1.2 is $[-\infty, X + 1.64]$, meaning that the posterior probability that μ lies in the interval from $-\infty$ to $X + 1.64$ is 0.95. Similar to frequentist construction of two-sided intervals, for given $\alpha \in (0, 1)$, a $100(1 - \alpha)\%$ two-sided

Bayesian credible interval for a single parameter θ with equal posterior tail probabilities is defined as

$$[\theta_{\alpha/2}, \theta_{1-\alpha/2}] \tag{1.9}$$

where the two end points are the $\alpha/2$ and $1-\alpha/2$ quantiles of the (marginal) posterior distribution of θ. For the the Gaussian Example 1.2, the two-sided 95% Bayesian credible interval is $[X-1.96, X+1.96]$.

In dealing simultaneously with more than one unknown quantity, the term credible region is used in place of credible interval. For a more general term, we refer to credible intervals and regions as *credible sets*. Constructing credible sets is somewhat subjective and usually depends on the problems of interest. A common way is to choose the region with highest posterior density (h.p.d.). The $100(1-\alpha)\%$ h.p.d. region is given by

$$R_{1-\alpha}^{(\pi)} = \{\theta : \pi(\theta|X) \geq \pi(\theta_{1-\alpha}|X)\} \tag{1.10}$$

for some $\theta_{1-\alpha}$ satisfying

$$\Pr\left(\theta \in R_{1-\alpha}^{(\pi)}|X\right) = 1-\alpha.$$

For the the Gaussian Example 1.2, the 95% h.p.d. interval is $[X-1.96, X+1.96]$, the same as the two-sided 95% Bayesian credible interval because the posterior of μ is unimodal and symmetric. We note that the concept of h.p.d. can also be used for functions of θ such as components of θ in high dimensional situations.

For a given probability content $(1-\alpha)$, the h.p.d. region has the smallest volume in the space of θ. This is attractive but depends on the functional form of unknown quantities, such as θ and θ^2. An alternative credible set is obtained by replacing the posterior density $\pi(\theta|X)$ in (1.10) with the likelihood $L(\theta|X)$:

$$R_{1-\alpha}^{(L)} = \{\theta : L(\theta|X) \geq L(\theta_{1-\alpha}|X)\} \tag{1.11}$$

for some $\theta_{1-\alpha}$ satisfying

$$\Pr\left(\theta \in R_{1-\alpha}^{(L)}|X\right) = 1-\alpha.$$

The likelihood based credible region does not depend on transformation of θ. This is appealing, in particular when no prior information is available on θ, that is, when the specified prior works merely as a working prior leading to inference having good frequency properties.

1.2.2 Hypothesis Testing: Bayes Factors

While the use of credible intervals is a Bayesian alternative to frequentist confidence intervals, the use of Bayes factors has been a Bayesian alternative to classical hypothesis testing. Bayes factors have also been used to develop Bayesian methods for model comparison and selection. Here we review the basics of Bayes factors. For more discussion on Bayes factors, including its history, applications, and difficulties, see Kass and Raftery (1995), Gelman *et al.* (2004), and references therein.

The concept of Bayes factors is introduced in the situation with a common observed data X and two competing hypotheses denoted by H_1 and H_2. A full Bayesian analysis requires

(i) specifying a prior distribution on H_1 and H_2, denoted by, $\Pr(H_1)$ and $\Pr(H_2)$, and

(ii) for each $k = 1$ and 2, specifying the likelihood $L_k(\theta_k|X) = f_k(X|\theta_k)$ and prior $\pi(\theta_k|H_k)$ for θ_k, conditioned on the truth of H_k, where θ_k is the parameter under H_k.

Integrating out θ_k yields

$$\Pr(X|H_k) = \int f_k(X|H_k)\pi(\theta_k|H_k)d\theta_k \tag{1.12}$$

for $k = 1$ and 2. The Bayes factor is the posterior odds of one hypothesis when the prior probabilities of the two hypotheses are equal. More precisely, the Bayes factor in favor of H_1 over H_2 is defined as

$$B_{12} = \frac{\Pr(X|H_1)}{\Pr(X|H_2)}. \tag{1.13}$$

The use of Bayes factors for hypothesis testing is similar to the likelihood ratio test, but instead of maximizing the likelihood, Bayesians in favor of Bayes factors average it over the parameters. According to the definition of Bayes factors, proper priors are often required. Thus, care must be taken in specification of priors so that inferential results are meaningful. In addition, the use of Bayes factors renders lack of probabilistic feature of Bayesian inference. In other words, it is consistent with the likelihood principle, but lacks of a metric or a probability scale to measure the strength of evidence. For a summary of evidence provided by data in favor of H_1 over H_2, Jeffreys (1961) (see also Kass and Raftery (1995)) proposed to interpret the Bayes factor as shown in Table 1.1.

The use of Bayes factor is illustrated by the following binomial example.

Table 1.1 Interpretation of Bayes factors.

$\log_{10}(B_{12})$	B_{12}	evidence against H_2
0 to 1/2	1 to 3.2	Barely worth mentioning
1.2 to 1	3.2 to 10	Substantial
1 to 2	10 to 100	Strong
>2	>100	Decisive

■ **Example 1.3 The Binomial Model (continued with a numerical example)**

Suppose we take a sample of $n = 100$ from Bernoulli(θ) with unknown θ, and observe $N = 63$ successes and $n - N = 37$ failures. Suppose that two competing hypotheses are

$$H_1 : \theta = 1/2 \quad \text{and} \quad H_2 : \theta \neq 1/2. \tag{1.14}$$

Under H_1, the likelihood is calculated according to the binomial distribution:

$$\Pr(N|H_1) = \binom{n}{N}\left(\frac{1}{2}\right)^N\left(\frac{1}{2}\right)^{n-N}$$

Under H_2, instead of the uniform over the unit interval we consider the Jeffreys prior

$$\pi(\theta) = \frac{\Gamma(1/2+1/2)}{\Gamma(1/2)\Gamma(1/2)}\theta^{1/2-1}(1-\theta)^{1/2-1} = \frac{1}{\pi}\theta^{1/2-1}(1-\theta)^{1/2-1}$$

the proper Beta distribution with shape parameters 1/2 and 1/2. Hence, we have

$$\Pr(N|H_2) = \frac{1}{\pi}\binom{n}{N}\text{Beta}(N+1/2, n-N+1/2).$$

The Bayes factor $\log_{10}(B_{12})$ is then -0.4, which is 'barely worth mentioning' even if it points very slightly towards H_2.

It has been recognized that Bayes factor can be sensitive to the prior, which is related to what is known as Lindley's paradox (see Shafer (1982)).

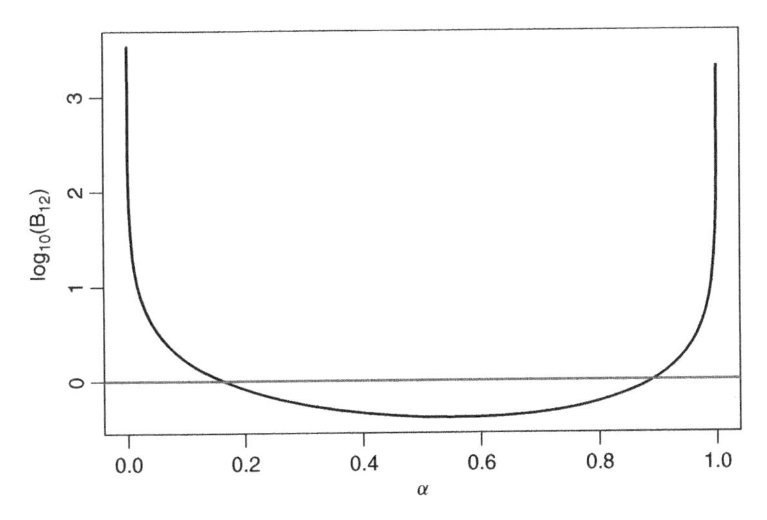

Figure 1.1 Bayes factors in the binomial example with $n = 100$, $N = 63$, and priors Beta($\alpha, 1-\alpha$) for $0 \leq \alpha \leq 1$.

This is shown in Figure 1.1 for a class of Beta priors Beta($\alpha, 1 - \alpha$) for $0 \leq \alpha \leq 1$. The Bayes factor is infinity at the two extreme priors corresponding to $\alpha = 0$ and $\alpha = 1$. It can be shown that this class of priors is necessary in the context of imprecise Bayes for producing inferential results that have desired frequency properties. This supports the idea that care must be taken in interpreting Bayesian factors in scientific inference.

Bayesian factors are not the same as a classical likelihood ratio test. A frequentist hypothesis test of H_1 considered as a *null* hypothesis would have produced a more dramatic result, saying that H_1 could be rejected at the 1% significance level, since the probability of getting 63 or more successes from a sample of 100 if $\theta = 1/2$ is 0.0060, and as a normal approximation based two-tailed test of getting a figure as extreme as or more extreme than 63 is 0.0093. Note that 63 is more than two standard deviations away from 50, the expected count under H_1.

1.3 Monte Carlo Integration

1.3.1 The Problem

Let ν be a probability measure over the Borel σ-field $\mathcal{X}$ on the sample space $\mathcal{X} \subseteq \mathbb{R}^d$, where $\mathbb{R}^d$ denotes the d-dimensional Euclidian space. A commonly encountered challenging problem is to evaluate integrals of the form

$$E_\nu[h(X)] = \int_{\mathcal{X}} h(x)\nu(dx) \tag{1.15}$$

where $h(x)$ is a measurable function. Suppose that ν has a pdf $f(x)$. Then (1.15) can be written as

$$E_f[h(X)] = \int_{\mathcal{X}} h(x)f(x)dx \tag{1.16}$$

For example, for evaluating the probability $\Pr(X \in \mathcal{S})$ for $S \subset \mathcal{X}$, $h(x)$ is the indicator function $h(x) = I_{x \in \mathcal{S}}$ with $h(x) = 1$ if $x \in \mathcal{S}$ and $h(x) = 0$ otherwise, and for computing the marginal distribution of $f_Y(y)$ from the joint distribution $f_{X,Y}(x, y)$, the representation in the form of (1.16) is $E_{f_X}\left[f_{Y|X}(y|x)\right]$, where $f_X(x)$ is the marginal pdf of X and $f_{Y|X}(y|x)$ is the conditional pdf of Y given X.

When the problem appears to be intractable analytically, the tool box of numerical integration methods is the next possible alternative, see, for example, Press *et al.* (1992) and references therein. For high dimensional problems, Monte Carlo methods have proved to be popular due to their simplicity and accuracy given limited computing power.

1.3.2 Monte Carlo Approximation

Suppose that it is easy to simulate a sample of size n, denoted by $X_1, \ldots, X_n$, from $f(x)$, the pdf involved in (1.16). Then the sample mean of $h(X)$,

$$\bar{h}_n = \frac{1}{n}\sum_{i=1}^{n} h(X_i), \tag{1.17}$$

can be used to approximate (1.16) because $\bar{h}_n$ converges to (1.16) almost surely by the Strong Law of Large Numbers. When $h(X)$ has a finite variance, the error of this approximation can be characterized by the central limit theorem, that is,

$$\frac{\bar{h}_n - E_f[h(X)]}{\sqrt{n\text{Var}(h(X))}} \sim N(0,1).$$

The variance term $\text{Var}(h(X))$ can be approximated in the same fashion, namely, by the sample variance

$$\frac{1}{n-1}\sum_{i=1}^{n}(h(X_i) - \bar{h}_n)^2.$$

This method of approximating integrals by simulated samples is known as the Monte Carlo method (Metropolis and Ulam, 1949).

1.3.3 Monte Carlo via Importance Sampling

When it is hard to draw samples from $f(x)$ directly, one can resort to importance sampling, which is developed based on the following identity:

$$E_f[h(X)] = \int_{\mathcal{X}} h(x)f(x)dx = \int_{\mathcal{X}} h(x)\frac{f(x)}{g(x)}g(x)dx = E_g[h(X)f(X)/g(X)],$$

where $g(x)$ is a pdf over $\mathcal{X}$ and is positive for every x at which $f(x)$ is positive. This identity suggests that samples from density functions different from $f(x)$ can also be used to approximate (1.16). The standard Monte Carlo theory in Section 1.3.2 applies here because of

$$E_f[h(X)] = E_g[h(X)f(X)/g(X)] = E_g[\tilde{h}(X)]$$

where $\tilde{h}(x) = h(x)\frac{f(x)}{g(x)}g(x)$. The estimator of $E_f[h(X)]$ now becomes

$$\bar{h} = \frac{1}{m}\sum_{i=1}^{m}\frac{f(x_i)}{g(x_i)}h(x_i), \tag{1.18}$$

where $x_1, \ldots, x_n$ are iid samples drawing from $g(x)$. Compared to (1.17), for each $i = 1, \ldots, m$, x_i enters with a weight $w_i = f(x_i)/g(x_i)$. For this reason, this method is called the *importance sampling* method. Most important for this method is to choose $g(x)$ for both simplicity in generating Monte Carlo samples and accuracy in estimating $E_f[h(X)]$ by controlling the associated Monte Carlo errors.

For Monte Carlo accuracy, a natural way is to choose $g(x)$ to minimize the variance of $\tilde{h}(X)$ with $X \sim g(x)$. Theoretical results on optimal $g(x)$ are also available. The following result is due to Rubinstein (1981); see also Robert and Casella (2004).

Theorem 1.3.1 *The choice of g that minimizes the variance of the estimator of $E_f[h(X)]$ in (1.18) is*

$$g^*(x) = \frac{|h(x)|f(x)}{\int_{\mathcal{X}} |h(y)|f(y)dy}.$$

The proof of Theorem 1.3.1 is left as an exercise. As always, theoretical results provide helpful guidance. In practice, balancing simplicity and optimality is more complex because human efforts and computer CPU time for creating samples from $g(x)$ are perhaps the major factors to be considered. Also, it is not atypical to evaluate integrals of multiple functions of $h(x)$, for example, in Bayesian statistics, with a common Monte Carlo sample.

1.4 Random Variable Generation

Monte Carlo methods rely on sampling from probability distributions. Generating a sample of iid draws on computer from the simplest continuous uniform Unif$(0, 1)$ is fundamentally important because all sampling methods depend on uniform random number generators. For example, for every continuous univariate distribution $f(x)$ with cdf $F(x)$ the so-called inverse-cdf method is given as follows.

Algorithm 1.1 *(Continuous Inverse-cdf)*

1. *Generate a uniform random variable U.*

2. *Compute and return $X = F^{-1}(U)$.*

where $F^{-1}(.)$ represents the inverse function of the cdf $F(.)$, provides an algorithm to create samples from $F(x)$. Similarly, for every discrete univariate distribution $p(x)$ with cdf $F(x)$ the inverse-cdf method becomes

Algorithm 1.2 *(Discrete Inverse-cdf)*

1. *Generate a uniform random variable U.*

2. *Find X such that $F(X-1) < U \leq F(X)$.*

3. *Return X.*

and provides an algorithm to create samples from $F(x)$. However, this algorithm is in general computationally expensive. More efficient methods are discussed in Sections 1.4.1 and 1.4.2, where a good and efficient uniform random generator is assumed to be available.

Unfortunately, computers are deterministic in nature and cannot be programmed to produce pure random numbers. Pseudo-random number generators are commonly used in Monte Carlo simulation. Pseudo-random number generators are algorithms that can automatically create long runs with good random properties but eventually the sequence repeats. We refer to Devroye (1986), Robert and Casella (2004), and Matsumoto and Nishimura (1998) for discussion on pseudo-random number generators, among which the Mersenne Twister of Matsumoto and Nishimura (1998) has been used as the default pseudo-random number generator in many softwares. In what follows, we give a brief review of the methods that are often used for sampling from distributions for which the inverse-cdf method does not work, including the transformation methods, acceptance rejection methods, ratio-of-uniform methods, adaptive direction sampling (Gilks, 1992), and perfect sampling (Propp and Wilson, 1996).

1.4.1 Direct or Transformation Methods

Transformation methods are those based on transformations of random variables. Algorithms 1.1 and 1.2 provide such examples. Except for a few cases, including the exponential and Bernoulli distributions, Algorithms 1.1 and 1.2 are often inefficient. Better methods based on transformations can be obtained, depending on the target distribution $f(x)$. Table 1.2 provides a few examples that are commonly used in practice.

1.4.2 Acceptance-Rejection Methods

Acceptance-Rejection (AR), or Accept-Reject, methods are very useful for random number generation, in particular when direct methods do not exist

Table 1.2 Examples of transformation methods for random number generation.

method	transformation	distribution
Exponential	$X = -\ln(U)$	$X \sim \text{Expo}(1)$
Cauchy	$X = \tan(\pi U - \pi/2)$	$X \sim \text{Cauchy}(0, 1)$
Box-Muller	$X_1 = \sqrt{-2\ln(U_1)}\cos(2\pi U_2)$	$X_i \overset{iid}{\sim} N(0, 1)$
(1958)	$X_2 = \sqrt{-2\ln(U_1)}\sin(2\pi U_2)$	
Beta	$X_i \overset{ind}{\sim} \text{Gamma}(\alpha_i)$, $i = 1, 2$	$\frac{X_1}{X_1+X_2} \sim \text{Beta}(\alpha_1, \alpha_2)$

where $U \sim \text{Unif}(0, 1)$ and $U_i \overset{ind}{\sim} \text{Unif}(0, 1)$ for $i = 1, 2$.

or are computationally inefficient. We discuss the AR methods via a geometric argument.

Suppose that the distribution to sample from is of d-dimension with the sample space $\mathcal{X} \subseteq \mathbb{R}^d$. According to the definition of pdf (or pmf), the region under the pdf curve/surface

$$\mathbb{C}_f = \{(x, u) : 0 \le u \le f(x)\} \subset \mathbb{R}^{d+1} \tag{1.19}$$

has unit volume. Thus, if (X, U) is uniform in the region $\mathbb{C}_f$ then $X \sim f(x)$. Note that $X \sim f(x)$ still holds when $f(x)$ in (1.19) is multiplied by an arbitrary positive constant, that is,

$$\mathbb{C}_h = \{(x, y) : 0 \le u \le h(x)\} \subset \mathbb{R}^{d+1}, \tag{1.20}$$

where $h(x) \propto f(x)$, because rescaling on U will not affect the marginal distribution of X.

This fact suggests a possible way of generating X by simulating points distributed uniformly over $\mathbb{C}_f$ or $\mathbb{C}_h$. When it is difficult to sample from $\mathbb{C}_h$ directly, samples from $\mathbb{C}_h$ can be obtained indirectly by (i) generating points uniformly over an enlarged and easy-to-sample region $\mathbb{D} \supseteq \mathbb{C}_h$ and (ii) collecting those falling inside of $\mathbb{C}_h$. Such an enlarged region $\mathbb{D}$ can be constructed by an easy-to-sample distribution with pdf $g(x)$ with the restriction that $f(x)/g(x)$ is bounded from above by some finite constant M so that $\mathbb{C}_h$ can be enclosed in the region

$$\mathbb{C}_g = \{(x, u) : 0 \le u \le g(x)\} \subset \mathbb{R}^{d+1}, \tag{1.21}$$

for some $h(x) \propto f(x)$. The distribution $g(x)$ is called the *envelope* or instrumental distribution, while $f(x)$ the *target*.

To summarize, we have the following AR algorithm to generate random numbers from $f(x)$ using an envelope distribution $g(x)$, where $\sup_x h(x)/g(x) \le M < \infty$.

Algorithm 1.3 *(Acceptance-Rejection)*

Repeat the following two steps until a value is returned in Step 2:

1. *Generate X from $g(x)$ and U from $Unif(0, 1)$.*

2. *If $U \le \frac{h(X)}{Mg(X)}$, return X (as a random deviate from $f(x)$).*

The acceptance rate is the ratio of the volume of the target region to the volume of the proposal region, that is,

$$r = \frac{1}{M} \frac{\int h(x)dx}{\int g(x)dx}.$$

In the case when both $h(x)$ and $g(x)$ are normalized, the acceptance ratio is $1/M$, suggesting the use of $M = \sup_x h(x)/g(x)$ when it is simple to compute.

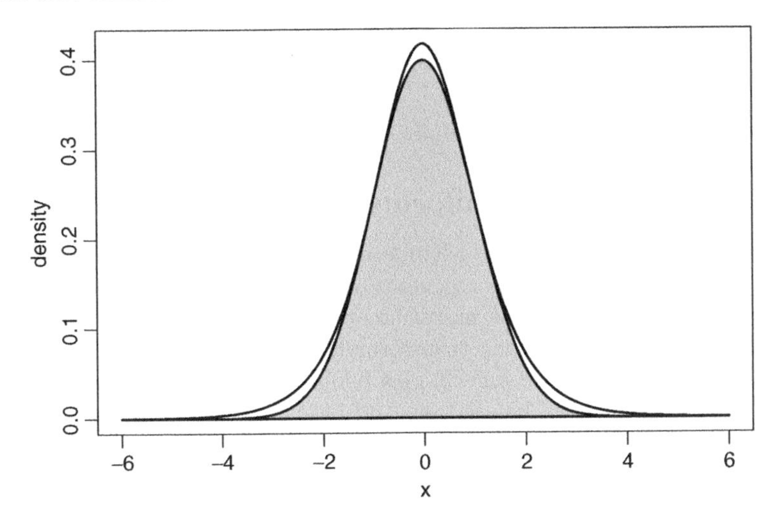

Figure 1.2 The target and proposal distributions $\phi(x)$ and $Mh_s(x)$ with $s = .648$ and $M = 1.081$ in Example 1.4.

■ **Example 1.4 The Standard Normal $N(0, 1)$**

This example illustrates the AR algorithm of using the logistic distribution with density

$$h_s(x) = \frac{e^{-x/s}}{s(1 + e^{-x/s})^2} = \frac{1}{4s\cosh^2(-x/s)} \qquad (-\infty < x < \infty)$$

where $s = .648$, as the proposal distribution to generate samples from the standard normal $N(0, 1)$. Note that $N(0, 1)$ has the pdf $\phi(x) = (2\pi)^{-1/2}e^{-x^2/2}$ $(-\infty < x < \infty)$. The maximum value

$$M = \max_{-\infty<x<\infty} \frac{\phi(x)}{h_s(x)},$$

obtained using the Newtow-Raphson method, is about 1.0808. Take $M = 1.081$. The target density and $Mh_s(x)$ are shown in Figure 1.2. The acceptance rate of the corresponding AR algorithm is given by $1/M = .925$.

When it is expensive to evaluate $h(x)$, an easy-to-compute function $s(x)$ $(0 \le s(x) \le h(x))$, called the squeeze function, can be used to reduce the frequency of computing $h(x)$. More specifically, the modified AR method is as follows.

Algorithm 1.4 *(Acceptance-Rejection With a Squeezer)*
Repeat the following two steps until a value is returned in Step 2:

1. *Generate X from $g(x)$ and U from Unif$(0, 1)$.*

2. *If* $U \leq \frac{s(X)}{Mg(X)}$ *or* $\frac{s(X)}{Mg(X)} < U \leq \frac{h(X)}{Mg(X)}$, *return* X *(as a random deviate from* $f(x)$*).*

Thus, in the case $U \leq \frac{s(X)}{Mg(X)}$, the algorithm does not evaluate $h(x)$.

1.4.3 The Ratio-of-Uniforms Method and Beyond

The ratio-of-uniforms method of Kinderman and Monahan (1977) is a popular method for random number generations of many standard distributions, including the Gamma, normal, and student-t. It can be considered as obtained from the rejection method via transformation subject to some kind of simplicity. Here we discuss the general idea behind the ratio-of-uniforms method and derive the method of Kinderman and Monahan (1977) and its extension proposed by Wakefield, Gelfand, and Smith (1991) as special cases.

The general idea of the ratio-of-uniforms method is to find a pair of differentiable transformations

$$U = u(Y) \quad \text{and} \quad X = x(Z, Y)$$

with $U = u(Y)$ strictly increasing to propagate the inequality in (1.20) and with constant Jacobian so that (Y, Z) is also uniform over the corresponding image of $\mathbb{C}_h$:

$$\mathbb{C}_h^{(Y,Z)} = \{(y, z) : u^{-1}(0) \leq y = u^{-1}(u) \leq u^{-1}(h(x(z, y)))\} \subset \mathbb{R}^{d+1}, \quad (1.22)$$

where $u^{-1}(.)$ denotes the inverse of $u(.)$. This leads to the following generic rejection algorithm to sample from $f(x)$ with a chosen easy-to-sample region $\mathbb{D}$ enclosing $\mathbb{C}_h^{(Y,Z)}$.

Algorithm 1.5 *(The Generic Acceptance-Rejection of Uniforms Algorithm)*

Repeat the following two steps until a value is returned in Step 2:

Step 1. Generate (Y, Z)*, uniform deviates over* $\mathbb{D} \supseteq \mathbb{C}_h^{(Y,Z)}$.

Step 2. If $(Y, Z) \in \mathbb{C}_h^{(Y,Z)}$*, return* $X = x(Y, Z)$ *as the desired deviate.*

This algorithm has the acceptance rate

$$r = \frac{\int_{\mathbb{C}_h^{(Y,Z)}} dydz}{\int_{\mathbb{D}} dydz} = \frac{\int_{\mathcal{X}} h(x)dx}{J_{x,u}(z, y)| \int_{\mathbb{D}} dydz},$$

where

$$J_{x,u}(z, y) = \begin{vmatrix} \frac{\partial x}{\partial z} & \frac{\partial x}{\partial y} \\ 0 & \frac{\partial u}{\partial y} \end{vmatrix} = u'(y) \left| \frac{\partial x}{\partial z} \right|$$

denotes the Jacobian of the transformations.

It is state-of-the-art to choose the pair of transformation and construct $\mathbb{D}$. Thus, simplicity plays an important role. Let x_i be a function of z_i and y, for example, the Jacobian of the transformation has a simple form

$$J_{x,u}(z,y) = u'(y)\left|\frac{\partial x}{\partial z}\right| = u'(y)\prod_{i=1}^{d}\left|\frac{\partial x_i(z_i,y)}{\partial z_i}\right| = \text{const.}$$

Hence, $x_i(z,y)$ is linear in z_i

$$x_i(z,y) = a_i(y)z_i + b_i(y) \qquad (i = 1,\ldots,d),$$

with the restriction $\prod_{i=1}^{d} a_i(y) = 1/u'(y)$. For example, the result of Wakefield *et al.* (1991) is obtained by letting $a_i(y) = [u'(y)]^{-1/d}$ and $b_i(y) = 0$ for $i = 1,\ldots,d$. The uniform region is

$$\{(y,z) : u^{-1}(0) \le y \le u^{-1}(h(z/[u'(y)]^{1/d}))\}.$$

The method of Wakefield *et al.* (1991) for generating multivariate distributions is obtained by taking the power transformation on Y: $u(y) = y^{r+1}/(r+1)$, $r \ge 0$. This results in the target region

$$\mathbb{C}_h^{(r)} = \left\{(y,z) : 0 \le y \le \left[(r+1)h\left(\frac{z}{y^{r/d}}\right)\right]^{1/(r+1)}\right\},$$

or equivalently

$$\mathbb{C}_h^{(r)} = \left\{(y,z) : 0 \le y \le \left[h\left(\frac{z}{y^{r/d}}\right)\right]^{1/(r+1)}\right\}, \tag{1.23}$$

because $h(x)$ is required to be known up to a proportionality constant. Wakefield *et al.* (1991) consider $(d+1)$-boxes $\mathbb{D}$ to bound $\mathbb{C}_h^{(r)}$, provided that $\sup_x h(x)$ and $\sup_x |x_i|[h(x)]^{r/(dr+d)}$, $i = 1,\ldots,d$, are all finite. Pérez *et al.* (2008) proposed to use ellipses as $\mathbb{D}$ in place of $(d+1)$-boxes.

In the univariate case, the algorithm based on (1.23) with the choice of $r = 1$ reduces to the famous ratio-of-uniforms method of Kinderman and Monahan (1977):

Algorithm 1.6 *(Ratio-of-Uniforms Algorithm of Kinderman and Monahan, 1977)*

Repeat the following two steps until a value is returned in Step 2:

1. *Generate* (y,z) *uniformly over* $\mathbb{D} \supseteq \mathbb{C}_h^{(1)}$.

2. *If* $(Y,Z) \in \mathcal{C}_h^{(1)}$ *return* $X = Z/Y$ *as the desired deviate.*

The uniform region is

$$\mathbb{C}_h^{(1)} = \left\{(y,z) : 0 \le y \le \left[h\left(\frac{z}{y}\right)\right]^{1/2}\right\}. \tag{1.24}$$

When $\sup_x h(x)$ and $\sup_x |x|[h(x)]^{1/2}$ are finite, the easy-to-sample bounding region $\mathbb{D}$ can be set to the tightest rectangle enclosing $\mathbb{C}_h^{(1)}$. For more efficient algorithms, more refined enclosing regions such as polygons could be used. This is potentially useful in simulating truncated distributions. Here we provide an illustrative example, which shows that efficient ratio-of-uniforms methods can be derived by considering transformations of the random variable X, including the relocation technique proposed by Wakefield *et al.* (1991), before applying the ratio-of-uniforms.

Example 1.5 Gamma(α) With $\alpha < 1$

Consider the ratio-of-uniforms method for generating a variate X from the Gamma density $f_\alpha(x) = \frac{1}{\Gamma(\alpha)} x^{\alpha-1} e^{-x}$, $x > 0$. To generate Gamma variates with density $f_\alpha(x)$, Kinderman and Monahan (1977) and Cheng and Feast (1979) used the ratio-of-uniforms method by setting $h(x) = x^{\alpha-1}e^{-x}$. The method is valid only for $\alpha > 1$ and gives a uniform region whose shape changes awkwardly when α is near 1. Cheng and Feast (1980) got around this problem by using the transformation $x = y^n$, $y > 0$, and setting $h(y) = y^{n\alpha-1}e^{-y^n}$. This effectively extends the range of α from 1 down to $1/n$. For more discussion on Gamma random number generators, see Tanizaki (2008) and the references therein.

We now use the transformation $X = e^{T/\alpha}$ (or $T = \alpha \ln X$), $-\infty < t < \infty$, for $\alpha < 1$. The random variable T has density $f(t) = e^{t-e^{t/\alpha}}/(\alpha\Gamma(\alpha))$. Let $h(t) = e^{t-e^{t/\alpha}}$. Then the uniform region is

$$\mathbb{C}_h^{(1)} = \left\{ (y, z) : 0 \le y \le e^{(t-e^{t/\alpha})/2}, t = \frac{z}{y} \right\}.$$

Hence, y has the upper bound $\max_t [h(t)]^{1/2} = (\alpha e^{-1})^{\alpha/2}$. The upper bound $\max_{t>0} t[h(t)]^{1/2}$ of z requires us to solve the equation

$$\frac{1}{t} - \frac{1}{2} - \frac{1}{2} e^{t/\alpha} = 0 \tag{1.25}$$

for $t > 0$. As a simple alternative, a slightly loose upper bound can be obtained by making use the inequality $\ln(t/\alpha) \le t/\alpha - 1$. That is, $e^{t/\alpha} \ge (e^1 t)/\alpha$ and, thereby,

$$t[h(t)]^{1/2} = te^{\frac{t}{2} - \frac{1}{2}e^{t/\alpha}} \le te^{\frac{t}{2} - \frac{et}{2\alpha}} \le \frac{2\alpha}{e(e-\alpha)} \qquad (t > 0)$$

The lower bound of z also exists and requires one to solve (1.25) for $t < 0$. A simple lower bound is obtained as follows:

$$t[h(t)]^{1/2} = te^{\frac{t}{2} - \frac{1}{2}e^{t/\alpha}} \ge te^{\frac{t}{2}} \ge -\frac{2}{e} \qquad (t < 0).$$

Although it is not very tight for α near 1 and better ones can be found, this lower bound works pretty well, as indicated by the boundary plots in Figure 1.3 for a selected sequence of α values. The following experimental computer code written in R (`http://www.r-project.com`) demonstrates the simplicity of this method.

Algorithm 1.7

```
EXP1 = exp(1) # define the constant
rou.gamma(n, shape, log=FALSE){ #arguments: n = sample size, shape = α
        # log = flag; if log=TRUE, it returns the deviates in log scale
    if(shape<=0 || shape>=1) stop("shape is not in (0, 1)")
    if(shape <0.01 && !log)
        warning("It is recommended to set log=TRUE for shape < 0.01")
    y.max = (shape/EXP1)^(shape/2)
    z.min = -2/EXP1
    z.max = 2*shape/EXP1/(EXP1-shape)
    s = numeric(n) #allocate space for the generated desired deviates
    for(i in 1:n) {
        repeat {
            y = runif(1, 0, y.max)          # y ~ Unif(0,y.max)
            t = runif(1, z.min, z.max)/y    # t = z/y
            x = exp(t/shape)
            if(2*log(y) <= t-x){
                s[i] = if(log) t/shape else x
                break
            }
        }
    }
    return(s)
}
```

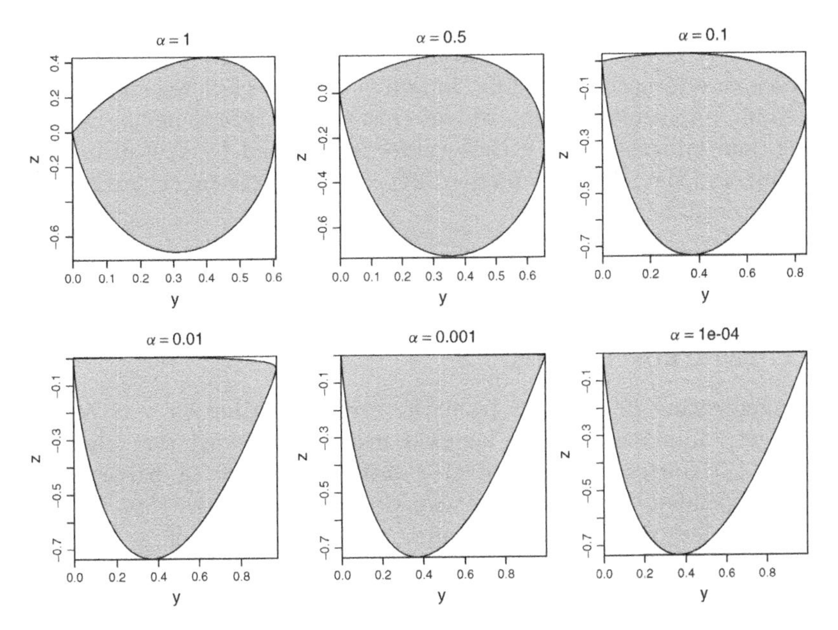

Figure 1.3 The uniform regions and their boundaries of the ratio-of-uniforms method for generating deviate X from Gamma(α), $\alpha < 1$, by letting $x = e^{t/\alpha}$, $-\infty < t < \infty$, and setting $h(t) = e^{t-e^{t/\alpha}}$.

We note that care must be taken in implementing Gamma deviate generators for small values of α. For example, we recommend in the above experimental code to return the output in a logarithmic scale when $\alpha < 0.01$.

1.4.4 Adaptive Rejection Sampling

Adaptive rejection sampling (ARS), introduced by Gilks (1992) (see also Gilks and Wild, 1992, and Wild, 1993), is a useful sampling method for log-concave densities. ARS works by constructing an envelope function of the log of the target density, which is then used in rejection sampling. For log-concave density functions, ARS is simple and efficient, especially when sampling from the same distributions occurs frequently. In this case, an adaptive squeezer can also be easily constructed to further improve its performance. For densities that are not log-concave, Gilks, Best, and Tan (1995) propose an Adaptive Metropolis sampling method.

1.4.5 Perfect Sampling

Propp and Wilson (1996) propose a perfect (exact) sampling MCMC algorithm, called *coupling from the past* (CFTP), for sampling from certain discrete distributions with finite number of states, for example, the Ising model. The algorithm uses a clever scheme to determine the time at which the Markov chain has reached its equilibrium. Later, it was extended by Murdoch and Green (1998) for sampling from a continuous state space. Fill (1998) proposes an alternative to CFTP, known as Fill's perfect rejection sampling algorithm. Interested readers are referred to the review paper by Djurić, Huang, and Ghirmai (2002), the website maintained by Wilson on perfect sampling, and the references therein (`http://www.dimacs.rutgers.edu/~/dbwilson/exact/`).

1.5 Markov Chain Monte Carlo

1.5.1 Markov Chains

When generating iid samples from the target distribution π is infeasible, dependent samples $\{X_i\}$ can be used instead, provided that the sample mean (1.17) converges to (1.16) at a satisfactory rate. A particular class of such dependent sequences that can be simulated is the class of Markov chains. A Markov chain, named after Andrey Markov, is a sequence of random variables $\{X_i : i = 0, 1, 2, \ldots\}$ with the Markov property that given the present state, the future and past states are independent, that is, for all measurable sets A in $\mathcal{X}$,

$$\Pr(X_{t+1} \in A | X_0 = x_0, \ldots, X_t = x_t) = \Pr(X_{t+1} \in A | X_t = x_t) \qquad (1.26)$$

holds for time $t = 0, 1, \ldots$

A convenient way of handling both discrete and continuous variables is to use the notation $\pi(dy)$ to denote the probability measure π on $(\mathbb{X}, \mathcal{X})$. For a continuous r.v. X, its pdf $f(x)$ is the Radon-Nikodym derivative of the probability measure $\pi(dx)$ with respect to the Lebesgue measure, while for discrete r.v. X, its pdf $f(x)$ is the derivative of $\pi(dx)$ with respect to the counting measure. Thus, we write $P_t(dx)$ for the marginal distribution of X_t over states $\mathcal{X}$ at time t. Starting with the distribution $P_0(dx)$, called the initial distribution, the Markov chain $\{X_t\}$ evolves according to

$$P_{t+1}(dy) = \int_{\mathbb{X}} P_t(dx) P_t(x, dy). \tag{1.27}$$

The distribution $P_t(x, dy)$ is the probability measure for X_{t+1} given $X_t = x$, called the transition kernel distribution at time t. In real life, this is the conditional density/mass function of X_{t+1} given $X_t = x$.

A primary class of Markov chains commonly used in MCMC is the class of time-homogeneous Markov chains or stationary Markov chains where

$$P_t(x, dy) = P(x, dy) \tag{1.28}$$

holds for $t = 1, 2, \ldots$. In this case, (1.27) becomes

$$P_{t+1}(dy) = \int_{\mathbb{X}} P_t(dx) P(x, dy) \tag{1.29}$$

and $P_t(dx)$ is uniquely determined by the initial distribution $P_0(dx)$ and the transition kernel $P(x, dy)$. For this reason, we write $P^n(x, .)$ for the conditional distribution of X_{t_0+n} given $X_{t_0} = x$. The basic idea of creating Markov chains for approximating $E_\pi(h(X))$ is to construct a transition kernel $P(x, dy)$ with $\pi(dx)$ as its invariant distribution, that is, $P(x, dy)$ and $\pi(dx)$ satisfy the *balance condition*

$$\pi(dy) = \int_{\mathbb{X}} \pi(dx) P(x, dy) \tag{1.30}$$

When the target distribution π has the density $f(x)$ and the transition kernel $P(x, dy)$ has the conditional density $p(y|x)$, this balance condition can be written as

$$f(y) = \int_{\mathbb{X}} p(y|x) f(x) dx.$$

The balance condition (1.30) can be viewed as obtained by requiring $P_{t+1}(dx) = P_t(dx) = \pi(dx)$ in (1.29). It says that if X_t is a draw from the target $\pi(x)$ then X_{t+1} is also a draw, possibly dependent on X_t, from $\pi(x)$. Moreover, for almost any $P_0(dx)$ under mild conditions $P_t(dx)$ converges to $\pi(dx)$. If for π-almost all x, $\lim_{t\to\infty} \Pr(X_t \in A | X_0 = x) = \pi(A)$ holds for all measurable sets A, $\pi(dx)$ is called the *equilibrium distribution* of the Markov chain. The relevant theoretical results, including those on the convergence behavior of the MCMC approximation $\bar{h}_n$ to $E(h(X))$, are summarized in Section 1.5.2.

1.5.2 Convergence Results

Except for rare cases where it is satisfactory to have one or few draws from the target distribution $f(x)$, most MH applications provide approximations to characteristics of $f(x)$, which can be represented by integrals of the form

$$E_\pi(h) = \int h(x)\pi(dx).$$

Assuming $E_\pi(|h|) < \infty$, this is to be approximated by

$$\bar{h}_{m,n} = \frac{1}{n}\sum_{i=1}^{n} h(X_{i+m}) \tag{1.31}$$

where $\{X_i\}$ denotes a simulated Markov chain and m is non-negative integer denoting the length of what is called the *burn-in* period; see Section 1.5.3.

This subsection includes theoretical results on (1.31) and is mainly based on Tierney (1994, Section 3). With some necessary preliminary theoretical results summarized in Section 1.5.2.2, Section 1.5.2.3 provides needed theoretical results concerning limiting behavior of (1.31) as $n \to \infty$ and m fixed. To proceed, Section 1.5.2.1 gives some key concepts and notations, in addition to those introduced in Section 1.5.1.

1.5.2.1 Notation and Definitions

The most important concepts are irreducibility and aperiodicity. A Markov chain with invariant distribution $\pi(dx)$ is *irreducible* if for any initial state, it has positive probability of entering any set to which $\pi(dx)$ assigns positive probability. A chain is *periodic* if there are portions of the state space $\mathcal{X}$ it can only visit at certain regularly spaced times; otherwise, the chain is *aperiodic*. A fundamental result is established in Theorem 1.5.1: If a chain has a proper invariant distribution $\pi(dx)$ and is irreducible and aperiodic, then $\pi(dx)$ is the unique invariant distribution and is also the equilibrium distribution of the chain.

Two additional crucial concepts in the theory of Markov chains are *recurrence* and *ergodicity*.

Definition 1.5.1 *Let X_n be a π-irreducible chain with invariant distribution $\pi(.)$ and let the notation $\{A_n\ i.o.\}$ mean that the sequence occurs infinitely often, that is, $\sum_i I_{A_i} = \infty$ (with probability one).*

(a) The chain is recurrent if, for every B with $\pi(B) > 0$,

$$Pr(X_n \in B\ i.o.|X_0 = x) > 0 \text{ for all } x$$

and

$$Pr(X_n \in B\ i.o.|X_0 = x) = 1 \text{ for } \pi\text{-almost all } x.$$

(b) The chain is Harris recurrent if $Pr(X_n \in B\ i.o.|X_0 = x) = 1$ for all x.

To define different forms of ergodicity, the total variation distance between two measures on $\mathcal{X}$ is used. The total variation distance between two measures on $(\mathbb{X}, \mathcal{X})$ is defined by the total variation norm of a signed measure λ on $(\mathbb{X}, \mathcal{X})$

$$\| \lambda \| = \sup_{A \in \mathcal{X}} \lambda(A) - \inf_{A \in \mathcal{X}} \lambda(A). \tag{1.32}$$

The concept of *hitting time* is also used. The *hitting time* of the subset $B \in \mathcal{X}$ is the random variable

$$H_B = \inf\{t \geq 0 : X_t \in B\}$$

where the infimum of the empty set is taken to be ∞.

Definition 1.5.2 *Different forms of ergodicity are as follows:*

(a) *A Markov chain is said to be ergodic if it is positive Harris recurrent and aperiodic.*

(b) *Let H_B denote the hitting time for the set B. An ergodic chain with invariant distribution $\pi(x)$ is said to be ergodic of degree 2 if*

$$\int_B E_x(H_B^2)\pi(dx) < \infty$$

holds for all $H \in \mathcal{X}$ with $\pi(H) > 0$.

(c) *An ergodic chain with invariant distribution $\pi(x)$ is said to be geometrically ergodic if there exists a nonnegative extended real-valued function M with $E(|M(X)|) < \infty$ and a positive constant $r < 1$ such that*

$$\| P^n(x, .) - \pi \| \leq M(x) r^n$$

for all x.

(d) *The chain in (c) is said to be uniformly ergodic if there exist a constant M and a positive constant $r < 1$ such that*

$$\| P^n(x, .) - \pi \| \leq M r^n.$$

1.5.2.2 Convergence of Distributions

The total variation distance between two measures on $(\mathbb{X}, \mathcal{X})$ is used to describe the convergence of a Markov chain in the following theorem (Theorem 1 of Tierney, 1994).

Theorem 1.5.1 *Suppose that $P(x, dy)$ is π-irreducible and π-invariant. Then $P(x, dy)$ is positive recurrent and $\pi(dx)$ is the unique invariant distribution of $P(x, dy)$. If $P(x, dy)$ is also aperiodic, then for π-almost all x,*

$$\| P^n(x, .) - \pi \| \to 0,$$

with $\| \,.\, \|$ denoting the total variation distance. If $P(x, dy)$ is Harris recurrent, then the convergence occurs for all x.

Tierney (1994) noted that the assumptions in Theorem 1.5.1 are essentially necessary and sufficient: if

$$||P_t(x,.) - \pi|| \to 0$$

for all x, then the chain is π-irreducible, aperiodic, positive Harris recurrent, and has the invariant distribution $\pi(dx)$. We refer to Tierney (1994) and Hernandez-Lerma and Lasserre (2001) for more discussion on sufficient conditions for Harris recurrence. Relevant theoretical results on the rate of convergence can also be found in Nummelin (1984), Chan (1989), and Tierney (1994).

1.5.2.3 Limiting Behavior of Averages

Tierney (1994) noted that a law of large numbers can be obtained from the ergodic theorem or the Chacon-Ornstein theorem. The following theorem is a corollary to Theorem 3.6 in Chapter 4 of Revuz (1975).

Theorem 1.5.2 *Suppose that X_n is ergodic with equilibrium distribution $f(x)$ and suppose $h(x)$ is real-valued and $E_f(|h(X)|) < \infty$. Then for any initial distribution, $\bar{h}_n \to E_f(h(X))$ almost surely.*

The central limit theorems that are available require more assumptions. Tierney (1994) gives the following central limit theorem.

Theorem 1.5.3 *Suppose that X_n is ergodic of degree 2 with equilibrium distribution $f(x)$ and suppose $h(x)$ is real-valued and bounded. Then there exists a real number σ_h such that the distribution of*

$$\sqrt{n}\left(\bar{h}_n - E_f(h(X))\right)$$

converges weakly to a normal distribution with mean 0 and variance σ_h^2 for any initial distribution.

The boundedness assumption on $h(x)$ can be removed if the chain is uniformly ergodic, provided $\mathrm{E}_f(h^2(X)) < \infty$.

Theorem 1.5.4 *Suppose that X_n is uniformly ergodic with equilibrium distribution $f(x)$ and suppose $h(x)$ is real-valued and $E_f(h^2(X)) < \infty$. Then there exists a real number σ_h such that the distribution of*

$$\sqrt{n}\left(\bar{h}_n - E_f(h(X))\right)$$

converges weakly to a normal distribution with mean 0 and variance σ_h^2 for any initial distribution.

1.5.3 Convergence Diagnostics

The theoretical results provide a useful guidance for designing practically valid and efficient MCMC (sampling) algorithms. It is difficult to make use of them to decide when it is safe to terminate MCMC algorithms and report MCMC approximations with their associated errors. More specifically, two critical issues arising in the context of computing $\bar{h}_{m,n}$ in (1.31) are (*i*) how to choose m, and (*ii*) how large should n be taken so that $\bar{h}_{m,n}$ has about the same precision as the corresponding Monte Carlo approximation based on an iid sample of a prespecified sample size, say, n_0. In theory, there are no definite answers to these two questions based on the simulated finite sequence $\{X_t : t = 0, \ldots, T\}$ because Markov chains can be trapped into local modes for arbitrarily long periods of time, if not indefinitely. While designing problem-specific efficient MCMC sampling algorithms is desirable, and is a major focus of this book, there have also been many proposed convergence diagnostic methods.

Gelman and Rubin (1992) is one of the most popular convergence diagnostic tools. The Gelman and Rubin method requires running multiple sequences $\{X_t^{(j)} : t = 0, 1, \ldots; j = 1, \ldots, J\}$, $J \geq 2$, with the starting sample $X_0^{(1)}, \ldots, X_0^{(J)}$ generated from an overdispersed estimate of the target distribution $\pi(dx)$. Let n be the length of each sequence after discarding the first half of the simulations. For each scalar estimand $\psi = \psi(X)$, write

$$\psi_i^{(j)} = \psi(X_i^{(j)}) \qquad (i = 1, \ldots, n; j = 1, \ldots, J).$$

Let

$$\bar{\psi}^{(j)} = \frac{1}{n}\sum_{i=1}^{n} \psi_i^{(j)} \quad \text{and} \quad \bar{\psi} = \frac{1}{J}\sum_{j=1}^{J} \bar{\psi}^{(j)},$$

for $j = 1, \ldots, J$. Then compute B and W, the between- and within-sequence variances:

$$B = \frac{n}{J-1}\sum_{j=1}^{J}\left(\bar{\psi}^{(j)} - \bar{\psi}\right)^2 \quad \text{and} \quad W = \frac{1}{J}\sum_{j=1}^{J} s_j^2,$$

where

$$s_j^2 = \frac{1}{n-1}\sum_{i=1}^{n}\left(\psi_i^{(j)} - \bar{\psi}^{(j)}\right)^2 \qquad (j = 1, \ldots, J).$$

Suppose that the target distribution of ψ is approximately normal and assume that the jumps of the Markov chains are local, as is often the case in practical iterative simulations. For any finite n, the within variance W underestimates the variance of ψ, σ_ψ^2; while the between variance B overestimates σ_ψ^2. In the limit as $n \to \infty$, the expectations of both B and W

approach σ^2_ψ. Thus, the Gelman and Rubin reduction coefficient,

$$\sqrt{\hat{R}} = \sqrt{\frac{\frac{n-1}{n}W + \frac{1}{n}B}{W}}, \tag{1.33}$$

should be expected to decline to one as $n \to \infty$. Gelman *et al.* (2004) recommend computing the reduction coefficient $\sqrt{\hat{R}}$ for all scalar estimands of interest; if $\sqrt{\hat{R}}$ is not near one for all of them, continue the simulation runs. Once $\sqrt{\hat{R}}$ is near one for all scalar estimands of interest, just collect the $J \times n$ samples from the second halves of the J sequences together and treat them as (dependent) samples from the target distribution $\pi(dx)$.

A number of criticisms of the GR method have been made in the literature. Readers are referred to Cowles and Carlin (1996) for these criticisms. Ideas of constructing overdispersed starting values can be found in Gelman and Rubin (1992) and Liu and Rubin (1996). For a review of other convergence diagnostic methods, see Cowles and Carlin (1996), Brooks and Gelman (1998), Mengersen *et al.* (1999), and Plummer *et al.* (2006) and the references therein.

Exercises

1.1 Suppose that a single observation $X = 2.0$ is considered to have been drawn from the Gaussian model $N(\theta, 1)$ with unknown θ. Consider the hypothesis $H_0 : \theta = 0$ versus the alternative hypothesis $H_a : \theta \neq 0$. Apply the Bayes approach using Bayes factors.

1.2 Consider inference about the binomial proportion θ in Binomial(n, θ) from an observed count X.

- **(a)** Show that the Jeffreys prior for the binomial proportion θ is the Beta distribution Beta $\left(\frac{1}{2}, \frac{1}{2}\right)$.
- **(b)** Derive the posterior $\pi(\theta|X)$.
- **(c)** For the case of $n = 1$, evaluate the frequency properties of the 95% credible interval for each of $\theta = .0001, 0.001, 0.01, 0.1, .25, .5, .75$, $.9$, $.99$, $.999$, and $.9999$.

1.3 Suppose that the sample density function of a single observation $X \in \mathcal{R}$ has the density of the form $f(x - \theta)$, where $\theta \in \mathcal{R}$ is unknown parameter to be estimated.

- **(a)** Show that the Jeffreys prior is $\pi(\theta) \propto 1$.
- **(b)** Consider frequency properties of one-sided credible intervals.
- **(c)** Discuss the case where θ is known to be on a sub-interval of $\mathcal{R}$.

1.4 Prove Theorem 1.3.1.

1.5 Verify the results in Table 1.2.

1.6 Extend the method of generating Beta to simulate the Dirichlet random variable.

1.7 Consider the problem of generating Poisson variates.

(a) Design an efficient discrete inverse-cdf algorithm using a mapping.

(b) Develop an Acceptance-Rejection method with a continuous envelope distribution.

(c) Investigate efficient ratio-of-uniforms methods; see Stadlober (1989).

1.8 Consider the problem of Example 1.5, that is, to generate a deviate from the Gamma distribution with shape parameter less than one.

(a) Create a mixture distribution as envelope for implementing the Acceptance-Rejection algorithm.

(b) Compare your methods against the ratio-of-uniforms methods in Example 1.5.

1.9 Develop the ratio-of-uniforms method for generating Gamma deviates (for all $\alpha > 0$) by replacing the transformation $T = \alpha \ln X$ in the problem of Example 1.5 with $T = \sqrt{\alpha} \ln \frac{X}{\alpha}$.

1.10 The standard student-t distribution with ν degrees of freedom has density

$$f_\nu(x) = \frac{\Gamma\left(\frac{\nu+1}{2}\right)}{\sqrt{\nu\pi}\Gamma\left(\frac{\nu}{2}\right)}\left(1 + \frac{x^2}{\nu}\right)^{-(\nu+1)/2} \qquad (-\infty < x < \infty)$$

where ν is the number of degrees of freedom and $\Gamma(.)$ is the Gamma function.

(a) Implement the ratio-of-uniforms method for generating random deviates from $f_\nu(x)$.

(b) Develop an efficient ratio-of-uniforms algorithm to generate random variables from interval-truncated Student-t distribution.

1.11 This is the same as Exercise 1.10 but for the standard normal distribution $N(0,1)$.

1.12 Suppose that $D = \{y_i = (y_{1i}, y_{2i})' : i = 1, \ldots, n\}$ is a random sample from a bivariate normal distribution $N_2(\mathbf{0}, \Sigma)$, where

$$\Sigma = \begin{pmatrix} 1 & \rho \\ \rho & 1 \end{pmatrix}$$

with unknown correlation coefficient $\rho \in (-1, 1)$ (Pérez *et al.*, 2008)

(a) Assuming the prior $\pi(\rho) \propto 1/(1-\rho^2)$, derive the posterior distribution $\pi(\rho|D)$.

(b) Implement the ratio-of-uniforms method to generate ρ from $\pi(\rho|D)$.

(c) Implement the ratio-of-uniforms method to generate η from $\pi(\eta|D)$, which is obtained from $\pi(\rho|D)$ via the one-to-one transformation $\eta = \ln \frac{1+\rho}{1-\rho}$.

(d) Conduct a simulation study to compare the two implementations in **(b)** and **(c)**.

1.13 Consider the simple random walk Markov chain with two reflecting boundaries on the space $\mathbb{X} = \{a, a+1, \ldots, b\}$ with the transition kernel distribution (matrix) $P = (p_{ij})$, where

$$p_{ij} = \Pr(X_{t+1} = j | X_t = i) = \begin{cases} p & \text{if } j = i+1 \text{ and } a < i < b; \\ q & \text{if } j = i-1 \text{ and } a < i < b; \\ 1 & \text{if } i = a \text{ and } j = a+1; \\ 1 & \text{if } i = b \text{ and } j = b-1, \end{cases}$$

with $0 < p < 1$ and $q = 1 - p$.

(a) Find the invariant distribution π.

(b) Show that the invariant distribution π is also the equilibrium distribution of the Markov chain.

1.14 Let π_i $(i = 1, 2)$ be the probability measure for $N(\mu_i, 1)$. Find the total variation distance between π_1 and π_2.

Hint: Let $\lambda = \pi_2 - \pi_1$ and let $\phi(x - \mu_i)$ be the density of π_i for $i = 1$ and 2.

Then $\sup_A \lambda(A) = \inf_{\phi(x-\mu_2)-\phi(x-\mu_1)>0} [\phi(x - \mu_2) - \phi(x - \mu_1)]\, dx$.

Chapter 2

The Gibbs Sampler

Direct sampling techniques discussed in Chapter 1 for generating multivariate variables are often practically infeasible for Bayesian inference, except for simple models. For example, for the Acceptance-Rejection or its variants such as the ratio-of-uniforms method, the acceptance rate often becomes effectively zero in high dimensional problems. This phenomenon is known as the curse of dimensionality. As an alternative to Monte Carlo methods using independent samples, dependent samples associated with target distributions can be used in two possible ways. The first is to generate a Markov chain with the target distribution as its stationary distribution. For this, the standard Monte Carlo theory is then extended accordingly for approximating integrals. The second is to create iid samples by using Markov chain Monte Carlo sampling methods; see Chapter 5. This chapter introduces the Gibbs sampling method also known as the Gibbs sampler. More discussion of MCMC is given in Chapter 3.

2.1 The Gibbs Sampler

The Gibbs sampler has become the most popular computational method for Bayesian inference. Known as the *heat bath* algorithm it was in use in statistical physics before the same method was used by Geman and Geman (1984) for analyzing Gibbs distributions on lattices in the context of image processing. Closely related to the EM algorithm (Dempster *et al.*, 1977), a similar idea was explored in the context of missing data problems by Tanner and Wong (1987; see also Li, 1988), who introduced the Data Augmentation algorithm. The paper by Gelfand and Smith (1990) demonstrated the value of the Gibbs sampler for a range of problems in Bayesian analysis and made the Gibbs sampler a popular computational tool for Bayesian computation.

Technically, the Gibbs sampler can be viewed as a special method for overcoming the curse of dimensionality via conditioning. The basic idea is the

Advanced Markov Chain Monte Carlo Methods: Learning from Past Samples
Faming Liang, Chuanhai Liu and Raymond J. Carroll

same as the idea behind iterative conditional optimization methods. Suppose that we want to generate random numbers from the target density $f(x)$, $x \in \mathcal{X} \subseteq \mathbb{R}^d$. Partition the d-vector x into K blocks and write $x = (x_1, \ldots, x_K)'$, where $K \leq d$ and $\dim(x_1) + \cdots + \dim(x_K) = d$ with $\dim(x_k)$ representing the dimension of x_k. Denote by

$$f_k(x_k|x_1, \ldots, x_{k-1}, x_{k+1}, \ldots, x_K) \qquad (k = 1, \ldots, K) \tag{2.1}$$

the corresponding full set of conditional distributions. Under mild conditions, this full set of conditionals, in turn, determines the target distribution $f(x)$; according to the Hammersley-Clifford theorem (Besag 1974; Gelman and Speed, 1993):

Theorem 2.1.1 (Hammersley-Clifford) *If $f(x) > 0$ for every $x \in \mathbb{X}$, then the joint distribution $f(x)$ is uniquely determined by the full conditionals (2.1). More precisely,*

$$f(x) = f(y) \prod_{k=1}^{K} \frac{f_{j_k}(x_{j_k}|x_{j_1}, \ldots, x_{j_{k-1}}, y_{j_{k+1}}, \ldots, y_{j_K})}{f_{j_k}(y_{j_k}|x_{j_1}, \ldots, x_{j_{k-1}}, y_{j_{k+1}}, \ldots, y_{j_K})} \qquad (x \in \mathcal{X}) \tag{2.2}$$

for every permutation j on $\{1, \ldots, n\}$ and every $y \in \mathbb{X}$.

Algorithmically, the Gibbs sampler is an iterative sampling scheme. Starting with an arbitrary point $x^{(0)}$ in $\mathcal{X}$ with the restriction that $f(x^{(0)}) > 0$, each iteration of the Gibbs sampler cycles through the full set of conditionals (2.1) to generate a random number from each $f_k(x_k|x_1, \ldots, x_K)$ by setting $x_1, \ldots, x_{k-1}, x_{k+1}, \ldots, x_K$ at their most recently generated values.

The Gibbs Sampler

Take $x^{(0)} = (x_1^{(0)}, \ldots, x_K^{(0)})$ from $f^{(0)}(x)$ with $f(x^{(0)}) > 0$, and iterate for $t = 1, 2, \ldots$

1. Generate $x_1^{(t)} \sim f_1(x_1|x_2^{(t-1)}, \ldots, x_K^{(t-1)})$.

$\vdots$

k. Generate $x_k^{(t)} \sim f_k(x_k|x_1^{(t)}, \ldots, x_{k-1}^{(t)}, x_{k+1}^{(t-1)}, \ldots, x_K^{(t-1)})$.

$\vdots$

K. Generate $x_K^{(t)} \sim f_K(x_K|x_1^{(t)}, \ldots, x_{K-1}^{(t)})$.

Under mild regularity conditions (see Section 1.5), the distribution of $x^{(t)} = (x_1^{(t)}, \ldots, x_K^{(t)})'$, denoted by $f^{(t)}(x)$, will converge to $f(x)$. Here, we provide an illustrative example.

Example 2.1 The Multivariate Normal Distribution

The p-dimensional normal distribution, denoted by $N_p(\mu, \Sigma)$, is defined on the sample space $\mathcal{X}^p$ with the parameter consisting of a mean vector $\mu \in \mathbb{R}^p$ and a positive defined $p \times p$ covariance matrix $\Sigma \in \mathbb{M}^+_{p\times p}$, where $\mathbb{M}^+_{p\times p}$ denotes the set of all positive definite $p \times p$ matrices. The density of $N_p(\mu, \Sigma)$ is given by

$$f_{\mu,\Sigma}(x) = \frac{1}{(2\pi)^{n/2}|\Sigma|} e^{-\frac{1}{2}(x-\mu)'\Sigma^{-1}(x-\mu)} \qquad (x \in \mathbb{R}^p) \tag{2.3}$$

Let $X \sim N_p(\mu, \Sigma)$. It is well known that the distribution of the linear combination of AX, where A is $q \times p$ matrix, is the q-variate normal with mean vector $A\mu$ and covariance matrix $A\Sigma A'$. This means that computation for marginal distributions is straightforward. Taking A to be permutation matrices, we see that reordering the components of X mounts to reordering elements of both μ and Σ accordingly. Thus, when X is partitioned into two blocks X_1 and X_2, we write

$$\begin{pmatrix} X_1 \\ X_2 \end{pmatrix} \sim N_p\left(\begin{pmatrix} \mu_1 \\ \mu_2 \end{pmatrix}, \begin{pmatrix} \Sigma_{11} & \Sigma_{12} \\ \Sigma_{21} & \Sigma_{22} \end{pmatrix} \right) \tag{2.4}$$

When conditioned on X_1, X_2 is also normal

$$X_2|X_1 \sim N_{\dim(X_2)} \left(\mu_2 + \Sigma_{21}\Sigma_{11}^{-1}(X_1 - \mu), \Sigma_{22} - \Sigma_{21}\Sigma_{11}^{-1}\Sigma_{12}\right) \tag{2.5}$$

The needed matrix computation in (2.5) can be summarized into a powerful operator, called the Sweep operator (see, e.g., Little and Rubin, 1987).

To illustrate the Gibbs sampler, we use a trivariate normal with mean vector $\mu = (\mu_1, \mu_2, \mu_3)'$ and the covariance

$$\Sigma(\rho) = \begin{pmatrix} 1 & \rho & \rho^2 \\ \rho & 1 & \rho \\ \rho^2 & \rho & 1 \end{pmatrix}.$$

The three-step Gibbs sampler with the partition of $X = (X_1, X_2, X_3)'$ into X_1, X_2, and X_3 is then implemented as follows.

The Gibbs sampler for $N_3(0, \Sigma(\rho))$: Set a starting value $x^{(0)} \in \mathbb{R}^3$, and iterate for $t = 1, 2, \ldots$

1. Generate $x_1^{(t)} \sim N(\mu_1 + \rho(x_2^{(t-1)} - \mu_2), 1 - \rho^2)$.
2. Generate $x_2^{(t)} \sim N\left(\mu_2 + \frac{\rho}{1+\rho^2}(x_1^{(t)} - \mu_1 + x_3^{(t-1)} - \mu_3), \frac{1-\rho^2}{1+\rho^2}\right)$.
3. Generate $x_3^{(t)} \sim N(\mu_3 + \rho(x_2^{(t)} - \mu_2), 1 - \rho^2)$.

Figure 2.1 displays trajectory plots of two Markov chains generated by this Gibbs sampler for the trivariate normal distributions with $\mu = (10, 10, 10)$ and $\rho = 0$ and $.99$. It shows that the performance of the Gibbs

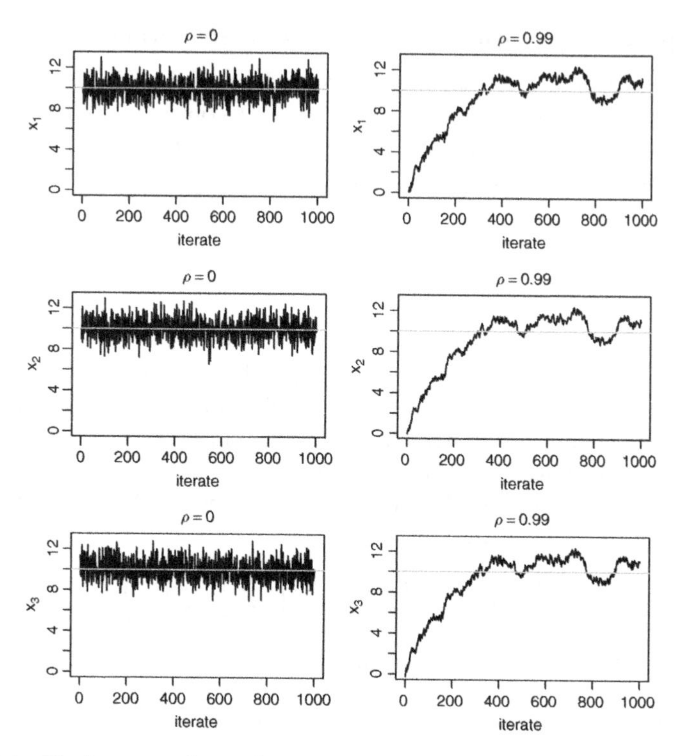

Figure 2.1 Trajectory plots of two Markov chains generated by the Gibbs sampler for the trivariate normal distributions in Example 2.1 with $\mu = (10, 10, 10)$ and $\rho = 0$ and .99.

sampler depends on the dependence structure among $X_1, \ldots, X_K$. In the case of $\rho = 0$, the sampler produces iid draws from the target distribution. In the case of $\rho = 0.99$, the Gibbs sequence is highly correlated.

2.2 Data Augmentation

The Data Augmentation (DA) algorithm (Tanner and Wong, 1987) can be viewed as a special case of the Gibbs sampler, the two-step Gibbs sampler. It is also viewed as the stochastic version of the EM algorithm (Dempster *et al.*, 1977); see Appendix 2A. In turn, the EM algorithm can be viewed as the deterministic version of the DA algorithm. Interest in DA is at least three-fold: first, it has applications in Bayesian analysis of incomplete data; second, DA is the simplest case of the Gibbs sampler and therefore can be extensively studied for understanding theoretical properties of the Gibbs sampler; third, ideas for developing efficient DA algorithms can usually be extended to the more general Gibbs sampler or even beyond, such as to the Metropolis-Hastings algorithm.

We describe the DA algorithm in the context of Bayesian analysis of incomplete data. Denote by X_{obs} the observed data and by X_{mis} ($\in \mathcal{X}_{\text{mis}}$) the missing-data. Write $X_{\text{com}} = (X_{\text{obs}}, X_{\text{mis}})$, called the complete data. Suppose that the complete-data model has density $g(X_{\text{obs}}, X_{\text{mis}}|\theta)$ with the parameter $\theta \in \Theta \subseteq \mathbb{R}^d$ for some positive integer d. The objective is to make Bayesian inference with a prior distribution $p(\theta)$ for the parameter θ. Let $f(X_{\text{obs}}|\theta)$ be the observed-data model, *i.e.*,

$$f(X_{\text{obs}}|\theta) = \int_{\mathbb{X}_{\text{mis}}} g(X_{\text{obs}}, X_{\text{mis}}|\theta)dX_{\text{mis}} \qquad (\theta \in \Theta) \tag{2.6}$$

For Bayesian inference about θ using MCMC methods, it is required to sample from the true or observed-data posterior

$$p(\theta|X_{\text{obs}}) \propto f(X_{\text{obs}}|\theta)p(\theta) \qquad (\theta \in \Theta) \tag{2.7}$$

or more generally, the joint distribution of θ and X_{mis},

$$p(\theta, X_{\text{mis}}|X_{\text{obs}}) \propto g(X_{\text{obs}}, X_{\text{mis}}|\theta)p(\theta) \qquad (\theta \in \Theta) \tag{2.8}$$

Let $h(X_{\text{mis}}|\theta, X_{\text{obs}})$ be the conditional distribution of X_{mis} given θ and X_{obs}. Suppose that both $h(X_{\text{mis}}|\theta, X_{\text{obs}})$ and $p(\theta|X_{\text{obs}}, X_{\text{mis}}) \propto g(X_{\text{obs}}, X_{\text{mis}}|\theta)p(\theta)$ are easy to draw samples from. The two-step Gibbs sampler based on these two conditionals is known as the DA algorithm and can be summarized as follows:

The DA Algorithm: A Two-Step Gibbs Sampler

Take $\theta^{(0)} \in \Theta$, and iterate for $t = 1, 2, \ldots$

I-Step. Generate $X_{\text{mis}}^{(t)} \sim f_{\text{mis}}(X_{\text{mis}}|\theta^{(t-1)}, X_{\text{obs}})$.

P-Step. Generate $\theta^{(t)} \sim p(\theta|X_{\text{obs}}, X_{\text{mis}}^{(t)})$.

As a two-step Gibbs sampler, DA creates two interleaving (marginal) Markov chains $\{\theta^{(t)} : t = 1, 2, \ldots\}$ and $\{X_{\text{mis}}^{(t)} : t = 1, 2, \ldots\}$. This explains why DA provides the mathematically simplest case of the Gibbs sampler. Example 2.2 demonstrates that DA is useful for missing data problems.

■ **Example 2.2 Multivariate Normal With Incomplete Data**

Consider a complete-data set consisting of a sample of n, $Y_1, \ldots, Y_n$, from the p-dimensional multivariate normal distribution $N_p(\mu, \Sigma)$ with unknown mean vector $\mu \in \mathbb{R}^p$ and $p \times p$ (positive definite) covariance matrix Σ. Each component of Y_i is either fully observed or missing for each $i = 1, \ldots, n$. Let $Y_{\text{obs}}^{(i)}$ denote the observed components and let $Y_{\text{mis}}^{(i)}$ the missing components of Y_i. Using the notation similar to that in Example 2.1, we write the conditional distribution of $Y_{\text{mis}}^{(i)}$ given $Y_{\text{obs}}^{(i)}$ and (μ, Σ) as

$$Y_{i,\text{mis}}|(Y_{i,\text{obs}}, \mu, \Sigma) \sim N\left(\mu_{\text{mis}}^{(i)} + \Sigma_{\text{mis,obs}}^{(i)}\left[\Sigma_{\text{obs,obs}}^{(i)}\right]^{-1}(Y_{i,\text{mis}} - \mu_{\text{mis}}^{(i)}),\right. \quad (2.9)$$
$$\left.\Sigma_{\text{mis,mis}}^{(i)} - \Sigma_{\text{mis,obs}}^{(i)}\left[\Sigma_{\text{obs,obs}}^{(i)}\right]^{-1}\Sigma_{\text{obs,mis}}^{(i)}\right)$$

Suppose that for Bayesian analysis, we use the prior distribution

$$p(\mu, \Sigma) \propto |\Sigma|^{-(q+1)/2}, \quad (2.10)$$

where q is a known integer. With $q = p$, this prior becomes the the Jeffreys prior for Σ.

Let $\bar{Y} = n^{-1}\sum_{i=1}^{n} Y_i$ and let $S = \sum_{i=1}^{n}(Y_i - \bar{Y})(Y_i - \bar{Y})'$. The complete-data posterior distribution $p(\mu, \Sigma|Y_1, \ldots, Y_n)$ can be characterized by

$$\Sigma|Y_1, \ldots, Y_n \sim \frac{1}{|\Sigma|^{(n+q)/2}} \exp\left\{-\frac{1}{2}\text{trace}\left(\Sigma^{-1}S\right)\right\}, \quad (2.11)$$

that is, the inverse Wishart distribution, and

$$\mu|\Sigma, Y_1, \ldots, Y_n \sim N_p(\bar{Y}, \Sigma/n). \quad (2.12)$$

Thus, the DA algorithm has the following I and P steps:

I-step. For $i = 1, \ldots, n$, draw $Y_{i,\text{mis}}$ from (2.9).

P-step. First draw Σ from (2.11) given $Y_1, \ldots, Y_n$ and then draw μ from (2.12) given $Y_1, \ldots, Y_n$ and Σ.

We note that the P-step can be split into two sub-steps, resulting in a three-step Gibbs sampler:

Step 1. This is the same as the I-step of DA.

Step 2. Draw μ from its conditional distribution given $Y_1, \ldots, Y_n$ and Σ.

Step 3. Draw Σ from its conditional distribution given $Y_1, \ldots, Y_n$ and μ.

Compared to the DA algorithm, a two-step Gibbs sampler, this three-step Gibbs sampler induces more dependence between the sequence $\{(\mu^{(t)}, \Sigma^{(t)}) : t = 1, 2, \ldots\}$ and, thereby, converges slower than the corresponding DA. In other words, DA can be viewed as obtained from the three-step Gibbs sampler by making μ and Σ into a single block. This grouping technique is referred to as 'blocking' by Liu *et al.* (1994); see Section 2.3.1. It should also be noted that more efficient DAs for incomplete multivariate normal data can be implemented by 'imputing less' missing data/information (see Rubin and Schafer, 1990; Liu, 1993; and Schafer, 1997).

In the case where computing the posterior distribution $p(\theta|X_{\text{obs}})$ is of interest, DA can be viewed as an MCMC sampling algorithm using auxiliary (random) variables, as illustrated in Example 2.3.

Example 2.3 A DA Alternative to the Acceptance-Rejection Method

Let $f(x)$ be a probability density function on $\mathbb{R}^d$. Consider the uniform distribution on the region

$$\{(x, u) : 0 \leq u \leq f(x)\} \subset \mathbb{R}^{d+1}.$$

The full conditionals consist of

$$U|\{X = x\} \sim \text{Unif}\,(0, f(x)) \quad \text{and} \quad X|\{U = u\} \sim \text{Unif}\,(\{x : f(x) = u\})$$

This leads to a two-step Gibbs sampler and is related to the slice sampler (see Chapter 4).

2.3 Implementation Strategies and Acceleration Methods

When applying the basic Gibbs sampler for creating a Markov chain with the target distribution $f(x)$ on $\mathcal{X} \subseteq \mathbb{R}^d$, we need to choose a Gibbs configuration that consists of (i) an appropriate coordinate system for X and (ii) a partition of the space X into (an ordered) K blocks $X = (X_1, \ldots, X_K)$. Some configurations are easier to implement and may be less efficient than others. The basic idea to improve efficiency is to weaken high correlations or dependence of the Gibbs sequence $\{X^{(t)} = (X_1^{(t)}, \ldots, X_K^{(t)}) : t = 1, 2, \ldots\}$. Thus, implementing the Gibbs sampler is the art of making intelligent tradeoffs between simplicity and efficiency. This section covers a number of selected simple and efficient methods. These include blocking and collapsing (Liu *et al.*, 1994; Liu, 1994), hierarchical centering (Hills and Smith, 1992; Gelfand *et al.*, 1995; Roberts and Sahu, 1997), conditional data augmentation and marginal data augmentation (Meng and van Dyk, 1999; van Dyk and Meng, 2001), parameter-expanded data augmentation (Liu and Wu, 1999), and alternating subspace-spanning resampling (Liu, 2003; Yu and Meng, 2008). Auxiliary variable methods are treated separately in Chapter 4.

2.3.1 Blocking and Collapsing

Liu *et al.* (1994) describe the benefits of 'blocking' and 'collapsing'. Blocking or grouping updates variables in large-dimensional groups. The idea of blocking is very intuitive and should be considered whenever the resulting Gibbs sampler is simple to implement. Suppose that a group of variables $\{X_{i_1}, \ldots, X_{i_k}\}$, where $\{i_1, \ldots, i_k\}$ is a subset of $\{1, \ldots, K\}$ in the original Gibbs setting, is to be considered to form a larger block. What is needed is to find a way of sampling the components $X_{i_1}, \ldots, X_{i_k}$ from their joint conditional distribution given the remaining variables, denoted by

$X_{-[i_1,\ldots,i_k]}$. A typical approach is to seek an ordering of $X_{i_1}, \ldots, X_{i_k}$, say, $X_{j_1}, \ldots, X_{j_k}$, among all possible permutations of $i_1, \ldots, i_k$ and draw samples sequentially from $f(X_{j_1}|X_{-[i_1,\ldots,i_k]})$, $f(X_{j_2}|X_{j_1}, X_{-[i_1,\ldots,i_k]}), \ldots$, and $f(X_{j_k}|X_{j_1}, \ldots, X_{j_{k-1}}, X_{-[i_1,\ldots,i_k]})$. This has been illustrated in Example 2.2.

Collapsing or *partial marginalizing* creates draws from partially marginalized distributions. Consider a three-step Gibbs sampler for (X_1, X_2, X_3). Suppose that it is simple to draw X_1 from $f(X_1|X_2)$ and X_2 from $f(X_2|X_1)$. That is, a two-step Gibbs sampler can be implemented to create Markov chain in the space of (X_1, X_2) with the marginal distribution $f(X_1, X_2)$ as the equilibrium distribution. These draws $\{(X_1^{(t)}, X_2^{(t)}) : t = 1, 2, \ldots\}$ are then augmented by imputing the missing component $X_3^{(t)}$ from $f(X_3|X_1^{(t)}, X_2^{(t)})$. This collapsing idea can also be applied to the general case of Gibbs sampler (Liu, 1994).

A sampling-based collapsing in this three-block Gibbs setting is to iterate between drawing $f(X_1, X_3|X_2)$ and $f(X_2, X_3|X_1)$. In this sense, parameter expansion and alternating subspace-spanning resampling (Section 2.3.3.2 and Section 2.3.4) can be viewed as special cases of the collapsed Gibbs sampler.

2.3.2 Hierarchical Centering and Reparameterization

Gelfand *et al.*, (1995) demonstrate that simple reparameterization such as hierarchical centering can be a useful method of breaking correlations for hierarchical models. We use the following toy example of Liu and Wu (1999) to explain the basic idea. This example can also be used to show that over-parameterization (Section 2.3.3.2) is often more efficient than reparameterization. Incidentally, we note that simple examples with closed-form solutions are used here for easy understanding without much loss of generality of the underlying basic ideas.

■ **Example 2.4 The Simple Random-Effects Model**

Consider a Bayesian approach to the simplest linear random-effects model with fixed effect θ and random effect Z

$$Z|\theta \sim N(0, v) \quad \text{and} \quad Y|(Z, \theta) \sim (\theta + Z, 1) \tag{2.13}$$

with the observed data $Y \in \mathbb{R}$, known $v > 0$, and the prior $\theta \sim \text{Unif}(\mathbb{R})$, *i.e.*, the flat prior on the real line $\mathbb{R}$, for the fixed effect θ. The complete-data model for Y and Z has the joint distribution

$$\mathrm{N}_2\left(\begin{bmatrix} \theta \\ 0 \end{bmatrix}, \begin{bmatrix} 1+v & v \\ v & v \end{bmatrix}\right), \tag{2.14}$$

leading to the joint posterior of (θ, Z)

$$\mathrm{N}_2\left(\begin{bmatrix} Y \\ 0 \end{bmatrix}, \begin{bmatrix} 1+v & v \\ v & v \end{bmatrix}\right).$$

In this case the target posterior distribution of θ given Y has a closed-form solution, $N(Y, 1 + v)$, because the observed-data model is $Y|\theta \sim N(\theta, 1 + v)$.

The basic idea of hierarchical centering amounts to comparing two different implementations of the DA algorithm with Z viewed as the missing data and θ the parameter. One is based on (2.13) and consists of the following two steps:

I-step. Draw Z from its conditional distribution $N(v(\theta - Y)/(1+v), v/(1+v))$, given Y and θ.

P-step. Draw θ from its conditional distribution $N(Y + Z, 1)$, given Y and Z.

The other is based on the reparameterized or recentered version of (2.13)

$$Z|\theta \sim N(\theta, v) \quad \text{and} \quad Y|(Z, \theta) \sim (Z, 1) \tag{2.15}$$

which has the same observed-data model. The corresponding DA has the following two steps:

I-step. Draw Z from its conditional distribution $N([vY + \theta]/(1+v), v/(1+v))$, given Y and θ.

P-step. Draw θ from its conditional distribution $N(Z, v)$, given Y and Z.

Each of the two DA implementations induces an AR series on θ. The first has the auto correlation coefficient $r = v/(1 + v)$; whereas the second has the auto-correlation coefficient $r = 1/(1 + v)$. Thus, the rate of convergence depends on the value of v, compared to the unit residual variance of Y conditioned on Z. For large values of v ($\gg 1$), the second scheme is more efficient than the first. For small values of v ($\ll 1$), the second scheme is very slow and the first is quite efficient. Nevertheless, this toy example indicates that hierarchical centering can be effective.

In general, reparameterization represents a transformation on the random variables, leading to a different implementation of the Gibbs sampler, which is by definition coordinate-dependent. Efficient implementation of the Gibbs sampler can be obtained if Gibbs components in the transformed coordinate system are approximately independent and the corresponding conditionals are easy to draw samples from. See Robert and Casella (2004) for more discussion on reparameterization and Liu *et al.* (1994) for a general method for comparing different sampling algorithms.

2.3.3 Parameter Expansion for Data Augmentation

2.3.3.1 Conditional and Marginal Data Augmentation

Weakening dependence among draws $\{(X_{\text{mis}}^{(t)}, \theta^{(t)}) : t = 1, 2, \ldots\}$ to speed up DA can be achieved by boosting the conditional variability (or variance if exists) of θ given X_{mis} and X_{obs}. A way of doing this is to consider different data augmentation schemes. Much work has been done in the EM world; see Meng and van Dyk (1997) and Liu *et al.* (1998). Meng and van Dyk (1997) seek within a given class of complete-data models to get the optimal one in terms of the fraction of missing information (see Little and Rubin, 1987). This idea of 'imputing less' is powerful. For example, it has motivated the Monotone-DA algorithm of Rubin and Schafer (1990) for incomplete multivariate normal models; see Liu (1993, 1995, 1996), Schafer (1997), Gelman *et al.* (1998), Liu and Rubin (1998) for more examples of Monotone-DA. Liu *et al.* (1998) activate hidden parameters that are identifiable in the complete-data model of the original EM algorithm but unidentifiable in the observed-data model. They use the standard EM algorithm to find maximum likelihood estimate of the original parameter from the parameter-expanded complete-data model. The resulting EM algorithm is called the PX-EM algorithm; a formal definition of PX-EM is included in Appendix 2A.

It is perhaps relatively straightforward to construct the DA version of Meng and van Dyk (1997) because once a complete-data model is chosen it defines a regular EM and, thereby, a regular DA. Unlike the other EM-type algorithms, such as ECM (Meng and Rubin, 1993), ECME (Liu and Rubin, 1994), and AECM (Meng and van Dyk, 1997), PX-EM has received considerable attention due to its fast convergence and the challenge for constructing its DA version. Constructing an exact DA version, if it exists, of PX-EM is still an open problem. Perhaps most important is that the research on this topic has produced a number of efficient DA algorithms with ideas that can be easily extended to the more general Gibbs sampler. The relevant algorithms selected in this book include conditional and marginal data augmentation of Meng and van Dyk (1999); see also van Dyk and Meng (2001); PX-DA of Liu and Wu (1999); and alternating subspace-spanning resampling of Liu (2003).

Suppose that a standard DA algorithm has the complete-data model $g(X_{\text{obs}}, X_{\text{mis}}|\theta)$ and the prior distribution $p(\theta)$ for $\theta \in \Theta$. The parameter-expanded (PXed) complete-data model is specified by augmenting the parameter to include an extra parameter $\alpha \in \mathbb{A} \subseteq \mathbb{R}^{d_x}$ for some positive integer d_x. The PXed complete-data model, denoted by $g_*(X_{\text{obs}}, X_{\text{mis}}|\theta_*, \alpha)$, $(\theta_*, \alpha) \in \Theta \times \mathbb{A}$, is required to preserve the observed-data model in the sense that for every $(\theta_*, \alpha) \in \Theta \times \mathbb{A}$, there exists a function $\theta = R_\alpha(\theta_*)$, called the reduction function, such that

$$\int_{\mathcal{X}_{\text{mis}}} g_*(X_{\text{obs}}, X_{\text{mis}}|\theta_*, \alpha) dX_{\text{mis}} = f(X_{\text{obs}}|\theta = R_\alpha(\theta_*)) \tag{2.16}$$

holds. Fixing α at its default value α_0 recovers the original complete-data model. Fixing α at different values results in different DA implementations. In other words, α indexes a class of DA schemes. This is explained by the following two toy examples.

Example 2.4 The Simple Random-Effects Model (continued)

Activate the mean of Z in (2.13) and write the original parameter θ as θ_*. This yields the PX version of (2.13)

$$Z|(\theta_*, \alpha) \sim N(\alpha, v) \quad \text{and} \quad Y|(Z, \theta_*, \alpha) \sim N(\theta_* + Z, 1) \tag{2.17}$$

where $\theta_* \in \mathbb{R}$ and $\alpha \in \mathbb{R}$ with its default value $\alpha_0 = 0$. The associated reduction function

$$\theta = R_\alpha(\theta_*) = \theta_* + \alpha.$$

is obtained by integrating out the missing data Z.

Example 2.5 is due to Lewandowski *et al.* (2010).

Example 2.5 The Simple Poisson-Binomial Random-Effects Model

Consider the complete-data model for the observed data $X_{\text{obs}} = X$ and the missing data $X_{\text{mis}} = Z$:

$$Z|\lambda \sim \text{Poisson}(\lambda) \quad \text{and} \quad X|(Z, \lambda) \sim \text{Binomial}(Z, \pi)$$

where $\pi \in (0, 1)$ is known and $\lambda > 0$ is the unknown parameter to be estimated. The observed-data model is $X|\lambda \sim \text{Poisson}(\pi\lambda)$. This provides another simple example for which analytical theoretical results can be easily derived.

Suppose that for Bayesian inference we take the prior $p(\lambda) \propto \lambda^{-\kappa}$, where $\kappa \in [0, 1]$ is a known constant. Then the posterior of λ given X is $\text{Gamma}(X + 1 - \kappa, \pi)$. Incidentally, we note that an imprecise Bayesian analysis can be carried out by considering the class of priors indexed by $\kappa \in [0, 1]$.

Activating the hidden parameter π gives the PXed complete-data model

$$Z|(\lambda_*, \alpha) \sim \text{Poisson}(\lambda_*) \quad \text{and} \quad X|(Z, \lambda_*, \alpha) \sim \text{Binomial}(Z, \alpha)$$

where the expansion parameter $\alpha \in (0, 1)$ takes its default value $\alpha_0 = \pi$. This PXed model has the reduction function

$$\lambda = R_\alpha(\lambda_*) = \frac{\alpha}{\pi}\lambda_*.$$

Conditional DA, termed by Meng and van Dyk (1999), implements the standard DA algorithm with a fixed α value, α_c, which is not necessarily its default value α_0. The preferred value α_c can be chosen according to certain criteria in the context of EM. Van Dyk and Meng (2001) suggest finding the α_c that minimizes the fraction of missing information. For complex models, the solution can be difficult to find analytically. Finding the optimal Conditional DA for each of the two toy examples is left as an exercise.

PX-EM treats expansion parameter α as a real parameter to be estimated for a better fit of the complete-data model to the imputed complete data, or more exactly, a larger value of (complete-data) log-likelihood. Liu *et al.* (1998) give a statistically intuitive explanation, in terms of covariance adjustment, of why PX-EM converges no slower than the original EM. For example, PX-EM converges in one iteration for the two toy examples, whereas EM can converge painfully slowly which gave rise to the idea of treating α as a real model parameter. What is needed for constructing the DA version of PX-EM is to specify a working prior for α, possibly conditioned on θ_*, so that α can be handled by standard DA updates.

Since the prior distribution for θ is available, it is sometimes convenient to consider the transformation, assumed to be one-to-one and differentiable,

$$\theta = R_{\alpha_*}(\theta_*) \quad \text{and} \quad \alpha = \alpha_*. \tag{2.18}$$

Let $\theta_* = R_\alpha^{-1}(\theta)$ denote the inverse mapping from θ to θ_* for fixed α. Then the complete-data likelihood function is given by

$$g_*(X_{\text{obs}}, X_{\text{mis}}|R_\alpha^{-1}(\theta), \alpha) \qquad (\theta \in \Theta; \alpha \in \mathbb{A})$$

■ Example 2.4 The Simple Random-Effects Model (continued)

In the space of (θ, α), the PXed model becomes

$$Z|(\theta, \alpha) \sim N(\alpha, v) \quad \text{and} \quad Y|(Z, \theta, \alpha) \sim N(\theta - \alpha + Z, 1) \tag{2.19}$$

Let $p_*(\theta, \alpha)$ be the joint prior for (θ, α), let $p_*(\theta)$ be the marginal prior for θ_*, and let $p_*(\alpha|\theta)$ be the conditional prior for α given θ. The following result on preserving the target posterior distribution $f(X_{\text{obs}}|\theta)p(\theta)$ is easy to prove; see Liu and Wu (1999).

Theorem 2.3.1 *Suppose that $p_*(\alpha|\theta)$ is a proper probability density function. Then the posterior distribution of θ is the same for both the original model and the parameter-expanded model if and only if the marginal prior distribution $p_*(\theta)$ agrees with $p(\theta)$, i.e., $p_*(\theta) \propto p(\theta)$, $\theta \in \Theta$.*

The following sampling algorithm is proposed by Meng and van Dyk (1999) and Liu and Wu (1999). Meng and van Dyk (1999) call it Marginal DA, whereas Liu and Wu (1999) call it PX-DA. For clarity, we refer to it as Marginal DA and to the version without drawing α in Step 1 as PX-DA.

The Marginal DA Algorithm: Marginalizing or Collapsing the Expansion Parameter

Take $\theta^{(0)} \in \Theta$, and iterate for $t = 1, 2, \ldots$

mI-Step. Draw (α, X_{mis}) from its conditional distribution given (X_{obs}, θ).

mP-Step. Draw (θ, α) from its conditional distribution given $(X_{\text{obs}}, X_{\text{mis}})$.

The mI-step can be implemented by first drawing α from $p_*(\alpha|\theta)$ and then drawing X_{mis} from its PXed predictive distribution $g_*(X_{\text{obs}}, X_{\text{mis}}|R_\alpha^{-1}(\theta), \alpha)$. The Marginal DA algorithm effectively marginalizes α out by drawing α in both mI-step and mP-step. In other words, the idea can be viewed as first introducing the expansion parameter as an auxiliary variable and then collapsing down this auxiliary variable.

This algorithm is very interesting in that it serves several purposes: Liu and Wu (1999) and Meng and van Dyk (1999) use proper priors $p^\omega(\alpha|\theta)$, indexed by ω, to establish a theory for using improper priors, with a limiting argument (discussed in Section 2.3.3.2); and, with some special degenerated cases of $p_*(\alpha|\theta)$, it provides a way of understanding other relevant sampling methods, such as conditional DA and hierarchical centering.

Taking a degenerated prior independent of θ, *i.e.*, $p_*(\alpha|\theta) \propto p_*(\alpha) = \delta_{\alpha_c}$, meaning $\Pr(\alpha = \delta_{\alpha_c}|\theta) = 1$, gives Conditional DA. Letting α be a function of θ, that is $\Pr(\alpha = \alpha(\theta)|\theta) = 1$, yields what we call the Reparameterized DA algorithm, which include hierarchical centering as a special case. For example, conditioned on θ, $\alpha = \alpha(\theta)$, and the observed data Y in Example 2.4, the missing data Z follows $N(\alpha(\theta) + v(Y-\theta)/(1+v), v/(1+v))$. With the degenerated prior having $\alpha(\theta) = c + (1+v)^{-1}v\theta$ for an arbitrary $c \in \mathbb{R}$ with probability one, the imputed missing data Z is independent of the current draw of θ. We note that for this example there exists a class of Reparameterized DAs, which are equivalent in the sense that they all converge in a single iteration.

■ Example 2.5 The Simple Poisson-Binomial Random-Effects Model (coutinued)

In the space of (λ, α), the PXed model becomes

$$Z|(\lambda, \alpha) \sim \text{Poisson}(\pi\lambda/\alpha) \quad \text{and} \quad X|(Z, \lambda, \alpha) \sim \text{Binomial}(Z, \alpha) \qquad (2.20)$$

Recall that $p(\lambda) \propto \lambda^{-\kappa}$. The pdf of the predictive distribution of Z given λ, $\alpha = \alpha(\lambda)$, and the observed data X is proportional to

$$\frac{1}{(Z-X)!}\left(\frac{\pi(1-\alpha(\lambda))}{\alpha(\lambda)}\lambda\right)^{Z-X} \qquad (Z \in \{X, X+1, \ldots\})$$

that is, $X + \text{Poisson}(\lambda\pi(1-\alpha(\lambda))/\alpha(\lambda))$. Let $\alpha(\lambda) = \lambda/(c+\lambda)$ for some constant $c > 0$. Then

$$Z|(X, \lambda, \alpha = \lambda/(c+\lambda)) \sim X + \text{Poisson}(c\pi),$$

independent of λ. Thus, we find a class of optimal Reparameterized DAs, indexed by c, all converge in one iteration. The target posterior of λ given X is derived from the conditional distribution of λ given (X, Z),

$$\lambda^{-\kappa}\left(\frac{\pi}{\alpha}\lambda\right)^Z \alpha^X(1-\alpha)^{Z-X}e^{-\frac{\pi}{\alpha}\lambda} \propto \lambda^{-\kappa}\lambda^X e^{-\pi\lambda} \qquad (\lambda > 0)$$

that is, the target posterior distribution $\text{Gamma}(X+1-\kappa, \pi)$.

2.3.3.2 The Data Transformation-Based PX-DA Algorithm

Suppose that a PXed model can be represented by a group of transformations of the missing data

$$X_{\text{mis}} = t_\alpha(T_{\text{mis}}) \qquad (\alpha \in \mathbb{A}) \tag{2.21}$$

where the α indexes the transformations. Assume that for any fixed $\alpha \in \mathbb{A}$, the transformation (2.21) is a one-to-one and differentiable mapping. Then the group of transformations induces a PXed complete-data model with the expansion parameter α

$$f(X_{\text{obs}}, X_{\text{mis}} = t_\alpha(T_{\text{mis}})|\theta)|J_\alpha(T_{\text{mis}})|, \tag{2.22}$$

where $J_\alpha(T_{\text{mis}}) = \det\{\partial t_\alpha(T_{\text{mis}})/\partial T_{\text{mis}}\}$ is the Jacobian evaluated at T_{mis}. This is the case for many augmented complete-data models; see, for example, Liu and Wu (1999), and van Dyk and Meng (2001), and Lawrence *et al.* (2008). It works for Example 2.4 but not for Example 2.5.

Example 2.4 The Simple Random-Effects Model (continued)

Take $Z = T - \alpha$. Then

$$Z|(\theta, \alpha) \sim N(\alpha, v) \quad \text{and} \quad Y|(Z, \theta, \alpha) \sim N(\theta - \alpha + Z, 1) \tag{2.23}$$

which is the same as (2.19) with the missing data Z now denoted by T.

To create most efficient Marginal DA, it is of interest to use very diffused prior α, which often corresponds to an improper prior. When an improper prior is used, the mI-step is not applicable. With the data transformation formulation (2.21) of PXed models, Liu and Wu (1999) and Meng and van Dyk (1999) take $p_*(\alpha|\theta) = p_*(\alpha)$, that is, α and θ are independent *a priori*, and consider a limiting procedure for using improper priors. Let $p_*^{(\omega)}(\alpha)$, indexed by ω. Suppose that $p_*^{(\omega)}(\alpha)$ converges to an improper prior $p_*^{(\omega_\infty)}(\alpha)$ as $\omega \to \omega_\infty$. For each $\omega \neq \omega_\infty$, Marginal DA results in a (Markov chain) transition kernel for θ. If the limiting prior $p_*^{(\omega_\infty)}(\alpha)$ produces the limit of the transition kernel for θ, then the Marginal DA sequence $\{\theta^{(t)} : t = 1, 2, \ldots\}$ has the target posterior distribution of θ as its stationary distribution. We refer to van Dyk and Meng (2001) for details and open issues on this method.

Assuming that the set of transformations (2.21) forms a locally compact group, Liu and Wu (1999; Scheme 2) propose an attractive DA algorithm, referred to here as the PX-DA algorithm because it has a similar structure and statistical interpretation to the PX-EM algorithm.

The PX-DA Algorithm

Take $\theta^{(0)} \in \Theta$, and iterate for $t = 1, 2, \ldots$

PX-I Step. Draw X_{mis} from its conditional distribution given (X_{obs}, θ).

PX-P Step. Draw (θ, α) according to

$$(\theta, \alpha)|(X_{\rm obs}, X_{\rm mis}) \sim f(X_{\rm obs}, t_\alpha(X_{\rm mis})|\theta)|J_\alpha(X_{\rm mis})|\frac{H(d\alpha)}{d\alpha}f(\theta).$$

where $H(d\alpha)$ represents the *right Haar measure* on the group $\mathbb{A}$ with density $\frac{H(d\alpha)}{d\alpha}$ with respect to the Lebesgue measure. For the translation group $\mathbb{A} = \{\alpha : \alpha \in \mathbb{R}^d\}$, corresponding to the transformations $t_\alpha(T_{\rm mis}) = T_{\rm mis} + \alpha$, $H(d\alpha)/(d\alpha) = 1$. For the scale group $\mathbb{A} = \{\alpha : \alpha > 0\}$, corresponding to the transformations $t_\alpha(T_{\rm mis}) = \alpha T_{\rm mis}$, $H(d\alpha)/(d\alpha) = \alpha^{-1}$. More generally, if $t_\alpha(T_{\rm mis}) = \alpha T_{\rm mis}$ with α being non-singular $k \times k$ matrix, the corresponding Haar measure is $H(d\alpha)/(d\alpha) = |\alpha|^{-k}$.

Example 2.4 The Simple Random-Effects Model (continued)

For this example, $|J_\alpha(Z)| = 1$ and $H(d\alpha)/(d\alpha) = 1$. The PX-DA algorithm share with the original DA algorithm its I-step, drawing Z from $N(v(\theta - Y)/(1+v), v/(1+v))$. The P-step of PX-DA draws (θ, α) and outputs $\theta \sim N(Y, 1+v)$, conditioned on Y and Z. That is, PX-DA generates independent draws of θ for this example.

For the simple Poisson-Binomial model, the PX-DA (or Marginal DA) is not applicable because the data transformation formulation does not work. A more general PX-DA algorithm can be defined. This is briefly discussed below.

2.3.3.3 The General PX-DA Algorithm

In a certain sense, data transformation introduced to construct PX-DA serves only as a technical device, but an important one due to its neat results when the set of transformations forms a locally compact group. A more general definition of PX-DA needs to be formulated without using data transformation. Such a general formulation would shed light on the essence of PX-EM, efficient inference via either efficient data augmentation or efficient data analysis. Suppose that $A_{\rm mis}$ is defined to represent a 'covariate' variable for covariance adjustment to be based on. Accordingly, write the missing data $X_{\rm mis}$ as

$$X_{\rm mis} = K(A_{\rm mis}, S_{\rm mis}) \tag{2.24}$$

to represent a one-to-one mapping between the missing data $X_{\rm mis}$ and its transformed representation by $A_{\rm mis}$ and $S_{\rm mis}$. The general PX-DA algorithm would be the one satisfying what we call the *efficiency condition*

$$A_{\rm mis} \perp \theta | \{X_{\rm obs}, S_{\rm mis}\} \tag{2.25}$$

that is, after the PX-P step, θ is independent of $A_{\rm mis}$, given $S_{\rm mis}$ and $X_{\rm obs}$. Provided in addition that the existence of $(A_{\rm mis}, S_{\rm mis})$ and an associated expanded prior $p_*(\theta_*, \alpha)$ that satisfies what we call the *validity condition*

$$\int_{\mathbb{A}} g_*(X_{\text{obs}}, X_{\text{mis}}|R_\alpha^{-1}(\theta), \alpha)p_*(\theta_*, \alpha) \left|\frac{\partial R_\alpha^{-1}(\theta)}{\partial \alpha}\right| d\alpha \propto f(X_{\text{obs}}|\theta)p(\theta) \quad (2.26)$$

we define the general PX-DA as follows:

The General PX-DA Algorithm

Take $\theta^{(0)} \in \Theta$, and iterate for $t = 1, 2, \ldots$ {

PX-I Step. Draw X_{mis} from its conditional distribution given (X_{obs}, θ).

PX-P Step. First draw (θ_*, α) according to

$$(\theta_*, \alpha)|(X_{\text{obs}}, X_{\text{mis}}) \sim g_*(X_{\text{obs}}, X_{\text{mis}}|\theta_*, \alpha)p_*(\theta_*, \alpha),$$

and then compute $\theta = R_\alpha(\theta_*)$.

Of course, mathematically, the PX-P step can be written as drawing (θ, α) without the need of the reduction step. While the existence of S_{mis} and $p_*(\alpha|\theta_*)$ remains as an open problem, it is illustrated by Example 2.5 that such a general definition of PX-DA can be important in maintaining the main feature of PX-EM.

■ **Example 2.5 The Simple Poisson-Binomial Random-Effects Model (continued)**

From the validity condition, $p_*(\lambda)p_*(\alpha|\lambda) = p(\lambda)p_*(\alpha|\lambda)$ and $p(\lambda) \propto \lambda^{-\kappa}$, $p_*(\lambda_*, \alpha)$ can be written as, with a slightly abuse of the notation for $\tilde{p}_*(\alpha|\lambda_*)$,

$$p(\lambda = R_\alpha(\lambda_*))p_*(\alpha|\lambda = R_\alpha(\lambda_*))|\partial R_\alpha(\lambda_*)/\partial\lambda_*| \propto \lambda_*^{-\kappa}\alpha^{1-\kappa}\tilde{p}_*(\alpha|\lambda_*)$$

Thus, the posterior is proportional to

$$\lambda_*^{Z-\kappa}e^{-\lambda_*}\alpha^X(1-\alpha)^{Z-X}\tilde{p}_*(\alpha|\lambda_*)$$

Suppose that constrained by the requirement that the resulting algorithm is simple to implement, take

$$\tilde{p}_*(\alpha|\lambda_*) \propto \alpha^{c_1-1}(1-\alpha)^{c_2-1} \qquad (\alpha \in (0,1))$$

A simple PX-DA implementation exists and is obtained by setting $c_1 = 1 - \kappa$ and $c_2 = 0$, because in this case

$$p(\lambda|X, Z) = \int p_*(\lambda_* = R_\alpha^{-1}(\lambda), \alpha|X, Z)|\partial R_\alpha^{-1}(\lambda)/\partial\lambda|d\alpha$$

is independent of Z. The PX-I step imputes $Z \sim X + \text{Poisson}(\lambda[1-\pi])$. The PX-P step generates (λ_*, α) by drawing $\lambda_* \sim \text{Gamma}(Z+1-\kappa)$ and

$\alpha \sim \text{Beta}(1-\kappa+X, Z-X)$, and then computes the $\lambda = R_\alpha(\lambda_*) = \lambda_*\alpha/\pi$. This results in

$$\lambda|(Z,X) \sim \pi^{-1}\text{Gamma}(X+1-\kappa)$$

after the PX-P step.

2.3.4 Alternating Subspace-Spanning Resampling

2.3.4.1 The CA-DA Algorithm

To avoid the need for an expanded prior in the (general) PX-DA algorithm, Liu (2003) considers an alternative to PX-DA, called CA-DA for 'covariance-adjusted data augmentation'. Using the notation in the discussion of the general PX-DA algorithm in Section 2.3.3.2, the CA-DA algorithm is given as follows.

The CA-DA Algorithm

Take $\theta^{(0)} \in \Theta$, and iterate for $t = 1, 2, \ldots$

CA-I Step. Draw $X_{\text{mis}} = K(A_{\text{mis}}, S_{\text{mis}}) \sim f(X_{\text{mis}}|\theta, \theta)$.

CA-P Step. Draw (θ, A_{mis}) from its conditional distribution given $(X_{\text{obs}}, S_{\text{mis}})$.

Since A_{mis} is effectively integrated out, the efficiency condition holds for the CA-DA algorithm. The CA-DA algorithm becomes trivial for the two toy examples because the CA-I step becomes redundant. We illustrate CA-DA in Section 2.4.

2.3.4.2 The ASSR Algorithm

The idea of *partial resampling* (drawing A_{mis} again in the CA-P step) in the CA-DA can be generalized to the more general Gibbs sampler or any other iterative simulation sampling methods, as formulated by Liu (2003). Assume that a Markov chain sampler such as the Gibbs sampler produces a sequence $\{X^{(t)} : t = 1, 2, \ldots\}$ with the equilibrium distribution $f(X)$ on the space $\mathbb{X}$. Write

$$X^{(t+1)} = \mathcal{M}(X^{(t)}) \qquad (t = 0, 1, \ldots)$$

Consider a one-to-one mapping from $\mathbb{X}$ onto the sampling space of a pair of variables, denoted by (A, S):

$$C = C(X) \quad \text{and} \quad D = D(X) \qquad (X \in \mathbb{X})$$

The ultimate goal of speeding up the sampler $\mathcal{M}$ is to be able to create iid draws $\{X^{(t)} : t = 1, 2, \ldots\}$ from $f(X)$. A much relaxed requirement is that

$$D^{(t)} \perp D^{(t+1)}|C^{(t)}$$

holds for $t = 1, 2, \ldots$. This is possible, provided that it can be implemented without much expense to draw D from its conditional distribution given C. One way of doing this to use what Liu (2003) calls the alternating subspace-spanning resampling (ASSR, pronounced as 'AS SIR') method.

The ASSR Algorithm

Take $X^{(0)} \in \mathbb{X}$, and iterate for $t = 1, 2, \ldots$

P-step. Complete an iteration of the parent sampler $\mathcal{M}$ to obtain $\tilde{X}^{(t)} = \mathcal{M}(X^{(t-1)})$.

A-step. Draw D from its conditional distribution given $C^{(t)} = C(\tilde{X}^{(t)})$ and obtain $X^{(t)}$ from $(C^{(t)}, D)$.

The supplemental A-step not only guarantees the correctness of the ASSR algorithm but also makes it faster than its parent sampler. This is evident from its special case, the CA-DA algorithm. Application of ASSR to accelerating ASSR algorithms leads to the multi-cycle ASSR algorithm, as a simple extension of ASSR. Liu (2003) demonstrates that ASSR is efficient for linear regression with interval-censored responses while the standard DA algorithm can be too slow to be useful; see Section 2.4.3.

Tierney (1994) briefly discusses the idea of combining different MCMC sampling schemes, which is certainly related to the idea of partial sampling in ASSR. In their discussion of Besag *et al.* (1995), Gelfand *et al.* (1995) consider combining multiple Gibbs samplers for a linear mixed-effects model. The basic idea is to get an 'average' performance of multiple samplers. In this sense, random scan Gibbs samplers (see, e.g., Levine and Casella, 2008), as opposed to the systematic scan Gibbs sampler of Section 2.1, share the same fundamental idea. The basic ASSR algorithm can be viewed as the partial combination of two samplers, one the parent sampler $\mathcal{M}$, the other a two-step Gibbs sampler that iterates between the two conditionals based on C and D. Note that C and D defines a different coordinate system or a different subspace-spanning scheme.

How to choose a new coordinate system for the A-step of ASSR is a challenging problem. The success of PX-EM for a complete-data model would suggest to consider grouping the model parameter and sufficient statistics for the expansion parameter as the D variable and defining the C variable that is approximately independent of the D variable. When the idea of parameter expansion is not used, the sufficient statistics for the expansion parameter are typically the ancillary statistics for θ. Yu and Meng (2008) investigate further along this line and provide an impressive example of speeding up the Gibbs sampler for complex Bayesian models. There are certainly other ways for finding subspaces worth resampling. For example, Liu and Rubin (2002) consider finding slow converging subspaces via Markov-normal analysis of multiple chains before convergence.

2.4 Applications

2.4.1 The Student-t Model

The Student-t distribution has served as a useful tool for robust statistical inference (Lange *et al.*, 1989; Liu and Rubin, 1995; Liu, 1996; Pinheiro *et al.*, 2001; and Zhang *et al.*, 2009). As noted in Liu (1997), maximum likelihood (ML) estimation of the t-distribution has also motivated several EM-type algorithms, including the ECME algorithm (Liu and Rubin, 1994), the AECM algorithm (Meng and van Dyk, 1997), and the PX-EM algorithm (Liu *et al.*, 1998). Here we consider the univariate t-distribution $t_\nu(\mu, \sigma^2)$ with the unknown center μ, the unknown scale parameter σ, and the known degrees of freedom ν (>0).

Suppose that we are interested in Bayesian fitting $t_\nu(\mu, \sigma^2)$ to the observed data $y_1, \ldots, y_n$ with the usual prior

$$p(\mu, \sigma^2) \propto \frac{1}{\sigma^2} \qquad (\mu \in \mathbb{R}; \sigma^2 > 0)$$

The commonly used complete-data model for the t-model is as follows: for $i = 1, \ldots, n$, (τ_i, y_i) are independent with

$$y_i|(\tau_i, \mu, \sigma) \sim N\left(\mu, \frac{\sigma^2}{\tau_i}\right) \quad \text{and} \quad \tau_i|(\mu, \sigma^2) \sim \text{Gamma}\left(\frac{\nu}{2}, \frac{\nu}{2}\right), \tag{2.27}$$

where τ_i are called the weights and the distribution Gamma (α, β) has density proportional to $\tau^{\alpha-1}\exp\{-\beta\tau\}$ for $\tau > 0$. Thus, conditioned on (μ, σ^2) and the observed data $y_1, \ldots, y_n$, τ_i's are independent with

$$\tau_i|(\mu, \sigma^2, y_1, \ldots, y_n) \sim \text{Gamma}\left(\frac{\nu+1}{2}, \frac{\nu + (y_i - \mu)^2/\sigma^2}{2}\right) \tag{2.28}$$

for $i = 1, \ldots, n$. This provides the conditional distribution for implementing the I-step of DA.

Let IG(α, β) denote the inverse Gamma distribution with shape parameter α and rate parameter β. Then the DA algorithm is given as follows:

I-step. Draw τ_i from (2.28) for $i = 1, \ldots, n$.

P-step. First draw σ^2 from IG$((n-1)/2, (1/2)\sum_{i=1}^n \tau_i(y_i - \bar{y})^2)$ and then draw μ from $N((\bar{y}, \sigma^2/\sum_{i=1}^n \tau_i)$, where $\bar{y} = \sum_{i=1}^n \tau_i y_i / \sum_{i=1}^n \tau_i)$.

The P-step can be verified via the routine algebraic operation. Note that this DA can be viewed as obtained via 'blocking' from the three-step Gibbs that splits the P-step into two steps.

Notice that the complete-data model (2.27) has a hidden parameter, denoted by α, for the scale of τ_i's. This hidden parameter α is fixed at $\alpha_0 = 1$ for identifiability of the observed-data model because α is confounded with σ_*^2

in the induced observed-data model, when α is to be estimated from the observed data,

$$y_i \stackrel{iid}{\sim} t_\nu(\mu, \sigma_*^2/\alpha) \qquad (i = 1, \ldots, n)$$

where σ_*^2 plays the same role of σ^2 in the original model, where $\alpha = \alpha_0 = 1$. When activated for a better fit of the imputed complete data, this expansion parameter α indexes a one-to-one mapping between σ_*^2 and σ^2:

$$\sigma^2 = R_\alpha(\sigma_*^2) = \sigma_*^2/\alpha \qquad (\sigma_*^2 > 0; \alpha > 0)$$

This is the basic setting for applying the PX-EM algorithm.

To apply the (data transformation-based) PX-DA, consider the data transformation

$$\tau \equiv (\tau_1, \ldots, \tau_n) = \lambda(w_1, \ldots, w_n).$$

The associated Jacobian is $|J_\lambda(w)| = \lambda^n$, where $w = (w_1, \ldots, w_n)$. The PX-P step is to draw (μ, σ^2, λ) from

$$\lambda^{\frac{n\nu}{2}-2} e^{-\frac{\lambda\nu}{2}\sum_{i=1}^n \tau_i} \left(\frac{\lambda}{\sigma^2}\right)^{1+n/2} e^{-\frac{\lambda}{2\sigma^2}\sum_{i=1}^n \tau_i (y_i - \mu)^2} \tag{2.29}$$

A simple way to draw (μ, σ^2, λ) from (2.29) is to make use of the transformation

$$\sigma^2 = \lambda \sigma_*^2, \tag{2.30}$$

conditional on (μ, λ), and to draw $(\mu, \sigma_*^2, \lambda)$ from

$$\lambda^{\frac{n\nu}{2}-1} e^{-\frac{\lambda\nu}{2}\sum_{i=1}^n \tau_i} (\sigma_*^2)^{-n/2} e^{-\frac{1}{2\sigma_*^2}\sum_{i=1}^n \tau_i (y_i - \mu)^2}$$

Note that conditioned on τ, λ and (σ_*^2, μ) are independent with

$$\lambda \sim \text{Gamma}\left(\frac{n\nu}{2}, \frac{\nu \sum_{i=1}^n \tau_i}{2}\right) \tag{2.31}$$

and (μ, σ_*^2) having the same distributed as (μ, σ^2) in the standard DA algorithm. To summarize, the PX-DA algorithm can be written as

PX-I step. This is the same as the I-step of DA.

PX-P step. Draw (μ, σ_*^2) in the exactly the same way of drawing (μ, σ^2) in DA; draw λ from (2.31); and compute $\sigma^2 = \lambda\sigma_*^2$, according to (2.30).

What PX-DA effectively does is to resample $A_{\text{mis}} = \sum_{i=1}^n \tau_i$ conditioned on the observed data and $S_{\text{mis}} = (w_1, \ldots, w_n)$ with $w_i = \tau_i / \sum_{i=1}^n \tau_i$ for $i = 1, \ldots, n$. This is made explicit in the CA-DA algorithm; see Exercise 2.6.

2.4.2 Robit Regression or Binary Regression with the Student-t Link

Consider the observed data consisting of n observations $X_{\rm obs} = \{(x_i, y_i) : i = 1, \ldots, n\}$ with a p-dimensional covariates vector x_i and binary response y_i that takes on values of 0 and 1. The binary regression model with Student-t link assumes that given the covariates, the binary responses y_i's are independent with the marginal probability distributions specified by

$$\Pr(y_i = 1|x_i, \beta) = 1 - \Pr(y_i = 0|x_i, \beta) = F_\nu(x_i'\beta) \qquad (i = 1, \ldots, n) \quad (2.32)$$

where $F_\nu(.)$ denotes the cdf of the Student-t distribution with center zero, unit scale, and ν degrees of freedom. With $\nu \approx 7$, this model provides a robust approximation to the popular logistic regression model for binary data analysis; see Mudholkar and George (1978), Albert and Chib (1993), and Liu (2004). Liu (2004) calls the binary regression model with Student-t link the *Robit regression*.

Here we consider the case with a known ν. A complete-data model for implementing EM to find the ML estimate of β is specified by introducing the missing data consisting of independent latent variables (τ_i, z_i) for each $i = 1, \ldots, n$ with

$$\tau_i|\beta \sim \text{Gamma}(\nu/2, \nu/2) \quad (2.33)$$

and

$$z_i|(\tau_i, \beta) \sim \text{N}(x_i'\beta, 1/\tau_i). \quad (2.34)$$

Let

$$y_i = \begin{cases} 1, & \text{if } z_i > 0; \\ 0, & \text{if } z_i \leq 0 \end{cases} \qquad (i = 1, \ldots, n) \quad (2.35)$$

then the marginal distribution of y_i is preserved and is given by (2.32). The complete-data model belongs to the exponential family and has the following sufficient statistics for β:

$$S_{\tau xx} = \sum_{i=1}^n \tau_i x_i x_i', \quad S_{\tau xz} = \sum_{i=1}^n \tau_i x_i z_i'. \quad (2.36)$$

For Bayesian estimation of the Robit regression model, here we use the multivariate $\mathbf{t}$-distribution

$$\text{pr}(\beta) = \mathbf{t}_p(0, S_0^{-1}, \nu_0) \quad (2.37)$$

as the prior distribution for the regression coefficients β, where S_0 is a known $(p \times p)$ non-negative definite scatter matrix and ν_0 is the known number of degrees of freedom. When S_0 is a positive definite matrix, the posterior distribution of β is proper because the likelihood is bounded. When $S_0 = 0$ the prior distribution for β is flat and β may have an improper posterior in the sense that $\int_\beta \text{pr}(\beta)\,\ell(\beta|Y_{\rm obs})d\beta = \infty$. Chen and Shao (1997) discuss this issue.

The **t**-distribution (2.37) can be represented as the marginal distribution of β in the following well-known hierarchical structure:

$$\tau_0 \sim \text{Gamma}(\nu_0/2, \nu_0/2) \quad \text{and} \quad \beta|\tau_0 \sim \text{N}_p(0, S_0^{-1}/\tau_0). \tag{2.38}$$

Like the missing weights τ_i ($i = 1, \ldots, n$), in the sequel τ_0 is treated as missing. The complete data for generating draws of β from its posterior distribution using the DA algorithm consist of Y_{obs}, $z = (z_1, \ldots, z_n)$ and $\tau = (\tau_0, \tau_1, \ldots, \tau_n)$.

The DA implementation for simulating the posterior of β consists of the following I-step and P-step:

I-step. Conditioning on the observed data and the current draw of β, draw $\{(z_i, \tau_i) : i = 1, \ldots, n\}$ by first taking a draw of z_i from the truncated $\mathbf{t}(\mu_i = x_i'\beta, 1, \nu)$, which is either left ($y_i = 1$) or right ($y_i = 0$) truncated at 0, and then taking a draw of τ_i from

$$\text{Gamma}\left(\frac{\nu+1}{2}, \frac{\nu+(z_i-\mu_i)^2}{2}\right)$$

for all $i = 1, \ldots, n$, and a draw of τ_0 from its distribution given in (2.38).

P-step. Conditioning on the current draws of $\{(z_i, \tau_i) : i = 1, \ldots, n\}$, draw β from the p-variate normal distribution

$$\text{N}_p\left(\hat{\beta}, (\tau_0 S_0 + S_{\tau xx})^{-1}\right),$$

where

$$\hat{\beta} = (\tau_0 S_0 + S_{\tau xx})^{-1} S_{\tau xz}, \tag{2.39}$$

and $S_{\tau xx}$ and $S_{\tau xz}$ are defined in (2.36).

For a more efficient DA algorithm, we take the CA-DA approach. Two versions are considered below. The first adjusts individual scores z_i for their common scale parameter, denoted by σ. The sufficient statistic for σ, after integrating out the regression coefficients β, is

$$s^2 = \sum_{i=1}^{n} \tau_i \left(z_i - x_i'\hat{\beta}\right)^2 + \hat{\beta}'\tau_0 S_0 \hat{\beta},$$

where $\hat{\beta} = (\tau_0 S_0 + S_{\tau xx})^{-1} S_{\tau xz}$. To draw (s^2, β) with z_i ($i = 1, \ldots, n$) fixed up to a proportionality constant (*i.e.*, the scale of z_is), take the re-scaling transformation

$$z_i^* = z_i/s \qquad (i = 1, \ldots, n). \tag{2.40}$$

with the constraint

$$\sum_{i=1}^{n} \tau_i \left(z_i^* - x_i'\hat{\beta}^*\right)^2 + (\hat{\beta}^*)'\tau_0 S_0 \hat{\beta}^* = 1, \tag{2.41}$$

where $\hat{\beta}^* = (\tau_0 S_0 + S_{\tau xx})^{-1} S_{\tau xz^*}$ with $S_{\tau xz^*}$ obtained from $S_{\tau xz}$ by substituting z_i^* for z_i. Since the transformation (2.40) from $(z*, s)$ to z with the constraint (2.41) is one-to-one, a CA-DA version can be obtained from DA by replacing the P-step of DA with a step that draws (β, s^2), conditioning on z^*. The Jacobian of the transformation from (z, β) onto $(z^*, s, \eta = \beta)$ with the constraints (2.41), as a function of (s, η), is proportional to s^{n-1}. The conditional distribution of (s, η) given z^* is then

$$\mathrm{pr}(s, \eta|\tau, z^*, Y_{\mathrm{obs}}) = \mathrm{pr}(s|\tau, z^*, Y_{\mathrm{obs}}) \cdot \mathrm{pr}(\eta|s, \tau, z^*, Y_{\mathrm{obs}}),$$

where $\mathrm{pr}(s^2|\tau, z^*, Y_{\mathrm{obs}}) = \mathrm{Gamma}(n/2, 1/2)$ and $\mathrm{pr}(\eta|s, \tau, z^*, Y_{\mathrm{obs}}) = \mathrm{N}(s\hat{\beta}^*, (\tau_0 S_0 + S_{\tau xx})^{-1})$. The resulting CA-DA, denoted by CA-DA1 is summarized as follows:

CA-I step. This is the same as the I-step of DA.

CA-P step. This is the same as the P-step of DA, except for rescaling $\hat{\beta}$ by a factor of $\chi_n / \left[\sum_{i=1}^n \tau_i (z_i - x_i'\hat{\beta})^2 + \hat{\beta}'\tau_0 S_0 \hat{\beta}\right]^{1/2}$, where χ_n^2 is a draw from the chi-square distribution with n degrees of freedom.

For the probit regression model, for example, $\nu = \infty$ and thereby $\tau_i = 1$ for all $i = 1, \ldots, n$, CA-DA 1 is equivalent to the PX-DA algorithm of Liu and Wu (1999), who considered a flat prior on β. The CA-P step of CA-DA1 implicitly integrates out the scale of z_is, which explains intuitively again why CA-DA converges faster than DA.

The second version adjusts both σ and the individual weights for their scale to obtain a DA sampling scheme that is even faster than CA-DA1. Let

$$w = \sum_{i=0}^{n} \nu_i \tau_i \quad \text{and} \quad w s^2 = \sum_{i=1}^{n} \tau_i (z_i - x_i'\hat{\beta})^2 + \hat{\beta}'\tau_0 S_0 \hat{\beta},$$

where $\nu_i = \nu$ for all $i = 1, \ldots, n$. Take the transformation

$$\tau_i = w\tau_i^* \quad (i = 0, \ldots, n; w > 0) \quad \text{and} \quad z_i = s z_i^* \quad (i = 1, \ldots, n; w > 0)$$

with the constraints

$$\sum_{i=0}^{n} \nu_i \tau_i^* = 1 \quad \text{and} \quad \sum_{i=1}^{n} \tau_i^* (z_i^* - x_i'\hat{\beta}^*)^2 + \hat{\beta}'\tau_0 S_0 \hat{\beta} = 1, \tag{2.42}$$

where $\hat{\beta}^* = (\tau_0^* S_0 + S_{\tau^* xx})^{-1} S_{\tau^* x z^*} = (\tau_0 S_0 + S_{\tau xx})^{-1} S_{\tau x z^*}$ with $S_{\tau^* xx}$ and $S_{\tau^* x z^*}$ obtained from $S_{\tau xx}$ and $S_{\tau xz}$, respectively, by replacing τ_i with τ_i^* and z_i with z_i^*. The Jacobian of the transformation from (τ, z, β) to $(\tau^*, z^*, w, s, \eta = \beta)$ with the constraints (2.42), as a function of (w, s, η) is proportional to $w^n s^{n-1}$. Thus, conditioning on z^*, τ^*, and Y_{obs}, $(w, s, \eta = \beta)$ is distributed as

$$\mathrm{pr}(w|z^*, \tau^*, Y_{\mathrm{obs}}) \cdot \mathrm{pr}(s|w, z^*, \tau^*, Y_{\mathrm{obs}}) \cdot \mathrm{pr}(\beta|w, s, z^*, \tau^*, Y_{\mathrm{obs}}),$$

where $\mathrm{pr}(w|z^*, \tau^*, Y_{\mathrm{obs}}) = \mathrm{Gamma}((\nu_0 + n\nu)/2, 1/2)$, $\mathrm{pr}(s^2|w, z^*, \tau^*, Y_{\mathrm{obs}}) = \mathrm{Gamma}(n/2, \ w/2)$, and $\mathrm{pr}(\beta|w, s, z^*, \tau^*, Y_{\mathrm{obs}}) = \mathrm{N}_p(s\hat{\beta}^*, w^{-1}(\tau_0^* S_0 + S_{\tau^* xx})^{-1})$. This leads to the following CA-DA algorithm:

CA-I step This is the same as the I-step of DA.

CA-P step. This is the same as the P-step of E-DA 1, except for rescaling the draw of β by a factor of $\left(\sum_{i=0}^{n} \nu_i \tau_i / \chi^2_{\nu_0+n\nu}\right)^{1/2}$, where $\chi^2_{\nu_0+n\nu}$ is a draw from the chi-square distribution with $\nu_0 + n\nu$ degrees of freedom.

This CA-DA has a P-step that implicitly integrates out both the scale of z_is and the scale of τ_is, which explains why it converges faster than both DA and CA-DA1.

2.4.3 Linear Regression with Interval-Censored Responses

Consider the experiment of improving the lifetime of fluorescent lights (Taguchi, 1987; Hamada and Wu, 1995; Liu and Sun, 2000; Liu 2003). The experiment was conducted over a time period of 20 days, with inspection every two days. It employed a 2^{5-2} fractional factorial design. The design matrix and the lifetime data are tabulated in Table 2.1, where the lifetime data are given as intervals representing the censored observations.

Let x_i be the k-vector of the factor levels, including the intercept, and let y_i be the logarithm of the corresponding lifetime for $i = 1, \ldots, n$, where

Table 2.1 The design matrix and lifetime data for the fluorescent-light experiment (Hamada and Wu, 1995).

	design								
run	intercept	A	B	C	D	E	AB	BD	lifetime (no. of days)
1	1	1	1	1	1	1	1	1	[14, 16)
2	1	1	1	−1	−1	−1	1	−1	[18, 20)
3	1	1	−1	1	1	−1	−1	−1	[8, 10)
4	1	1	−1	−1	−1	1	−1	1	[18, 20)
5	1	−1	1	1	−1	1	−1	−1	[20, ∞)
6	1	−1	1	−1	1	−1	−1	1	[12, 14)
7	1	−1	−1	1	−1	−1	1	1	[16, 18)
8	1	−1	−1	−1	1	1	1	−1	[12, 14)
9	1	1	1	1	1	1	1	1	[20, ∞)
10	1	1	1	−1	−1	−1	1	−1	[20, ∞)
11	1	1	−1	1	1	−1	−1	−1	[10, 12)
12	1	1	−1	−1	−1	1	−1	1	[20, ∞)
13	1	−1	1	1	−1	1	−1	−1	[20, ∞)
14	1	−1	1	−1	1	−1	−1	1	[20, ∞)
15	1	−1	−1	1	−1	−1	1	1	[20, ∞)
16	1	−1	−1	−1	1	1	1	−1	[14, 16)

$k = 8$ and $n = 2 \times 2^{5-2} = 16$, and denoting by $[Y_i^{(\ell)}, Y_i^{(r)})$ the observed censoring interval for y_i, that is, $y_i \in [Y_i^{(\ell)}, Y_i^{(r)})$. Write $Y = (y_1, \ldots, y_n)$, $Y_{\text{obs}} = \{[Y_i^{(\ell)}, Y_i^{(r)}) : i = 1, \ldots, n\}$, and X for the $(n \times k)$ design matrix, that is, the i-th row of X is x_i for $i = 1, \ldots, n$. We consider the following model (Hamada and Wu, 1995):

$$y_i|(\beta, \sigma^2) \overset{\text{iid}}{\sim} N(x_i'\beta, \sigma^2), \quad i = 1, \ldots, n$$

with the prior distribution

$$\text{pr}(\beta, \sigma^2) = \text{pr}(\sigma^2)\text{pr}(\beta|\sigma^2)$$

for (β, σ^2) with $\text{pr}(\sigma^2) = \text{IG}(\nu_0/2, \nu_0 s_0/2)$ and $\text{pr}(\beta|\sigma^2) = N_k(\beta_0, \sigma^2 I_k/\tau_0)$, where IG denotes the inverted Gamma distribution, that is,

$$\text{pr}(\sigma^{-2}) \propto (\sigma^{-2})^{\nu_0/2-1} \exp\{-\nu_0 s_0/(2\sigma^2)\},$$

$\nu_0 = 1$, $s_0 = 0.01$, $\beta_0 = (3, 0, \ldots 0)$, I_k is the $(k \times k)$ identity matrix, and $\tau_0 = 0.0001$. For convenience, we replace β_0 with 0 and, accordingly, $y_i \leftarrow (y_i - 3)$, $Y_i^{(\ell)} \leftarrow (Y_i^{(\ell)} - 3)$, and $Y_i^{(r)} \leftarrow (Y_i^{(r)} - 3)$ for $i = 1, \ldots, n$.

The DA algorithm for this model is straightforward. Each iteration of DA consists of an I-step that imputes the missing values given the current draw of the parameters $\theta = (\beta, \sigma^2)$ and the observed data and a P-step that draws θ from its posterior given the currently imputed complete data. More precisely:

The DA Algorithm

I-step. Draw Y from $\text{pr}(Y|Y_{\text{obs}}, \beta, \sigma^2) = \prod_{i=1}^n \text{pr}(y_i|Y_{\text{obs}}, \beta, \sigma^2)$, where $\text{pr}(y_i|Y_{\text{obs}}, \beta, \sigma^2)$ is the truncated normal

$$\text{pr}(y_i|Y_{\text{obs}}, \beta, \sigma^2) \propto \frac{1}{\sigma} \exp\{-(y_i - x_i\beta)^2/(2\sigma^2)\} \qquad (y_i \in (Y_i^{(\ell)}, Y_i^{(r)}))$$

for $i = 1, \ldots, n$.

P-step. Draw (β, σ^2) from $\text{pr}(\beta, \sigma^2|Y, Y_{\text{obs}}) = \text{pr}(\beta, \sigma^2|Y) = \text{pr}(\sigma^2|Y)\text{pr}(\beta|Y, \sigma^2)$, where

$$\text{pr}(\sigma^2|Y) = \text{IG}\left((\nu_0 + n + k)/2, (\nu_0 s_0 + \tau_0\beta'\beta + \textstyle\sum_{i=1}^n (y_i - x_i\beta)^2)/2\right)$$

and

$$\text{pr}(\beta|Y, \sigma^2) = \text{N}_k\left((X'X + \tau_0 I_k)^{-1}X'Y, \sigma^2(X'X + \tau_0 I_k)^{-1}\right).$$

Unfortunately, although simple, this algorithm is so slow that the DA sequences are of little use in estimating posterior distributions. The trajectory plots of the DA sequence, Figure 2.2, show that the slowly converging subspace is associated with the regression coefficients β. To remedy the slow convergence of DA for this example, we add a CA-step that redraws $\alpha \equiv X_{[1:8]}\beta$ jointly with its complete-data sufficient statistics $d \equiv (y_1 + y_9, \ldots, y_8 + y_{16})'$

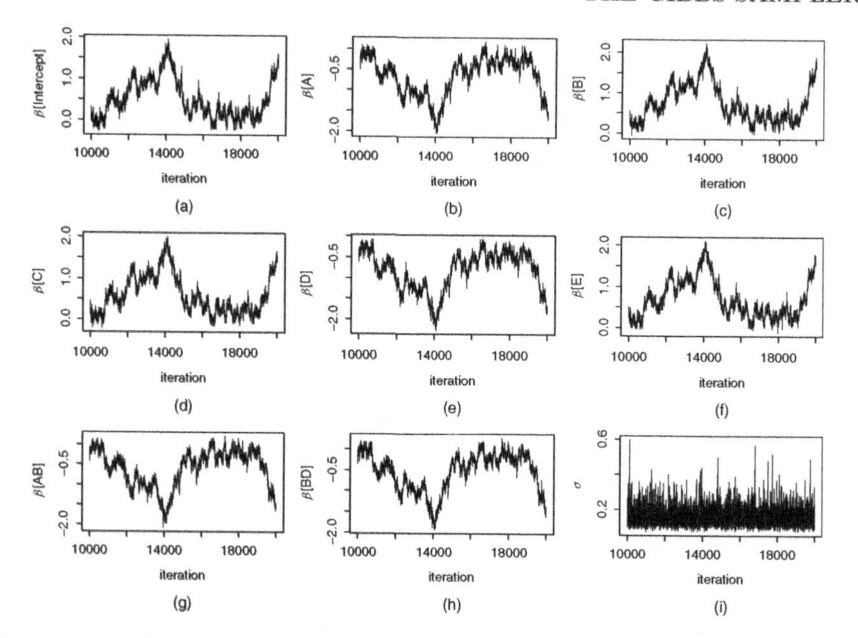

Figure 2.2 The DA sequences of β and σ in the model for the lifetime data in the fluorescent-light lifetime experiment.

given σ^2 and $c \equiv (y_1 - y_9, \ldots, y_8 - y_{16})'$, where $X_{[1:8]}$ is the (8×8) matrix consisting of the first 8 rows of X. The CA-step is given as follows:

CA-step. Draw $D = (\alpha, d)$ from $\text{pr}(\alpha, d|\sigma^2, c)$, the conditional distribution of D given $C = (c, \sigma^2)$, and then adjust (β, Y) by $\beta \leftarrow (1/8)X'_{[1:8]}\alpha$ and $y_i \leftarrow (d_i + c_i)/2$ and $y_{i+8} \leftarrow (d_i - c_i)/2$ for $i = 1, \ldots, k$.

Thus, each iteration of the adjusted DA algorithm consists of three steps: the I-step, the P-step, and the CA-step.

To implement the CA-step, we need the conditional distribution of $D \equiv (\alpha, d)$ given $C \equiv (c, \sigma^2)$. From the conditional distribution of β and Y given σ^2, $\text{pr}(\beta, Y|\sigma^2)$, is proportional to

$$\exp\left\{-\frac{1}{2\sigma^2}\left[(Y - X\beta)'(Y - X\beta) + \tau_0\beta'\beta\right]\right\} \qquad \left(y_i \in [Y_i^{(\ell)}, Y_i^{(r)})\right)$$

and the fact that the Jacobian of the transformation from (β, Y) to (α, d, c) is constant, we obtain the conditional distribution of α, d, and c given σ^2

$$\text{pr}(\alpha, d, c|\sigma^2) \propto \exp\left\{-\frac{1}{2\sigma^2}\left[\frac{2k + \tau_0}{k}\alpha'\alpha - 2d'\alpha + \frac{1}{2}(d'd + c'c)\right]\right\}$$

where $\alpha_i \in (-\infty, +\infty)$, $d_i + c_i = 2y_i \in (2Y_i^{(\ell)}, 2Y_i^{(r)})$, and $d_i - c_i = 2y_{k+i} \in (2Y_{k+i}^{(\ell)}, 2Y_{k+i}^{(r)})$ for $i = 1, \ldots, k = 8$. Thus, we have the conditional distribution of $D \equiv (\alpha, d)$ given $C \equiv (c, \sigma^2)$

$$\text{pr}(\alpha, d|c, \sigma^2) \propto \exp\left\{-\frac{1}{2\sigma^2}\left[\frac{2k+\tau_0}{k}\alpha'\alpha - 2d'\alpha + \frac{1}{2}d'd\right]\right\}$$

where $\alpha_i \in (-\infty, +\infty)$ and

$$d_i \in R_i \equiv \left(\max\{2Y_i^{(\ell)} - c_i, 2Y_{i+8}^{(\ell)} + c_i\}, \min\{2Y_i^{(r)} - c_i, 2Y_{i+8}^{(r)} + c_i\}\right)$$

for $i = 1, \ldots, k$. That is, $\text{pr}(\alpha, d|\sigma^2, c) = \prod_{i=1}^{k} \text{pr}(\alpha_i, d_i|\sigma^2, c)$, where

$$\text{pr}(\alpha_i, d_i|\sigma^2, c) = \text{pr}(d_i|\sigma^2, c)\text{pr}(\alpha_i|d_i, \sigma^2, c)$$

with $\text{pr}(d_i|\sigma^2, c)$ the univariate truncated normal $\text{N}\left(0, \sigma^2(2 + 4k/\tau_0)\right)$ constrained on the interval R_i, and $\text{pr}(\alpha_i|d_i, \sigma^2, c)$ the univariate normal

$$N\left(kd_i/(2k+\tau_0), \sigma^2 k/(2k+\tau_0)\right)$$

for $i = 1, \ldots, k$. This provides a simple way of implementing the CA-step.

The trajectory plots of the CA-DA sequences displayed in Figure 2.3 clearly show the dramatically improved efficiency; compare, for example, the ranges (along Y-axes) that DA sequences wandered during the 10 000 iterations with the corresponding ranges of DA sequences.

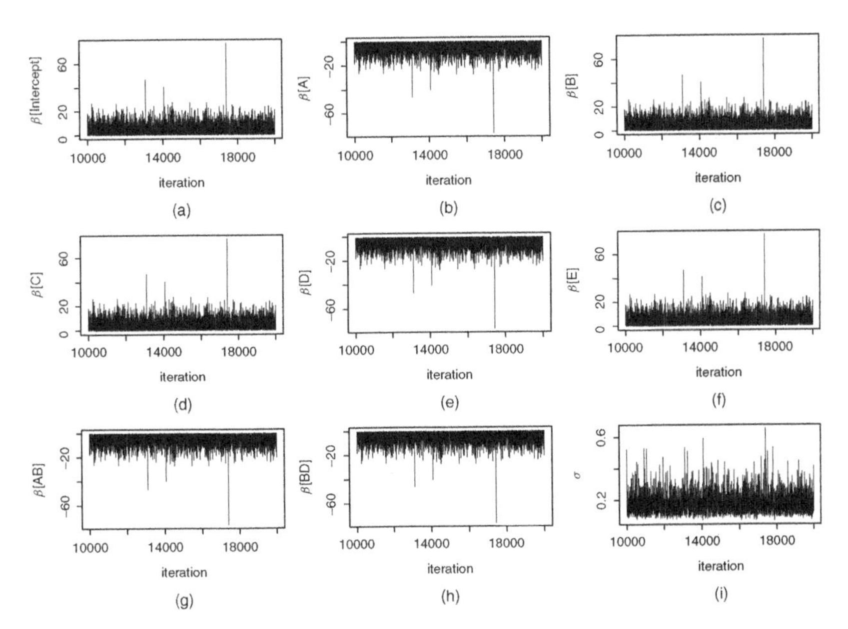

Figure 2.3 The CA-DA sequences of β and σ in the lifetime data in the fluorescant light lifetime experiment.

Exercises

2.1 Prove the Hammersley-Clifford theorem for the case of $K = 2$.

2.2 Verify the conditions in the Gibbs sampler for $N_3(0, \Sigma(\rho))$ in Example 2.1.

2.3 A collection of candidate conditionals (or conditional distributions) is said to be *compatible* if they are associated with a common joint distribution.

(a) Find the necessary and sufficient conditionals, in terms of the parameters $\alpha_{1|2}$, $\beta_{1|2}$, $\sigma^2_{1|2}$, $\alpha_{2|1}$, $\beta_{2|1}$, and $\sigma^2_{2|1}$, for the two conditionals

$$X_1|X_2 \sim N(\alpha_{1|2} + \beta_{1|2}X_2, \sigma^2_{1|2})$$

and

$$X_2|X_1 \sim N(\alpha_{2|1} + \beta_{2|1}X_1, \sigma^2_{2|1})$$

to be compatible.

(b) Generalize the result for (a) to the multivariate case.

2.4 Let $X \sim N_p(\mu, \Sigma)$ and take the prior (2.10) with $q = 2$. Consider the multiple linear regression of any X_i on the other components X_{-i}: $X_i = \tilde{X}'_{-i}\beta + e$, where $e \sim N(0, \sigma^2)$ and $\tilde{X}_{-i}$ denotes the vector obtained from X by replacing its i component with constant one. Show that these (prior) conditionals are compatible and that the induced prior for (β, σ^2) is $p(\beta, \sigma^2) \propto \sigma^{-2}$.

2.5 Consider the Robit linear regression in Section 2.4.2.

(a) Derive the PX-DA algorithm for Bayesian inference using the prior specified in Section 2.4.2.

(b) The probit regression model can be obtained from the Robit regression model. Simplify the above PX-DA implementation for the probit regression model.

2.6 Consider the Student-t model in Section 2.4.1.

(a) Let $A_{\text{mis}} = \sum_{i=1}^n \tau_i$ and let $S_{\text{mis}} = (w_1, \ldots, w_n)$ with $w_i = \tau_i / \sum_{i=1}^n \tau_i$ for $i = 1, \ldots, n$. Show that the CA-DA algorithm is the same as the PX-EM algorithm for the Student-t model.

(b) Assume $\mu = 0$ and $n = 1$. Characterize the dependence of the standard DA sequence and prove that in this simple case, PX-DA converges in one iteration.

2.7 Consider Example 2.1, taking the Gibbs sampler as the basic sampler.

(a) Implement ASSR with $D = \bar{X}$ and $C = (X_1 - \bar{X}, X_2 - \bar{X}, X_3 - \bar{X})$, where $\bar{X} = (X_1 + X_2 + X_3)/3$. Note that C lies in a 2-dimensional space.

(b) Compare the performance of the parent sampler and the ASSR algorithm.

2.8 Consider the Gibbs sampler based on the conditionals of the n-dimensional multivariate normal with the zero means and the exchangeable covariance matrix $(1 - \rho)I + \rho J$, where I is the $n \times n$ identity matrix, J is the $n \times n$ matrix of ones, and $\rho \in (-1/(n-1), 1)$.

(a) Show that

$$X_i|X_{-i} \sim N\left(\frac{\rho}{1 + (n-2)\rho}\sum_{i \neq j} X_j, \frac{(1-\rho)[1 + (n-1)\rho]}{1 + (n-2)\rho}\right) \tag{2.43}$$

for $i = 1, \ldots, n$.

(b) Implement the ASSR algorithm with the transformation $D = \bar{X} = n^{-1}\sum_{i=1}^{n} X_i$ and $C_i = X_i - \bar{X}$ for $i = 1, \ldots, n$.

2.9 Many Bayesian models are hierarchical. Specify appropriate prior (or hyperprior) distributions and implement the Gibbs sampler for the following Bayesian hierarchical models to fit some datasets available on the web.

(a) Multiple normal models with their unknown means modeled by a normal model and variances models by an inverse Gamma model.

(b) The linear mixed-effects models.

(c) Multiple logistic regression models with regression coefficients modeled by a multivariate normal model.

(d) The mixture of univariate normal models.

(e) Examples of Gelfand and Smith (1990).

2.10 Develop the Gibbs sampler for fitting multivariate probit models.

2.11 Develop the Gibbs sampler for fitting factor analysis models.

2.12 Develop efficient Gibbs sampling algorithms for DNA sequence alignment.

Appendix 2A: The EM and PX-EM Algorithms

The EM algorithm is an iterative algorithm for ML estimation from incomplete data. Let X_{obs} be the observed data and let $f(X_{obs};\theta)$ denote the observed-data model with unknown parameter θ, where $X_{obs} \in \mathbb{X}_{obs}$ and $\theta \in \Theta$. Suppose that the observed-data model can be obtained from a complete-data model, denoted by $g(X_{obs}, X_{mis};\theta)$, where $X_{obs} \in \mathbb{X}_{obs}$, $X_{mis} \in \mathbb{X}_{mis}$ and $\theta \in \Theta$. That is,

$$f(X_{obs};\theta) = \int_{\mathbb{X}_{mis}} g(X_{obs}, X_{mis};\theta)dX_{mis}.$$

Given a starting point $\theta^{(0)} \in \Theta$, the EM algorithm iterates for $t = 0, 1, \ldots$ between the E step and M step:

E step. Compute the expected complete-data log-likelihood

$$Q(\theta|\theta^{(t)}) = \mathrm{E}\left(\ln g(X_{obs}, X_{mis};\theta)|X_{obs}, \theta = \theta^{(t)}\right)$$

as a function of $\theta \in \Theta$; and

M step. Maximize $Q(\theta|\theta^{(t)})$ to obtain $\theta^{(t+1)} = \arg\max_{\theta\in\Theta} Q(\theta|\theta^{(t)})$.

Suppose that the complete-data model can be embedded in a larger model

$$g_*(X_{obs}, X_{mis};\theta_*, \alpha)$$

with the expanded parameter $(\theta_*, \alpha) \in \Theta \times \mathbb{A}$. Assume that the observed-data model is preserved in the sense that for every $(\theta_*, \alpha) \in \Theta \times \mathbb{A}$,

$$f(X_{obs};\theta) = f_*(X_{obs};\theta_*, \alpha), \tag{2.44}$$

holds for some $\theta \in \Theta$, where $f_*(X_{obs};\theta_*, \alpha) = \int_{\mathbb{X}_{mis}} g_*(X_{obs}, X_{mis};\theta_*, \alpha)dX_{mis}$. The condition (2.44) defines a mapping $\theta = R_\alpha(\theta_*)$, called the reduction function, from the expanded parameter space $\Theta \times \mathbb{A}$ to the original parameter space Θ. For convenience, assume that there exists a *null* value of α, denoted by α_0, such that $\theta = R_{\alpha_0}(\theta)$ for every $\theta \in \Theta$, that is, the original complete-data and observed-data models are recovered by fixing α at α_0. When applied to the parameter-expanded complete-data model $g_*(X_{obs}, X_{mis};\theta_*, \alpha)$, the EM algorithm, called the PX-EM algorithm, creates a sequence $\{(\theta_*^{(t)}, \alpha^{(t)})\}$ in $\Theta \times \mathbb{A}$. In the original parameter space Θ, PX-EM generates a sequence $\{\theta^{(t)} = R(\theta_*^{(t)}, \alpha^{(t)})\}$ and converges no slower than the corresponding EM based on $g(X_{obs}, X_{mis};\theta)$.

For simplicity and stability, Liu *et al.* (1998) use $(\theta^{(t)}, \alpha_0)$ instead of $(\theta_*^{(t)}, \alpha^{(t)})$ for the E step. As a result, PX-EM shares with EM its E-step and modifies its M step by mapping $(\theta_*^{(t+1)}, \alpha^{(t+1)})$ to the original space: $\theta^{(t)} = R_{\alpha^{(t+1)}}(\theta_*^{(t+1)})$. More precisely, the PX-EM algorithm is defined replacing the E and M steps of EM with the following PX-E and PX-M steps:

PX-E step. Compute

$$Q(\theta_*, \alpha|\theta^{(t)}, \alpha_0) = \mathrm{E}\left(\ln g_*(X_{\text{obs}}, X_{\text{mis}}; \theta_*, \alpha)|X_{\text{obs}}, \theta_* = \theta^{(t)}, \alpha = \alpha_0\right)$$

as a function of $(\theta_*, \alpha) \in \Theta \times \mathbb{A}$.

PX-M step. Find $(\theta_*^{(t+1)}, \alpha^{(t+1)}) = \arg\max_{\theta_*,\alpha} Q(\theta_*, \alpha|\theta^{(t)}, \alpha_0)$ and update $\theta^{(t+1)} = R_{\alpha^{(t+1)}}(\theta_*^{(t+1)})$.

Liu *et al.* (1998) provide a statistical interpretation of PX-M step in terms of covariance adjustment. This is reviewed in Lewandowski *et al.* (2009).

Chapter 3

The Metropolis-Hastings Algorithm

Although powerful in Bayesian inference for many statistical models, the Gibbs sampler cannot be applied to Bayesian model selection problems which involve multiple parameter spaces of different dimensionality. In addition, the Gibbs sampler is not convenient for sampling from distributions for which the conditional distributions of some or all components are not standard. For these problems, the Metropolis-Hastings (MH) algorithm (Metropolis *et al.*, 1953; Hastings, 1970), which can be viewed as a generalized version of the Gibbs sampler, is needed.

In this chapter, we first give a description of the basic MH algorithm, and then consider its variants, such as the Hit-and-Run algorithm (Boneh and Golan, 1979; Smith, 1980; Chen and Schmeiser, 1993), the Langevin algorithm (Besag, 1994; Grenander and Miller, 1994), the multiple-try MH algorithm (Liang *et al.*, 2000), the reversible jump MH algorithm (Green, 1995), and the Metropolis-within-Gibbs sampler (Müller, 1991, 1993). Finally, we consider two applications, the change-point identification and ChIP-chip data analysis.

3.1 The Metropolis-Hastings Algorithm

Consider the target distribution $\pi(dx)$ with pdf $f(x)$ on the sample space $\mathcal{X}$ with σ-field $\mathcal{B}_{\mathcal{X}}$. The basic idea of creating a Markov chain with a transition kernel $P(x, dy)$ is to have $\pi(dx)$ as its invariant distribution such that

$$\pi(dy) = \int_{\mathcal{X}} \pi(dx) P(x, dy). \tag{3.1}$$

Advanced Markov Chain Monte Carlo Methods: Learning from Past Samples
Faming Liang, Chuanhai Liu and Raymond J. Carroll

The indeterminacy of the transition kernel $P(x, dy)$ in (3.1) allows for attractive flexibility, but does not help much in constructing $P(x, dy)$ for given $\pi(dx)$. To create practically useful transition kernels, a commonly used strategy is to impose the restriction called the reversibility condition. A Markov chain with the probability transition kernel $P(x, dy)$ and invariant distribution $\pi(dx)$ is said to be *reversible* if it satisfies the *detailed balance condition*

$$\int_B \int_A \pi(dx)P(x, dy) = \int_A \int_B \pi(dy)P(y, dx), \forall A, B \in \mathcal{B}_{\mathcal{X}}, \tag{3.2}$$

In terms of $f(x)$ and $p(x, y)$, the pdf of $P(x, dy)$ given x, the detailed balance condition (3.2) can be equivalently written as

$$f(x)p(x, y) = f(y)p(y, x). \tag{3.3}$$

It can be shown that the detailed balance condition implies the balance condition; see Exercise 3.1.

Let $x = X^{(t)}$ be the state of the Markov chain at time t. To construct transition kernels satisfying (3.3), Metropolis *et al.* (1953) consider a two-step approach: (i) specifying a symmetric proposal distribution with pdf $q(y|x)$, that is, $q(y|x) = q(x|y)$, and (ii) adjusting draws from $q(y|x)$ via acceptance-rejection in such a way that the resulting Markov chain is reversible. More precisely, the Metropolis algorithm can be summarized as follows:

The Metropolis Sampler

1. Draw y from $q(y|x_t)$.
2. Compute the acceptance ratio

$$\alpha(x_t, y) = \min\left\{1, \frac{f(y)}{f(x_t)}\right\}.$$

   Set $x_{t+1} = y$ with probability $\alpha(x_t, y)$ and $x_{t+1} = x_t$ with the remaining probability $1 - \alpha(x, y)$.

Hastings (1970) generalizes the Metropolis sampler by allowing proposal distributions to be asymmetric. A commonly used such algorithm, known as the MH algorithm, is obtained from the Metropolis sampler by allowing $q(y|x)$ in Step 1 to be asymmetric and modifying Step 2 so that (3.3) holds:

The Metropolis-Hastings (MH) Sampler

1. Draw y from $q(y|x_t)$.
2. Compute the acceptance ratio

$$\alpha(x_t, y) = \min\left\{1, \frac{f(y)q(x_t|y)}{f(x_t)q(y|x_t)}\right\},$$

   and set $x_{t+1} = y$ with the probability $\alpha(x_t, y)$ and $x_{(t+1)} = x_t$ with the remaining probability $1 - \alpha(x_t, y)$.

The acceptance ratio $\alpha(x_t, y)$ in Step 2 of MH is not uniquely determined by the reversibility condition; alternatives exist. For example, consider the case of finite state space $\mathcal{X} = \{k : k = 1, \ldots, K\}$. Write $q_{ij} = q(j|i), \pi_i = f(i), p_{ij} = p(i, j)$, α_{ij} for the acceptance probability/ratio to be found for all $i, j \in \mathcal{X}$. Then for all states $i \neq j$

$$p_{ij} = q_{ij}\alpha_{ij} \tag{3.4}$$

with $p_{ii} = 1 - \sum_{j \neq i} p_{ij}$. Since p_{ii} imposes no problem with the reversibility condition, the solution to the problem of finding α_{ij} is obtained from the system of equations

$$\pi_i q_{ij} \alpha_{ij} = \pi_j q_{ji} \alpha_{ji}$$

subject to $0 \leq \alpha_{ij} \leq 1$. That is, for each pair of i and $j\,(i \neq j)$

$$\alpha_{ij} = \frac{c_{ij}}{1 + \frac{\pi_i q_{ij}}{\pi_j q_{ji}}} \tag{3.5}$$

with a symmetric positive constants $c_{ij} \leq 1 + \frac{\pi_i q_{ij}}{\pi_j q_{ji}}$, where it should be noted that $\pi_j q_{ji} = 0$ implies $\alpha_{ij} = 0$. It follows that for the fixed proposal distribution $q(y|x)$, the resulting Markov chain with the fastest mixing rate is the one that has the largest possible α_{ij} (and α_{ji}):

$$\alpha_{ij} = \begin{cases} 1, & \text{if } \frac{\pi_i q_{ij}}{\pi_j q_{ji}} \geq 1; \\ \frac{\pi_i q_{ij}}{\pi_j q_{ji}}, & \text{otherwise,} \end{cases}$$

or, equivalently,

$$\alpha_{ij} = \min\left\{1, \frac{\pi_i q_{ij}}{\pi_j q_{ji}}\right\} \tag{3.6}$$

This acceptance probability, known as the Metropolis (M)-ratio, is the one used in the MH algorithm. Another well-known specific solution (Barker, 1965) is given by

$$\alpha_{ij}^{(B)} = \frac{\pi_j q_{ji}}{\pi_i q_{ij} + \pi_j q_{ji}}.$$

Peskun (1973) provides an argument for the optimality of (3.6) in terms of minimal variance of MCMC approximations to integrals.

In the general case, the MH-ratio $\alpha(x, y)$ is chosen so that the resulting transition kernel obeys the reversibility condition

$$f(x)q(y|x)\alpha(x, y) = f(y)q(x|y)\alpha(y, x) \tag{3.7}$$

assuming that $\alpha(y, x)$ is measurable with respect to the proposal distribution $Q(dy|x) = q(y|x)\nu(dy)$. The MH kernel can be written as, for all $A \in \mathcal{B}_{\mathcal{X}}$,

$$\begin{aligned} P(x, A) &= \int_A Q(dy|x)\alpha(x, y) + I_{\{x \in A\}} \int_{\mathcal{X}} Q(dy|x)(1 - \alpha(x, y)) \\ &= \int_A Q(dy|x)\alpha(x, y) + I_{\{x \in A\}} \left[1 - \int_{\mathcal{X}} Q(dy|x)\alpha(x, y)\right], \end{aligned}$$

that is,

$$P(x, dy) = Q(dy|x)\alpha(x, y) + \delta_x(dy)r(x)$$

where $\delta_x(dy)$ stands for the probability measure with unit mass at x and $r(x) = 1 - \int_{\mathcal{X}} Q(dy|x)\alpha(x, y)$, the (average) rejection probability at the state x. The reversibility condition can be verified by the standard algebraic operations, for any $A, B \in \mathcal{B}_{\mathcal{X}}$,

$$\begin{aligned}
& \int_B \int_A \pi(dx)P(x, dy) \\
= & \int_B \int_A f(x)q(y|x)\alpha(x, y)\nu(dx)\nu(dy) + \int_{A\cap B} r(x)f(x)\nu(dx) \\
\stackrel{(3.7)}{=} & \int_B \int_A f(y)q(x|y)\alpha(y, x)\nu(dx)\nu(dy) + \int_{A\cap B} r(x)f(x)\nu(dx) \\
= & \int_A \int_B f(y)q(x|y)\alpha(y, x)\nu(dy)\nu(dx) + \int_{B\cap A} r(y)f(y)\nu(dy) \\
= & \int_A \int_B \pi(dy)P(y, dx).
\end{aligned}$$

Thus, under the assumption that the transition kernel $P(x, dy)$ determined by $q(y|x)$ and $\alpha(x, y)$ is irreducible and aperiodic, $\pi(dx)$ is the unique equilibrium distribution.

The efficiency of the MH algorithm depends largely on its proposal distribution. In Sections 3.1.1 and 3.1.2, we describe two popular choices of the proposal distribution, the independence proposal and the random walk proposal.

3.1.1 Independence Sampler

For the independence sampler, we have $q(y|x) = q(y)$; that is, the candidate state y is drawn independently of the current state of the Markov chain. The MH ratio becomes

$$r(x, y) = \frac{f(y)q(x)}{f(x)q(y)} = \frac{f(y)/q(y)}{f(x)/q(x)},$$

the importance ratio, where $x = x_t$.

The independent sampler can be viewed as a generalization of the acceptance-rejection algorithm. It is easy to see that the independence chain is irreducible and aperiodic if the support of $q(x)$ contains the support of $f(x)$, that is,

$$\{x : x \in \mathcal{X}, f(x) > 0\} \subseteq \{x : x \in \mathcal{X}, g(x) > 0\}.$$

An important consideration in specifying $q(x)$ is that $q(x)$ should resemble $f(x)$ and have longer tails than $f(x)$. This is supported by the following theoretical result of Mengerson and Tweedie (1996); see also Robert and Casella (2004).

Theorem 3.1.1 *The independence chain is uniformly ergodic if there exists a constant M such that*

$$f(x) \leq Mg(x) \qquad (x \in \{x : f(x) > 0\}).$$

3.1.2 Random Walk Chains

Random walk MH chains are created by taking the proposal distributions of the form

$$q(x, y) = q(y - x).$$

That is, the proposed jump direction and distance from the current state x_t is independent of x_t.

The most common choices for $q(\cdot)$ include simple spherical distributions such as a scaled standard normal distribution, a scaled Student-t distribution, a uniform distribution over a ball centered at the origin, and ellipsoidal distributions. An important consideration is the specification of the scale parameter for the proposal distribution $q(\cdot)$. Small values tend to produce proposals to be accepted with large probabilities, but may result in chains that are highly dependent. In contrast, large scales tend to produce desirable large steps, but result in very small acceptance rates. Thus, it is often worthwhile to select appropriate scales by controlling acceptance rates in a certain range, say 20–40% as suggested by Gelman *et al.* (1996) and Roberts and Rosenthal (2001), via some pilot runs. Choosing proposals with a desired acceptance rate in the context of adaptive Metropolis algorithms (Haario *et al.*, 2001; Andrieu and Thoms, 2008) is discussed in Section 8.1.

3.1.3 Problems with Metropolis-Hastings Simulations

The MH algorithm has proven to be fundamental and plays a central role in Monte Carlo computation. However, as pointed out by many researchers, it suffers from two difficulties:

- the local-trap problem;
- inability to sample from distributions with intractable integrals.

In simulations of a complex system whose energy landscape is rugged, the local-trap problem refers to the sampler getting trapped in a local energy minimum indefinitely, rendering the simulation useless. Figure 3.1 shows sampling paths of the MH sampler for a mixture normal distribution

$$\frac{1}{3}N(\boldsymbol{\mu}_1, \Sigma) + \frac{2}{3}N(\boldsymbol{\mu}_2, \Sigma), \tag{3.8}$$

where $\boldsymbol{\mu}_1 = (0, 0)^T, \boldsymbol{\mu}_2 = (5, 5)^T$, and $\Sigma = \text{diag}(1/4, 2)$. In each run, the sampler started with a random point drawn from $N_2((2.5, 2.5)^T, I_2)$, and

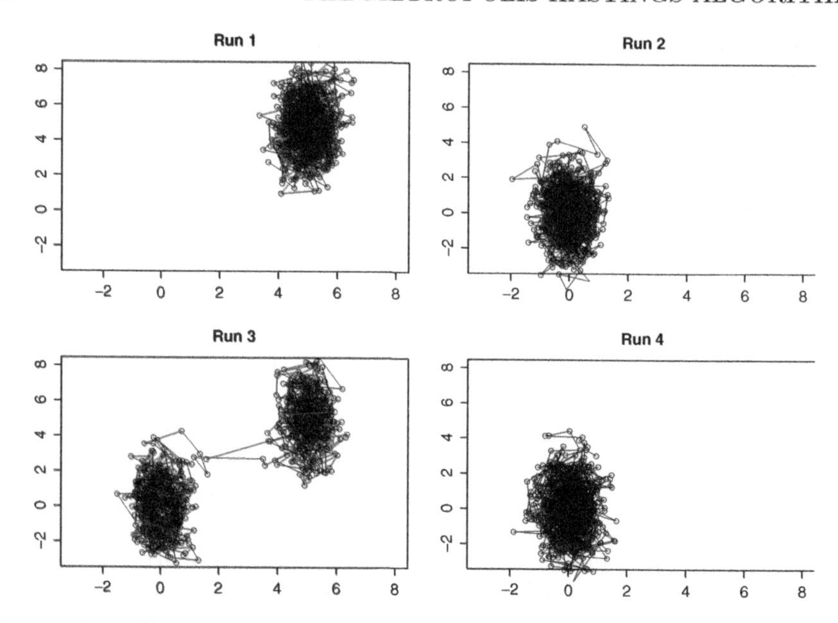

Figure 3.1 Sampling paths of the MH sampler for the mixture distribution (3.8).

then iterated for 10 000 steps. Employed here is a Gaussian random walk proposal with an identity covariance matrix. Among the four runs, two runs got trapped in the left mode, and one run got trapped in the right mode. The third run visited both modes, but there are only two transitions between the two modes. Clearly, the MH algorithm fails for this example; the parameters of (3.8) cannot be correctly estimated from the samples produced in any of the four runs.

Suppose that the target distribution of interest has density of the form $f(x) \propto c(x)\psi(x)$, where $c(x)$ denotes an intractable integral. Clearly, the MH algorithm cannot be applied to sample from $f(x)$, as the acceptance probability would involve an unknown ratio $c(x')/c(x)$, where x' denotes the proposed value. This difficulty naturally arises in Bayesian inference for many statistical models, such as spatial statistical models, generalized linear mixed models, and exponential random graph models.

To alleviate or overcome these two difficulties, various advanced MCMC methods have been proposed in the literature, including auxiliary variable-based methods, population-based methods, importance weight-based methods, and stochastic approximation-based methods, which will be described, respectively, in Chapters 4, 5, 6, and 7. Chapter 8 is dedicated to MCMC methods with adaptive proposals, which aim to provide an online learning for an appropriate proposal distribution such that the MCMC method can perform optimally under certain criteria.

3.2 Variants of the Metropolis-Hastings Algorithm

In this section, we describe several MH schemes that improve the mixing of the MH chain in certain scenarios.

3.2.1 The Hit-and-Run Algorithm

The Hit-and-Run algorithm of Boneh and Golan (1979) and Smith (1980), with further refinements and extensions by Smith (1984), Bélisle *et al.* (1993), and Chen and Schmeiser (1996), can be obtained by separating the process of creating a proposed jump in MH into two subprocesses: (i) Generate a direction d from a distribution on the surface of the unit sphere $\mathbb{O}$, and (ii) Generate a signed distance λ along the direction d in the constrained space

$$\mathcal{X}_{x,d} = \{\lambda : \lambda \in \mathbb{R}, x + \lambda d \in \mathcal{X}\},$$

where $x = X^{(t)}$. That is, the proposed jump is $y = X^{(t)} + \lambda d \in \mathcal{X}$. In the MH framework, this algorithm can be summarized as follows: take X_0 from the starting distribution $f_0(X)$ with $f(X_0) > 0$, and iterate for $t = 0, 1, 2, \ldots$

Hit-and-Run Algorithm

1. Draw $d \sim g(d)(d \in \mathbb{O})$ and $\lambda \sim l(\lambda|d, x)$ over $\mathcal{X}_{x,d}$, and compute an MH acceptance probability $\alpha(x, y)$, where $x = X^{(t)}$.
2. Generate U from Unif $(0, 1)$ and set

$$X^{(t+1)} = \begin{cases} x + \lambda d, & \text{if } U \leq \alpha(x, y); \\ x, & \text{otherwise.} \end{cases}$$

Chen *et al.* (2000) note that the most common choice of $g(d)$ is the uniform distribution on $\mathbb{O}$, and discuss common choices for $g(.|x, d)$ and $\alpha(x, y)$. Berger (1993) also recognizes that Hit-and-Run is particularly useful for problems with a sharply constrained parameter space.

3.2.2 The Langevin Algorithm

The Langevin algorithm is rooted in the Langevin diffusion process, which is defined by the stochastic differential equation

$$dX_t = dB_t + \frac{1}{2}\nabla \log f(X_t), \tag{3.9}$$

where B_t is the standard Brownian motion. This process leaves f as the stationary distribution. The implementation of the diffusion algorithm involves a discretization step which replaces (3.9) by a random-walk style transition

$$x^{(t+1)} = x^{(t)} + \frac{\sigma^2}{2}\nabla \log f(x^{(t)}) + \sigma\epsilon_t, \tag{3.10}$$

where $\epsilon_t \sim N_d(0, I_d)$ and σ is the step size of discretization. However, as shown by Roberts and Tweedie (1995), the discretized process may be transient and no longer leaves f as the stationary distribution.

To correct this negative behavior, Besag (1994) suggests moderating the discretization step by applying the MH acceptance-rejection rule; that is, treating (3.10) as a conventional proposal, and accepting it according to the MH rule. In summary, one iteration of the Langevin algorithm can be described as follows:

Langevin Algorithm

1. Propose a new state

$$x^* = x^{(t)} + \frac{\sigma^2}{2}\nabla \log f(x^{(t)}) + \sigma\epsilon_t,$$

where σ is user-specified parameter.

2. Calculate the MH ratio

$$r = \frac{f(x^*)}{f(x^{(t)})} \frac{\exp(-\|x^{(t)} - x^* - \frac{\sigma^2}{2}\nabla \log f(x^*)\|^2/2\sigma^2)}{\exp(-\|x^* - x^{(t)} - \frac{\sigma^2}{2}\nabla \log f(x^{(t)})\|^2/2\sigma^2)}.$$

Set $x^{(t+1)} = x^*$ with probability $\min(1, r)$, and set $x^{(t+1)} = x^{(t)}$ with the remaining probability.

This algorithm, Grenander and Miller (1994), is useful in problems for which the gradient of $f(x)$ is available, for example, when training a feedforward neural network (Rumelhart *et al.*, 1986). Roberts and Rosenthal (1998) show that the optimal convergence rate of the algorithm can be attained at an acceptance rate of 0.574, which can be achieved by choosing an appropriate value of σ. Roberts and Tweedie (1996) show that the Langevin algorithm is not geometrically ergodic when $\nabla \log f(x)$ goes to zero at infinity, but the basic ergodicity is still ensured. See Stramer and Tweedie (1999a, b) and Roberts and Stramer (2002) for further study of the Langevin algorithm.

3.2.3 The Multiple-Try MH Algorithm

As implied by the Langevin algorithm, use of the gradient information for the target distribution can be used to accelerate the convergence of the MH simulation. However, gradient information is not available for most target distributions. One way of approximating the gradient of the target distribution is to use Monte Carlo samples. Liu *et al.* (2000) propose using the multiple-try Metropolis (MTM) algorithm, which modifies the standard MH by replacing the single proposal y with a set of k iid proposals $y_1, \ldots, y_k$ from $q(y|x)$ and selecting a good candidate (in importance sampling) to jump to, according to the MH rule.

Suppose that $q(y|x) > 0$ if and only if $q(x|y) > 0$. Let $\lambda(x, y)$ be a non-negative symmetric function in x and y. Suppose that $\lambda(x, y) > 0$ whenever $q(y|x) > 0$. Define

$$w(x, y) = f(x)q(y|x)\lambda(x, y). \tag{3.11}$$

Let $x = X^{(t)}$. The MTM transition is then defined as follows:

Multiple-Try Metropolis

1. Draw k iid candidates $y_1, \ldots, y_k$ from $q(y|x)$, and compute $w_i = w(y_i, x)$ for $i = 1, \ldots, k$.

2. Select $y = y_j$ from $\{y_1, \ldots, y_k\}$ with probability proportional to w_i, $i = 1, \ldots, k$, draw $x_1^*, \ldots, x_{k-1}^*$ from $q(.|y)$, set $x_k^* = x$, and compute $w_i^* = w(x_i^*, y)$ for $i = 1, \ldots, k$.

3. Accept y with probability

$$a_m = \min\left\{1, \frac{w_1 + \ldots + w_k}{w_1^* + \ldots + w_k^*}\right\} \tag{3.12}$$

and reject it (or set $X^{(t+1)} = x$) with probability $1 - a_m$.

Liu *et al.* (2000) give some simple choices of $\lambda(x, y)$, including $\lambda(x, y) = 1$, $\lambda(x, y) = [q(y|x) + q(x|y)]^{-1}$, and $\lambda(x, y) = (q(y|x)q(x|y))^{-\alpha}$ for a constant α. When $q(x|y)$ is symmetric, for example, one can choose $\lambda(x, y) = 1/q(y|x)$, and then $w(x, y) = f(x)$. In this case, the MTM algorithm is reduced to the orientational bias Monte Carlo algorithm used in the field of molecular simulation.

MTM can be used in combination with other methods, such as conjugate gradient Monte Carlo, the Hit-and-Run algorithm, and the griddy Gibbs sampler (Ritter and Tanner, 1992). Augmenting a set of $k - 1$ auxiliary variables $x_1^*, \ldots, x_{k-1}^*$ in Step 2 of MTM is an interesting technique for matching dimensionality so that the standard MH rule can be applied easily. The technique of dimension matching plays an important role in the reversible jump MCMC method (Green, 1995).

3.3 Reversible Jump MCMC Algorithm for Bayesian Model Selection Problems

3.3.1 Reversible Jump MCMC Algorithm

Consider the problem of Bayesian model selection. Let $\{\mathcal{M}_k : k \in \mathcal{K}\}$ denote a countable collection of models to be fitted to the observed data Y. Each model

$\mathcal{M}_k$ has its own parameter space $\Theta_k \subseteq \mathbb{R}^{d_k}$. Without loss of generality, we assume here that different models have different dimensions. A full Bayesian model can be written as

$$p(k)p(\theta_k|k)p(Y|k, \theta_k), \tag{3.13}$$

where $p(k)$ is the prior probability imposed on the model, $\mathcal{M}_k$, $p(\theta_k|k)$ is the prior distribution specified for the parameter θ_k, and $p(Y|k, \theta_k)$ represents the sampling model for the observed data Y, conditional on

$$(k, \theta_k) \in \{k\} \times \Theta_k \qquad (k \in \mathcal{K}) \tag{3.14}$$

Let $X = (k, \theta_k)$ and let $\mathcal{X}_k = \{k\} \times \Theta_k$. Thus the Markov chain $\{X_t\}$ jumps between models with parameters changing over the space $\mathcal{X} = \cup_{k \in \mathcal{K}} \mathcal{X}_k$. Since the subspaces $\mathcal{X}_k$'s are of different dimensionality, the Gibbs sampler cannot be applied. Green (1995) proposes a reversible jump MCMC (RJMCMC) sampling method that generates a Markov chain that jumps between models of different dimensions. The RJMCMC has become a very popular Bayesian tool for problems involving multiple parameter spaces of different dimensionality.

The basic idea of RJMCMC is to match dimensions with the resulting chain reserving

$$f(k, \theta_k|Y) \propto p(k)p(\theta_k|k)p(Y|k, \theta_k)$$

as its invariant distribution. RJMCMC takes an auxiliary variable approach to the problem of 'dimension matching'. Let $x_t = (k^{(t)}, \theta_k^{(t)})$ denote the current state. Suppose that $x^* = (k^*, \theta^*_{k^*})$ is the proposed state for $X^{(t+1)}$. If $k^* = k$, the proposed move explores different locations within the same subspace $\mathcal{X}_k$ and therefore the dimension-matching problem does not exist. If $k^* \neq k$, generate s random variables $\boldsymbol{u} = (u_1, \ldots, u_s)$ from a distribution $\psi_{k^{(t)} \to k^*}(u)$, and consider a bijection

$$(\theta^*_{k^*}, \boldsymbol{u}^*) = T(\theta_k^{(t)}, \boldsymbol{u}), \tag{3.15}$$

where $\boldsymbol{u}^* = (u_1, \ldots, u_{s^*})$ is a random vector of s^*-dimension, and s and s^* satisfy the dimension-matching condition $s + d_k = s^* + d_{k^*}$. The general idea of 'augmenting less' suggests that $s^* = 0$ be taken if $d_{k^{(t)}} \leq d_{k^*}$ and vice versa. In summary, RJMCMC is a special type of MH algorithm whose proposal distribution includes auxiliary variables for dimension matching when needed. In the above notations, the RJMCMC algorithm can be itemized as follows:

Reversible Jump MCMC Algorithm

1. Select model $\mathcal{M}_{k^*}$ with probability $q(k^{(t)}, k^*)$.

2. Generate $u_1, \ldots, u_s \sim \psi_{k^{(t)} \to k^*}(u)$.

3. Set $(\theta_{k^*}, u^*) = T(\theta_k^{(t)}, u)$.

4. Compute the MH ratio

$$r = \frac{f(k^*, \theta^*_{k^*}|Y)q(k^*, k^{(t)})\psi_{k^* \to k^{(t)}}(\boldsymbol{u}^*)}{f(k^{(t)}, \theta^{(t)}_k|Y)q(k^{(t)}, k^*)\psi_{k^{(t)} \to k^*}(\boldsymbol{u})} \left| \frac{\partial(\theta^*_{k^*}, \boldsymbol{u}^*)}{\partial(\theta^{(t)}_k, \boldsymbol{u})} \right| \tag{3.16}$$

   where $\partial(\theta^*_{k^*}, \boldsymbol{u}^*)/\partial(\theta^{(t)}_k, \boldsymbol{u})$ is the Jacobian of the transformation of (3.15).

5. Set $X^{(t+1)} = (k^*, \theta^*_{k*})$ with probability $\min(1, r)$ and $X^{(t+1)} = X_t$ with the remaining probability.

For many problems, the Jacobian can be reduced to 1 by choosing an identity transformation in (3.15); that is, proposing new samples in the space $\mathcal{X}_{k^*}$ directly, as illustrated by Example 3.1.

■ **Example 3.1 Bayesian Analysis of Mixture Models**

Let $\boldsymbol{Z} = (z_1, \ldots, z_n)$ denote a sequence of independent observations drawn from a mixture distribution with the likelihood function

$$f(\boldsymbol{Z}|m, \boldsymbol{p}_m, \Phi_m, \eta) = \prod_{i=1}^{n} \left[p_1 f(z_i; \phi_1, \eta) + \cdots + p_m f(z_i; \phi_m, \eta)\right],$$

where m is the unknown number of components, $\boldsymbol{p}_m = (p_1, \ldots, p_m)$, $\Phi_m = (\phi_1, \ldots, \phi_m)$, and η is a common parameter vector for all components.

To conduct a Bayesian analysis, we take the prior distribution $\pi(k, \boldsymbol{p}_k, \Phi_k, \eta)$ with k varying in a pre specified range $K_{\min} \leq k \leq K_{\max}$. The posterior distribution of $(k, \boldsymbol{p}_k, \Phi_k, \eta)$ is then

$$\pi(k, \boldsymbol{p}_k, \Phi_k, \eta|\boldsymbol{Z}) \propto f(\boldsymbol{Z}|k, \boldsymbol{p}_k, \Phi_k, \eta)\pi(k, \boldsymbol{p}_k, \Phi_k, \eta). \tag{3.17}$$

For simulating from (3.17), RJMCMC consists of three types of moves, 'birth', 'death', and 'parameter updating', which can be prescribed as in Stephens (2000). In the 'birth' move, a new component is generated and $(\boldsymbol{p}_k, \Phi_k)$ is updated to be

$$\{(p_1(1-p), \phi_1), \ldots, (p_k(1-p), \phi_k), (p, \phi)\}$$

In the 'death' move, a randomly chosen component i is proposed to be removed with $(\boldsymbol{p}_k, \Phi_k)$ updated to

$$\left\{\left(\frac{p_1}{1-p_i}, \phi_1\right), \ldots, \left(\frac{p_{i-1}}{1-p_i}, \phi_{i-1}\right), \left(\frac{p_{i+1}}{1-p_i}, \phi_{i+1}\right), \ldots, \left(\frac{p_k}{1-p_i}, \phi_k\right)\right\}.$$

In the 'parameter updating' move, the parameters $(\boldsymbol{p}_k, \Phi_k, \eta)$ are updated using the MH algorithm. In summary, RJMCMC works as follows for this example. Let $(k, \boldsymbol{p}_k, \Phi_k, \eta)$ denote the current state of the Markov chain.

- Propose a value of k^* according to a stochastic matrix Q, where, for example, we set $Q_{k,k+1} = Q_{k,k-1} = Q_{k,k} = 1/3$ for $K_{\min} < k < K_{\max}$, $Q_{K_{\min},K_{\min}+1} = Q_{K_{\max},K_{\max}-1} = 1/3$, and $Q_{K_{\min},K_{\min}} = Q_{K_{\max},K_{\max}} = 2/3$.

- According to the value of k^*, do step (a), (b) or (c):

  (a) If $k^* = k+1$, make a 'birth' move: Draw $p \sim Unif[0,1]$ and draw ϕ from a proposal distribution $g(\phi|\boldsymbol{p}_k, \Phi_k, \eta)$, and accept the new component with probability

$$\min\left\{1, \frac{\pi(k+1, \boldsymbol{p}_{k+1}, \Phi_{k+1}, \eta|\boldsymbol{Z})}{\pi(k, \boldsymbol{p}_k, \Phi_k, \eta|\boldsymbol{Z})} \frac{Q_{k+1,k}}{Q_{k,k+1}} \frac{1}{(k+1)g(\phi|\boldsymbol{p}_k, \Phi_k, \eta)}\right\}.$$

  (b) If $k^* = k-1$, make a 'death' move: Randomly choose a component, say component i, to remove, and accept the new state with probability

$$\min\left\{1, \frac{\pi(k-1, \boldsymbol{p}_{k-1}, \Phi_{k-1}, \eta|\boldsymbol{Z})}{\pi(k, \boldsymbol{p}_k, \Phi_k, \eta|\boldsymbol{Z})} \frac{Q_{k-1,k}}{Q_{k,k-1}} \frac{kg(\phi_i|\boldsymbol{p}_{k-1}, \Phi_{k-1}, \eta)}{1}\right\}.$$

  (c) If $k^* = k$, make a 'parameter updating' move, updating the parameters $(\boldsymbol{p}_k, \Phi_k, \eta)$ using the MH algorithm: Generate $(\boldsymbol{p}_k^*, \Phi^*, \eta^*)$ from a proposal distribution $q(\boldsymbol{p}_k^*, \Phi^*, \eta^*|\boldsymbol{p}_k, \Phi, \eta)$, and accept the proposal with probability

$$\min\left\{1, \frac{\pi(k, \boldsymbol{p}_k^*, \Phi_k^*, \eta^*|\boldsymbol{Z})}{\pi(k, \boldsymbol{p}_k, \Phi_k, \eta|\boldsymbol{Z})} \frac{q(\boldsymbol{p}_k, \Phi_k, \eta|\boldsymbol{p}_k^*, \Phi^*, \eta^*)}{q(\boldsymbol{p}_k^*, \Phi^*, \eta^*|\boldsymbol{p}_k, \Phi, \eta)}\right\}.$$

In steps (a) and (b), the Jacobian term is 1 is used as an identity transformation in the 'birth' and 'death' moves. Step (c) can be split into several substeps, updating the parameters $\boldsymbol{p}_k$, Φ_k and η separately, as prescribed by the Metropolis-within-Gibbs sampler (Müller, 1991, 1993).

A major difficulty with RJMCMC noted by many researchers (e.g., Richardson and Green, 1997; Brooks *et al.*, 2003) is that the reversible jump move often has a very low acceptance probability, rendering simulation inefficient. This difficulty can be alleviated by stochastic approximation Monte Carlo (Liang *et al.*, 2007), whereby the self-adjusting mechanism helps the system transmit between different models (see Chapter 7 for details).

3.3.2 Change-Point Identification

Consider the following application of RJMCMC to the change-point identification problem. Let $Z = (z_1, z_2, \cdots, z_n)$ denote a sequence of independent observations. The change-point identification problem is to identify a partition of the sequence such that the observations follow the same distribution within each block of the partition. Let $\boldsymbol{\vartheta} = (\vartheta_1, \cdots, \vartheta_{n-1})$ be the change-point indicator, a binary vector with $\vartheta_{c_1} = \cdots = \vartheta_{c_k} = 1$ and 0 elsewhere. That is,

$$0 = c_0 < c_1 < \cdots < c_k < c_{k+1} = n$$

and

$$z_i \sim p_r(\cdot), \quad c_{r-1} < i \leq c_r$$

for $r = 1, 2, \cdots, k+1$. Our task is to identify the change-point positions $c_1, \cdots, c_k$.

Consider the case where $p_r(\cdot)$ is Gaussian with unknown mean μ_r and variance σ_r^2. Let $\boldsymbol{\vartheta}^{(k)}$ denote a configuration of $\boldsymbol{\vartheta}$ with k ones, which represents a model of k change points. Let $\eta^{(k)} = (\boldsymbol{\vartheta}^{(k)}, \mu_1, \sigma_1^2, \cdots, \mu_{k+1}, \sigma_{k+1}^2)$, $\mathcal{X}_k$ denote the space of models with k change points, $\boldsymbol{\vartheta}^{(k)} \in \mathcal{X}_k$, and let $\mathcal{X} = \cup_{k=0}^{n} \mathcal{X}_k$. The log-likelihood of $\eta^{(k)}$ is

$$L(Z|\eta^{(k)}) = -\sum_{i=1}^{k+1} \left\{ \frac{c_i - c_{i-1}}{2} \log \sigma_i^2 + \frac{1}{2\sigma_i^2} \sum_{j=c_{i-1}+1}^{c_i} (z_j - \mu_i)^2 \right\}. \quad (3.18)$$

Assume that the vector $\boldsymbol{\vartheta}^{(k)}$ has the prior distribution

$$P(\boldsymbol{\vartheta}^{(k)}) = \frac{\lambda^k}{\sum_{j=0}^{n-1} \frac{\lambda^j}{j!}} \frac{(n-1-k)!}{(n-1)!}, \quad k = 0, 1, \cdots, n-1,$$

which is equivalent to assuming that $\mathcal{X}_k$ has a truncated Poisson distribution with parameter λ, and each of the $(n-1)!/[k!(n-1-k)!]$ models in $\mathcal{X}_k$ is a priori equally likely. Assume that the component mean μ_i is subject to an improper prior, and that the component variance σ_i^2 is subject to an inverse-Gamma $IG(\alpha, \beta)$. Assuming that all the priors are independent, the log-prior density can be written as

$$\log P(\eta^{(k)}) = a_k - \sum_{i=1}^{k+1} \left[(\alpha - 1) \log \sigma_i^2 + \frac{\beta}{\sigma_i^2} \right], \quad (3.19)$$

where $a_k = (k+1)[\alpha \log \beta - \log \Gamma(\alpha)] + \log(n-1-k)! + k \log \lambda$, and α, β and λ are hyperparameters to be specified by the user. The log-posterior of $\eta^{(k)}$ (up to an additive constant) can be obtained by adding (3.18) and (3.19). Integrating out the parameters $\mu_1, \sigma_1^2, \cdots, \mu_{k+1}, \sigma_{k+1}^2$ from the full posterior distribution, we have

$$\log P(\boldsymbol{\vartheta}^{(k)}|Z) = a_k + \frac{k+1}{2}\log 2\pi - \sum_{i=1}^{k+1}\left\{ \frac{\log(c_i - c_{i-1})}{2} \right.$$

$$- \log\Gamma\left(\frac{c_i - c_{i-1} - 1}{2} + \alpha\right) + \left(\frac{c_i - c_{i-1} - 1}{2} + \alpha\right)$$

$$\left. \log\left[\beta + \frac{1}{2}\sum_{j=c_{i-1}+1}^{c_i} z_j^2 - \frac{(\sum_{j=c_{i-1}+1}^{c_i} z_j)^2}{2(c_i - c_{i-1})}\right]\right\}. \tag{3.20}$$

The MAP (maximum a posteriori) estimate of $\boldsymbol{\vartheta}^{(k)}$ is often a reasonable solution to the problem. However, of more interest to a Bayesian, may be the marginal posterior distribution $P(\mathcal{X}_k|Z)$, which can be estimated using the reversible jump MCMC algorithm. Without loss of generality, we restrict our considerations to the models with $k_{\min} \le k \le k_{\max}$. For the change-point identification problem, the dimensional jumping moves can be performed as follows.

Let $\boldsymbol{\vartheta}_t^{(k,l)}$ denote the lth sample generated at iteration t, where k indicates the number of change-points of the sample. The next sample can be generated according to the following procedure:

(a) Set $j = k-1, k$, or $k+1$ according to the probabilities $q_{k,j}$, where $q_{k,k} = \frac{1}{3}$ for $k_{\min} \le k \le k_{\max}$, $q_{k_{\min},k_{\min}+1} = q_{k_{\max},k_{\max}-1} = \frac{2}{3}$, and $q_{k,k+1} = q_{k,k-1} = \frac{1}{3}$ for $k_{\min} < k < k_{\max}$.

(b) If $j = k$, update $\boldsymbol{\vartheta}_t^{(k,l)}$ by a 'simultaneous' move (described below); if $j = k+1$, update $\boldsymbol{\vartheta}_t^{(k,l)}$ by a 'birth' move (described below); and if $j = k-1$, update $\boldsymbol{\vartheta}_t^{(k,l)}$ by a 'death' move (described below).

In the 'birth' move, a random number, say u, is first drawn uniformly from the set $\{0, 1, \cdots, k\}$; then another random number, say v, is drawn uniformly from the set $\{c_u + 1, \cdots, c_{u+1} - 1\}$, and it is proposed to set $\vartheta_v = 1$. The resulting new sample is denoted by $\boldsymbol{\vartheta}_*^{(k+1)}$. In the 'death' move, a random number, say u, is drawn uniformly from the set $\{1, 2, \cdots, k\}$, and it is proposed to set $\vartheta_{c_u} = 0$. The resulting new sample is denoted by $\boldsymbol{\vartheta}_*^{(k-1)}$. In the 'simultaneous' move, a random number, say u, is first randomly drawn from the set $\{1, 2, \cdots, k\}$; then another random number, say v, is uniformly drawn from the set $\{c_{u-1} + 1, \cdots, c_u - 1, c_u + 1, \cdots, c_{u+1} - 1\}$, and it is proposed to set $\vartheta_{c_u} = 0$ and $\vartheta_v = 1$. The resulting new sample is denoted by $\boldsymbol{\vartheta}_*^{(k)}$. The acceptance probabilities of these moves are as follows. For the 'birth' move, the acceptance probability is

$$\min\left\{1, \frac{P(\boldsymbol{\vartheta}_*^{(k+1)}|X)}{P(\boldsymbol{\vartheta}_t^{(k,l)}|X)} \frac{q_{k+1,k}}{q_{k,k+1}} \frac{c_{u+1} - c_u - 1}{1}\right\}. \tag{3.21}$$

The Jacobian term is 1, as the new sample is directly drawn in the space $\mathcal{X}_{k+1}$. Similarly, for the 'death' move, the acceptance probability is

$$\min\left\{1, \frac{P(\boldsymbol{\vartheta}_*^{(k-1)}|X)}{P(\boldsymbol{\vartheta}_t^{(k,l)}|X)} \frac{q_{k-1,k}}{q_{k,k-1}} \frac{1}{c_{u+1} - c_{u-1} - 1}\right\}. \tag{3.22}$$

For the 'simultaneous' move, the acceptance probability is

$$\min\left\{1, \frac{P(\boldsymbol{\vartheta}_*^{(k)}|X)}{P(\boldsymbol{\vartheta}_t^{(k,l)}|X)}\right\}, \tag{3.23}$$

for which the proposal is symmetric in the sense $T(\boldsymbol{\vartheta}_t^{(k,l)} \to \boldsymbol{\vartheta}_*^{(k)}) = T(\boldsymbol{\vartheta}_*^{(k)} \to \boldsymbol{\vartheta}_t^{(k,l)}) = 1/(c_{u+1} - c_{u-1} - 2)$.

For numerical illustration, we consider a dataset simulated by Liang (2009a), which consists of 1000 observations with $z_1, \cdots, z_{120} \sim N(-0.5, 1)$, $z_{121}, \cdots, z_{210} \sim N(0.5, 0.5)$, $z_{211}, \cdots, z_{460} \sim N(0, 1.5)$, $z_{461}, \cdots, z_{530} \sim N(-1, 1)$, $z_{531}, \cdots, z_{615} \sim N(0.5, 2)$, $z_{616}, \cdots, z_{710} \sim N(1, 1)$, $z_{711}, \cdots, z_{800} \sim N(0, 1)$, $z_{801}, \cdots, z_{950} \sim N(0.5, 0.5)$, and $z_{951}, \cdots, z_{1000} \sim N(1, 1)$. The time plot of the data is shown in Figure 3.2. RJMCMC was run 20 times independently with the parameters: $\alpha = \beta = 0.05, \lambda = 1, k_{\min} = 1$ and $k_{\max} = 20$. Each run consists of 2×10^6 iterations and costs about 130 seconds on a 3.0 GHz personal computer.

Figure 3.2 compares the true change-point pattern and its MAP estimate – (120, 210, 460, 530, 615, 710, 800, 950) and (120, 211, 460, 531, 610, 709, 801, 939) respectively, with the largest discrepancy occuring at the last change-point position. A detailed exploration of the original data gives strong support to the MAP estimate. The last ten observations of the second last cluster have a larger mean value than expected, and thus tend to be in the last cluster. The log-posterior probability of the MAP estimate is 5.33 higher than that of the true pattern.

Figure 3.3 shows the histogram of the change-points sampled from the posterior distribution. It indicates that there are most likely 8 change-points contained in the data. Table 3.1 shows an estimate of the marginal posterior distribution produced by RJMCMC. The estimate is consistent with Figure 3.3; the marginal posterior distribution attains its mode at $k = 8$.

For this problem, the posterior (3.20) can also be simulated using the Gibbs sampler, as in Barry and Hartigan (1993). For the binary state space,

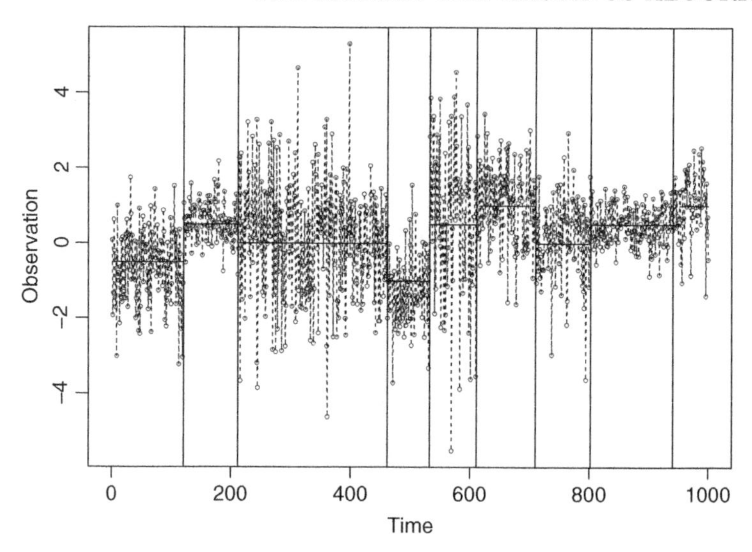

Figure 3.2 Comparison of the true change-point pattern (horizontal lines) and its MAP estimate (vertical lines).

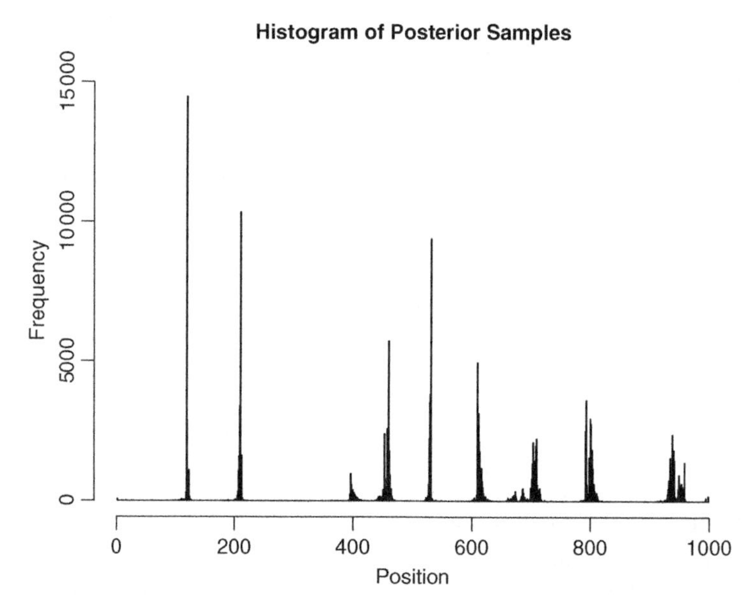

Figure 3.3 Histogram of change-points sampled by RJMCMC for the simulated example.

Table 3.1 RJMCMC estimate of the marginal posterior distribution $P(\mathcal{X}_k|Z)$.

K	≤ 7	8	9	10	11	12	13	≥ 14
$P(\mathcal{X}_k\|Z)$	.103	55.332	33.324	9.458	1.587	.180	.015	.001
SD	.024	1.400	.894	.567	.199	.037	.007	.002

$P(\mathcal{X}_k|Z)$: Estimates of the posterior probabilities (normalized to 100%). SD: Standard deviation of the estimate.

the Gibbs sampler is essentially the method used by Barker (1965), which is generally inferior to the MH algorithm as shown by Peskun (1973). Further discussion on this issue can be found in the next section.

3.4 Metropolis-Within-Gibbs Sampler for ChIP-chip Data Analysis

3.4.1 Metropolis-Within-Gibbs Sampler

Consider the Gibbs sampling algorithm described in Section 2.1. When some components cannot be easily simulated, rather than resorting to a customized algorithm such as the acceptance-rejection algorithm, Müller (1991, 1993) suggests a compromised Gibbs algorithm – the Metropolis-within-Gibbs sampler. For any step of the Gibbs sampler that has difficulty in sampling from $f_k(x_k|x_i, i \neq k)$, substitute a MH simulation.

Metropolis-Within-Gibbs Sampler (Steps)

For $i = 1, \ldots, K$, given $(x_1^{(t+1)}, \ldots, x_{i-1}^{(t+1)}, x_i^{(t)}, \ldots, x_K^{(t)})$:

1. Generate $x_i^* \sim q_i(x_i|x_1^{(t+1)}, \ldots, x_{i-1}^{(t+1)}, x_i^{(t)}, \ldots, x_K^{(t)})$.
2. Calculate

$$r = \frac{f_i(x_i^*|x_1^{(t+1)}, \ldots, x_{i-1}^{(t+1)}, x_{i+1}^{(t)}, \ldots, x_K^{(t)})}{f_i(x_i^{(t)}|x_1^{(t+1)}, \ldots, x_{i-1}^{(t+1)}, x_{i+1}^{(t)}, \ldots, x_K^{(t)})} \times \frac{q_i(x_i^{(t)}|x_1^{(t+1)}, \ldots, x_{i-1}^{(t+1)}, x_i^*, x_{i+1}^{(t)}, \ldots, x_K^{(t)})}{q_i(x_i^*|x_1^{(t+1)}, \ldots, x_{i-1}^{(t+1)}, x_i^{(t)}, x_{i+1}^{(t)}, \ldots, x_K^{(t)})}.$$

3. Set $x_i^{(t+1)} = x_i^*$ with probability $\min(1, r)$ and $x_i^{(t+1)} = x_i^{(t)}$ with the remaining probability.

In this algorithm, the MH step is performed only once at each iteration; $f(x_1, \ldots, x_K)$ is still admitted as the stationary distribution (see Exercise 3.5). As to whether one should run multiple MH steps to produce a precise approximation of $f_i(\cdot)$, Chen and Schmeiser (1998) note that this is not necessary. A precise approximation of $f_i(\cdot)$ does not necessarily lead to a better approximation of $f(\cdot)$, and a single step substitution may be beneficial for the speed of excursion of the chain on the sample space of $f(\cdot)$.

When all components' x_i's are discrete, the Metropolis-within-Gibbs sampler can be more efficient than the Gibbs sampler. For discrete state Markov chains, Liu (1996) shows that the Gibbs sampler can be improved by MH steps in terms of a smaller variance of the empirical mean of $h(x^{(t)})$, where $h(\cdot)$ denotes an integrable function. Given a conditional distribution $f_i(x_i|x_j, j \neq i)$ defined on a discrete state space and the current state $x_i^{(t)}$, Liu's modification can be described as follows:

Metropolized Gibbs Sampler

1. Draw a state y_i, different from x_i, with probability

$$\frac{f(y_i|x_j^{(t)}, j \neq i)}{1 - f(x_i^{(t)}|x_j^{(t)}, j \neq i)}.$$

2. Set $x_i^{(t+1)} = y_i$ with a MH acceptance probability

$$\min\left\{1, \frac{1 - f(x_i^{(t)}|x_j^{(t)}, j \neq i)}{1 - f(y_i|x_j^{(t)}, j \neq i)}\right\},$$

and set $x_i^{(t+1)} = x_i^{(t)}$ with the remaining probability.

The efficiency of the modified algorithm derives from its increased mobility around the state space. Regarding this issue, Peskun (1973) established the following general result:

Theorem 3.4.1 *Consider two reversible Markov chains on a countable state space, with transition matrices P_1 and P_2 such that the non-diagonal elements of P_2 are larger than that of P_1. Then the chain induced by T_2 dominates the chain induced by T_1 in terms of variance.*

The Metropolized Gibbs sampler can be viewed as an application of this theorem.

3.4.2 Bayesian Analysis for ChIP-chip Data

Chromatin immunoprecipitation (ChIP) coupled with microarray (chip) analysis provides an efficient way of mapping protein-DNA interactions across a whole genome. ChIP-chip technology has been used in a wide range of biomedical studies, for instance, in identification of human transcription factor binding sites (Cawley *et al.*, 2004), investigation of DNA methylation (Zhang *et al.*, 2006), and investigation of histone modifications in animals (Bernstein *et al.*, 2005) and plants (Zhang *et al.*, 2007). Data from ChIP-chip experiments encompass DNA-protein interaction measurements on millions of short oligonucleotides (also known as probes), which often tile one or several chromosomes or even the whole genome. The data analysis consists of two steps: (*i*) identifying the bound regions where DNA and the protein are cross-linked in the experiments, and (*ii*) identifying the binding sites through sequence analysis of the bound regions. Our goal here is to provide an effective method for the first step analysis.

Analysis of ChIP-chip data is very challenging due to the large amount of probes and the small number of replicates. The existing methods in the literature can be roughly grouped into three categories: sliding window methods (Cawley *et al.*, 2004; Bertone *et al.*, 2004; Keles *et al.*, 2006); hidden Markov Model (HMM) methods (Li *et al.*, 2005; Ji and Wong, 2005; Munch *et al.*, 2006; Humburg *et al.*, 2008); and Bayesian methods (Qi *et al.*, 2006; Keles, 2007; Gottardo *et al.*, 2008; Wu *et al.*, 2009; Mo and Liang, 2009). Other methods have been suggested by Zheng *et al.* (2007), Huber *et al.* (2006) and Reiss *et al.* (2008), but are less commonly used.

Sliding window methods test a hypothesis for each probe using information from the probes within a certain genomic distance sliding window, then try to correct for multiple hypothesis tests. The test statistics used are varied: Cawley *et al.* (2004) use Wilcoxon's rank sum test; Keles *et al.* (2006) use a scan statistic which is the average of t-statistics within the sliding window; Ji and Wong (2005) use a scan statistic which is the average of empirical Bayesian t-statistics within the sliding window. Since each test uses information from neighboring probes, the tests are not independent, rendering adjustment difficult in the multiple hypothesis testing step.

The power of the sliding window test is usually low, as there is only limited neighboring information. In the ChIP-chip experiments, DNA samples hybridized to the microarrays are prepared by PCR, which is known to perform independently of the form of DNA, and the far probes should have similar intensity patterns as long as they have similar positions to their nearest bound regions. This provides a basis for devising powerful methods that make use of information from all probes. HMM methods have the potential to make use of all probe information where model parameters are estimated using all available data. However, as pointed out by Humburg *et al.* (2008), both the Baum-Welch algorithm and the Viterbi training algorithm (Rabiner, 1989) tend to converge to a local maximum of the likelihood function, rendering the parameter estimation and model inference suboptimal.

Bayesian methods also have the potential to make use of all probe information. Like HMM methods, Bayesian methods estimate model parameters using all available data. However, these methods usually require multiple replicates or some extra experimental information to parameterize the model. For example, the joint binding deconvolution model (Qi *et al.*, 2006) requires one to know the DNA fragment lengths, measured separately for each sample via extrophoretic analysis; and the hierarchical Gamma mixture model (HGMM) (Keles, 2007) requires one to first divide the data into genomic regions containing at most one bound region, information which is, in general, unavailable. The Bayesian hierarchical model of Gottardo *et al.* (2008) essentially models the probe intensities using a mixture of normal distributions, and models the spatial structure of the probes using a Gaussian intrinsic auto-regression model (Besag and Kooperberg, 1995). This model does not need extra experimental information, but the algorithm (the software, called BAC, is available at `http://www.bioconductor.org/packages/2.2/bioc`) is extremely slow, taking about 10 hours for datasets consisting of 300 000 probes on a personal computer. The reason is that the model has too many parameters.

In this subsection, we describe the latent variable method developed by Wu *et al.* (2009). It directly works on the difference between the averaged treatment and control samples. This allows the use of a simple model which avoids the probe-specific effect and the sample (control/treatment) effect. This enables an efficient MCMC simulation of the posterior distribution, and also makes the model robust to the outliers.

3.4.2.1 Bayesian Latent Model

The Model. Consider a ChIP-chip experiment with two conditions, treatment and control. Let X_1 and X_2 denote, respectively, the samples measured under the treatment and control conditions. Each sample has $m_l, l = 1, 2$, replicates providing measurements for n genomic locations along a chromosome or the genome. Suppose that these samples have been normalized and log-transformed. Summarize the measurements for each probe by the differences

$$Y_i = \bar{X}_{1i} - \bar{X}_{2i}, \tag{3.24}$$

where $\bar{X}_{li}$ is the intensity measurement of probe i averaged over m_l replicates. The underlying assumption for this summary statistic is that the intensity measurements for each probe have a variance independent of its genomic position. The rationale is that the DNA samples used in the experiments are prepared by PCR, which is known to perform independently of the form of DNA, and therefore it is the amount of the DNA samples that provides the main source for the variation of probe intensities.

Suppose that the dataset consists of a total of K bound regions, and that region k consists of $n_k (k = 1, \ldots, K)$ consecutive probes. For convenience, we refer to all the non bound regions as region 0 and denote by n_0 the total number of probes contained in all the non bound regions, although the probes in which may be non consecutive. Thus, we have $\sum_{k=0}^{K} n_k = n$. Let

$\boldsymbol{z} = (z_1, \ldots, z_n)$ be a latent binary vector associated with the probes, where $z_i = 1$ indicates that probe i belongs to a bound region and 0 otherwise. Given $\boldsymbol{z}$, we rewrite $(Y_1, \ldots, Y_n)$ as $\{Y_{kj}\}, k = 0, \ldots, K, j = 1, \ldots, n_k$, which are modeled as

$$y_{kj} = \mu_0 + \nu_k + \epsilon_{kj}, \tag{3.25}$$

where μ_0 is the overall mean, which models the difference of sample effects (between the treatment samples and the control samples); $\nu_0 = 0$ and $\nu_k > 0$, $k = 1, \ldots, K$ accounts for the difference in probe intensities in different bound regions; ϵ_{kj}'s are random errors independently and identically distributed as $N(0, \sigma^2)$. Conditional on $\boldsymbol{z}$, the likelihood of the model can be written as

$$\begin{aligned} f(\boldsymbol{y}|\boldsymbol{z}, \mu_0, \nu_1, \ldots, \nu_K, \sigma^2) = & \prod_{j=1}^{n_0} \left(\frac{1}{\sqrt{2\pi}\sigma} e^{-\frac{1}{2\sigma^2}(y_{0j}-\mu_0)^2} \right) \\ & \times \prod_{k=1}^{K} \prod_{j=1}^{n_k} \left(\frac{1}{\sqrt{2\pi}\sigma} e^{-\frac{1}{2\sigma^2}(y_{kj}-\mu_0-\nu_k)^2} \right). \end{aligned} \tag{3.26}$$

To conduct a Bayesian analysis, we specify the following prior distributions:

$$\sigma^2 \sim IG(\alpha, \beta), \quad f(\mu_0) \propto 1, \quad \nu_k \sim \text{Uniform}(\nu_{min}, \nu_{max}), \tag{3.27}$$

where $IG(\cdot, \cdot)$ denotes an inverse Gamma distribution, and $\alpha, \beta, \nu_{min}, \nu_{max}$ are hyperparameters. For example, the hyperparameters can be set as follows: $\alpha = \beta = 0.05, \nu_{min} = 2s_y$ and $\nu_{max} = \max_i y_i$, where s_y is the sample standard error of y_i's. Wu *et al.* (2009) experimented with different choices of ν_{min}, for example s_y and $1.5s_y$, and found that the results are similar. The latent vector $\boldsymbol{z}$ has the prior distribution

$$f(\boldsymbol{z}|\lambda) \propto \frac{\lambda^K e^{-\lambda}}{K!}, \quad K \in \{0, 1, \ldots, K_{\max}\}, \tag{3.28}$$

where $K = |\boldsymbol{z}|$ denotes the total number of bound regions specified by $\boldsymbol{z}$, λ is a hyperparameter, and $K_{\max}$ is the largest number of bounded regions allowed by the model. Since the length of each bound region is very short compared to the chromosome or the whole genome, it is reasonable to view each bound region as a single point, and thus, following standard Poisson process theory, the total number of bound regions can be modeled as a Poisson random variable. Conditioning on the total number of bound regions, as implied by (3.28), we impose the uniform prior on all possible configurations of $\boldsymbol{z}$. The prior (3.28) penalizes a large value of K, where the parameter λ represents the strength of penalty. We do not recommend using a large value of λ, as the number of true bound regions is usually small and a large value of λ may lead to a high false discovery rate. Our experience shows that a value of λ around

0.01 usually works well for most data. In this subsection, we set $\lambda = 0.01$ and $K_{\max} = 5000$ in all simulations.

If $\nu_1, \ldots, \nu_K \in (\nu_{\min}, \nu_{\max})$, combining the likelihood and prior distributions, integrating out σ^2, and taking the logarithm, we get the following log-posterior density function:

$$\log f(\boldsymbol{z}, \mu_0, \nu_1, \ldots, \nu_K|\boldsymbol{y}) = Constant - \Big(\frac{n}{2} + \alpha\Big) \log\left(\frac{1}{2} \sum_{j=1}^{n_0} (y_{0j} - \mu_0)^2 + \frac{1}{2} \sum_{k=1}^{K} \sum_{j=1}^{n_k} (y_{kj} - \mu_0 - \nu_k)^2 + \beta \right) - \log(K!) + K\Big(\log(\lambda) - \log(\nu_{max} - \nu_{min})\Big), \tag{3.29}$$

otherwise, the posterior is equal to 0.

MCMC Simulations. To simulate from the posterior distribution (3.29), the Metropolis-within-Gibbs sampler is used as follows:

1. Conditioned on $\boldsymbol{z}^{(t)}$, update $\mu_0^{(t)}, \nu_1^{(t)}, \ldots, \nu_K^{(t)}$ using the MH algorithm.

2. Conditioned on $\mu_0^{(t)}, \nu_1^{(t)}, \ldots, \nu_K^{(t)}$, update each component of $\boldsymbol{z}^{(t)}$: change $z_i^{(t)}$ to $z_i^{(t+1)} = 1 - z_i^{(t)}$ according to the MH rule.

In updating a component of $\boldsymbol{z}$, the sum of square terms in the posterior (3.29) can be calculated in a recursive manner. This greatly simplifies the computation of the posterior distribution.

Inference of Bound Regions. Let $p_i = P(z_i = 1|\boldsymbol{y})$ be the marginal posterior probability that probe i belongs to a bound region. Since the bound regions are expected to consist of several consecutive probes with positive IP-enrichment effects, the regions that consist of several consecutive probes with high marginal posterior probabilities are likely to be bound regions. To identify such regions, we calculate the joint posterior probability

$$\rho_i(w, m|\boldsymbol{y}) = P\left(\sum_{j=i-w}^{i+w} z_j \geq m|\boldsymbol{y} \right), \tag{3.30}$$

where i is the index of the probes, w is a pre specified half-window size, and m is the minimum number of probes belonging to the bound region. This removes false bound regions, which usually consist of only a few isolated probes with high intensities. We find that the choice $w = 5$ and $m = 5$ often works well in practice. Estimation of ρ_i is trivial using the MCMC samples simulated from the posterior distribution. The value of ρ_i depends on many parameters, such as w, m, and the hyperparameters of the model. However, the order of ρ_i's appears robust to these parameters. This suggests treating ρ_i as a conventional testing p-value, and, to control the false discovery rate (FDR) of the bound regions, using a FDR control method, such as the empirical Bayes method (Efron, 2004) or the stochastic approximation method (Liang and Zhang, 2008), both of which allow for dependence between testing statistics.

Although a strict control of FDR is important to the detection of bound regions, it is not our focus here. We will simply set a cut-off value of ρ_i, classifying probe i to be in bound regions if $\rho_i \geq 0.5$ and in the nonbound region otherwise.

3.4.2.2 Numerical Examples

As seen in the numerical examples, the joint posterior probability can lead to satisfactory detection of true bound regions.

The Estrogen Receptor Data. The estrogen receptor (ER) data generated by Carroll *et al.* (2005) mapped the association of the estrogen receptor on chromosomes 21 and 22. We used a subset of the data, available from the BAC software, to illustrate how the Bayesian latent model works. It consists of intensity measurements for 30 001 probes under the treatment and control conditions, with three replicates each.

The Bayesian latent method was run 5 times for the data. Each run consisted of 11 000 iterations, and took about 4.4 minutes CPU time on a personal computer. For comparison, BAC (Gottardo *et al.*, 2008) and tileHMM (Humburg *et al.*, 2008; the software is available at `http://www.cran.r-project.org/web/packages`) were also applied to this dataset. Both BAC and tileHMM produced a probability measure for each probe, similar to ρ_i, on how likely it was to belong to a bound region. As shown in Figure 3.4, all three methods produce very similar results in this example. However, the results produced by the Bayesian latent model are neater; the resulting joint posterior probabilities tend to be dichotomized, either close to 1 or 0. To provide some numerical evidence for this statement, we calculated the ratio $\#\{i : P_i > 0.5\}/\#\{i : P_i > 0.05\}$, where P_i refers to the joint posterior probability for the Bayesian latent model and BAC, and the conditional probability for tileHMM. The ratios resultant from the Bayesian latent model, BAC and tileHMM are 0.816, 0.615 and 0.674, respectively.

p53 Data. In a ChIP-chip experiment, Cawley *et al.* (2004) mapped the binding sites of four human transcription factors Sp1, cMyc, p53-FL, and p53-DO1 on chromosomes 21 and 22. The experiment consisted of 6 treatment and 6 input control arrays, and the chromosomes spanned over three chips A, B and C. For our illustration, analysis of p53-FL data on chips A, B and C, which contain 14 quantitative PCR verified regions, was conducted. The data were pre processed by filtering out local repeats, quantile-normalized (Bolstad

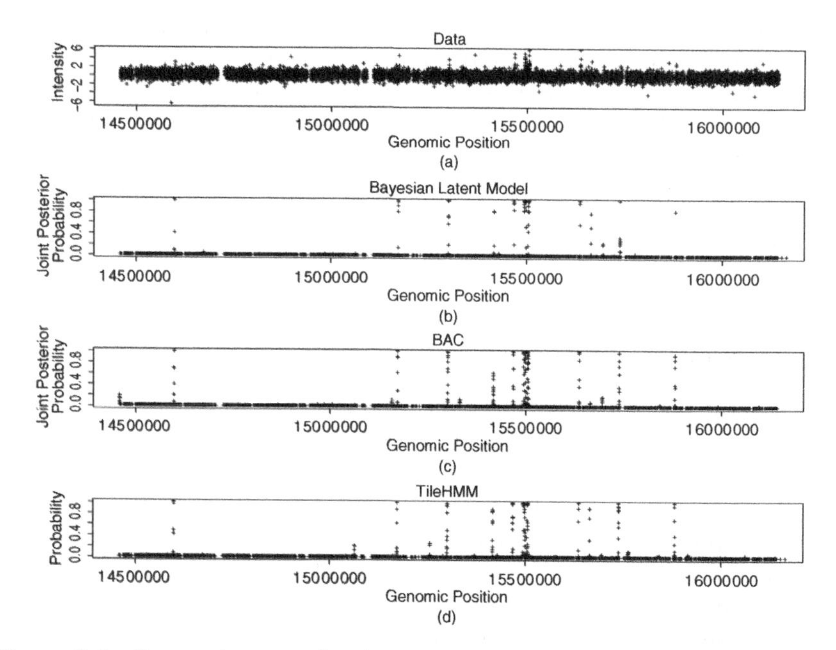

Figure 3.4 Comparison results for the ER data: (a) original data; (b) the joint posterior probability produced by the Bayesian latent model; (c) the joint posterior probability produced by BAC; and (d) the posterior probability produced by tileHMM (Wu *et al.*, 2009).

et al., 2003), rescaled to have a median feature intensity of 1000, and then log-transformed.

The Bayesian latent method, BAC and tileHMM were applied to the p53 data. The data on chip A, chip B, and chip C were analyzed separately. Given the posterior probabilities, a cut-off of 0.5 was used for all methods used to detect bound regions. All resultant regions having less than 3 probes or 100 bps were considered to be spurious and removed, and those regions separated by 500 bps or less were merged together to form a predicted bound region. The results are summarized in Table 3.2. Although tileHMM detected all the 14 validated regions, it essentially failed for this data. It identified a total of 33 796 bound regions, indicating too many false bound regions. We suspect that the failure of tileHMM for this example is due to its training algorithm; it is very likely that tileHMM converged to a local maximum of the likelihood function.

The Bayesian latent method and BAC work well for this example. At a cut-off of 0.5, BAC identified 100 bound regions, covering 12 out of 14 experimentally validated bound regions. The Bayesian latent method works even better. At the same cut-off, it only identified 70 bound regions, but which cover 12 out of 14 experimentally validated bound regions. For further

Table 3.2 Computational results for the p53-FL data with a cut-off of 0.5 (Wu *et al.*, 2009).

	chip A		chip B		chip C		p53	
method	V	total	V	total	V	total	V	total
Latent	2	15	2	28	8	27	12	70 (127)
BAC	2	38	1	29	9	33	12	100 (1864)
tileHMM	2	29 708	3	1944	9	2144	14	33 796

The column 'V' reports the number of bound regions that were found by the method and have been verified by PCR experiments. It is known there are 2, 3, and 9 PCR verified bound regions on chips A, B and C, respectively. The column 'Total' reports the total number of bound regions identified by the method. The columns under 'p53' summarize the results on chips A, B and C. In the parentheses is the number of clusters needed to cover all 14 experimentally validated bound regions (Wu *et al.*, 2009).

comparison, we relaxed the cut-off value and counted the total number of regions needed to cover all experimentally validated regions. The Bayesian latent method only needs to increase the total number of regions to 127, while BAC needs to increase it to 1864 regions. The BAC and tileHMM results reported here may be slightly different from those reported by other authors due to the difference of normalization methods used in data preparation.

Exercises

3.1 Consider the discrete state space $\mathcal{X} = \{i : i = 1, 2, \ldots\}$. Show that if the transition probability satisfy the detailed balance condition $\pi_i P_{i,j} = \pi_j P_{j,i}$ for all i and j, then π_i is the invariant distribution of the Markov chain. Also show that in general the detailed balance or reversibility condition (3.2) implies the balance condition.

3.2 Detailed balance also implies that around any closed cycle of states, there is no net flow of probability. Show

$$P(a, b)P(b, c)P(c, a) = P(a, c)P(c, b)P(b, a), \quad \forall a, b, c.$$

3.3 Consider the random walk MH scheme. Show that if the support region of $f(x)$ is connected and $h(\cdot)$ is positive in a neighborhood of 0, the resulting chain is irreducible and aperiodic (see Robert and Casella, 2004).

3.4 Show that the Multiple-try Metropolis in Section 3.2.3 satisfies the detailed balance condition.

3.5 Show that the Metropolis-within-Gibbs sampler leads to a Markov chain which admits $f(x_1, \ldots, x_K)$ as the stationary distribution.

3.6 Implement the reversible jump MCMC algorithm given in Example 3.1 for a one-dimensional mixture normal model.

3.7 Let $\pi(x)$ denote a density function. Suppose that the sample space $\mathcal{X}$ has been partitioned into m subspaces $A_1, \ldots, A_m$. Liang and Liu (2005) propose the following equation-solving method to estimate the probabilities $\pi(A_i)$'s:

(a) Draw a random sample from a proposal distribution $T(x_t, y)$.

(b) Calculate the ratio

$$r(x_t, y) = \frac{\pi(y)}{\pi(x_t)} \frac{T(y, x_t)}{T(x_t, y)}.$$

Set $x_{t+1} = y$ with probability $\min\{1, r(x_t, y)\}$ and set $x_{t+1} = x_t$ with the remaining probability.

(c) Suppose $x_t \in A_i$ and $y \in A_j$, set $\tilde{P}_{ij} \leftarrow \tilde{P}_{ij} + \frac{r(x_t,y)}{1+r(x_t,y)}$ and $\tilde{P}_{ii} \leftarrow \tilde{P}_{ii} + \frac{1}{1+r(x_t,y)}$.

Let $\widehat{P}$ denote the normalized matrix of $\tilde{P}$, and solve the systems of equations:

$$\sum_{i=1}^{m} \widehat{\pi}_n(A_i)\widehat{P}_{ij} = \widehat{\pi}_n(A_j), \quad j = 1, \ldots, m$$

$$\text{subject to} \quad \sum_{i=1}^{m} \widehat{\pi}_n(A_i) = 1,$$

where the subscript n denotes the number of iterations performed in the MCMC simulation, and $\widehat{\pi}_n(A_i)$ is called the equation solving estimator of $\pi(A_i)$. Show that

$$\widehat{\pi}_n(A_i) \longrightarrow \pi(A_i), \quad \text{almost surely},$$

as $n \to \infty$.

3.8 Implement the equation-solving method (see Exercise 3.7) for the change-point identification problem studied in Section 3.3.2, and compare the results with those presented in Table 3.1.

Chapter 4

Auxiliary Variable MCMC Methods

Consider the problem of sampling from a multivariate distribution with density function $f(x)$. It is known that *Rao-Blackwellization* (Bickel and Doksum, 2000) is the first principle of Monte Carlo simulation: in order to achieve better convergence of the simulation, one should try to integrate out as many components of x as possible. However, sometimes one can include one or more additional variables in simulations to accelerate or facilitate the simulations. This often occurs in the following two scenarios:

- *The target distribution $f(x)$ is multimodal*: An auxiliary variable, such as temperature or some unobservable measurement, is included in simulations to help the system to escape from local-traps.
- *The target distribution $f(x)$ includes an intractable normalizing constant*: an auxiliary realization of X is included in simulations to have the normalizing constants canceled.

The Metropolis-Hastings (MH) algorithm involves two basic components, the target distribution and the proposal distribution. Accordingly, the auxiliary variable methods can also be performed in two ways, namely augmenting auxiliary variables to the target distribution and augmenting auxiliary variables to the proposal distribution. We refer to the former as the target distribution augmentation method and to the latter as the proposal distribution augmentation method. The target distribution augmentation method can be implemented as follows:

- Specify an auxiliary variable u and the conditional distribution $f(u|x)$ to form the joint distribution $f(x, u) = f(u|x)f(x)$.
- Update (x, u) using the MH algorithm or the Gibbs sampler.

Advanced Markov Chain Monte Carlo Methods: Learning from Past Samples
Faming Liang, Chuanhai Liu and Raymond J. Carroll

The samples of $f(x)$ can then be obtained from the realizations, $(x_1, u_1), \ldots, (x_N, u_N)$, of (X, U) through marginalization or conditioning.

The method of proposal distribution augmentation can be implemented as follows:

- Specify a proposal distribution $T(x', u|x)$ and its reversible version $T(x, u|x')$ such that $\int T(x', u|x)du = T(x'|x)$ and $\int T(x, u|x')du = T(x|x')$.

- Generate a candidate sample x' from the proposal $T(x', u|x)$, and accept it with probability $\min\{1, r(x, x', u)\}$, where

$$r(x, x', u) = \frac{f(x')}{f(x)} \frac{T(x, u|x')}{T(x', u|x)}.$$

  Repeat this step to generate realizations $x_1, \ldots, x_N$, which will be approximately distributed as $f(x)$ when N becomes large.

The validity of this method can be shown as follows. Let

$$K(x'|x) = \int_{\mathcal{U}} s(x, x', u)du + I(x = x') \left[1 - \int_{\mathcal{X}} \int_{\mathcal{U}} s(x, x^*, u)dudx^*\right], \quad (4.1)$$

denote the integrated transitional kernel from x to x', where $s(x, x', u) = T(x', u|x)\, r(x, x', u)$, and $I(\cdot)$ is the indicator function. Then

$$f(x) \int_{\mathcal{U}} s(x, x', u)du = \int_{\mathcal{U}} \min\{f(x')T(x, u|x'), f(x)T(x', u|x)\}du, \quad (4.2)$$

which is symmetric about x and x'. This implies $f(x)K(x'|x) = f(x')K(x|x')$; that is, the detailed balance condition is satisfied for the transition $x \to x'$.

In this chapter, we review the existing auxiliary MCMC methods. Methods of target distribution augmentation include simulated annealing (Kirkpatrick *et al.*, 1983), simulated tempering (Marinari and Parisi, 1992; Geyer and Thompson, 1995), slice sampler (Higdon, 1998), the Swendsen-Wang algorithm (Swendsen and Wang, 1987), and the Møller algorithm (Møller *et al.*, 2006). Technically, the data augmentation algorithm (Tanner and Wong, 1987), described in Section 2.2, also belongs to this class because it views missing observations as auxiliary variables. Methods classed as proposal distribution augmentation include the Wolff algorithm (1989), the exchange algorithm (Murray *et al.*, 2006), and the double MH algorithm (Liang, 2009c).

4.1 Simulated Annealing

In industry, annealing refers to a process used to harden steel. First, the steel is heated to a high temperature, almost to the point of transition to its liquid phase. Subsequently, the steel is slowly cooled down enough to make the atoms self-arranged in an ordered pattern. The highest ordered pattern, which corresponds to the global minimum of the free energy of the steel, can be achieved only if the cooling process proceeds slowly enough. Otherwise, the frozen status will fall into a local minimum of free energy.

Realizing that the MH algorithm can be used to simulate the evolution of a solid at various temperature towards thermal equilibrium, Kirkpatrick *et al.* (1983) proposed the simulated annealing algorithm, which mimics the thermal annealing process, for combinational optimization problems. The algorithm can be described as follows.

Suppose that one aims to find the global minimum of an objective function $H(x)$, which is also called the energy function in the standard terms of simulated annealing. By augmenting to the system an auxiliary variable, the so-called temperature T, minimizing $H(x)$ is equivalent to sampling from the Boltzmann distribution $f(x,T) \propto \exp(-H(x)/T)$ at a very small value (closing to 0) of T. [When T is close to 0, most of the mass of the distribution $f(x,T)$ concentrates on the global minimizers of $H(x)$.] In this sense, we say that *sampling is more basic than optimization*. In order to sample successfully from $f(x,T)$ at a very small value of T, Kirkpatrick *et al.* (1983) suggested simulating from a sequence of Boltzmann distributions, $f(x,T_1),\ldots,f(x,T_m)$, in a sequential manner, where the temperatures form a decreasing ladder $T_1 > T_2 > \cdots > T_m$ with $T_m \approx 0$ and T_1 being reasonably large such that most uphill MH moves at that level can be accepted. Simulation at high temperature levels aims to provide a good initial sample, hopefully a point in the attraction basin of the global minimum of $H(x)$, for the simulation at low temperature levels. In summary, the simulated annealing algorithm can be described as follows:

Simulated Annealing Algorithm

1. Initialize the simulation at temperature T_1 and an arbitrary sample x_0.

2. At each temperature T_i, simulate of the distribution $f(x,T_i)$ for N_i iterations using a MCMC sampler. Pass the final sample to the next lower temperature level as the initial sample.

From the viewpoint of auxiliary MCMC methods, simulated annealing simulates the augmented target distribution $f(x,T)$ with the auxiliary variable T taking values from a finite set $\{T_1,\ldots,T_m\}$ and in a fixed order from high to low.

The main difficulty of using simulated annealing is in choosing the cooling schedule. One cooling schedule of theoretical interest is the 'logarithmic' cooling, in which T_i is set at the order of $O(1/\log(M_i))$ with $M_i = N_1 + \cdots + N_i$. This cooling schedule ensures the simulation to converge to the global minimum of $H(x)$ in probability 1 (Geman and Geman, 1984). In practice, however, it is so slow that no one can afford to have such a long running time.

A linearly or geometrically decreasing cooling schedule is commonly used, but, as shown by Holley *et al.* (1989), these schedules can no longer guarantee the global minima to be reached.

Simulated annealing has many successful applications in optimization, such as the traveling salesman problem (Bonomi and Lutton, 1984; Rossier *et al.*, 1986; Golden and Skiscim, 1986) and the VLSI design (Wong *et al.*, 1988; Nahar *et al.*, 1989). During the past 25 years, simulated annealing has been widely accepted as a general purpose optimization algorithm. See Tan (2008) for recent developments of this algorithm.

4.2 Simulated Tempering

Suppose that it is of interest to sample from the distribution $f(x) \propto \exp(-H(x))$, $x \in \mathcal{X}$. As in simulated annealing, simulated tempering (Marinari and Parisi, 1992; Geyer and Thompson, 1995) augments the target distribution to $f(x, T) \propto \exp(-H(x)/T)$ by including an auxiliary variable T, called temperature, which takes values from a finite set pre specified by the user. Simulated tempering is obtained from simulated annealing by treating the temperature T as an auxiliary random variable to be simulated jointly with x:

- Simulated tempering updates the joint state (x, T) in a Gibbs sampling fashion; that is, updating x and T in an alternative manner.
- In simulated tempering, the lowest temperature is set to 1, as the purpose is to sample from $f(x)$.

Suppose that the temperature T takes m different values, $T_1 > T_2 > \cdots > T_m$, where $T_m \equiv 1$ and is called the target temperature. Let $f(x, T_i) = \exp(-H(x)/T_i)/Z_i$ denote the trial distribution defined on the temperature level T_i, where Z_i is the normalizing constant of the distribution. Let q_{ij} denote the proposal probability of transition from level T_i to T_j. Typically, one sets $q_{i,i+1} = q_{i,i-1} = q_{i,i} = 1/3$ for $1 < i < m$, $q_{1,2} = 1/3$, $q_{m,m-1} = 1/3$, $q_{1,1} = 2/3$, and $q_{m,m} = 2/3$. Starting with $i_0 = 1$ and an initial sample $x_0 \in \mathcal{X}$, simulating tempering iterates between the following three steps:

Simulated Tempering

1. Draw a random number $U \sim$ Uniform$[0, 1]$, and determine the value of j according to the proposal transition matrix (q_{ij}).
2. If $j = i_t$, let $i_{t+1} = i_t$ and let x_{t+1} be drawn from a MH kernel $K_{i_t}(x, y)$ which admits $f(x, T_{i_t})$ as the invariant distribution.
3. If $j \neq i_t$, let $x_{t+1} = x_t$ and accept the proposal with probability

$$\min\left\{1, \frac{\widehat{Z}_j}{\widehat{Z}_{i_t}} \exp\left\{-H(x)\left(\frac{1}{T_j} - \frac{1}{T_{i_t}}\right)\right\} \frac{q_{j,i_t}}{q_{i_t,j}}\right\},$$

where $\widehat{Z}_i$ denotes an estimate of Z_i. If it is accepted, set $i_{t+1} = j$. Otherwise, set $i_{t+1} = i_t$.

The intuition underlying simulated tempering is that simulation at high temperature levels provides a good exploration of the energy landscape of the target distribution, and the low energy samples generated thereby are transmitted to the target level through a sequence of temperature updating operations. As reported by many researchers, simulated tempering can converge substantially faster than the MH algorithm, especially for distributions for which the energy landscape is rugged. For effective implementation of simulated tempering, two issues must be considered:

- *Choice of the temperature ladder.* The highest temperature T_1 should be set such that most of the uphill moves can be accepted at that level. The intermediate temperatures can be set in a sequential manner: Start with T_1, and sequentially set the next lower temperature such that

$$\mathrm{Var}_i(H(x))\delta^2 = O(1), \tag{4.3}$$

  where $\delta = 1/T_{i+1} - 1/T_i$, and $\mathrm{Var}_i(\cdot)$ denotes the variance of $H(x)$ taken with respect to $f(x, T_i)$. This condition is equivalent to requiring that the distributions $f(x, T_i)$ and $f(x, T_{i+1})$ have considerable overlap. In practice, $\mathrm{Var}_i(H(x))$ can be estimated roughly through a preliminary run of the sampler at level T_i.

- *Estimation of Z_i's.* This is the key to the efficiency of simulated tempering. If the pseudo-normalizing constants Z_i's are well estimated, simulated tempering will perform like a 'symmetric random walk' along the temperature ladder (if ignoring the x-updating steps). Otherwise, it may get stuck at a certain temperature level, rendering the simulation a failure. In practice, the Z_i's can be estimated using the stochastic approximation Monte Carlo method (Liang *et al.*, 2007; Liang, 2005b), which is described in Chapter 7. Alternatively, the Z_i's can be estimated using the reverse logistic regression method, as suggested by Geyer and Thompson (1995).

We note that the mixing of simulated tempering suffers from a waiting time dilemma: To make simulated tempering work well, the adjacent distributions $f(x, T_i)$ and $f(x, T_{i+1})$ should have considerable overlap, requiring many intermediate temperature levels to be used. On the other hand, even in the ideal case that simulated tempering performs like a 'symmetric random walk' along the temperature ladder, the expected waiting time for a traversal of the temperature ladder will be of the order $O(m^2)$. This puts a severe limit on the number of temperature levels that one can afford to use in simulations. The same criticism also applies to the tempering-based population MCMC algorithms, such as parallel tempering (Geyer, 1991) and evolutionary Monte Carlo (Liang and Wong, 2000, 2001a), which are described in Chapter 5.

Simulated tempering has been applied successfully to many complex systems, such as protein folding (Hansmann and Okamoto, 1997), and VLSI floorplan design (Cong *et al.*, 1999). In Li *et al.* (2004), the authors discuss at length how to fine tune the proposal distributions used at different temperature levels, and how to lean the temperature ladder to have more samples generated at low temperature levels, while maintaining frequent visits to high temperature levels.

4.3 The Slice Sampler

Suppose that one is interested in sampling from a density $f(x)$, $x \in \mathcal{X}$. Recall that sampling $x \sim f(x)$ is equivalent to sampling uniformly from the area under the graph $f(x)$:

$$\mathcal{A} = \{(x, u) : 0 \leq u \leq f(x)\},$$

which is the basis of the acceptance-rejection algorithm described in Section 1.4.2. To achieve this goal, one can augment the target distribution by an auxiliary variable U, which, conditional on x, is uniformly distributed on the interval $[0, f(x)]$. Therefore, the joint density function of (X, U) is

$$f(x, u) = f(x)f(u|x) \propto 1_{(x,u)\in\mathcal{A}},$$

which can be sampled using the Gibbs sampler as follows:

Slice Sampler

1. Draw $u_{t+1} \sim \text{Uniform}[0, f(x_t)]$.
2. Draw x_{t+1} uniformly from the region $\{x : f(x) \geq u_{t+1}\}$.

This sampler, called the slice sampler by Higdon (1998), potentially can be more efficient than the simple MH algorithm for multimodal distributions, due to the free between-mode-transitions within a slice (as illustrated by Figure 4.1). However, the slice sampler is often difficult to implement in practice. For many distributions, sampling uniformly from the region $\{x : f(x) \geq u\}$ is almost as difficult as sampling from the original distribution $f(x)$.

Edwards and Sokal (1988) note that when $f(x)$ can be decomposed into a product of k distribution functions, that is, $f(x) \propto f_1(x) \times f_2(x) \times \cdots \times f_k(x)$, the slice sampler can be easily implemented. To sample from such a distribution, they introduce k auxiliary variables, $U_1, \ldots, U_k$, and propose the following algorithm:

1. Draw $u_{t+1}^{(i)} \sim \text{Uniform}[0, f_i(x_t)]$, $i = 1, \ldots, k$.
2. Draw x_{t+1} uniformly from the region $\mathcal{S} = \cap_{i=1}^{k}\{x : f_i(x) \geq u_{t+1}^{(i)}\}$.

This algorithm is a generalization of the Swendsen-Wang algorithm (Swendsen and Wang, 1987) described in Section 4.4. Damien *et al.* (1999) explore the use of this algorithm as a general MCMC sampler for Bayesian inference problems.

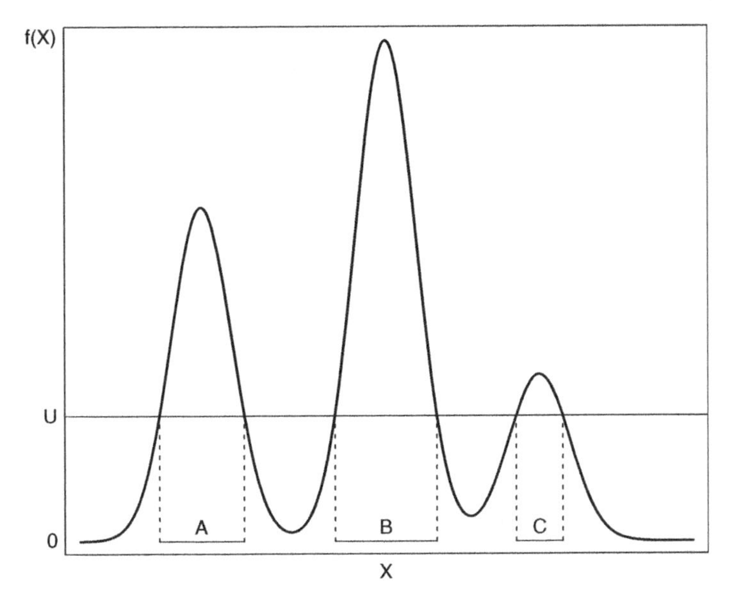

Figure 4.1 The slice sampler for a multimodal distribution: Draw $X|U$ uniformly from the sets labeled by A, B and C.

They illustrate how one can decompose $f(x)$ into a product of k functions for which the joint set $\mathcal{S}$ is easy to compute. For example, for a model of k iid data points, the likelihood function can be decomposed into a product of k distribution functions, one for each data point. However, the convergence of this algorithm may be slow due to the presence of multiple auxiliary variables. Recently, the implementation of the slice sampler was further improved by Neal (2003), who suggested a Gibbs style implementation: update each component of x in turn using a single-variable slicing sampling procedure.

The slice sampler has attractive theoretical properties. Under certain conditions, as Roberts and Rosenthal (1999) show, the slice sampler is geometrically ergodic. Under slightly stronger conditions, Mira and Tierney (2002) show that the slice sampler is uniformly ergodic. These theories assume that the new sample $x^{(t+1)}$ is sampled uniformly from the joint region $\mathcal{S}$, independently of the old sample $x^{(t)}$. However, this is often difficult in practice.

4.4 The Swendsen-Wang Algorithm

Consider a 2-D Ising model with the Boltzmann density

$$f(\boldsymbol{x}) \propto \exp\left\{K \sum_{i \sim j} x_i x_j\right\}, \tag{4.4}$$

where the spins $x_i = \pm 1$, K is the inverse temperature, and $i \sim j$ represents the nearest neighbors on the lattice. An interesting physical property of this model is the phase transition: When the temperature is high, all the spins behave nearly independently, whereas when the temperature is below a critical point ($K_0 \approx 0.44$), all the spins tend to stay in the same state, either '+1' or '−1'.

When the temperature is high, this model can be easily simulated using the Gibbs sampler (Geman and Geman, 1984): iteratively reset the value of each spin according to the conditional distribution

$$\begin{aligned} P(x_i = 1 | x_j, j \in n(i)) &= \frac{1}{1 + \exp\left\{-2K \sum_{j \in n(i)}\right\}}, \\ P(x_i = -1 | x_j, j \in n(i)) &= 1 - P(x_i = 1 | x_j, j \in n(i)), \end{aligned} \tag{4.5}$$

where $n(i)$ denotes the set of neighbors of spin i. For the Ising model, this sampler is also known as the heat bath method (Creutz, 1980; Rubinstein, 1981). However, the Gibbs sampler slows down rapidly when the temperature is approaching or below the critical temperature. This is the so-called 'critical slowing down'. Swendsen and Wang (1987) proposed a powerful auxiliary variable MCMC method, which can eliminate much of the critical slowing down. The Swendsen-Wang algorithm can be described as follows.

As in slice sampling, the density (4.4) is rewritten in the form

$$f(\boldsymbol{x}) \propto \prod_{i \sim j} \exp\{K(1 + x_i x_j)\} = \prod_{i \sim j} \exp\{\beta I(x_i = x_j)\}, \tag{4.6}$$

where $\beta = 2K$ and $I(\cdot)$ is the indicator function, as $1 + x_i x_j$ is either 0 or 2. If we introduce auxiliary variables $\boldsymbol{u} = (u_{i \sim j})$, where each component $u_{i \sim j}$, conditional on x_i and x_j, is uniformly distributed on $[0, \exp\{\beta I(x_i = x_j)\}]$, then

$$f(\boldsymbol{x}, \boldsymbol{u}) \propto \prod_{i \sim j} I(0 \leq u_{i \sim j} \leq \exp\{\beta I(x_i = x_j)\}).$$

In Swendsen and Wang (1987), $u_{i \sim j}$ is called a 'bond variable', which can be viewed as a variable physically sitting on the edge between spin i and spin j.

If $u_{i \sim j} > 1$, then $\exp\{\beta I(x_i = x_j) > 1$ and there must be $x_i = x_j$. Otherwise, there is no constraint on x_i and x_j. Let $b_{i \sim j}$ be an indicator variable for the constraint. If x_i and x_j are constrained to be equal, we set $b_{i \sim j} = 1$ and 0 otherwise. Note that for any two 'like-spin' neighbors (that is, the two spins have the same values), they are bonded with probability $1 - \exp(-\beta)$. Based on the configurations of $\boldsymbol{u}$, we can 'cluster' the spins according to whether they are connected via a 'mutual bond' ($b_{i \sim j} = 1$). Then all the spins within the same cluster will have identical values, and flipping all the spins in a cluster simultaneously will not change the equilibrium of $f(\boldsymbol{x}, \boldsymbol{u})$. The

Swendsen-Wang algorithm iterates between the steps of sampling $f(\boldsymbol{b}|\boldsymbol{x})$ and sampling $f(\boldsymbol{x}|\boldsymbol{u})$:

The Swendsen-Wang Algorithm

1. *Update bond values.* Check all 'like-spin' neighbors, and set $b_{i\sim j} = 1$ with probability $1 - \exp(-\beta)$.

2. *Update spin values.* Cluster spins by connecting neighboring sites with a mutual bond, and then flip each cluster with probability 0.5.

For the Ising model, the introduction of the auxiliary variable $\boldsymbol{u}$ has the dependence between neighboring spins partially decoupled, and the resulting sampler can thus converge substantially faster than the single site updating algorithm. As demonstrated by Swendsen and Wang (1987), this algorithm can eliminate much of the critical slowing down.

Motivated by the success of the Swendsen-Wang algorithm on Ising models, much effort has been spent on generalizing it to a wider class of models, such as the Potts model and the continuous-spin model (de Meo and Oh, 1992). A generalized form of this algorithm was given by Edwards and Sokal (1988), which is also referred to as a slice sampler, as described in Section 4.3. Later, the Edwards-Sokal algorithm was further generalized by Higdon (1998); the joint density $f(\boldsymbol{x}, \boldsymbol{u})$ is replaced by

$$f(\boldsymbol{x}, \boldsymbol{u}) \propto f_0(\boldsymbol{x}) \prod_k f_k(\boldsymbol{x})^{1-\delta_k} I\left[0 \le u_k \le f_k(\boldsymbol{x})^{\delta_k}\right]. \tag{4.7}$$

The simulation of $f(\boldsymbol{x}, \boldsymbol{u})$ can be done, as in the previous case, by iterating between the steps of sampling $f(\boldsymbol{x}|\boldsymbol{u})$ and sampling $f(\boldsymbol{u}|\boldsymbol{x})$. This algorithm is called the partial decoupling algorithm. Because (4.7) includes an extra term $f_0(x)$, the partial decoupling algorithm is potentially useful for models which do not possess the nice symmetry of the Ising model. The partial decoupling algorithm has been successfully applied to image analysis problems (Higdon, 1998).

4.5 The Wolff Algorithm

Wolff (1989) introduces a modification for the Swendsen-Wang algorithm, which can almost completely eliminate the critical slowing down for the Ising model as defined in (4.4).

The Wolff Algorithm

1. For a given configuration of $\boldsymbol{x}$, one randomly picks a spin, say x, as the first point of a cluster C to be built, and then grows the cluster as follows:

   – Visit all links connecting $x \in C$ to its nearest neighbors y. The bond $x \sim y$ is activated with probability $p_b = 1 - \exp(-2K)$ as in the Swendsen-Wang algorithm, and if this happens, y is adjoined to c. The p_b is called the bonding probability.

– Continue iteratively in the same way for all bonds leading to neighbors of newly adjoined spins until the process stops.

2. Flip all the spins in C to their opposites.

The Wolff algorithm and the Swendsen-Wang algorithm are very similar, the only difference being that in the Wolff algorithm, at each iteration, only one cluster is formed and all spins in the cluster are flipped. However, these two algorithms have very different auxiliary-variable augmentation schemes: The Swendsen-Wang algorithm augments its target distribution, whereas the Wolff algorithm augments its proposal distribution. This can be explained as follows. Suppose that the cluster C one has grown has $m+n$ neighbors, among which m are $+1$ spins and n are -1 spins. If the current state of C is all $+1$, by flipping all the spins in C to -1, then the probability ratio is

$$\frac{f(\boldsymbol{x}^*)}{f(\boldsymbol{x})} = \exp\{2K(n-m)\}, \tag{4.8}$$

where $\boldsymbol{x}^*$ denotes the new configuration of $\boldsymbol{x}$. The proposal probability for the update is given by

$$T(\boldsymbol{x} \to (\boldsymbol{x}^*, \boldsymbol{b})) = p_b^{c_1}(1-p_b)^{c_0+m}, \tag{4.9}$$

where $\boldsymbol{b}$ denotes the configuration of mutual bonds formed for the cluster C and it plays the role of auxiliary variables, c_1 denotes the number of interior links of C that is formed as a mutual bond, and c_0 denotes the number of interior links of C that is not formed as a mutual bond. Similarly, we have

$$T(\boldsymbol{x}^* \to (\boldsymbol{x}, \boldsymbol{b})) = p_b^{c_1}(1-p_b)^{c_0+n}. \tag{4.10}$$

Combining equations (4.8), (4.9) and (4.10) yields

$$\frac{f(\boldsymbol{x}^*)}{f(\boldsymbol{x})}\frac{T(\boldsymbol{x}^* \to (\boldsymbol{x}, \boldsymbol{b}))}{T(\boldsymbol{x} \to (\boldsymbol{x}^*, \boldsymbol{b}))} = 1,$$

which implies that the proposed change should be always accepted.

The reason why the Wolff algorithm outperforms the Swendsen-Wang algorithm is clear: by randomly flipping all clusters simultaneously, the Swendsen-Wang algorithm introduces too much randomness at each iteration. The Wolff algorithm avoids this drawback by flipping a single cluster at each iteration. The recursivity of Wolff's flipping improves the convergence of the simulation.

It is apparent that the growth of the cluster is not necessarily restricted to the 'like-spins'. More generally, Niedermayer (1988) suggests that one can bond two neighboring spins with opposite values, and then flip each spin in the cluster to its opposite. In this scheme, the acceptance probability for a new configuration of spins can be calculated in a similar way to the Wolff algorithm, but not necessarily equal to 1, depending on the choice of bonding probabilities.

4.6 The Møller Algorithm

Spatial models, for example, the autologistic model, the Potts model, and the autonormal model (Besag, 1974), have been used in modeling of many scientific problems. Examples include image analysis (Hurn *et al.*, 2003), disease mapping (Green and Richardson, 2002), geographical genetic analysis (Francois *et al.*, (2006)), among others. A major problem with these models is that the normalizing constant is intractable. The problem can be described as follows. Suppose we have a dataset $\boldsymbol{X}$ generated from a statistical model with the likelihood function

$$f(\boldsymbol{x}|\theta) = \frac{1}{Z(\theta)} \exp\{-U(\boldsymbol{x}, \theta)\}, \quad \boldsymbol{x} \in \mathcal{X}, \quad \theta \in \Theta, \tag{4.11}$$

where θ is the parameter, and $Z(\theta)$ is the normalizing constant which depends on θ and is not available in closed form. Let $f(\theta)$ denote the prior density of θ. The posterior distribution of θ given $\boldsymbol{x}$ is then given by

$$f(\theta|\boldsymbol{x}) \propto \frac{1}{Z(\theta)} \exp\{-U(\boldsymbol{x}, \theta)\} f(\theta). \tag{4.12}$$

The MH algorithm cannot be directly applied to simulate from $f(\theta|\boldsymbol{x})$, because the acceptance probability would involve a computationally intractable ratio $Z(\theta)/Z(\theta')$, where θ' denotes the proposed value. To circumvent this difficulty, various methods of approximating the likelihood function or the normalizing constant function have been proposed in the literature. Besag (1974) proposed to approximating the likelihood function by a pseudo-likelihood function which is tractable. This method is easy to use, but it typically performs less well for the models with strong neighboring dependence. It is further discussed and generalized by Dryden *et al.* (2002) and Huang and Ogata (2002). Geyer and Thompson (1992) propose an importance sampling-based approach to approximating $Z(\theta)$, for which Liang *et al.* (2007) suggest refining the choice of the trial density function using the stochastic approximation Monte Carlo algorithm. Liang (2007) proposes an alternative Monte Carlo approach to approximating $Z(\theta)$, where $Z(\theta)$ is viewed as a marginal distribution of the unnormalized distribution $g(\boldsymbol{x}, \theta) = \exp\{-U(\boldsymbol{x}, \theta)\}$ and is estimated by an adaptive kernel density estimator using Monte Carlo draws.

A significant step made by Møller *et al.* (2006) proposes augmenting the distribution $f(\theta|\boldsymbol{x})$ by an auxiliary variable such that the normalizing constant ratio $Z(\theta)/Z(\theta')$ can be canceled in simulations. The Møller algorithm can be described as follows.

Let $\boldsymbol{y}$ denote the auxiliary variable, which shares the same state space with $\boldsymbol{x}$. Let

$$f(\theta, \boldsymbol{y}|\boldsymbol{x}) = f(\boldsymbol{x}|\theta)f(\theta)f(\boldsymbol{y}|\theta, \boldsymbol{x}), \tag{4.13}$$

denote the joint distribution of θ and $\boldsymbol{y}$ conditional on $\boldsymbol{x}$, where $f(\boldsymbol{y}|\theta, \boldsymbol{x})$ is the distribution of the auxiliary variable $\boldsymbol{y}$. To simulate from (4.13) using the MH algorithm, one can use the proposal distribution

$$q(\theta', \boldsymbol{y}'|\theta, \boldsymbol{y}) = q(\theta'|\theta, \boldsymbol{y})q(\boldsymbol{y}'|\theta'), \tag{4.14}$$

which corresponds to the usual change on the parameter vector $\theta \to \theta'$, followed by an exact sampling (Propp and Wilson, 1996) step of drawing $\boldsymbol{y}'$ from $q(\cdot|\theta')$. If $q(\boldsymbol{y}'|\theta')$ is set as $f(\boldsymbol{y}'|\theta)$, then the MH ratio can be written as

$$r(\theta, \boldsymbol{y}, \theta', \boldsymbol{y}'|\boldsymbol{x}) = \frac{f(\boldsymbol{x}|\theta')f(\theta')f(\boldsymbol{y}'|\theta', \boldsymbol{x})q(\theta|\theta', \boldsymbol{y}')f(\boldsymbol{y}|\theta)}{f(\boldsymbol{x}|\theta)f(\theta)f(\boldsymbol{y}|\theta, \boldsymbol{x})q(\theta'|\theta, \boldsymbol{y})f(\boldsymbol{y}'|\theta')}, \tag{4.15}$$

where the unknown normalizing constant $Z(\theta)$ can be canceled. To ease computation, Møller *et al.* (2006) further suggest setting the proposal distributions $q(\theta'|\theta, \boldsymbol{y}) = q(\theta'|\theta)$ and $q(\theta|\theta', \boldsymbol{y}') = q(\theta|\theta')$, and the auxiliary distributions

$$f(\boldsymbol{y}|\theta, \boldsymbol{x}) = f(\boldsymbol{y}|\widehat{\theta}), \quad f(\boldsymbol{y}'|\theta', \boldsymbol{x}) = f(\boldsymbol{y}'|\widehat{\theta}), \tag{4.16}$$

where $\widehat{\theta}$ denotes an estimate of θ, for example, which can be obtained by maximizing a pseudo-likelihood function. In summary, the Møller algorithm starts with an arbitrary point $\theta^{(0)}$ and an exact sample $\boldsymbol{y}^{(0)}$ drawn from $f(\boldsymbol{y}|\widehat{\theta})$, and then iterates between the following three steps:

Møller Algorithm

1. Generate θ' from the proposal distribution $q(\theta'|\theta_t)$.

2. Generate an exact sample $\boldsymbol{y}'$ from the distribution $f(\boldsymbol{y}|\theta')$.

3. Accept $(\theta', \boldsymbol{y}')$ with probability $\min(1, r)$, where

$$r = \frac{f(\boldsymbol{x}|\theta')f(\theta')f(\boldsymbol{y}'|\widehat{\theta})q(\theta_t|\theta')f(\boldsymbol{y}|\theta_t)}{f(\boldsymbol{x}|\theta_t)f(\theta_t)f(\boldsymbol{y}|\widehat{\theta})q(\theta'|\theta_t)f(\boldsymbol{y}'|\theta')}.$$

   If it is accepted, set $(\theta_{t+1}, \boldsymbol{y}_{t+1}) = (\theta', \boldsymbol{y}')$. Otherwise, set $(\theta_{t+1}, \boldsymbol{y}_{t+1}) = (\theta_t, \boldsymbol{y}_t)$.

This algorithm is applied by Møller *et al.* (2006) to estimate parameters for the Ising model. A problem with it is that the acceptance rate of the exact samples is usually very low.

4.7 The Exchange Algorithm

Like the Møller algorithm (Møller *et al.*, 2006), the exchange algorithm (Murray *et al.*, 2006) is dedicated to sample from the distribution $f(\theta|x)$ given in (4.12). The exchange algorithm is motivated by the parallel tempering algorithm (Geyer, 1991; Hukushima and Nemoto, 1996), and can be described as follows. Consider the augmented distribution

$$f(\boldsymbol{y}_1, \ldots, \boldsymbol{y}_m, \theta|\boldsymbol{x}) = \pi(\theta) f(\boldsymbol{x}|\theta) \prod_{j=1}^{m} f(\boldsymbol{y}_j|\theta_j), \tag{4.17}$$

where θ_i's are instantiated and fixed, and $\boldsymbol{y}_1, \ldots, \boldsymbol{y}_m$ are independent auxiliary variables with the same state space as $\boldsymbol{x}$ and the joint distribution $\prod_{j=1}^{m} f(\boldsymbol{y}_j|\theta_j)$. Suppose that a change to θ is proposed with probability $q(\theta_i|\theta)$. To ensure that $\boldsymbol{y}_i = \boldsymbol{x}$, we swap the settings of $\boldsymbol{x}$ and $\boldsymbol{y}_i$. The resulting MH ratio for the change is

$$\begin{aligned} r(\theta, \theta_i, \boldsymbol{y}_i|\boldsymbol{x}) &= \frac{\pi(\theta_i) f(\boldsymbol{x}|\theta_i) f(\boldsymbol{y}_i|\theta) \prod_{j \neq i} f(\boldsymbol{y}_j|\theta_j) q(\theta|\theta_i)}{\pi(\theta) f(\boldsymbol{x}|\theta) f(\boldsymbol{y}_i|\theta_i) \prod_{j \neq i} f(\boldsymbol{y}_j|\theta_j) q(\theta_i|\theta)} \\ &= \frac{\pi(\theta_i) f(\boldsymbol{x}|\theta_i) f(\boldsymbol{y}_i|\theta) q(\theta|\theta_i)}{\pi(\theta) f(\boldsymbol{x}|\theta) f(\boldsymbol{y}_i|\theta_i) q(\theta_i|\theta)}. \end{aligned} \tag{4.18}$$

Based on the above arguments, Murray *et al.* (2006) propose the following algorithm:

Exchange Algorithm

1. Propose $\theta' \sim q(\theta'|\theta, \boldsymbol{x})$.
2. Generate an auxiliary variable $\boldsymbol{y} \sim f(\boldsymbol{y}|\theta')$ using an exact sampler.
3. Accept θ' with probability $\min\{1, r(\theta, \theta', \boldsymbol{y}|\boldsymbol{x})\}$, where

$$r(\theta, \theta', \boldsymbol{y}|\boldsymbol{x}) = \frac{\pi(\theta') f(\boldsymbol{x}|\theta') f(\boldsymbol{y}|\theta) q(\theta|\theta')}{\pi(\theta) f(\boldsymbol{x}|\theta) f(\boldsymbol{y}|\theta') q(\theta'|\theta)}. \tag{4.19}$$

Since a swapping change between $(\theta, \boldsymbol{x})$ and $(\theta', \boldsymbol{y})$ is involved, the algorithm is called the exchange algorithm. It generally improves the performance of the Møller algorithm, as it removes the need to estimate the parameter before sampling begins. Murray *et al.* (2006) report that the exchange algorithm tends to have a higher acceptance probability for the exact samples than the Møller algorithm.

The exchange algorithm can also be viewed as an auxiliary variable MCMC algorithm with the proposal distribution being augmented, for which the proposal distribution can be written as

$$T(\theta \to (\theta', \boldsymbol{y})) = q(\theta'|\theta) f(\boldsymbol{y}|\theta'), \quad T(\theta' \to (\theta, \boldsymbol{y})) = q(\theta|\theta') f(\boldsymbol{y}|\theta).$$

This simply validates the algorithm, following the arguments for auxiliary variable Markov chains made around (4.1) and (4.2).

4.8 The Double MH Sampler

Although the Møller algorithm and the exchange algorithm work for the Ising model, they cannot be applied to many other models for which exact sampling is not available. In addition, even for the Ising model, exact sampling may be very expensive when the temperature is near or below the critical point. To overcome this difficulty, Liang (2009c) proposes the double MH algorithm, which avoids the requirement for exact sampling, the auxiliary variables being generated using MH kernels, and thus can be applied to many statistical models for which exact sampling is not available or is very expensive. While for the models for which exact sampling is available, such as the Ising model and the autologistic model, it can produce almost the same accurate results as the exchange algorithm, but using much less CPU time.

Suppose that one is interested in simulating a sample $\boldsymbol{y}$ from $f(\boldsymbol{y}|\theta')$. If the sample is generated through m MH updates starting with the current state $\boldsymbol{x}$, the transition probability, $P_{\theta'}^{(m)}(\boldsymbol{y}|\boldsymbol{x})$, is

$$P_{\theta'}^{(m)}(\boldsymbol{y}|\boldsymbol{x}) = K_{\theta'}(\boldsymbol{x} \to \boldsymbol{x}_1) \cdots K_{\theta'}(\boldsymbol{x}_{m-1} \to \boldsymbol{y}), \tag{4.20}$$

where $K(\cdot \to \cdot)$ is the MH transition kernel. Then

$$\begin{aligned} \frac{P_{\theta'}^{(m)}(\boldsymbol{x}|\boldsymbol{y})}{P_{\theta'}^{(m)}(\boldsymbol{y}|\boldsymbol{x})} &= \frac{K_{\theta'}(\boldsymbol{y} \to \boldsymbol{x}_{m-1}) \cdots K_{\theta'}(\boldsymbol{x}_1 \to \boldsymbol{x})}{K_{\theta'}(\boldsymbol{x} \to \boldsymbol{x}_1) \cdots K_{\theta'}(\boldsymbol{x}_{m-1} \to \boldsymbol{y})} \\ &= \frac{f(\boldsymbol{x}|\theta')}{f(\boldsymbol{y}|\theta')} \frac{f(\boldsymbol{y}|\theta')}{f(\boldsymbol{x}|\theta')} \frac{K_{\theta'}(\boldsymbol{y} \to \boldsymbol{x}_{m-1}) \cdots K_{\theta'}(\boldsymbol{x}_1 \to \boldsymbol{x})}{K_{\theta'}(\boldsymbol{x} \to \boldsymbol{x}_1) \cdots K_{\theta'}(\boldsymbol{x}_{m-1} \to \boldsymbol{y})} \\ &= \frac{f(\boldsymbol{x}|\theta')}{f(\boldsymbol{y}|\theta')}, \end{aligned} \tag{4.21}$$

where the last equality follows from the detailed balance equality $f(\boldsymbol{x}|\theta') K_{\theta'}(\boldsymbol{x} \to \boldsymbol{x}_1) \cdots K_{\theta'}(\boldsymbol{x}_{m-1} \to \boldsymbol{y}) = f(\boldsymbol{y}|\theta') K_{\theta'}(\boldsymbol{y} \to \boldsymbol{x}_{m-1}) \cdots K_{\theta'}(\boldsymbol{x}_1 \to \boldsymbol{x})$.

Returning to the problem of simulating from the posterior distribution (4.12). By (4.21), the MH ratio (4.19) can be re-expressed as

$$r(\theta, \theta', \boldsymbol{y}|\boldsymbol{x}) = \frac{\pi(\theta')q(\theta|\theta')}{\pi(\theta)q(\theta'|\theta)} \frac{f(\boldsymbol{y}|\theta)P_{\theta'}^{(m)}(\boldsymbol{x}|\boldsymbol{y})}{f(\boldsymbol{x}|\theta)P_{\theta'}^{(m)}(\boldsymbol{y}|\boldsymbol{x})}. \tag{4.22}$$

It is easy to see that if one chooses $q(\theta'|\theta)$ as a MH transition kernel which satisfies the detailed balance condition, then $\pi(\theta')q(\theta|\theta') = \pi(\theta)q(\theta'|\theta)$, and the exchange update is reduced to a simple MH update for which $f(\boldsymbol{x}|\theta)$ works as the target distribution and $P_{\theta'}^{(m)}(\boldsymbol{y}|\boldsymbol{x})$ works as the proposal distribution. In summary, the double MH algorithm proceeds as follows.

Double MH sampler

1. Simulate a new sample θ' from $\pi(\theta)$ using the MH algorithm starting with θ_t.

2. Generate an auxiliary variable $\boldsymbol{y} \sim P_{\theta'}^{(m)}(\boldsymbol{y}|\boldsymbol{x})$, and accept it with probability $\min\{1, r(\theta_t, \theta', \boldsymbol{y}|\boldsymbol{x})\}$, where, by (4.21),

$$r(\theta_t, \theta', \boldsymbol{y}|\boldsymbol{x}) = \frac{f(\boldsymbol{y}|\theta_t)P_{\theta'}^{(m)}(\boldsymbol{x}|\boldsymbol{y})}{f(\boldsymbol{x}|\theta_t)P_{\theta'}^{(m)}(\boldsymbol{y}|\boldsymbol{x})} = \frac{f(\boldsymbol{y}|\theta_t)f(\boldsymbol{x}|\theta')}{f(\boldsymbol{x}|\theta_t)f(\boldsymbol{y}|\theta')}. \tag{4.23}$$

3. Set $\theta_{t+1} = \theta'$ if the auxiliary variable is accepted in step (b), and set $\theta_{t+1} = \theta_t$ otherwise.

Since two types of MH updates are performed in step (b), one for drawing the auxiliary variable $\boldsymbol{y}$ and one for acceptance of θ', the algorithm is called the double MH sampler. The MH update performed in step (a) is not essential, which can be incorporated into step (b) by changing (4.23) to (4.22). Also, (4.23) holds regardless of the value of m. The double MH sampler avoids the requirement for exact sampling, so it can be applied to a wide range of problems for which exact sampling is infeasible.

It is apparent that the double MH sampler will converge to the correct posterior distribution for a large value of m. In practice, to get good samples from the posterior distribution, m is not necessarily large. The key to the efficiency of the MH kernel is of starting with $\boldsymbol{x}$. One can expect that the posterior will put most of its mass in the region of the parameter space where the likelihood $f(\boldsymbol{x}|\theta)$ is high, provided that the prior is noninformative. Therefore, most proposed values of θ' will lie in this region and the likelihood $f(\boldsymbol{y}|\theta')$ will put most of its mass in the vicinity of $\boldsymbol{x}$. A more theoreically justified algorithm can be found in the next section.

4.8.1 Spatial Autologistic Models

The autologistic model (Besag, 1974) has been widely used for spatial data analysis (see, e.g., Preisler, 1993; Wu and Huffer, 1997; and Sherman *et al.* 2006). Let $\boldsymbol{x} = \{x_i : i \in D\}$ denote the observed binary data, where x_i is called a spin and D is the set of indices of the spins. Let $|D|$ denote the total number of spins in D, and let $n(i)$ denote the set of neighbors of spin i. The likelihood function of the model is

$$f(\boldsymbol{x}|\alpha, \beta) = \frac{1}{Z(\alpha, \beta)} \exp\left\{\alpha \sum_{i \in D} x_i + \frac{\beta}{2} \sum_{i \in D} x_i \left(\sum_{j \in n(i)} x_j \right) \right\}, \quad (\alpha, \beta) \in \Theta, \tag{4.24}$$

where the parameter α determines the overall proportion of $x_i = +1$, the parameter β determines the intensity of interaction between x_i and its neighbors, and $Z(\alpha, \beta)$ is the intractable normalizing constant defined by

$$Z(\alpha, \beta) = \sum_{\text{for all possible } \boldsymbol{x}} \exp\left\{\alpha \sum_{j \in D} x_j + \frac{\beta}{2} \sum_{i \in D} x_i \left(\sum_{j \in n(i)} x_j \right) \right\}.$$

An exact evaluation of $Z(\alpha, \beta)$ is prohibited even for a moderate system.

To conduct a Bayesian analysis for the model, assume a uniform prior

$$(\alpha, \beta) \in \Theta = [-1, 1] \times [0, 1]$$

for the parameters in this section. Then the double MH sampler can be applied to simulate from the posterior distribution $\pi(\alpha, \beta|\boldsymbol{x})$. In step (a), (α_t, β_t), the current state of the Markov chain, is updated by a single MH step with a random walk proposal $N_2((\alpha_t, \beta_t)', s^2 I_2)$, where s is the step size, and I_2 is the 2×2 identity matrix. In step (b), the auxiliary variable $\boldsymbol{y}$ is generated by a single cycle of Gibbs updates. Two or more cycles have also been tried for the examples, the results are similar. The acceptance rate of the double MH moves can be controlled by the choice of s. In this subsection, we set $s = 0.03$ for all examples.

4.8.1.1 US Cancer Mortality Data

United States cancer mortality maps have been compiled by Riggan *et al.* (1987) for investigating possible association of cancer with unusual demographic, environmental, industrial characteristics, or employment patterns. Figure 4.2 shows the mortality map of liver and gallbladder (including bile ducts) cancers for white males during the decade 1950–1959, which indicates some apparent geographic clustering. See Sherman *et al.* (2006) for further description of the data. As in Sherman *et al.* (2006), Liang (2009c) modeled the data by a spatial autologistic model. The total number of spins is $|D| = 2293$. A free boundary condition is assumed for the model, under which the boundary points have less neighboring points than the interior points. This assumption is natural to this dataset, as the lattice has an irregular shape.

Liang (2009c) compares the double MH sampler and the exchange algorithm for this example. The double MH sampler started with the initial value $(\alpha_0, \beta_0) = (0, 0)$ and was run 5 times independently. Each run consisted of 10 500 iterations. The CPU time cost by each run was 4.2 s on a 2.8 GHz computer. The overall acceptance rate the double MH moves was about 0.23. Averaging over the 5 runs produced the following estimate: $(\widehat{\alpha, \beta}) = (-0.3028, 0.1228)$ with the standard error $(8.2 \times 10^{-4}, 2.7 \times 10^{-4})$.

The exchange algorithm was run for this example in a similar way to the double MH sampler, except that the auxiliary variable $\boldsymbol{y}$ was generated using an exact sampler, the summary state algorithm by Childs *et al.* (2001), which is known to be suitable for high dimensional binary spaces. The exchange algorithm was also run 5 times, and each run consisted of 10 500 iterations. The CPU time cost by each run was 111.5 s, about 27 times longer than that cost by the double MH sampler. The overall acceptance rate of the exact auxiliary variables was 0.2. Averaging over the 5 runs produced the estimate $(\widehat{\alpha, \beta}) = (-0.3030, 0.1219)$ with the standard error $(1.1 \times 10^{-3}, 6.0 \times 10^{-4})$.

It can be seen that the double MH sampler and the exchange algorithm produced almost identical estimates for this example. These estimates are

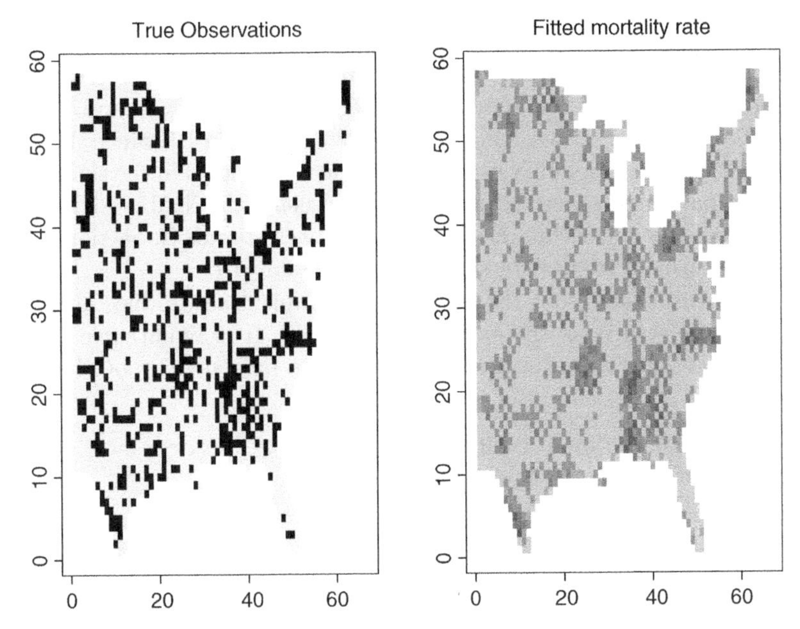

Figure 4.2 US cancer mortality data. Left: The mortality map of liver and gallbladder cancers (including bile ducts) for white males during the decade 1950–1959. Black squares denote counties of high cancer mortality rate, and white squares denote counties of low cancer mortality rate. Right: Fitted cancer mortality rates by the autologistic model with the parameters being replaced by its approximate Bayesian estimates. The cancer mortality rate of each county is represented by the gray level of the corresponding square (Liang, 2009c).

also very close to the estimate $(-0.3008, 0.1231)$ obtained by Liang (2007a) using contour Monte Carlo, and the estimate $(-0.2999, 0.1234)$ obtained by Liang *et al.* (2007) using stochastic approximation Monte Carlo. Note that both contour Monte Carlo and stochastic approximation Monte Carlo try first to approximate the unknown normalizing constant function, and then estimate the parameters based on the approximated normalizing constant function. As reported by the authors, both the algorithms took hours of CPU time to approximate the normalizing constant function. This data is analyzed by Sherman *et al.* (2006) using the MCMC maximum likelihood algorithm (Geyer and Thompson, 1992), producing a similar estimate of $(-0.304, 0.117)$.

For this example, the double MH sampler produced almost identical results with the exchange algorithm, but using much less CPU time. This advantage can be seen clearer from the next example, where for some cases the exact sampler does not work while the double MH sampler still works well.

4.8.1.2 Simulation Studies

To assess general accuracy of the estimates produced by the double MH sampler, Liang (2009c) simulate 50 independent samples for the US cancer mortality data under each setting of (α, β) given in Table 4.1. Since the lattice is irregular, the free boundary condition was again assumed in the simulations. The given parameters were then re-estimated using the double MH sampler and the exchange algorithm. Both algorithms were run as in Section 4.8.1.1. The computational results are summarized in Table 4.1.

Table 4.1 indicates that the double MH sampler can produce almost the same accurate results as the exchange algorithm. It is remarkable that the CPU time cost by the double MH sampler is independent of the values

Table 4.1 Computational results for the simulated US cancer mortality data (recompiled from Liang, 2009c).

	double MH sampler			exchange algorithm		
(α, β)	$\widehat{\alpha}$	$\widehat{\beta}$	CPU[a]	$\widehat{\alpha}$	$\widehat{\beta}$	CPU[b]
(0,0.1)[c]	−.0038 (.0024)	.1010 (.0018)	4.2	−.0038 (.0024)	.1002 (.0018)	103
(0,0.2)[c]	−.0026 (.0021)	.2018 (.0019)	4.2	−.0025 (.0020)	.2007 (.0019)	251
(0,0.3)[c]	−.0018 (.0014)	.2994 (.0018)	4.2	−.0014 (.0014)	.2971 (.0018)	821
(0,0.4)[c]	.0013 (.0009)	.4023 (.0015)	4.2	−.0007 (.0004)	.3980 (.0012)	7938
(0.1,0.1)[c]	.1025 (.0025)	.0993 (.0022)	4.2	.1030 (.0025)	.0986 (.0022)	110
(0.3,0.3)[c]	.2944 (.0098)	.3032 (.0043)	4.2	.3012 (.0098)	.3008 (.0043)	321
(0.5,0.5)[d]	.5040 (.0227)	.5060 (.0085)	4.2	– –	– –	–

The numbers in the parentheses denote the standard error of the estimates. [a]The CPU time (in seconds) cost by a single run of the double MH sampler. [b]The CPU time (in seconds) cost by a single run of the exchange algorithm. [c]The data were simulated using the exact sampler. [d]The data were simulated using the Gibbs sampler, starting with a random configuration and then iterating for 100 000 Gibbs cycles (Liang, 2009c).

of (α, β). Whereas, for the exchange algorithm, when β increases, the CPU time increases exponentially. Childs *et al.* (2006) study the behavior of the exact sampler for the Ising model, a simplified autologistic model with α being constrained to 0. For the Ising model, they fitted an exponential law for the convergence time, and reported that the exact sampler may diverge at a value of β lower than the critical value ($\approx$0.44). Childs *et al.*'s finding is consistent with the results reported in Table 4.1. It takes an extremely long CPU time for the exact sampler to generate a sample under the settings $(0, 0.4)$ and $(0.5, 0.5)$. Note that due to the effect of α, it usually takes a longer CPU time for the exact sampler to generate a sampler under the setting $(0, \beta)$ than under the setting (α, β); and that when α and β are large, the accuracy of the estimates tend to be reduced by their correlation.

4.9 Monte Carlo MH Sampler

In this section, we present the Monte Carlo Metropolis-Hastings (MCMH) algorithm proposed by Liang and Jin (2010) for sampling from distributions with intractable normalizing constants. The MCMH algorithm is a Monte Carlo version of the Metropolis-Hastings algorithm. At each iteration, it replaces the unknown normalizing constant ratio $Z(\theta)/Z(\theta')$ by a Monte Carlo estimate. Under mild conditions, it is shown that the MCMH algorithm still admits $f(\theta|x)$ as its stationary distribution. Like the double MH sampler, the MCMH algorithm also avoids the requirement for perfect sampling, and thus can be applied to many statistical models for which perfect sampling is not available or very expensive.

4.9.1 Monte Carlo MH Algorithm

Consider the problem of sampling from the distribution (4.12). Let θ_t denote the current draw of θ by the algorithm. Let $y_1^{(t)}, \ldots, y_m^{(t)}$ denote the auxiliary samples simulated from the distribution $f(y|\theta_t)$, which can be drawn by either a MCMC algorithm or an automated rejection sampling algorithm (Booth and Hobert, 1999). The MCMH algorithm works by iterating between the following steps:

Monte Carlo MH Algorithm I

1. Draw ϑ from some proposal distribution $Q(\theta_t, \vartheta)$.
2. Estimate the normalizing constant ratio $R(\theta_t, \vartheta) = \kappa(\vartheta)/\kappa(\theta_t)$ by

$$\widehat{R}_m(\theta_t, \boldsymbol{y}_t, \vartheta) = \frac{1}{m} \sum_{i=1}^{m} \frac{g(y_i^{(t)}, \vartheta)}{g(y_i^{(t)}, \theta_t)},$$

where $g(y,\theta) = \exp\{-U(y,\theta)\}$, and $\boldsymbol{y}_t = (y_1^{(t)},\ldots,y_m^{(t)})$ denotes the collection of the auxiliary samples.

3. Calculate the Monte Carlo MH ratio

$$\tilde{r}_m(\theta_t,\boldsymbol{y}_t,\vartheta) = \frac{1}{\widehat{R}_m(\theta_t,\boldsymbol{y}_t,\vartheta)} \frac{g(x,\vartheta)f(\vartheta)}{g(x,\theta_t)f(\theta_t)} \frac{Q(\vartheta,\theta_t)}{Q(\theta_t,\vartheta)},$$

where $f(\theta)$ denotes the prior distribution imposed on θ.

4. Set $\theta_{t+1} = \vartheta$ with probability $\tilde{\alpha}(\theta_t,\boldsymbol{y}_t,\vartheta) = \min\{1,\tilde{r}_m(\theta_t,\boldsymbol{y}_t,\vartheta)\}$, and set $\theta_{t+1} = \theta_t$ with the remaining probability.

5. If the proposal is rejected in step 4, set $\boldsymbol{y}_{t+1} = \boldsymbol{y}_t$. Otherwise, draw samples $\boldsymbol{y}_{t+1} = (y_1^{(t+1)},\ldots,y_m^{(t+1)})$ from $f(y|\theta_{t+1})$ using either a MCMC algorithm or an automated rejection sampling algorithm.

Since the algorithm involves a Monte Carlo step to estimate the unknown normalizing constant ratio, it is termed as 'Monte Carlo MH'. Clearly, the samples $\{(\theta_t,\boldsymbol{y}_t)\}$ forms a Markov chain, for which the transition kernel can be written as

$$\begin{aligned}\tilde{P}_m(\theta,\boldsymbol{y};d\vartheta,d\boldsymbol{z}) &= \tilde{\alpha}(\theta,\boldsymbol{y},\vartheta)Q(\theta,d\vartheta)f_\vartheta^m(d\boldsymbol{z}) \\ &\quad + \delta_{\theta,\boldsymbol{y}}(d\vartheta,d\boldsymbol{z})\left[1-\int_{\Theta\times\mathbb{Y}}\tilde{\alpha}(\theta,\boldsymbol{y},\vartheta)Q(\theta,d\vartheta)f_\theta^m(d\boldsymbol{y})\right],\end{aligned} \tag{4.25}$$

where $f_\theta^m(\boldsymbol{y}) = f(y_1,\ldots,y_m|\theta)$ denotes the joint density of $y_1,\ldots,y_m$. However, since here we are mainly interested in the marginal law of θ, in what follows we will consider only the marginal transition kernel

$$\begin{aligned}\tilde{P}_m(\theta,d\vartheta) &= \int_{\mathbb{Y}}\tilde{\alpha}(\theta,\boldsymbol{y},\vartheta)Q(\theta,d\vartheta)f_\theta^m(d\boldsymbol{y}) \\ &\quad + \delta_\theta(d\vartheta)\left[1-\int_{\Theta\times\mathbb{Y}}\tilde{\alpha}_m(\theta,\boldsymbol{y},\vartheta)Q(\theta,d\vartheta)f_\theta^m(d\boldsymbol{y})\right],\end{aligned} \tag{4.26}$$

showing that $\tilde{P}_m(\theta,d\vartheta)$ will converge to the posterior distribution $f(\theta|x)$ when m is large and the number of iterations goes to infinity.

The MCMH algorithm requires the auxiliary samples to be drawn at equilibrium, if a MCMC algorithm is used for generating the auxiliary samples. To ensure this requirement to be satisfied, Liang and Jin (2010) propose to choose the initial auxiliary sample at each iteration through an importance resampling procedure; that is, to set $y_0^{(t+1)} = y_i^{(t)}$ with a probability proportional to the importance weight

$$w_i = g(y_i^{(t)},\theta_{t+1})/g(y_i^{(t)},\theta_t). \tag{4.27}$$

As long as $y_0^{(t+1)}$ follows correctly from $f(y|\theta_{t+1})$, this procedure ensures that all samples, $\boldsymbol{y}_{t+1},\boldsymbol{y}_{t+2},\boldsymbol{y}_{t+3},\ldots$, will be drawn at equilibrium in the followed iterations, provided that θ does not change drastically at each iteration.

Regarding the choice of m, we note that m may not necessarily be very large in practice. In our experience, a value between 20 and 50 may be good for most problems. It seems that the random errors introduced by the Monte Carlo estimate of $\kappa(\theta_t)/\kappa(\vartheta)$ can be smoothed out by path averaging over iterations. This is particularly true for parameter estimation.

The MCMH algorithm can have many variants. A simple one is to draw auxiliary samples at each iteration, regardless of acceptance or rejection of the last proposal. This variant be described as follows:

Monte Carlo MH Algorithm II

1. Draw ϑ from some proposal distribution $Q(\theta_t, \vartheta)$.
2. Draw auxiliary samples $\boldsymbol{y}_t = (y_1^{(t)}, \ldots, y_m^{(t)})$ from $f(y|\theta_t)$ using a MCMC algorithm or an automated rejection algorithm.
3. Estimate the normalizing constant ratio $R(\theta_t, \vartheta) = \kappa(\vartheta)/\kappa(\theta_t)$ by
$$\widehat{R}_m(\theta_t, \boldsymbol{y}_t, \vartheta) = \frac{1}{m} \sum_{i=1}^{m} \frac{g(y_i^{(t)}, \vartheta)}{g(y_i^{(t)}, \theta_t)}.$$
4. Calculate the Monte Carlo MH ratio
$$\tilde{r}_m(\theta_t, \boldsymbol{y}_t, \vartheta) = \frac{1}{\widehat{R}_m(\theta_t, \boldsymbol{y}_t, \vartheta)} \frac{g(x, \vartheta) f(\vartheta)}{g(x, \theta_t) f(\theta_t)} \frac{Q(\vartheta, \theta_t)}{Q(\theta_t, \vartheta)}.$$
5. Set $\theta_{t+1} = \vartheta$ with probability $\tilde{\alpha}(\theta_t, \boldsymbol{y}_t, \vartheta) = \min\{1, \tilde{r}_m(\theta_t, \boldsymbol{y}_t, \vartheta)\}$ and set $\theta_{t+1} = \theta_t$ with the remaining probability.

MCMH II has a different Markovian structure from MCMH I. In MCMH II, $\{\theta_t\}$ forms a Markov chain with the transition kernel as given by (4.26). Note that MCMH II may be less efficient than MCMH I, as the latter recycles the auxiliary samples when rejection occurs. Due to the recycle of the auxiliary samples, one may expect that the successive samples generated by MCMH I have significantly higher correlation than those generated by MCMH II. However, our numerical results for one example show that this may not be true (see Table 4.2 and Figure 4.3 for the details). This is understandable, as the random error of $\widehat{R}_m(\theta_t, \boldsymbol{y}_t, \vartheta)$ depends mainly on θ_t and ϑ instead of $\boldsymbol{y}_t$, especially when m is large.

Similar to MCMH II, we can propose another variant of MCMH, which in Step 2 draws auxiliary samples from $f(y|\vartheta)$ instead of $f(y|\theta_t)$. Then
$$\widehat{R}_m^*(\theta_t, \boldsymbol{y}_t, \vartheta) = \frac{1}{m} \sum_{i=1}^{m} \frac{g(y_i^{(t)}, \theta_t)}{g(y_i^{(t)}, \vartheta)},$$
forms an unbiased estimator of the ratio $\kappa(\theta_t)/\kappa(\vartheta)$, and the Monte Carlo MH ratio can be calculated as
$$\tilde{r}_m^*(\theta_t, \boldsymbol{y}_t, \vartheta) = \widehat{R}_m^*(\theta_t, \boldsymbol{y}_t, \vartheta) \frac{g(x, \vartheta)\pi(\vartheta)}{g(x, \theta_t)\pi(\theta_t)} \frac{Q(\vartheta, \theta_t)}{Q(\theta_t, \vartheta)}.$$

Table 4.2 Computational results for the US cancer mortality data (Liang and Jin, 2010).

algorithm	setting	$\widehat{\alpha}$	$\widehat{\beta}$	CPU
MCMH I	$m = 20$	-0.3018 (1.2×10^{-3})	0.1232 (6.0×10^{-4})	11
	$m = 50$	-0.3018 (1.1×10^{-3})	0.1230 (5.3×10^{-4})	24
	$m = 100$	-0.3028 (6.7×10^{-4})	0.1225 (3.8×10^{-4})	46
MCMH II	$m = 20$	-0.3028 (1.2×10^{-3})	0.1226 (5.9×10^{-4})	26
	$m = 50$	-0.3019 (1.0×10^{-3})	0.1228 (5.3×10^{-4})	63
	$m = 100$	-0.3016 (8.2×10^{-4})	0.1231 (3.8×10^{-4})	129
Exchange	–	-0.3015 (4.3×10^{-4})	0.1229 (2.3×10^{-4})	33

The CPU time (in seconds) is recorded for a single run on a 3.0 GHz personal computer. The numbers in the parentheses denote the standard error of the estimates.

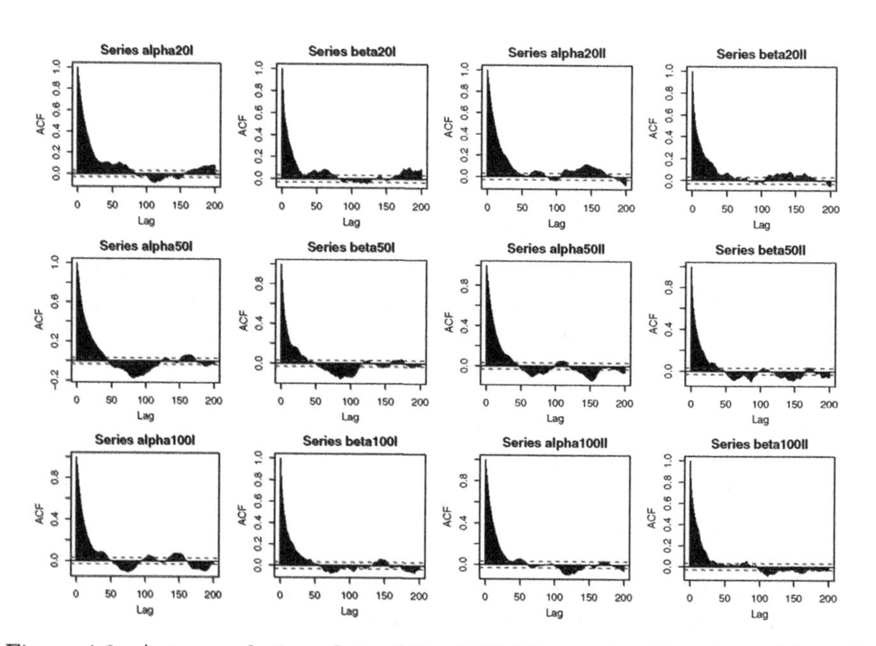

Figure 4.3 Autocorrelation plots of the MCMH samples. The title 'alpha20I' stands for the α samples generated by MCMH I with $m = 20$, 'alpha20II' stands for the α samples generated by MCMH II with $m = 20$, and other titles can be interpreted similarly (Liang and Jin, 2010).

It is interesting to point out that if $m = 1$, this variant is reduced to the double MH sampler described in Section 4.8.

In addition to $f(y|\theta_t)$ or $f(y|\vartheta)$, the auxiliary samples can also be generated from a third distribution which has the same support set as $f(y|\theta_t)$ and $f(y|\vartheta)$. In this case, the ratio importance sampling method (Torrie and Valleau, 1977; Chen and Shao, 1997) can be used for estimating the normalizing constant ratio $\kappa(\theta_t)/\kappa(\vartheta)$. The existing normalizing constant ratio estimation techniques, such as bridge sampling (Meng and Wong, 1996) and path sampling (Gelman and Meng, 1998), are also applicable to MCMH with an appropriate strategy for generating auxiliary samples.

4.9.2 Convergence

Here, we prove the convergence of MCMH I; that is, showing

$$\|\tilde{P}_m^k(\theta_0, \cdot) - f(\cdot|x)\| \to 0, \quad \text{as } k \to \infty,$$

where k denotes the number of iterations, $f(\cdot|x)$ denotes the posterior distribution of θ, and $\|\cdot\|$ denotes the total variation norm. Extension of these results to MCMH II and other variants is straightforward.

Define

$$\gamma_m(\theta, \boldsymbol{y}, \vartheta) = \frac{R(\theta, \vartheta)}{\widehat{R}(\theta, \boldsymbol{y}, \vartheta)}.$$

In the context where confusion is impossible, we denote $\gamma_m = \gamma_m(\theta, \boldsymbol{y}, \vartheta)$, and $\lambda_m = |\log(\gamma_m(\theta, \boldsymbol{y}, \vartheta))|$. Define

$$\rho(\theta) = 1 - \int_{\Theta \times \mathbb{Y}} \tilde{\alpha}_m(\theta, \boldsymbol{y}, \vartheta) Q(\theta, d\vartheta) f_\theta^m(d\boldsymbol{y}),$$

which represents the mean rejection probability a MCMH transition from θ.

To show the convergence of the MCMH algorithm, we also consider the transition kernel

$$P(\theta, \vartheta) = \alpha(\theta, \vartheta) Q(\theta, \vartheta) + \delta_\theta(d\vartheta) \left[1 - \int_\Theta \alpha(\theta, \vartheta) Q(\theta, \vartheta) d\vartheta\right],$$

which is induced by the proposal $Q(\cdot, \cdot)$. In addition, we assume the following conditions:

(A_1) Assume that P defines an irreducible and aperiodic Markov chain such that $\pi P = \pi$, and for any $\theta_0 \in \Theta$, $\lim_{k\to\infty} \|P^k(\theta_0, \cdot) - \pi(\cdot)\| = 0$.

(A_2) For any $(\theta, \vartheta) \in \Theta \times \Theta$,

$$\gamma_m(\theta, \boldsymbol{y}, \vartheta) > 0, \quad f_\theta^m(\cdot) - a.s.$$

(A_3) For any $\theta \in \Theta$ and any $\epsilon > 0$,

$$\lim_{m\to\infty} Q\left(\theta, f_\theta^m(\lambda_m(\theta, \boldsymbol{y}, \vartheta) > \epsilon)\right) = 0,$$

where $Q\left(\theta, f_\theta^m(\lambda_m(\theta, \boldsymbol{y}, \vartheta) > \epsilon)\right) = \int_{\{(\vartheta,\boldsymbol{y}):\lambda_m(\theta,\boldsymbol{y},\vartheta) > \epsilon\}} f_\theta^m(d\boldsymbol{y}) Q(\theta, d\vartheta)$.

The condition (A_1) can be simply satisfied by choosing an appropriate proposal distribution $Q(\cdot,\cdot)$, following from the standard theory of the Metropolis-Hastings algorithm (Tierney, 1994). The condition (A_2) assumes that the distributions $f(y|\theta)$ and $f(y|\vartheta)$ have a reasonable overlap such that $\widehat{R}$ forms a reasonable estimator of R. The condition (A_3) is equivalent to assuming that for any $\theta \in \Theta$ and any $\epsilon > 0$, there exists a positive integer M such that for any $m > M$,

$$Q\left(\theta, f_\theta^m(\lambda_m(\theta, \boldsymbol{y}, \vartheta) > \epsilon)\right) \le \epsilon.$$

Lemma 4.9.1 states that the marginal kernel $\tilde{P}_m$ has a stationary distribution. It is proved in a similar way to Theorem 1 of Andrieu and Roberts (2009). The relation between this work and Beaumont (2003) and Andrieu and Roberts (2009) will be discussed in Section 4.9.4.

Lemma 4.9.1 *Assume (A_1) and (A_2) hold. Then for any $N \in \mathcal{N}$ such that for any $\theta \in \Theta$, $\rho(\theta) > 0$, $\tilde{P}_m$ is also irreducible and aperiodic, and hence there exists a stationary distribution $\tilde{\pi}(\theta)$ such that*

$$\lim_{k\to\infty} \|\tilde{P}_m^k(\theta_0, \cdot) - \tilde{\pi}(\cdot)\| = 0.$$

Proof: Since P defines an irreducible and aperiodic Markov chain, to show $\tilde{P}_m$ has the same property, it suffices to show that the accessible sets of P are included in those of $\tilde{P}_m$. More precisely, we show by induction that for any $k \in \mathcal{N}$, $\theta \in \Theta$ and $A \in \mathcal{B}(\Theta)$ such that $P^k(\theta, A) > 0$, then $\tilde{P}_m^k(\theta, A) > 0$. First, for any $\theta \in \Theta$ and $A \in \mathcal{B}(\Theta)$,

$$\tilde{P}_m(\theta, A) \ge \int_A \left[\int_{\mathbb{Y}} (1 \wedge \gamma_m) f_\theta^m(d\boldsymbol{y})\right] \alpha(\theta, \vartheta) Q(\theta, d\vartheta) + \mathbb{I}(\theta \in A)\rho(\theta),$$

where $\mathbb{I}(\cdot)$ is the indicator function. By condition (A_2), we deduce that the implication is true for $k = 1$. Assume the induction assumption is true up to some $k = n \ge 1$. Now, for some $\theta \in \Theta$, let $A \in \mathcal{B}(\Theta)$ be such that $P^{n+1}(\theta, A) > 0$ and assume that

$$\int_\Theta \tilde{P}_m^n(\theta, d\vartheta) \tilde{P}_m(\vartheta, A) = 0,$$

which implies that $\tilde{P}_m(\vartheta, A) = 0$, $\tilde{P}_m^n(\theta, \cdot)$-a.s. and hence that $P(\vartheta, A) = 0$, $\tilde{P}_m^n(\theta, \cdot)$-a.s. from the induction assumption for $k = 1$. From this and the induction assumption for $k = n$, we deduce that $P(\vartheta, A) = 0$, $P^n(\theta, \cdot)$-a.s. (by contradiction), which contradicts the fact that $P^{n+1}(\theta, A) > 0$. ∎

Lemma 4.9.2 considers the distance between the kernel $\tilde{P}_m$ and the kernel P. It states that the two kernels can be arbitrarily close to each other, provided that m is large enough.

Lemma 4.9.2 *Assume* (A_3) *holds. Let* $\epsilon \in (0,1]$. *Then there exists a value* $M \in \mathcal{N}$ *such that for any* $\psi : \Theta \to [-1,1]$ *and any* $m > M$,

$$|\tilde{P}_m\psi(\theta) - P\psi(\theta)| \leq 4\epsilon.$$

Proof: Let

$$\begin{aligned} S &= P\psi(\theta) - \tilde{P}_m\psi(\theta) \\ &= \int_{\Theta\times\mathbb{Y}} \psi(\vartheta)\Big[1 \wedge r(\theta,\vartheta) - 1 \wedge \gamma_m r(\theta,\vartheta)\Big]Q(\theta,d\vartheta)f_\theta^m(d\boldsymbol{y}) \\ &\quad - \psi(\theta)\int_{\Theta\times\mathbb{Y}} \Big[1 \wedge r(\theta,\vartheta) - 1 \wedge \gamma_m r(\theta,\vartheta)\Big]Q(\theta,d\vartheta)f_\theta^m(d\boldsymbol{y}), \end{aligned}$$

and we therefore focus on the quantity

$$\begin{aligned} S_0 &= \int_{\Theta\times\mathbb{Y}} \Big|1 \wedge r(\theta,\vartheta) - 1 \wedge \gamma_m r(\theta,\vartheta)\Big|Q(\theta,d\vartheta)f_\theta^m(d\boldsymbol{y}) \\ &= \int_{\Theta\times\mathbb{Y}} \Big|1 \wedge r(\theta,\vartheta) - 1 \wedge \gamma_m r(\theta,\vartheta)\Big|\mathbb{I}(\lambda_m > \epsilon)Q(\theta,d\vartheta)f_\theta^m(d\boldsymbol{y}) \\ &\quad + \int_{\Theta\times\mathbb{Y}} \Big|1 \wedge r(\theta,\vartheta) - 1 \wedge \gamma_m r(\theta,\vartheta)\Big|\mathbb{I}(\lambda_m \leq \epsilon)Q(\theta,d\vartheta)f_\theta^m(d\boldsymbol{y}). \end{aligned}$$

Since, for any $(x,y) \in \mathcal{R}^2$,

$$|1 \wedge e^x - 1 \wedge e^y| = 1 \wedge |e^{0\wedge x} - e^{0\wedge y}| \leq 1 \wedge |x - y|,$$

we deduce that

$$S_0 \leq Q(\theta, f_\theta^m(\mathbb{I}(\lambda_m > \epsilon))) + Q(\theta, f_\theta^m(1 \wedge \lambda_m \mathbb{I}(\lambda_m \leq \epsilon))).$$

Consequently, we have

$$|S| \leq 2Q(\theta, f_\theta^m(\mathbb{I}(\lambda_m > \epsilon))) + 2Q(\theta, f_\theta^m(1 \wedge \lambda_m \mathbb{I}(\lambda_m \leq \epsilon))) \leq 2\epsilon + 2\epsilon = 4\epsilon.$$

This completes the proof of Lemma 4.9.2. ■

Theorem 4.9.1 concerns the convergence of the MCMH algorithm. It states that the kernel $\tilde{P}_m$ shares the same stationary distribution with the MH kernel P.

Theorem 4.9.1 *Assume* (A_1), (A_2) *and* (A_3) *hold. For any* $\varepsilon \in (0,1]$, *there exist* $M \in \mathcal{N}$ *and* $K \in \mathcal{N}$ *such that for any* $m > M$ *and* $k > K$

$$\|\tilde{P}_m^k(\theta_0,\cdot) - \pi(\cdot|x)\| \leq \varepsilon,$$

where $\pi(\cdot|x)$ *denotes the posterior density of* θ.

Proof: Dropping m for notational simplicity, we have that for any $k \geq 1$ and any $\psi : \Theta \to [-1, 1]$,

$$\tilde{P}^k \psi(\theta_0) - \pi(\psi) = S_1(k) + S_2(k),$$

where $\pi(\psi) = \pi(\psi(\theta))$ for notational simplicity, and

$$S_1(k) = P^k \psi(\theta_0) - \pi(\psi), \qquad S_2(k) = \tilde{P}^k \psi(\theta_0) - P^k \psi(\theta_0).$$

The magnitude of $S_1(k)$ can be controlled following from the convergence of the transition kernel P. For any $\epsilon > 0$, there exists $k_0 = k(\epsilon, \theta_0)$ such that for any $k > k_0$,

$$|S_1(k)| \leq \epsilon,$$

and, by Lemma 4.9.1 and condition (A_1),

$$\begin{aligned} |S_2(k)| &\leq |S_2(k_0)| + 2\epsilon \\ &= \left| \sum_{l=0}^{k_0 - 1} [P^l \tilde{P}^{k_0 - l} \psi(\theta_0) - P^{l+1} \tilde{P}^{k_0 - (l+1)} \psi(\theta_0)] \right| + 2\epsilon \\ &= \left| \sum_{l=0}^{k_0 - 1} P^l (\tilde{P} - P) \tilde{P}^{k_0 - (l+1)} \psi(\theta_0) \right| + 2\epsilon. \end{aligned}$$

For any $l > 1$, we have for any $\bar{\psi} : \Theta \to [-1, 1]$,

$$P^l \bar{\psi}(\theta_0) = \rho(\theta_0)^l \bar{\psi}(\theta_0) + \sum_{j=1}^{l} P^{j-1} \{Q(\theta_{j-1}, \alpha(\theta_{j-1}, \theta_j) \rho(\theta_j)^{l-j} \bar{\psi})\}(\theta_0).$$

We apply Lemma 4.9.2 k_0 times to show that there exist $M(\epsilon, \theta_0)$ such that for any $m > M(\epsilon, \theta_0)$

$$|S_2| \leq (4k_0 + 2)\epsilon.$$

Summarizing the results of S_1 and S_2, we conclude the proof by choosing $\epsilon = \varepsilon/(4k_0 + 3)$. ■

Theorem 4.9.1 implies, by standard MCMC theory (Theorem 1.5.3), that for an integrable function $h(\theta)$, the path averaging estimator $\sum_{k=1}^n h(\theta_k)/n$ will converge to its posterior mean almost surely; that is,

$$\frac{1}{n} \sum_{k=1}^{n} h(\theta_k) \to \int h(\theta) f(\theta | x) d\theta, \quad a.s.,$$

provided that $\int |h(\theta)| f(\theta|x) d\theta < \infty$.

4.9.3 Spatial Autologistic Models (Revisited)

Consider again the spatial autologistic model and the US Cancer Mortality data studied in Section 4.8.1. MCMH I and MCMH II were applied to this data with different choices of $m = 20$, 50, and 100. For each value of m, each algorithm was run 50 times independently. Each run started with a random point drawn uniformly on the region $[-1,1] \times [0,1]$, and consisted of 5000 iterations with the first 1000 iterations being discarded for the burn-in process and the remaining iterations being used for inference. The overall acceptance rates of the MCMH I moves are 0.408, 0.372 and 0.356 for the runs with $m = 20$, $m = 50$, and $m = 100$, respectively. For MCMH II, they are 0.423, 0.380 and 0.360, respectively. This is reasonable, extra randomness introduced by MCMH II in drawing auxiliary samples helps the system escape from the current point. As m increases, this improvement decays. The numerical results were summarized in Table 4.2. MCMH I and MCMH II produced almost identical results for this example, while MCMH I only cost less than 50% CPU times than MCMH II. Figure 4.3 shows the autocorrelation of the samples generated by MCMH I and MCMH II. From the plots, it can be seen that the autocorrelation of the samples generated by MCMH I is not significantly higher than that generated by MCMH II. This is consistent with the numerical results presented in Table 4.2.

From Table 4.2, it is easy to see a pattern that when m increases, MCMH tends to produce more accurate estimates, of course, at the price of longer CPU times. It is worth noting that the MCMH estimator seems unbiased even with a value of m as small as 20.

To assess the validity of the MCMH algorithms, the exchange algorithm (Murray *et al.*, 2006) was applied to this data. It was also run 50 times independently, and each run consisted of 5000 iterations with the first 1000 iterations being discarded for the burn-in process. The overall acceptance rate was 0.2. The numerical results, summarized in Table 4.2, indicate that the MCMH algorithms are valid.

4.9.4 Marginal Inference

In the literature, there are two algorithms, namely, grouped independence MH (GIMH) and Monte Carlo within Metropolis (MCWM) (Beaumont, 2003), which are similar in spirit to the MCMH algorithm. Both GIMH and MCWM are designed for marginal inference for joint distributions.

Let $\pi(\theta, y)$ denote a joint distribution. Suppose that one is mainly interested in the marginal distribution $\pi(\theta)$. For example, in Bayesian statistics, θ could represent a parameter of interest and y a set of missing data or latent variables. As implied by the Rao-Blackwellization theorem (Bickel and Doksum, 2000), a basic principle in Monte Carlo computation is to carry out analytical computation as much as possible. Motivated by this principle, Beaumont (2003) proposes to replace $\pi(\theta)$ by its Monte Carlo estimate in simulations when an analytical form $\pi(\theta)$ is not available. Let $\boldsymbol{y} = (y_1, \ldots, y_m)$ denote a set of independently identically distributed (iid) samples drawn from a trial distribution $q_\theta(y)$. By the standard theory of importance sampling, we know

$$\widetilde{\pi}(\theta) = \frac{1}{m}\sum_{i=1}^{m}\frac{\pi(\theta, y_i)}{q_\theta(y_i)}, \tag{4.28}$$

forms an unbiased estimate of $\pi(\theta)$. In simulations, GIMH treats $\widetilde{\pi}(\theta)$ as a known target density, and the algorithm can be summarized as follows. Let θ_t denote the current draw of θ, and let $\boldsymbol{y}_t = (y_1^{(t)}, \ldots, y_m^{(t)})$ denote a set of iid auxiliary samples drawn from $q_\theta(y)$. One iteration of GIMH consists of the following steps:

Group Independence MH Algorithm

1. Generate a new candidate point θ' from a proposal distribution $T(\theta'|\theta_t)$.
2. Draw m iid samples $\boldsymbol{y}' = (y_1', \ldots, y_m')$ from the trial distribution $q_{\theta'}(y)$.
3. Accept the proposal with probability

$$\min\left\{1, \frac{\widetilde{\pi}(\theta')}{\widetilde{\pi}(\theta_t)}\frac{T(\theta_t|\theta')}{T(\theta'|\theta_t)}\right\}.$$

   If it is accepted, set $\theta_{t+1} = \theta'$ and $\boldsymbol{y}_{t+1} = \boldsymbol{y}'$. Otherwise, set $\theta_{t+1} = \theta_t$ and $\boldsymbol{y}_{t+1} = \boldsymbol{y}_t$.

Like GIMH, MCWM also treats $\widetilde{\pi}(\theta)$ as a known target density, but it refreshes the auxiliary samples $\boldsymbol{y}_t$ at each iteration. One iteration of the MCMW algorithm consists of the following steps:

Monte Carlo within Metropolis Algorithm

1. Generate a new candidate point θ' from a proposal distribution $T(\theta'|\theta_t)$.
2. Draw m iid samples $\boldsymbol{y} = (y_1, \ldots, y_m)$ from the trial distribution $q_{\theta_t}(y)$ and draw m iid samples $\boldsymbol{y}' = (y_1', \ldots, y_m')$ from the trial distribution $q_{\theta'}(y)$.
3. Accept the proposal with probability

$$\min\left\{1, \frac{\widetilde{\pi}(\theta')}{\widetilde{\pi}(\theta_t)}\frac{T(\theta_t|\theta')}{T(\theta'|\theta_t)}\right\}.$$

   If it is accepted, set $\theta_{t+1} = \theta'$. Otherwise, set $\theta_{t+1} = \theta_t$.

Although GIMH and MCMW look very similar, they have different Markovian structures. In GIMH, $\{\theta_t, \boldsymbol{y}_t\}$ forms a Markov chain; whereas, in MCMW, $\{\theta_t\}$ forms a Markov chain. The convergence of these two algorithms has been studied by Andrieu and Roberts (2009) under similar conditions to those assumed for MCMH in this section. In the context of marginal inference, MCMH can be described as follows:

Monte Carlo MH Algorithm (For Marginal Inference)

1. Generate a new candidate point θ' from a proposal distribution $T(\theta'|\theta_t)$.
2. Accept the proposal with probability
$$\min\left\{1, \widetilde{R}(\theta_t, \theta')\frac{T(\theta_t|\theta')}{T(\theta'|\theta_t)}\right\},$$
where $\widetilde{R}(\theta_t, \theta') = \frac{1}{m}\sum_{i=1}^{m} \pi(\theta', y_i^{(t)})/\pi(\theta_t, y_i^{(t)})$ forms an unbiased estimate of the marginal density ratio $R(\theta_t, \theta') = \int \pi(\theta', y)dy / \int \pi(\theta_t, y)dy$. If it is accepted, set $\theta_{t+1} = \theta'$; otherwise, set $\theta_{t+1} = \theta_t$.
3. Set $\boldsymbol{y}_{t+1} = \boldsymbol{y}_t$ if a rejection occurs in the previous step. Otherwise, generate auxiliary samples $\boldsymbol{y}_{t+1} = (y_1^{(t+1)}, \ldots, y_m^{(t+1)})$ from the conditional distribution $\pi(y|\theta_{t+1})$. The auxiliary samples $y_1^{(t+1)}, \ldots, y_m^{(t+1)}$ can be generated via a MCMC simulation.

Taking a closer look at MCMH, we can find that MCMH is designed with a very different rule in comparison with GIMH and MCMW. In GIMH and MCMW, one estimates the marginal distributions; whereas, in MCMH, one directly estimates the ratio of marginal distributions. In addition, MCMH recycles the auxiliary samples when a proposal is rejected, and thus is potentially more efficient than GIMH and MCMW, especially for the problems for which generation of auxiliary samples is expensive.

MCMH can potentially be applied to many statistical models for which marginal inference is our main interest, such as generalized linear mixed models (see, e.g., McCulloch *et al.*, 2008) and hidden Markov random field models (Rue and Held, 2005). MCMH can also be applied to Bayesian analysis for the missing data problems that are traditionally treated with the EM algorithm (Dempster *et al.*, 1977) or the Monte Carlo EM algorithm (Wei and Tanner, 1990). Since the EM and Monte Carlo EM algorithms are local optimization algorithms, they tend to converge to suboptimal solutions. MCMH may perform better in this respect. Note that one may run MCMH under the framework of parallel tempering (Geyer, 1991) to help it escape from suboptimal solutions.

4.10 Applications

In this section, we consider two applications of the double MH sampler, Bayesian analysis for autonormal models and social networks.

4.10.1 Autonormal Models

Consider a second-order zero-mean Gaussian Markov random field $\boldsymbol{X} = (X_{ij})$ defined on an $M \times N$ lattice, whose conditional density function is given by

$$f(x_{ij}|\boldsymbol{\beta}, \sigma^2, x_{uv}; (u,v) \neq (i,j)) = \frac{1}{\sqrt{2\pi}\sigma} \exp\left\{-\frac{1}{2\sigma^2}\left(x_{ij} - \beta_h \sum_{(u,v)\in n_h(i,j)} x_{uv} - \beta_v \sum_{(u,v)\in n_v(i,j)} x_{uv} - \beta_d \sum_{(u,v)\in n_d(i,j)} x_{uv}\right)^2\right\}, \tag{4.29}$$

where $\boldsymbol{\beta} = (\beta_h, \beta_v, \beta_d)$ and σ^2 are parameters, $n_h(i,j) = \{(i,j-1),(i,j+1)\}$, $n_v(i,j) = \{(i-1,j),(i+1,j)\}$ and $n_d(i,j) = \{(i-1,j-1),(i-1,j+1),$ $(i+1,j-1),(i+1,j+1)\}$ are neighbors of (i,j). This model is stationary when $|\beta_h|+|\beta_v|+2|\beta_d| < 0.5$ (Balram and Moura, 1993). The joint likelihood function of this model can be written as

$$f(\boldsymbol{x}|\boldsymbol{\beta}, \sigma^2) = (2\pi\sigma^2)^{-MN/2}|B|^{1/2} \exp\left\{-\frac{1}{2\sigma^2}\boldsymbol{x}'B\boldsymbol{x}\right\},$$

where B is an $(MN \times MN)$-dimensional matrix, and $|B|$ is intractable except for some special cases (Besag and Moran, 1975).

To conduct a Bayesian analysis for the model, the following prior was assumed for the parameters:

$$f(\boldsymbol{\beta}) \propto I(|\beta_h| + |\beta_v| + 2|\beta_d| < 0.5), \qquad \pi(\sigma^2) \propto \frac{1}{\sigma^2}, \tag{4.30}$$

where $I(\cdot)$ is the indicator function. Under the free boundary condition, the posterior distribution can be expressed as

$$f(\boldsymbol{\beta}, \sigma^2|\boldsymbol{x}) \propto (\sigma^2)^{-\frac{MN}{2}-1}|B|^{1/2} \exp\left\{-\frac{MN}{2\sigma^2}\left(S_x - 2\beta_h X_h - 2\beta_v X_v - 2\beta_d X_d\right)\right\} \times I(|\beta_h| + |\beta_v| + 2|\beta_d| < 0.5), \tag{4.31}$$

where $S_x = \sum_{i=1}^{M}\sum_{j=1}^{N} x_{ij}^2/MN$, $X_h = \sum_{i=1}^{M}\sum_{j=1}^{N-1} x_{ij}x_{i,j+1}/MN$, $X_v = \sum_{i=1}^{M-1}\sum_{j=1}^{N} x_{ij}x_{i+1,j}/MN$, and $X_d = \left(\sum_{i=1}^{M-1}\sum_{j=1}^{N-1} x_{ij}x_{i+1,j+1} + \sum_{i=1}^{M-1}\sum_{j=2}^{N} x_{ij}x_{i+1,j-1}\right)/MN$. Although σ^2 can be integrated out from the posterior, it is not suggested here. Working on the joint posterior will ease the generation of auxiliary variables for the double MH sampler.

To ease implementation of sampling from the prior distribution, σ^2 is reparameterized by $\tau = \log(\sigma^2)$. Then

$$f(\boldsymbol{\beta}, \tau) \propto I(|\beta_h| + |\beta_v| + 2|\beta_d| < 0.5).$$

Liang (2009c) applies the double MH sampler to this model. In step (a), $(\boldsymbol{\beta}_t, \tau_t)$ is updated by a single MH step with a random walk proposal $N((\boldsymbol{\beta}_t, \tau_t)', s^2 I_4)$, where s is set to 0.02. In step (b), the auxiliary variable $\boldsymbol{y}$ is generated by a single cycle of Gibbs updates; that is,

$$y_{ij}|\boldsymbol{y}_{(u,v)\in n(i,j)} \sim N\left(\beta_h \sum_{(u,v)\in n_h(i,j)} y_{uv} + \beta_v \sum_{(u,v)\in n_v(i,j)} y_{uv} \right.$$
$$\left. + \beta_d \sum_{(u,v)\in n_d(i,j)} y_{uv}, e^{\tau_t}\right),$$

for $i = 1, \ldots, M$ and $j = 1, \ldots, N$, starting with $\boldsymbol{y} = \boldsymbol{x}$.

The exchange algorithm is not applicable to the autonormal model, as no exact sampler is available for it. However, under the free boundary condition, the log-likelihood function of the model admits the following analytic form (Balram and Moura, 1993):

$$\begin{aligned} l(\boldsymbol{X}|\boldsymbol{\beta}, \sigma^2) = \text{Constant} &- \frac{MN}{2}\log(\sigma^2) \\ &- \frac{MN}{2\sigma^2}\Big(S_x - 2\beta_h X_h - 2\beta_v X_v - 2\beta_d X_d\Big) \\ &+ \frac{1}{2}\sum_{i=1}^{M}\sum_{j=1}^{N}\log\left(1 - 2\beta_v \cos\frac{i\pi}{M+1} - 2\beta_h \cos\frac{j\pi}{N+1}\right. \\ &\left. - 4\beta_d \cos\frac{i\pi}{M+1}\cos\frac{j\pi}{N+1}\right), \end{aligned} \tag{4.32}$$

where S_x, X_h, X_v and X_d are as defined in (4.31). The Bayesian inference for the model is then standard, with the priors as specified in (4.30). The Bayesian analysis based on this analytic likelihood function is called the true Bayesian analysis, and the resulting estimator is called the true Bayesian estimator.

In Liang (2009c), the double MH sampler and the true Bayesian analysis were compared on the wheat yield data, which was collected on a 20×25 rectangular lattice (Table 6.1, Andrews and Herzberg, 1985). The data has been analyzed by a number of authors, including Besag (1974), Huang and Ogata (1999) and Gu and Zhu (2001), among others.

The double MH sampler was run 5 times independently. Each run started with the point (0,0,0,0) and consisted of 50 500 iterations. In each run, the first 500 iterations were discarded for burn-in, and 10 000 samples were then collected at equally spaced time points. The overall acceptance rate of the double MH moves was about 0.23. In the true Bayesian analysis, the posterior was simulated using the MH algorithm 5 times. Each run also consisted of 50 500 iterations, with the first 500 iterations being discarded for burn-in and then 10 000 samples being collected from the remaining iterations at equally-spaced time points. The proposal adopted here was a random walk proposal with the

Table 4.3 Computational results for the wheat yield data (Liang, 2009c).

algorithm	β_h	β_v	β_d	σ^2
True Bayes	0.102(4e-4[a])	0.355(3e-4)	0.006(2e-4)	0.123(2e-4)
DMH	0.099(6e-4)	0.351(5e-4)	0.006(3e-4)	0.126(3e-4)

[a]The numbers in the parentheses denote the standard error of the estimates (Liang, 2009c).

variance-covariance matrix $0.02^2 I_4$. The overall acceptance rate of the MH moves was about 0.22. The numerical results presented in Table 4.3. indicate that for this example the double MH sampler produced almost identical results with the true Bayesian analysis.

4.10.2 Social Networks

Social network analysis has emerged as a key technique in modern sociology, and it has also gained significant followings in biology, communication studies, economics, etc. The exponential family of random graphs is among the most widely-used, flexible models for social network analysis, which includes edge and dyadic independence models, Markov random graphs (Frank and Strauss, 1986), exponential random graphs (also known as p^* models) (Snijders *et al.*, 2006), and many other models. The model that is of particular interest to current researchers is the exponential random graph model (ERGM), which is a generalization of the Markov random graph model by incorporating some higher order specifications. Not only does the ERGM show improvements in goodness of fit for various datasets, but also it helps avoid the problem of near degeneracy that often afflicts the fitting of Markov random graphs. Refer to Robins *et al.* (2007) for an overview of ERGMs.

Consider a social network with n actors. Let X_{ij} denote a network tie variable; $X_{ij} = 1$ if there is a network tie from i to j and 0 otherwise. Then, one can specify $\boldsymbol{X}$ as a total set of X_{ij}'s in a matrix, the so-called adjacency matrix. Here, $\boldsymbol{X}$ can be directed or non-directed. Let x_{ij} denote the (i,j)-th entry of an observed matrix $\boldsymbol{x}$. The likelihood function of the ERGM is given by

$$f(x|\theta) = \frac{1}{\kappa(\theta)} \exp\left\{ \sum_{a \in A} \theta_a s_a(x) \right\}, \tag{4.33}$$

where A denotes the set of configuration types, different sets of configuration types representing different models; $s_a(x)$ is an explanatory variable/statistic, θ_a is the coefficient of $s_a(x)$; $\theta = \{\theta_a : a \in A\}$; and $\kappa(\theta)$ is the normalizing constant which makes (4.33) a proper probability distribution.

In the literature, two methods are usually used for estimation of θ, namely, the pseudo-likelihood method (Strauss and Ikeda, 1990) and Markov chain Monte Carlo maximum likelihood estimation (MCMCMLE) method (Snijders, 2002; Hunter and Handcock, 2006). The pseudo-likelihood method originates in Besag (1974), which works on a simplified form of the likelihood function under the dyadic independence assumption. Since this assumption is often violated by real networks, the maximum pseudo-likelihood estimator can perform very badly in practice (Handcock, 2003). The MCMCMLE method originates in Geyer and Thompson (1992), and works as follows. Let $\theta^{(0)}$ denote an arbitrary point in the parameter space of (4.33), and let $z_1, \ldots, z_m$ denote a sample of random networks simulated from $f(z|\theta^{(0)})$, which can be obtained via a MCMC simulation. Then

$$\log f_m(x|\theta) = \sum_{a \in A} \theta_a s_a(x) - \log\left(\frac{1}{m}\sum_{i=1}^{m} \exp\left\{\sum_{a \in A} \theta_a S_a(z_i) - \sum_{a \in A} \theta_a^{(0)} s_a(z_i)\right\}\right) - \log(\kappa(\theta^{(0)})),$$

approaches to $\log f(x|\theta)$ as $m \to \infty$. The estimator $\widehat{\theta} = \arg\max_\theta \log P_m(x|\theta)$ is called the MCMCMLE of θ. It is known that the performance of the method depends on the choice of $\theta^{(0)}$. If $\theta^{(0)}$ lies in the attractive region of the MLE, the method usually produces a good estimation of θ. Otherwise, the method may converge to a local optimal solution. To resolve the difficulty in choosing $\theta^{(0)}$, Geyer and Thompson (1992) recommended an iterative approach, which drew new samples at the current estimate of θ and then re-estimate:

(a) Initialize with a point $\theta^{(0)}$, usually taking to be the maximum pseudo-likelihood estimator.

(b) Simulate m samples from $f(z|\theta^{(t)})$ using MCMC, for example, the MH algorithm.

(c) Find $\theta^{(t+1)} = \arg\max_\theta \log f_m(x|\theta^{(t)})$.

(d) Stop if a specified number of iterations has reached, or the termination criterion $\max_\theta f_m(x|\theta^{(t)}) - f_m(x|\theta^{(t)}) < \epsilon$ has reached. Otherwise, go back to Step (b).

Even with this iterative procedure, as pointed out by Bartz *et al.* (2008), nonconvergence is still quite common in the ERGMs. The reason is that the starting point, the maximum pseudo-likelihood estimator, is often too far from the MLE.

4.10.2.1 Exponential Random Graph Models

To define explicitly the ERGM, the explanatory statistics $s_a(x)$ need to be specified. Since the number of possible specifications is large, only a few key statistics are considered here, including the edge, degree distribution and shared partnership distribution. The edge, denoted by $e(x)$, counts the number of edges in the network. The other two statistics are defined below.

Degree Distribution. Let $D_i(x)$ denote the number of nodes in the network x whose degree, the number of edges incident to the node, equals i. For example, $D_{n-1}(x) = n$ when x is the complete graph and $D_0(x) = n$ when x is the empty graph. Note that $D_0(x), \ldots, D_{n-1}(x)$ satisfy the constraint $\sum_{i=0}^{n-1} D_i(x) = n$, and the number of edges in x can be expressed as

$$e(x) = \frac{1}{2}\sum_{i=1}^{n-1} iD_i(x).$$

The degree distribution statistic (Snijders *et al.*, 2006; Hunter and Handcock, 2006; Hunter, 2007) is defined as

$$u(x|\tau) = e^{\tau}\sum_{i=1}^{n-2}\left\{1 - \left(1 - e^{-\tau}\right)^i\right\}D_i(x), \tag{4.34}$$

where the additional parameter τ specifies the decreasing rate of the weights put on the higher order terms. This statistic is also called the geometrically weighted degree (GWD) statistic. Following Hunter *et al.* (2008), τ is fixed to 0.25 throughout this subsection. Fixing τ to be a constant is sensible, as $u(x|\tau)$ plays a role of explanatory variable for the ERGMs.

Shared Partnership. Following Hunter and Handcock (2006) and Hunter (2007), we define one type of shared partner statistics, the edgewise shared partner statistic, which are denoted by $EP_0(x), \ldots, EP_{n-2}(x)$. The $EP_k(x)$ is the number of unordered pairs (i, j) such that $X_{ij} = 1$ and i and j have exactly k common neighbors. The geometrically weighted edgewise shared partnership (GWESP) statistic is defined as

$$v(x|\tau) = e^{\tau}\sum_{i=1}^{n-2}\left\{1 - \left(1 - e^{-\tau}\right)^i\right\}EP_i(x), \tag{4.35}$$

where the parameter τ specifies the decreasing rate of the weights put on the higher order terms. Again, following Hunter *et al.* (2008), τ is fixed to 0.25.

Based on the above summary statistics, we now consider three ERGMs whose likelihood functions are as follows:

$$\begin{aligned}
f(\boldsymbol{x}|\theta) &= \frac{1}{\kappa(\theta)}\exp\{\theta_1 e(x) + \theta_2 u(x|\tau)\} \quad \text{(Model 1)}\\
f(\boldsymbol{x}|\theta) &= \frac{1}{\kappa(\theta)}\exp\{\theta_1 e(x) + \theta_2 v(x|\tau)\} \quad \text{(Model 2)}\\
f(\boldsymbol{x}|\theta) &= \frac{1}{\kappa(\theta)}\exp\{\theta_1 e(x) + \theta_2 u(x|\tau) + \theta_3 v(x|\tau)\} \quad \text{(Model 3)}
\end{aligned}$$

To conduct a Bayesian analysis for the models, the prior $f(\theta) = N_d(0, 10^2 I_d)$ was imposed on θ, where d is the dimension of θ, and I_d is an identity matrix of size $d \times d$. Then, the DMH sampler can be applied to simulate from the posterior. In Step (a), θ_t can be updated by a Gaussian random walk proposal $N_d(\theta_t, s^2 I_d)$, where s is called the step size. In Step (b), the auxiliary variable y is generated through a single cycle of Metropolis-within-Gibbs updates.

4.10.2.2 AddHealth Network

Jin and Liang (2009) compared the performance of the double MH sampler and the MCMCMLE method on the AddHealth network, which was collected during the first wave (1994–1995) of National Longitudinal Study of Adolescent Health (AddHealth). The data were collected through a stratified sampling survey in US schools containing grades 7 through 12. School administrators made a roster of all students in each school and asked students to nominate five close male and female friends. Students were allowed to nominate friends who were not in their school, or not to nominate if they did not have five close male or female students. A detailed description of the dataset can be found in Resnick *et al.* (1997), Udry and Bearman (1998), or at `http://www.cpc.unc.edu/projects/addhealth`. The full dataset contains 86 schools and 90 118 students. In Jin and Liang (2009), only the subnetwork for school 10, which has 205 students, was analyzed. Also, only the undirected network for the case of mutual friendship was considered.

The DMH sampler was run 5 times for the network. Each run was started with (0,0) and iterated for 10 500 iterations, where the first 500 iterations were discarded for the burn-in process and the samples collected from the remaining iterations were used for estimation. During the simulations, the step size s was fixed to 0.2. The results are summarized in Table 4.4. For comparison, the MCMCMLEs were also included in the table, where the estimates

Table 4.4 Parameter estimation for the AddHealth data (Jin and Liang, 2009).

methods	statistics	model 1	model 2	model 3
	edges	−3.895(0.003)	−5.545(0.004)	−5.450(0.015)
DMH	GWD	−1.563(0.006)		−0.131(0.011)
	GWESP		1.847(0.004)	1.797(0.008)
	edges	−1.423(0.50)	−5.280(0.10)	−5.266(0.070)
MCMCMLE[a]	GWD	−1.305(0.20)		−0.252(0.173)
	GWESP		1.544(0.10)	1.635(0.022)

[a]The results of model 1 and model 2 are from Hunter *et al.* (2008), and the results of model 3 are from the software ERGM.

for the models 1 and 2 were from Hunter *et al.* (2008) and the estimates for model 3 were calculated using the package `statnet`, which is available at `http://cran.r-project.org/web/packages/statnet/index.html`. For models 2 and 3, the double MH sampler and the MCMCMLE method produced similar estimates, except that the estimates produced by the double MH sampler have smaller standard errors. However, for model 1, the estimates produced by the two methods are quite different. As shown later, the MCMCMLE method fails for model 1; the estimates produced by it may be very far from the true values.

To assess accuracy of the MCMCMLE and double MH estimates, Jin and Liang (2009) propose the following procedure, in a similar spirit to the parametric bootstrap method (Efron and Tibshirani, 1993). Since the statistics $\{S_a(x) : a \in A\}$ are sufficient for θ, if an estimate $\widehat{\theta}$ is accurate, then $S_a(x)$'s can be reversely estimated by simulated networks from the distribution $f(x|\widehat{\theta})$. The proposed procedure calculated the root mean square errors of the estimates of $S_a(x)$'s:

(a) Given the estimate $\hat{\theta}$, simulate m networks, $x_1, \ldots, x_m$, independently using the Gibbs sampler.

(b) Calculate the statistics $S_a(x)$, $a \in A$ for each of the simulated networks.

(c) Calculate RMSE by following equation.

$$RMSE(S_a) = \sqrt{\sum_{i=1}^{m} \left[S_a(x_i) - S_a(x)\right]^2 / m}, \quad a \in A, \tag{4.36}$$

where $S_a(x)$ is the corresponding statistic calculated from the network x.

For each of the estimates shown in Table 4.4, the RMSEs were calculated with $m = 1000$ and summarized in Table 4.5. The results indicate that the

Table 4.5 Root mean square errors of the MCMCMLE and DMH estimates for the ADDHealth School 10 data (Jin and Liang, 2009).

methods	statistics	model 1	model 2	model 3
Double MH	edges	22.187	19.204	20.449
	GWD	10.475		9.668
	GWESP		22.094	22.820
MCMCMLE	edges	4577.2	20.756	22.372
	GWD	90.011		10.045
	GWESP		40.333	30.308

DMH sampler produced much more accurate estimates than the MCMCMLE method for all the three models. For model 1, the MCMCMLE method even failed; the corresponding estimates have large standard errors and very large RMSEs, as shown in Table 4.4 and Table 4.5. In addition to parameter estimation, Jin and Liang (2009) also applied the double MH sampler to the variable selection for ERGMs. See Jin and Liang (2009) for the details.

Exercises

4.1 Implement the simulated annealing algorithm for minimizing the function

$$\begin{aligned} H(x,y) = & -(x\sin(20y) + y\sin(20x))^2 \cos h(\sin(10x)x) \\ & -(x\cos(10y) - y\sin(10x))^2 \cos h(\cos(20y)y), \end{aligned}$$

where $(x,y) \in [-1.1, 1.1]^2$. It is known that the global minimum of $H(x,y)$ is -8.12465, and attained at $(-1.0445, -1.0084)$ and $(1.0445, -1.0084)$.

4.2 Given a set of cities, the traveling salesman problem (TSP) is to find the shortest tour which goes through each city once and only once. Implement the simulated annealing algorithm for a traveling salesman problem for which 100 cities are uniformly distributed on a square region of length 100 miles. Find the average length of the shortest tour.

4.3 Implement the simulated tempering algorithm for the traveling salesman problem described in Exercise 4.2.

4.4 Consider a simplified witch's hat distribution (Geyer and Thompson, 1995), which is defined on a unit hypercube $[0,1]^d$ of dimension d and has the density function

$$f(x) \propto \begin{cases} (3^d - 1)/2, & x \in [0, \frac{1}{3}]^d, \\ 1, & \text{otherwise}. \end{cases}$$

Let $d = 30$, and simulate this density using the simulated tempering algorithm.

4.5 Implement the slice sampler for the mixture distribution

$$\frac{1}{3}N(-5,1) + \frac{2}{3}N(5,1).$$

4.6 Implement the Swendsen-Wang algorithm and the Wolff algorithm for the Ising model, and compare their convergence speeds.

4.7 Let $\boldsymbol{x}$ denote a configuration of an Ising model of size 100×100 and inverse temperature 0.4. Re-estimate the inverse temperature of the model using the exchange algorithm and the double MH algorithm.

4.8 The likelihood function of the very-soft-core model is defined as

$$f(\boldsymbol{x}|\theta) = \frac{1}{Z(\theta)} \exp\left\{ -\sum_{i=1}^{n} \sum_{j>i} \phi(\|x_i - x_j\|, \theta) \right\}, \quad \theta > 0,$$

where $x_1, \ldots, x_n$ denote n different points in a planar region $\mathcal{A}$, $\phi(t, \theta) = -\log\{1 - \exp(-nt^2/(\theta|\mathcal{A}|))\}$ is the pairwise potential function, and $|\mathcal{A}|$ denotes the area of the region $\mathcal{A}$. The normalizing constant function of this model is also intractable. Assuming the prior distribution $f(\theta) \propto 1/\theta$, implement the double MH sampler for this model.

4.9 Discuss the theoretical property of the double MH algorithm.

Chapter 5

Population-Based MCMC Methods

An actively pursued research direction for alleviating the local-trap problem suffered by the Metropolis-Hastings (MH) algorithm (Metropolis *et al.*, 1953; Hastings, 1970) is the population-based MCMC, where a population of Markov chains are run in parallel, each equipped with possibly different but related invariant distributions. Information exchange between different chains provides a means for the target chains to learn from past samples, and this in turn improves the convergence of the target chains.

Mathematically, the population-based MCMC may be described as follows. In order to simulate from a target distribution $f(x)$, one simulates an augmented system with the invariant distribution

$$f(x_1, \ldots, x_N) = \prod_{i=1}^{N} f_i(x_i), \tag{5.1}$$

where $(x_1, \ldots, x_N) \in \mathcal{X}^N, N$ is called the population size, $f(x) \equiv f_i(x)$ for at least one $i \in \{1, 2, \ldots, N\}$, and those different from $f(x)$ are called the trial distributions in terms of importance sampling. Different ways of specifying the trial distributions and updating the population of Markov chains lead to different algorithms, such as adaptive direction sampling (Gilks *et al.*, 1994), conjugate gradient Monte Carlo (Liu *et al.*, 2000), parallel tempering (Geyer, 1991; Hukushima and Nemoto, 1996)), evolutionary Monte Carlo (Liang and Wong, 2000, 2001a), sequential parallel tempering (Liang, 2003), and equi-energy sampler (Kou *et al.*, 2006), which will be described in the following sections.

Advanced Markov Chain Monte Carlo Methods: Learning from Past Samples
Faming Liang, Chuanhai Liu and Raymond J. Carroll

5.1 Adaptive Direction Sampling

Adaptive direction sampling (ADS) (Gilks *et al.*, 1994) is an early population-based MCMC method, in which each distribution $f_i(x)$ is identical to the target distribution, and at each iteration, one sample is randomly selected from the current population to undergo an update along a line passed through another sample randomly selected from the remaining set of the current population. An important form of the ADS is the *snooker algorithm*, described below.

At each iteration, the snooker algorithm keeps a population of samples. Let $\boldsymbol{x}^{(t)} = (x_1^{(t)}, \ldots, x_N^{(t)})$ denote the population obtained at iteration t, where $x_i^{(t)}$ is called an individual of the population. One iteration of the algorithm involves the following steps.

The Snooker Algorithm (Steps)

1. Select one individual, say $x_c^{(t)}$, at random from the current population $\boldsymbol{x}^{(t)}$. The $x_c^{(t)}$ is called the current point.

2. Select another individual, say $x_a^{(t)}$, from the remaining set of the current population, (i.e., $\{x_i^{(t)} : i \neq c\}$), and form a direction $e_t = x_c^{(t)} - x_a^{(t)}$. The individual $x_a^{(t)}$ is called the anchor point.

3. Set $y_c = x_a^{(t)} + r_t e_t$, where r_t is a scalar sampled from the density

$$f(r) \propto |r|^{d-1} f(x_a^{(t)} + r_t e_t), \tag{5.2}$$

   where d is the dimension of x, and the factor $|r|^{d-1}$ is derived from a transformation Jacobian (Roberts and Gilks, 1994).

4. Form the new population $\boldsymbol{x}^{(t+1)}$ by replacing $x_c^{(t)}$ by y_c and leaving all other individuals unchanged (i.e., set $x_i^{(t+1)} = x_i^{(t)}$ for $i \neq c$).

To show the sampler is proper, we need to show that at the equilibrium the new sample y_c is independent of the $x_i^{(t)}$ for $i \neq a$ and is distributed as $f(x)$. This fact follows directly from the following lemma, a generalized version of Lemma 3.1 of Roberts and Gilks (1994), proved by Liu *et al.* (2000).

Lemma 5.1.1 *Suppose $x \sim \pi(x)$ and y is any fixed point in a d-dimensional space. Let $e = x - y$. If r is drawn from distribution $f(r) \propto |r|^{d-1}\pi(y+re)$, then $x' = y + re$ follows the distribution $\pi(x)$. If y is generated from a distribution independent of x, then x' is independent of y.*

Proof: Without loss of generality, we assume that y is the origin, and then $e = x$. If r is drawn from $f(r) \propto |r|^{d-1}\pi(rx)$, then for any measurable function $h(x)$,

$$E\{h(x')\} = E[E\{h(x')|x\}] = \int\int h(rx)\frac{|r|^{d-1}\pi(rx)}{\int |r'|^{d-1}\pi(r'x)dr'}\pi(x)drdx.$$

Let $g(x) = \int |r'|^{d-1}\pi(r'x)dr'$. Then $g(x)$ has the property that $g(sx) = |s|^{-d}g(x)$. Let $z = rx$, then

$$\begin{aligned} E\{h(x')\} &= \int\int h(z)\pi(z)|r|^{-1}\pi(r^{-1}z)/g(r^{-1}z)drdz \\ &= \int h(z)\pi(z)/g(z)\int |r|^{-d-1}\pi(r^{-1}z)drdz \\ &= \int h(z)\pi(z)dz = E_\pi\{h(z)\}. \end{aligned}$$

Thus, the sample x' follows the distribution $\pi(x)$. Because the expectation $E\{h(x')\}$ does not depend on a particular value of y, x' and y are independent. ∎

Note that the directional sampling step can be replaced by one or several MH moves or the griddy Gibbs sampler (Ritter and Tanner, 1992) in cases where sampling directly from $f(r)$ is difficult.

Although the ADS provides a useful framework enabling different chains to learn from each other, the algorithm alone is not very effective in improving sampling efficiency, mainly because the sampling direction generated by the ADS is too arbitrary. The question of how a learnt/efficient sampling direction for the ADS can be generated is discussed in the next section: we describe the *conjugate gradient Monte Carlo* (CGMC) algorithm (Liu *et al.*, 2000), which, at each iteration, employs a local optimization method, such as the conjugate gradient method, to construct a sampling direction for the ADS.

5.2 Conjugate Gradient Monte Carlo

Let $\boldsymbol{x}^{(t)} = (x_1^{(t)}, \ldots, x_N^{(t)})$ denote the current population of samples. One iteration of the CGMC sampler consists of the following steps.

Conjugate Gradient Monte Carlo (Steps)

1. Select one individual, say $x_c^{(t)}$, at random from the current population $\boldsymbol{x}^{(t)}$.

2. Select another individual, say $x_a^{(t)}$, at random from the remaining set of the population (i.e. $\{x_i^{(t)} : i \neq c\}$). Starting with $x_a^{(t)}$, conduct a deterministic search, using the conjugate gradient method or the steepest descent method, to find a local mode of $f(x)$. Denote the local mode by $z_a^{(t)}$, which is called the anchor point.

3. Set $y_c = z_a^{(t)} + r_t e_t$, where $e_t = x_c^{(t)} - z_a^{(t)}$, and r_t is a scalar sampled from the density

$$f(r) \propto |r|^{d-1} f(z_a^{(t)} + r_t e_t), \tag{5.3}$$

where d is the dimension of x, and the factor $|r|^{d-1}$ is derived from the transformation Jacobian.

4. Form the new population $\boldsymbol{x}^{(t+1)}$ by replacing $x_c^{(t)}$ by y_c and leaving other individuals unchanged (i.e., set $x_i^{(t+1)} = x_i^{(t)}$ for $i \neq c$).

The validity of this algorithm follows directly from Lemma 5.1.1. Note that y_c is independent of $x_a^{(t)}$, as the latter can be viewed as a function of the anchor point $z_a^{(t)}$.

The gradient-based optimization procedure performed in Step 2 can be replaced by some other optimization procedures, for example, a short run of simulated annealing (Kirkpatrick *et al.*, 1983). Since the local optimization step is usually expensive in computation, Liu *et al.* (2000) proposed the multiple-try MH algorithm (described in Section 3.2.3) for the line sampling step, which enables effective use of the local modal information of the distribution and thus improve the convergence of the algorithm. The numerical results indicate that the CGMC sampler offers significant improvement over the MH algorithm, especially for the hard problems. Note that for CGMC, the population size N is not necessarily very large. As shown by Liu *et al.* (2000), $N = 4$ or 5 works well for most problems they tried.

5.3 Sample Metropolis-Hastings Algorithm

In adaptive direction sampling and conjugate gradient Monte Carlo, when updating the population, one first selects an individual from the population and then updates the selected individual using the standard Metropolis-Hastings procedure. If the candidate state is of high quality relative to the whole population, one certainly wants to keep it in the population. However, the acceptance of the candidate state depends on the quality of the individual that is selected for updating. To improve the acceptance rate of high quality candidates and to improve the set $\{x_i^{(t)} : i = 1, \ldots, N\}$ as a sample of size N from $f(x)$, Lewandowski and Liu (2008) proposed the sampling Metropolis-Hastings (SMH) algorithm. A simple version of this algorithm can be described as follows.

Let $\boldsymbol{x}_t = \left\{x_1^{(t)}, \ldots, x_N^{(t)}\right\}$ denote the current population. One iteration of SMH consists of the following two steps:

Sample MH algorithm

1. Take one candidate draw $x_0^{(t)}$ from a proposal distribution $g(x)$ on $\mathcal{X}$, and compute the acceptance probability

$$\alpha_0^{(t)} = \frac{\sum_{i=1}^{N} \frac{g(x_i^{(t)})}{f(x_i^{(t)})}}{\sum_{i=0}^{N} \frac{g(x_i^{(t)})}{f(x_i^{(t)})} - \min_{0 \le k \le N} \frac{g(x_k^{(t)})}{f(x_k^{(t)})}}.$$

2. Draw $U \sim \text{Unif}(0,1)$, and set

$$\begin{aligned} \mathcal{S}_{t+1} &= \left\{x_1^{(t+1)}, \ldots, x_n^{(t+1)}\right\} \\ &= \begin{cases} \mathcal{S}_t, & \text{if } U > \alpha_0^{(t)}; \\ \left\{x_1^{(t)}, \ldots, x_{i-1}^{(t)}, x_0^{(t)}, x_{i+1}^{(t)}, \ldots, x_n^{(t)}\right\}, & \text{if } U \le \alpha_0^{(t)}, \end{cases} \end{aligned}$$

where i is chosen from $(1, \ldots, n)$ with the probability weights

$$\left(\frac{g(x_1^{(t)})}{f(x_1^{(t)})}, \ldots, \frac{g(x_n^{(t)})}{f(x_n^{(t)})}\right).$$

Thus, $\boldsymbol{x}_{t+1}$ and $\boldsymbol{x}_t$ differ by one element at most.

It is easy to see that in the case of $N = 1$, SMH reduces to the traditional MH with independence proposals. The merit of SMH is that to accept a candidate state, it compares the candidate with the whole population, instead of a single individual randomly selected from the current population. Lewandowski and Liu (2008) show that SMH will converge under mild conditions to the target distribution $\prod_{i=1}^{N} f(x_i)$ for $\{x_1, \ldots, x_N\}$, and can be more efficient than the traditional MH and adaptive direction sampling. Extensions of SMH to the case where x_i's are not identically distributed are of interest.

5.4 Parallel Tempering

Another research stream of population-based MCMC is pioneered by the parallel tempering algorithm (Geyer, 1991), also known as exchange Monte Carlo Hukushima and Nemoto (1996). In parallel tempering, each chain is equipped with an invariant distribution powered by an auxiliary variable, the so-called inverse temperature. Let $f(x) \propto \exp(-H(x))$, $x \in \mathcal{X}$ denote the distribution of interest, where $H(x)$ is called the energy function in terms of physics. In Bayesian statistics, $H(x)$ corresponds to the negative log-posterior distribution of the parameters. Parallel tempering simulates in parallel a sequence of distributions

$$f_i(x) \propto \exp(-H(x)/T_i), \quad i = 1, \ldots, n, \tag{5.4}$$

where T_i is the temperature associated with the distribution $f_i(x)$. The temperatures form a ladder $T_1 > T_2 > \cdots > T_{n-1} > T_n \equiv 1$, so $f_n(x) \equiv f(x)$ corresponds to the target distribution. The idea underlying this algorithm can be explained as follows: Raising temperature flattens the energy landscape of the distribution and thus eases the MH traversal of the sample space, the high density samples generated at the high temperature levels can be transmitted to the target temperature level through the exchange operations (described below), and this in turn improves convergence of the target Markov chain.

Let $\boldsymbol{x}^{(t)} = (x_1^{(t)}, \ldots, x_N^{(t)})$ denote the current population of samples. One iteration of parallel tempering consists of the following steps.

1. *Parallel MH step*: Update each $x_i^{(t)}$ to $x_i^{(t+1)}$ using the MH algorithm.

2. *State swapping step*: Try to exchange $x_i^{(t+1)}$ with its neighbors: Set $j = i - 1$ or $i + 1$ according to probabilities $q_e(i, j)$, where $q_e(i, i + 1) = q_e(i, i - 1) = 0.5$ for $1 < i < N$ and $q_e(1, 2) = q_e(N, N - 1) = 1$, and accept the swap with probability

$$\min\left\{1, \exp\left(\left[H(x_i^{(t+1)}) - H(x_j^{(t+1)})\right]\left[\frac{1}{T_i} - \frac{1}{T_j}\right]\right)\right\}. \tag{5.5}$$

In practice, to have a reasonable acceptance rate of the proposed exchange, the temperatures need to be chosen carefully. As for simulated tempering, we recommend the following method to set the temperature ladder in a sequential manner: It starts with the highest temperature T_1, and set the next lower temperature such that

$$\text{Var}_i(H(x))\delta^2 = O(1), \tag{5.6}$$

where $\delta = 1/T_{i+1} - 1/T_i$, and the variance $\text{Var}_i(\cdot)$ is taken with respect to $f_i(x)$ and can be estimated through a preliminary run. The condition (5.6) essentially requires that the distributions on neighboring temperature levels have a considerable overlap.

Parallel tempering is very powerful for simulating complicated systems, such as spin-glasses (Hukushima and Nemoto, 1996; de Candia and Coniglio, 2002) and polymer simulations (Neirotti *et al.*, 2000; Yan and de Pablo, 2000). Compared to simulated tempering (Marinari and Parisi, 1992), parallel tempering has the apparent advantage that it avoids the problem of normalizing constant estimation for the distributions $f_i(x)$'s.

5.5 Evolutionary Monte Carlo

The genetic algorithm (Holland, 1975) has been successfully applied to many hard optimization problems, such as the traveling salesman problem (Chatterjee *et al.*, 1996), protein folding (Patton *et al.*, 1995), and machine learning (Goldberg, 1989), among others. It is known that its crossover operator is the key to its power; this makes it possible for the genetic algorithm to explore a far greater range of potential solutions to a problem than conventional optimization algorithms. Motivated by the genetic algorithm, Liang and Wong (2000, 2001a) propose the evolutionary Monte Carlo algorithm (EMC), which incorporates most attractive features of the genetic algorithm into the framework of Markov chain Monte Carlo. EMC works in a fashion similar to parallel tempering: A population of Markov chains are simulated in parallel with each chain having a different temperature. The difference between the two algorithms is that EMC includes a genetic operator, namely, the crossover operator in its simulation. The numerical results indicate that the crossover operator improves the convergence of the simulation and that EMC can outperform parallel tempering in almost all scenarios.

To explain how the crossover operator can be used effectively in EMC, we suppose the target distribution of interest is written in the form

$$f(x) \propto \exp\{-H(x)\}, \quad x \in \mathcal{X} \subset \mathbb{R}^d,$$

where the dimension $d > 1$, and $H(x)$ is called the fitness function in terms of genetic algorithms. Let $\boldsymbol{x} = \{x_1, \ldots, x_N\}$ denote a population of size N with x_i from the distribution with density

$$f_i(x) \propto \exp\{-H(x)/T_i\}.$$

In terms of genetic algorithms, x_i is called a chromosome or an individual, each element of x_i is called a gene, and a realization of the element is called a genotype. As in parallel tempering, the temperatures form a decreasing ladder $T_1 > T_2 > \cdots > T_N \equiv 1$, with $f_N(x)$ being the target distribution.

5.5.1 Evolutionary Monte Carlo in Binary-Coded Space

We describe below how EMC works when x_i is coded by a binary vector. For simplicity, let $\boldsymbol{x} = \{x_1, \ldots, x_N\}$ denote the current population, where $x_i = (\beta_{i,1}, \ldots, \beta_{i,d})$ is a d-dimensional binary vector, and $\beta_{i,j} \in \{0, 1\}$. The mutation, crossover and exchange operators employed by EMC can be described as follows.

Mutation. In mutation, a chromosome, say x_k, is first randomly selected from the current population $\boldsymbol{x}$, then mutated to a new chromosome y_k by reversing the value ($0 \leftrightarrow 1$) of some genotypes which are also randomly selected. A new population is formed as $\boldsymbol{y} = \{x_1, \cdots, x_{k-1}, y_k, x_{k+1}, \cdots, x_N\}$, and it is accepted with probability $\min(1, r_m)$ according to the Metropolis rule, where

$$r_m = \frac{f(\boldsymbol{y})}{f(\boldsymbol{x})} \frac{T(\boldsymbol{x}|\boldsymbol{y})}{T(\boldsymbol{y}|\boldsymbol{x})} = \exp\left\{ -\frac{H(y_k) - H(x_k)}{T_k} \right\} \frac{T(\boldsymbol{x}|\boldsymbol{y})}{T(\boldsymbol{y}|\boldsymbol{x})}, \tag{5.7}$$

and $T(\cdot|\cdot)$ denotes the transition probability between populations. If the proposal is accepted, replace the current population $\boldsymbol{x}$ by $\boldsymbol{y}$; otherwise, keep $\boldsymbol{x}$ as the current population.

In addition to the 1-point and 2-point mutations, one can also use the uniform mutation in which each genotype of x_k is mutated with a nonzero probability. All these operators are symmetric, that is, $T(\boldsymbol{x}|\boldsymbol{y}) = T(\boldsymbol{y}|\boldsymbol{x})$.

Crossover. One chromosome pair, say x_i and x_j $(i \neq j)$, are selected from the current population $\boldsymbol{x}$ according to some selection procedure, for example, a roulette wheel selection or a random selection. Without loss of generality, we assume that $H(x_i) \geq H(x_j)$. Two 'offspring' are generated according to some crossover operator (described below), the offspring with a smaller fitness value is denoted by y_j and the other denoted by y_i. A new population is formed as $\boldsymbol{y} = \{x_1, \cdots, x_{i-1}, y_i, x_{i+1}, \cdots, x_{j-1}, y_j, x_{j+1}, \cdots, x_N\}$. According to the Metropolis rule, the new population is accepted with probability $\min(1, r_c)$,

$$r_c = \frac{f(\boldsymbol{y})}{f(\boldsymbol{x})} \frac{T(\boldsymbol{x}|\boldsymbol{y})}{T(\boldsymbol{y}|\boldsymbol{x})} = \exp\left\{-\frac{H(y_i) - H(x_i)}{T_i} - \frac{H(y_j) - H(x_j)}{T_j}\right\} \frac{T(\boldsymbol{x}|\boldsymbol{y})}{T(\boldsymbol{y}|\boldsymbol{x})}, \tag{5.8}$$

where $T(\boldsymbol{y}|\boldsymbol{x}) = P(x_i, x_j|\boldsymbol{x})P(y_i, y_j|x_i, x_j)$, $P(x_i, x_j|\boldsymbol{x})$ denotes the selection probability of (x_i, x_j) from the population $\boldsymbol{x}$, and $P(y_i, y_j|x_i, x_j)$ denotes the generating probability of (y_i, y_j) from the parental chromosomes (x_i, x_j).

Liang and Wong (2000, 2001a) recommended the following procedure for selection of parental chromosomes:

1. Select the first chromosome x_i according to a roulette wheel procedure with probability

$$p(x_i) = \frac{\exp(-H(x_i)/T_s)}{\sum_{i=1}^N \exp(-H(x_i)/T_s)},$$

   where T_s is called the selection temperature, and is not necessarily the same with any T_i. Intuitively, if T_s is low, then a high quality chromosome will be likely selected from the current population to mate with others.

2. Select the second chromosome x_j randomly from the remaining of the population.

Then, the selection probability of (x_i, x_j) is

$$P((x_i, x_j)|\boldsymbol{x}) = \frac{1}{(N-1)Z(\boldsymbol{x})}\left\{\exp\left(-\frac{H(x_i)}{T_s}\right) + \exp\left(-\frac{H(x_j)}{T_s}\right)\right\}, \tag{5.9}$$

where $Z(\boldsymbol{x}) = \sum_{i=1}^N \exp\{-H(x_i)/T_s\}$. The $P((y_i, y_j)|\boldsymbol{y})$ can be calculated similarly.

There are many possible crossover operators which can leave the joint distribution (5.1) invariant. The 1-point crossover operator is perhaps the simplest one. Given the two parental chromosomes, x_i and x_j, the 1-point crossover operator proceeds as follows. First an integer crossover point k is drawn randomly from the set $\{1, 2, \cdots, d\}$, then x_i' and x_j' are constructed by swapping the genotypes to the right of the crossover point between the two parental chromosomes:

$$\begin{array}{ccc} (x_{i,1}, \cdots, x_{i,d}) & & (x_{i,1}, \cdots, x_{i,k}, x_{j,k+1}, \cdots, x_{j,d}) \\ & \Longrightarrow & \\ (x_{j,1}, \cdots, x_{j,d}) & & (x_{j,1}, \cdots, x_{j,k}, x_{i,k+1}, \cdots, x_{i,d}). \end{array}$$

If there are k $(k > 1)$ crossover points, it is called the k-points crossover. One extreme case is the uniform crossover, in which each genotype of x_i' is randomly chosen from the two parental genotypes and the corresponding genotype of x_j' is assigned to the parental genotype not chosen by x_i'. With the operation, the beneficial genes of parental chromosomes can combine together and possibly produce some high quality individuals.

In addition to the k-point crossover operator, Liang and Wong (2000) introduce an adaptive crossover operator, in which two offspring are generated. For each position, if x_i and x_j have the same value, y_i and y_j copy the value, and independently reverse it with probability p_0; if x_i and x_j have different values, y_i copies the value of x_i and reverses it with probability p_2, y_j copies the value of x_j and reverses it with probability p_1. Usually one sets $0 < p_0 \leq p_1 \leq p_2 < 1$. The adaptive crossover tends to preserve the good genotypes of a population, and thus enhance the ability of EMC learning from existing samples. The generating probability of the new offspring is

$$P((y_i, y_j)|(x_i, x_j)) = p_{11}^{n_{11}} p_{12}^{n_{12}} p_{21}^{n_{21}} p_{22}^{n_{22}}, \tag{5.10}$$

where p_{ab} and n_{ab} $(a, b = 1, 2)$ denote, respectively, the probability and frequency of the cell (a, b) of Table 5.1. Note that $n_{11} + n_{12} + n_{21} + n_{22} = d$.

Exchange. This operation is the same as that used in parallel tempering (Geyer, 1991; Hukushima and Nemoto, 1996). Given the current population $\boldsymbol{x}$

Table 5.1 Generating probability for the single position of the new offspring in the adaptive crossover operator.

Parents:	offspring: y_i and y_j	
x_i and x_j	same	different
Same	$p_0^2 + (1 - p_0)^2$	$2p_0(1 - p_0)$
Different	$p_1(1 - p_2) + p_2(1 - p_1)$	$p_1 p_2 + (1 - p_1)(1 - p_2)$

and the temperature ladder $\boldsymbol{t}$, $(\boldsymbol{x},\boldsymbol{t}) = (x_1, T_1, \cdots, x_N, T_N)$, one tries to make an exchange between x_i and x_j without changing the t's. The new population is accepted with probability min(1,r_e),

$$r_e = \frac{f(\boldsymbol{x}')}{f(\boldsymbol{x})} \frac{T(\boldsymbol{x}|\boldsymbol{x}')}{T(\boldsymbol{x}'|\boldsymbol{x})} = \exp\left\{ (H(x_i) - H(x_j)) \left(\frac{1}{T_i} - \frac{1}{T_j} \right) \right\}. \tag{5.11}$$

Typically, the exchange is only performed on neighboring temperature levels.

The Algorithm. Based on the operators described above, the algorithm can be summarized as follows. Given an initial population $\boldsymbol{x} = \{x_1, \cdots, x_N\}$ and a temperature ladder $\boldsymbol{t} = \{T_1, T_2, \cdots, T_N\}$, EMC iterates between the following two steps:

1. Apply either mutation or crossover operator to the population with probability q_m and $1 - q_m$, respectively. The q_m is called the mutation rate.

2. Try to exchange x_i with x_j for N pairs (i,j) with i being sampled uniformly on $\{1, \cdots, N\}$ and $j = i \pm 1$ with probability $q_e(i,j)$, where $q_e(i, i+1) = q_e(i, i-1) = 0.5$ and $q_e(1,2) = q_e(N, N-1) = 1$.

5.5.2 Evolutionary Monte Carlo in Continuous Space

Let $\boldsymbol{x} = \{x_1, \ldots, x_N\}$ denote the current population. If x_i is coded as a real vector, that is, $x_i = (\beta_{i,1}, \ldots, \beta_{i,d})$ with $\beta_{i,j} \in \mathbb{R}$, then the mutation and crossover operators can be defined as follows. (The description for the exchange operator is skipped below, as it is the same as that used for the binary-coded vectors.)

Mutation. The mutation operator is defined as an additive Metropolis-Hastings move. One chromosome, say x_k, is randomly selected from the current population $\boldsymbol{x}$. A new chromosome is generated by adding a random vector e_k so that

$$y_k = x_k + e_k, \tag{5.12}$$

where the scale of e_k is chosen such that the operation has a moderate acceptance rate, for example, 0.2 to 0.5, as suggested by Gelman *et al.* (1996). The new population $\boldsymbol{y} = \{x_1, \cdots, x_{k-1}, y_k, x_{k+1}, \cdots, x_N\}$ is accepted with probability min(1,r_m), where

$$r_m = \frac{f(\boldsymbol{y})}{f(\boldsymbol{x})} \frac{T(\boldsymbol{x}|\boldsymbol{y})}{T(\boldsymbol{y}|\boldsymbol{x})} = \exp\left\{ -\frac{H(y_k) - H(x_k)}{T_k} \right\} \frac{T(\boldsymbol{x}|\boldsymbol{y})}{T(\boldsymbol{y}|\boldsymbol{x})}, \tag{5.13}$$

and $T(\cdot|\cdot)$ denotes the transition probability between populations.

Crossover. One type of crossover operator that works for the real-coded chromosomes is the so-called 'real crossover', which includes the k-point and uniform crossover operators as described in Section 5.5.1. They are called real crossover by Wright (1991) to indicate that they are applied to real-coded chromosomes.

In addition to the real crossover, Liang and Wong (2001a) propose the snooker crossover operator:

1. Randomly select one chromosome, say x_i, from the current population $\boldsymbol{x}$.

2. Select the other chromosome, say x_j, from the sub-population $\boldsymbol{x} \setminus \{x_i\}$ with a probability proportional to $\exp\{-H(x_j)/T_s\}$, where T_s is called the selection temperature.

3. Let $e = x_i - x_j$, and $y_i = x_j + re$, where $r \in (-\infty, \infty)$ is a random variable sampled from the density

$$f(r) \propto |r|^{d-1} f(x_j + r\mathbf{e}). \tag{5.14}$$

4. Construct a new population by replacing x_i with the 'offspring' y_i, and replace $\boldsymbol{x}$ by $\boldsymbol{y}$.

The validity of this operator follows directly from Lemma 5.1.1. As mentioned in Section 5.1, the line sampling step can be replaced by one or a few MH moves or the griddy Gibbs when sampling directly from $f(r)$ is difficult.

5.5.3 Implementation Issues

EMC includes three free parameters, namely the population size N, the temperature ladder $\boldsymbol{t}$, and the mutation rate q_m. Regarding the setting of these parameters, we first note that both parameters, N and $\boldsymbol{t}$, are related to the diversity of the population, and that a highly diversified population is always preferred for the system mixing. In EMC, one has two ways to increase the diversity of a population. One is to increase the population size, and the other is to steepen the temperature ladder by increasing the value of T_1. A certain balance is needed between the two ways. A small population size may result in a steeper temperature ladder and a low acceptance rate of exchange operations. A large population size may result in the target chain being less likely to be updated in per-unit CPU time, and thus converges slowly. Neither is an attractive choice. A heuristic guideline is to choose the population size comparable (at least) with the dimension of the problem, choose the highest temperature such that the (anticipated) energy barriers can be easily overcome by a MH move, and set the temperature ladder to produce a moderate acceptance rate of exchange operations. In practice, N and $\boldsymbol{t}$ can be set as in parallel tempering. Starting with the highest temperature, add temperature level by level to the ladder, until the lowest (target) temperature level has been reached. When a new temperature level is added to the current ladder, a moderate acceptance rate should be maintained for the exchange operations between neighboring levels.

Regarding the choice of q_m, we note that the mutation operation usually provides a local exploration around a local mode, and in contrast, the crossover operation usually provides a much more global exploration over the entire sample space and has often a low acceptance rate. To balance the two kinds of operations, we suggest to set q_m to a value between 0.25 and 0.4, which may not be optimal but usually works well.

5.5.4 Two Illustrative Examples

Below, we illustrate the performance of EMC with two examples and make comparisons with parallel tempering.

A Highway Data Example. This dataset relates the automobile accident rate (in accidents per million vehicle miles) to 13 potential independent variables. It includes 39 sections of large highways in the state of Minnesota in 1973. Weisberg (1985) uses this data to illustrate variable selection for a multiple regression. According to Weisberg's analysis, variable 1, the length of the segment in miles, should be included in the regression. Variables 11, 12 and 13 are dummy variables that taken together indicate the type of highway, so they should be regarded as one variable, including all or none of them in a model. After this simplification, the data consists of 10 independent variables, and the number of all possible subset models is reduced to 1024, which is manageable even for an exhaustive search.

To illustrate EMC as a simulation approach, it was applied to simulate from the Boltzmann distribution defined on C_p (Mallows, 1973) as follows:

$$f(x) = \frac{1}{Z(\tau)} \exp\{-C_p(x)/\tau\}, \tag{5.15}$$

where x denotes a model, τ denotes the temperature, $C_p(x)$ denotes the C_p value of the model x, and $Z(\tau) = \sum_x \exp\{-C_p(x)/\tau\}$ is the normalizing constant. The interest of this distribution comes from the connection between Bayesian and non-Bayesian methods for regression variable selection problems. Liang *et al.* (2001) showed that under an appropriate prior setting, sampling from the posterior distribution of a linear regression model is approximately equivalent to sampling from (5.15) with $\tau = 2$.

For this example, the chromosome was coded as a 10-dimensional binary vector, with each component indicating the inclusion/exclusion of the corresponding variable. The highest temperature was set to 5, the lowest temperature was set to 1, and the intermediate temperatures were equally spaced between the two limits. A 1-point mutation and a uniform crossover were used in the simulations, and the mutation rate was 0.25.

In the first run, the population size was set to $N = 5$, and EMC was run for 15 000 iterations. The CPU time was 105s on an Ultra Sparc2 workstation (all computations reported in this subsection were done on the same computer). The overall acceptance rates of the mutation, crossover and exchange operations were 0.59, 0.60, and 0.76, respectively. During the run, 457 different C_p values were sampled at the lowest temperature $t = 1$. These C_p values have included the smallest 331 C_p values and covered 99.9% probability of the distribution (5.15). The histogram of the samples is shown in Figure 5.1(b). For comparison, Figure 5.1(a) shows the true distribution $f(x)$, which was calculated for all 1024 models at temperature $t = 1$. This indicates that the distribution $f(x)$ has been estimated accurately by the EMC samples. In

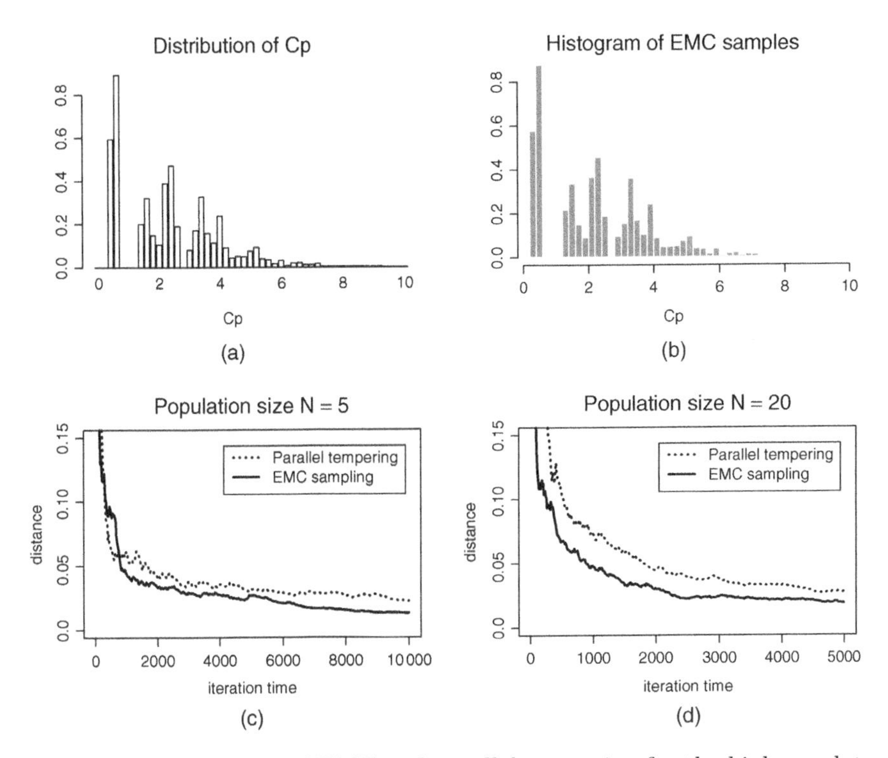

Figure 5.1 Comparison of EMC and parallel tempering for the highway data example. (a) A histogram representation of the distribution (5.15). (b) The histogram of C_p produced in a run of EMC. (c) Comparison of the convergence rates of EMC and parallel tempering with the population size $N = 5$. (d) Comparison of the convergence rates of EMC and parallel tempering with the population size $N = 20$ (Liang and Wong, 2000).

another run, the population size was set to 20 and the other parameters were kept unchanged. EMC was run for 10 000 iterations with CPU 179s. The overall acceptance rates of the mutation, crossover and exchange operations were 0.62, 0.51 and 0.95, respectively. The results are summarized in Figure 5.1.

Parallel tempering was also applied to this example. A 1-point mutation was used for the local updating. In one run, the population size was set to 5 and 10 000 iterations were produced within 105s. The overall acceptance rates of the local updating and exchange operations were 0.59 and 0.77, respectively. Parallel tempering cost longer CPU time per iteration, because each chain needs to undergo an update step at each iteration, whereas, in EMC, only 40% of samples were selected to mate in the crossover step. In another run, the population size was 20, and 5000 iterations were produced within 193s. The overall acceptance rates of the local updating and exchange operations were 0.62 and 0.94, respectively. The results are summarized in Figure 5.1.

Figure 5.1(c) and 5.1(d) compare the convergence rates of parallel tempering and EMC, where the 'distance' is defined as the L^2 distance between the estimated and true distributions of C_p. In the two plots, EMC and parallel tempering have been adjusted to have the same time scale. The comparison indicates that EMC has a faster convergence rate than parallel tempering, regardless of the population size used in simulations.

A Multimodal Example. Consider simulating from a 2D mixture normal distribution

$$f(x) = \frac{1}{\sqrt{2\pi}\sigma} \sum_{k=1}^{20} w_k \exp\left\{-\frac{1}{2\sigma^2}(x - \mu_k)'(x - \mu_k)\right\}, \qquad (5.16)$$

where $\sigma = 0.1$, $w_1 = \cdots = w_{20} = 0.05$. The mean vectors $\mu_1, \mu_2, \cdots, \mu_{20}$ (given in Table 5.2) are uniformly drawn from the rectangle $[0, 10] \times [0, 10]$ (Liang and Wong, 2001a). Among them, components 2, 4, and 15 are well separated from the others. The distance between component 4 and its nearest neighboring component is 3.15, and the distance between component 15

Table 5.2 Mean vectors of the 20 components of the mixture normal distribution (Liang and Wong, 2001a).

k	μ_{k1}	μ_{k2}	k	μ_{k1}	μ_{k2}	k	μ_{k1}	μ_{k2}	k	μ_{k1}	μ_{k2}
1	2.18	5.76	6	3.25	3.47	11	5.41	2.65	16	4.93	1.50
2	8.67	9.59	7	1.70	0.50	12	2.70	7.88	17	1.83	0.09
3	4.24	8.48	8	4.59	5.60	13	4.98	3.70	18	2.26	0.31
4	8.41	1.68	9	6.91	5.81	14	1.14	2.39	19	5.54	6.86
5	3.93	8.82	10	6.87	5.40	15	8.33	9.50	20	1.69	8.11

and its nearest neighboring component (except component 2) is 3.84, which are 31.5 and 38.4 times of the standard deviation, respectively. Mixing the components across such long distances puts a great challenge on EMC.

Liang and Wong (2001a) applied EMC to this example with the following settings: Each chromosome was coded by a real 2D vector; the population size was set to 20, the highest temperature to 5, the lowest temperature to 1; the intermediate temperatures were equally spaced between 1 and 5; and the mutation rate was set to 0.2. The population was initialized by some random vectors drawn uniformly from the region $[0,1]^2$. EMC was run for 100 000 iterations. In the mutation step, $e_k \sim N_2(0, 0.25^2 T_k)$ for $k = 1, \cdots, N$, where T_k is the temperature at level k. Figure 5.2(a) shows the sample paths (at level $t = 1$) of the first 10 000 iterations, and Figure 5.3(a) shows the whole samples (at level $t = 1$) obtained in the run. It can be seen that EMC has sampled all components in the first 10 000 iterations, although the population was initialized at one corner.

For comparison, Liang and Wong (2001a) also applied parallel tempering to this example with the same parameter setting and initialization. Within the same computational time, parallel tempering produced 73 500 iterations. The local updating step was done with the mutation operator used in EMC, and

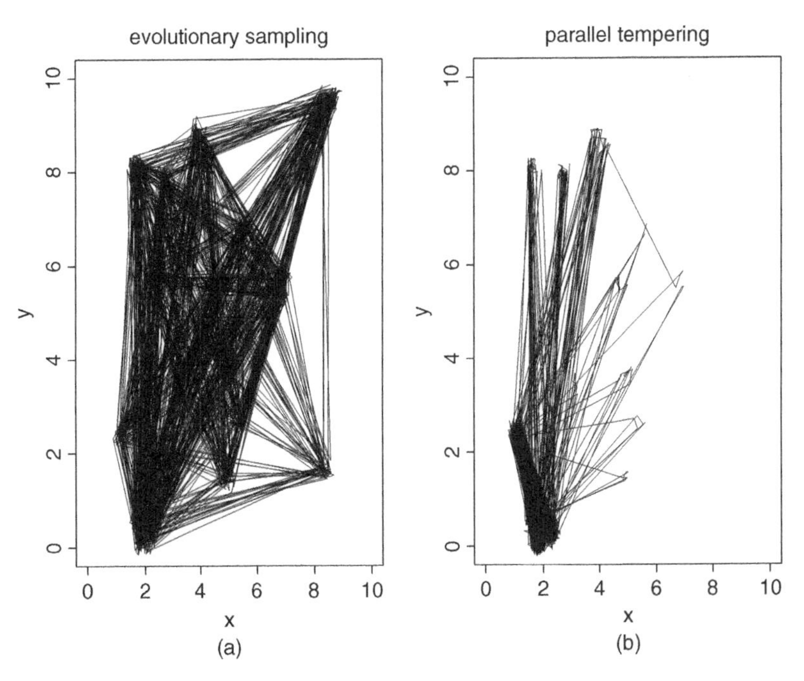

Figure 5.2 The sample path of the first 10 000 iterations at temperature $t = 1$. (a) EMC. (b) Parallel tempering (Liang and Wong, 2001a).

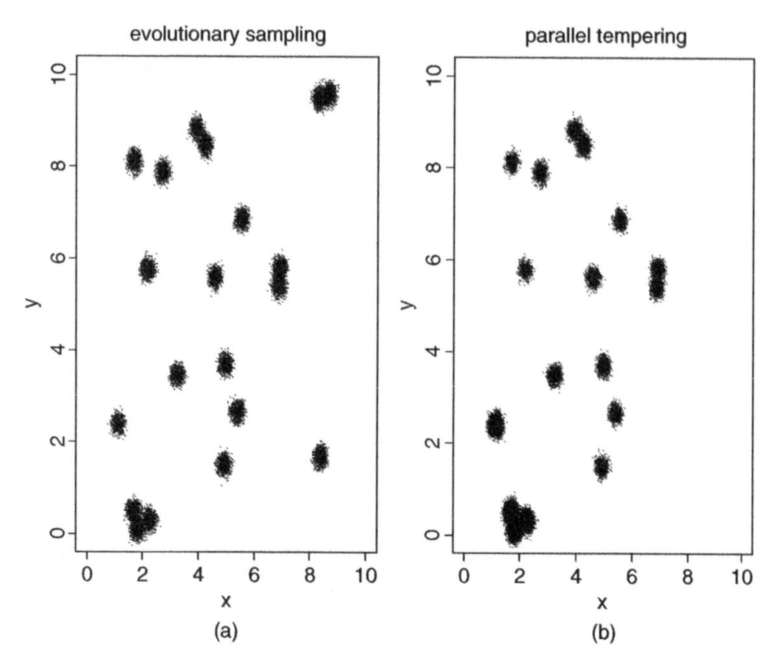

Figure 5.3 The plot of whole samples. (a) EMC. (b) Parallel tempering (Liang and Wong, 2001a).

the overall acceptance rate was 0.28, which suggests that parallel tempering has been implemented effectively. Figure 5.2(b) shows the sample path (at the level $t = 1$) of the first 10 000 iterations, and Figure 5.3(b) shows the whole samples. It can be seen that parallel tempering was not able to sample all components of the mixture even even with 73 500 iterations; the components 2, 4, and 15 were never sampled.

Table 5.3 shows the estimates of the mean and variance of (5.16) produced by EMC (EMC-A) and parallel tempering (PT), where all estimates were calculated based on 20 independent runs. Each run of EMC consists of 10^6 iterations, and each run of parallel tempering consists of 7.35×10^5 iterations, such that they cost about the same CPU time. Clearly, EMC outperforms parallel tempering in terms of smaller biases and variances.

In another experiment, Liang and Wong (2001a) examined the effect of the snooker crossover. The EMC algorithm with only the real crossover operator was run 20 times. Each run consisted of 10^6 iterations, and the CPU time was about the same as that used by the previous experiment. The estimation results are reported in Table 5.3 (EMC-B). The comparison shows that the snooker crossover is superior to the real crossover and can lead to a faster convergence rate of EMC.

Table 5.3 Comparison of EMC and parallel tempering for the mixture normal example (Liang and Wong, 2001a).

		EMC-A		EMC-B		PT	
parameter	true value	est.	SD	est.	SD	est.	SD
μ_1	4.48	4.48	0.004	4.44	0.026	3.78	0.032
μ_2	4.91	4.91	0.008	4.86	0.023	4.34	0.044
Σ_{11}	5.55	5.55	0.006	5.54	0.051	3.66	0.111
Σ_{22}	9.86	9.84	0.010	9.78	0.048	8.55	0.049
Σ_{12}	2.61	2.59	0.011	2.58	0.043	1.29	0.084

EMCA: EMC with both the real and snooker crossover operators; EMC-B: EMC with only the real crossover operator; PT: parallel tempering. The $\boldsymbol{\mu}_1$ and $\boldsymbol{\mu}_2$ denote, respectively, the first and second component of the mean vector of (5.16); Σ_{11}, Σ_{22}, and Σ_{12} denote, respectively, three components of the covariance matrix of (5.16); and SD denotes the standard deviation of the corresponding estimate.

5.5.5 Discussion

In the EMC algorithm presented above, each sample is coded as a fixed dimensional vector, even for some varying dimensional problems, for example, the variable selection problem considered in the highway data example. For many problems, such as the change-point identification problem (Liang and Wong, 2000), Bayesian curve fitting (Liang *et al.*, 2001), and Bayesian neural network learning (Liang and Kuk, 2004; Liang, 2005a), this kind of coding scheme is very natural. However, we note that fixing the dimension of each sample is not a intrinsic constraint of EMC. Extending EMC to the transdimensional case, where each sample is coded as a varying dimensional vector, is straightforward. In this case, the k-point crossover operator described above still works as shown by Jasra *et al.* (2007), who, for the snooker algorithm, develop a substitute for mixture model problems.

The exchange operator plays an important role in EMC, which provides the main way for the Markov chains to interact with each other. In the EMC algorithm presented above, only the plain, nearest neighboring exchange operators were prescribed. In practice, it can be improved in various ways. For example, Jasra *et al.* (2007) propose improving it by using the delayed rejection method (Green and Mira, 2001), and Goswami and Liu (2007) propose several different types of exchange operators with the idea of finding at each iteration the most similar samples (in fitness values) to swap. All these methods allow for an increased interaction within the population, and thus accelerate the convergence of the algorithm.

Regarding crossover operations, we note that many different crossover operators have been developed for the genetic algorithm (Goldberg, 1989). However, only a few of them can be applied to EMC, due to the stringent requirement of detailed balance (or reversible) condition of Markov chain. The crossover operation has often a low acceptance rate. In EMC, one has to retain two offspring produced by two parental individuals through a crossover operation, otherwise, the detailed balance condition will be violated. However, it is often the case that one offspring is of bad quality, so the acceptance rate of the operation is low. A worthwhile topic to study would be how to improve the acceptance rate of crossover operations, or more generally, how to create efficient crossover operations, for a population-based MCMC sampler. This issue will be further addressed in Section 7.6.3.2, where we show that the acceptance rate of crossover operations can be generally improved under the framework of population-based stochastic approximation Monte Carlo (Liang, 2009d).

5.6 Sequential Parallel Tempering for Simulation of High Dimensional Systems

The development of science and technology means we increasingly need to deal with high dimensional systems, in order to, for example, align a group of protein or DNA sequences to infer their homology (Durbin *et al.*, 1998); identify a single-nucleotide polymorphism (SNP) associated with certain disease from millions of SNPs (Emahazion *et al.*, 2001); estimate the volatility of asset returns to understand the price trend of the option market (Hull and White, 1987); or simulate from spin systems to understand their physical properties (Swendsen and Wang, 1987). In such problems, the dimensions of the systems often range from several hundreds to several thousands or even higher, and the solution spaces are so huge that Monte Carlo has been an indispensable tool for making inference for them. How to efficiently sample from these high dimensional systems puts a great challenge on the existing MCMC methods.

As discussed in other parts of this book, many advanced MCMC algorithms have been proposed during the past two decades for accelerating the convergence of simulations. These include simulated tempering (Marinari and Parisi, 1992), parallel tempering (Geyer, 1991), evolutionary Monte Carlo (Liang and Wong, 2000, 2001a), dynamic weighting (Wong and Liang, 1997), multicanonical sampling (Berg and Neuhaus, 1991), Wang-Landau algorithm (Wang and Landau, 2001), stochastic approximation Monte Carlo (Liang *et al.*, 2007), among others. A central goal of these algorithms is to overcome the local-trap problem caused by multimodality – that on the energy landscape of the system, there are many local energy minima separated by high barriers. For example, in the tempering-based algorithms, the energy barriers are flattened by increasing the 'temperature' of the system such that the sampler can move across them freely. In the multicanonical sampling, dynamic weighting, and stochastic approximation Monte Carlo, the sampler is equipped with an importance weight such that it can move across the energy barriers freely.

However, for many problems slow convergence is not due to the multimodality, but the curse of dimensionality; that is, the number of samples increases exponentially with dimension to maintain a given level of accuracy. For example, the witch's hat distribution (Matthews, 1993) has only a single mode, but the convergence time of the Gibbs sampler on it increases exponentially with dimension. For this kind of problems, although the convergence can be improved to some extent by the tempering-based or importance weight-based algorithms, the curse of dimensionality cannot be much reduced, as these samplers always work in the same sample space.

To eliminate the curse of dimensionality, Liang (2003) provides a sequential parallel tempering (SPT) algorithm, which makes use of the sequential structure of high dimensional systems. As an extension of parallel tempering, SPT works by simulating from a sequence of systems of different dimensions. The idea is to use the information provided by the simulation of low dimensional systems as a clue for the simulation of high dimensional systems.

5.6.1 Build-up Ladder Construction

A build-up ladder (Wong and Liang, 1997) comprises a sequence of systems of different dimensions. Consider m systems with density $f_i(x_i)$, where $x_i \in \mathcal{X}_i$ for $i = 1, \ldots, m$. Typically,

$$dim(\mathcal{X}_1) < dim(\mathcal{X}_2) < \cdots < dim(\mathcal{X}_m),$$

The principle of build-up ladder construction is to approximate the original system by a system with a reduced dimension, the reduced system is again approximated by a system with a further reduced dimension, until a system of a manageable dimension is reached; that is, the corresponding system can be easily sampled using a local updating algorithm, such as the MH algorithm or the Gibbs sampler. The solution of the reduced system is then extrapolated level by level until the target system is reached. For many problems, the build-up ladder can be constructed in a simple way. For example, for both the traveling salesman problem (Wong and Liang, 1997) and the phylogenetic tree reconstruction problem (Cheon and Liang, 2008), the build-up ladders are constructed by marginalization, this method being illustrated in Section 5.6.3.

Note that both the temperature ladder used in simulated tempering, parallel tempering, and EMC and the energy ladder used in the equi-energy sampler (Kou *et al.*, 2006) can be regarded as special kinds of build-up ladders. Along with the ladder, the complexity of the systems increases monotonically.

5.6.2 Sequential Parallel Tempering

As a member of population-based MCMC methods, SPT also works on a joint distribution of the form

$$f(\boldsymbol{x}) = \prod_{i=1}^{N} f_i(x_i),$$

where $\boldsymbol{x} = \{x_1, x_2, \cdots, x_N\}$, and x_i represents a sample from f_i. The simulation consists of two steps: local updating and between-level transitions. In the local updating step, each f_i is simulated by a local updating algorithm, such as the MH algorithm or the Gibbs sampler. The between-level transition involves two operations, namely, projection and extrapolation. Two different levels, say, i and j, are proposed to make the between-level transition. Without loss of generality, we assume that $\boldsymbol{X}_i \subset \boldsymbol{X}_j$. The transition is to extrapolate x_i ($\in \boldsymbol{X}_i$) to x'_j ($\in \boldsymbol{X}_j$), and simultaneously project x_j ($\in \boldsymbol{X}_j$) to x'_i ($\in \boldsymbol{X}_i$). The extrapolation and projection operators are chosen such that the pairwise move (x_i, x_j) to (x'_i, x'_j) is reversible. The transition is accepted with probability

$$\min\left\{1, \frac{f_i(x'_i)f_j(x'_j)}{f_i(x_i)f_j(x_j)} \frac{T_e(x'_i \to x_j)T_p(x'_j \to x_i)}{T_e(x_i \to x'_j)T_p(x_j \to x'_i)}\right\}, \tag{5.17}$$

where $T_e(\cdot \to \cdot)$ and $T_p(\cdot \to \cdot)$ denote the extrapolation and projection probabilities, respectively. For simplicity, the between-level transitions are only restricted to neighboring levels, that is, $|i - j| = 1$. In summary, each iteration of SPT proceeds as follows.

Sequential Parallel Tempering

1. *Local Updating*: Update each x_i independently by a local updating algorithm for a few steps.

2. *Between-level transition*: Try between-level transitions for N pairs of neighboring levels, with i being sampled uniformly on $\{1, 2, \cdots, N\}$ and $j = i \pm 1$ with probability $q_e(i, j)$, where $q_e(i, i+1) = q_e(i, i-1) = 0.5$ for $1 < i < N$ and $q_e(1, 2) = q_e(N, N-1) = 1$.

5.6.3 An Illustrative Example: the Witch's Hat Distribution

The witch's hat distribution has the density

$$\pi_d(x) = (1-\delta)\left(\frac{1}{\sqrt{2\pi}\sigma}\right)^d \exp\left\{-\frac{\sum_{i=1}^{d}(\tilde{x}_i - \theta_i)^2}{2\sigma^2}\right\} + \delta I_{x \in C},$$

where $x = (\tilde{x}_1, \ldots, \tilde{x}_d)$, d is dimension, C denotes the open d-dimensional hypercube $(0,1)^d$, and δ, σ and θ_i's are known constants. In the case of $d = 2$, the density shapes with a broad flat 'brim' and a high conical peak – the witch's hat distribution. This distribution is constructed by Matthews (1993) as a counter-example to the Gibbs sampler, on which the mixing time of the Gibbs sampler increases exponentially with dimension. The slow convergence can be understood intuitively as follows: as dimension increases, the volume of the peak decreases exponentially, thus, the time for the Gibbs sampler to locate the peak will increase exponentially. For example, when $d = 100$ and $\delta = 0.05$, 95% mass of the distribution is contained in a hypercube of volume $3.4e - 19$, and the remaining 5% mass is almost uniformly distributed in the part of C outside the hypercube. Hence, sampling from such a distribution is like searching for a needle in a haystack, and other advanced Gibbs techniques, such as grouping, collapsing (Liu *et al.*, 1994), and reparameterizations (Hills and Smith, 1992), will also fail, as they all try to sample from $\pi_d(\cdot)$ directly.

SPT works well for this example with the use of a build-up ladder. As a test, Liang (2003) applied to SPT to simulate from $\pi_d(x)$, with $d = 5, 6, \ldots, 15$, $\delta = 0.05$, $\sigma = 0.05$, and $\theta_1 = \cdots = \theta_d = 0.5$. For each value of d, the build-up ladder was constructed by setting $f_i(x) = \pi_i(x)$ for $i = 1, 2, \cdots, d$, where $\pi_i(\cdot)$ is the i-dimensional witch's hat distribution, which has the same parameter as $\pi_d(\cdot)$ except for the dimension. In the local updating step, each x_i is updated iteratively by the MH algorithm for i steps. At each MH step, one coordinate is randomly selected and then proposed to be replaced by a random draw from uniform(0,1). The between-level transition, say, between level i and level $i+1$, proceeds as follows:

1. Extrapolation: draw $u \sim unif(0,1)$ and set $x'_{i+1} = (x_i, u)$.

2. Projection: set x'_i to be the first i coordinates of x_{i+1}. The corresponding extrapolation and projection probabilities are $T_e(\cdot \to \cdot) = T_p(\cdot \to \cdot) = 1$.

For each value of d, SPT was run 10 times independently. Each run consisted of $2.01e + 6$ iterations, where the first 10 000 iterations were discarded for the burn-in process, and the remaining iterations were used for the inference. For $d = 10$, the acceptance rates of the local updating and between-level transitions are 0.2 and 0.17, respectively. It is interesting to point out that, for this example, the acceptance rate of between-level transitions is independent of levels. This suggests that the simulation can be extended to a very large value of d. To characterize the mixing of the simulation, the probability that the first coordinate of x_i lie in the interval $(\theta_1 - \sigma, \theta_1 + \sigma)$ was estimated for $i = 1, \ldots, d$. It is easy to calculate, under the above setting, the true value of the probability is $\alpha = 0.6536$. The standard deviations of the estimates were calculated using the batch mean method (Roberts, 1996) with a batch number of 50. The computational results are summarized in Table 5.4.

Table 5.4 Comparison of SPT and parallel tempering for the witch's hat distribution (Liang, 2003).

	SPT			PT		
d^1	time^2(s)	$\hat{\alpha}$	SD($\times 10^{-4}$)	time^2(s)	$\hat{\alpha}$	SD($\times 10^{-4}$)
5	72.5	0.6546	8.5	58.8	0.6529	9.7
6	94.9	0.6540	9.1	84.7	0.6530	10.5
7	118.6	0.6541	9.2	115.6	0.6525	11.2
8	145.8	0.6530	9.3	152.4	0.6530	13.2
9	174.6	0.6534	9.2	190.8	0.6538	15.8
10	206.0	0.6533	9.4	236.7	0.6517	20.5
11	239.3	0.6528	9.3	711.7	0.6531	17.7
12	275.5	0.6525	9.9	847.7	0.6530	21.3
13	312.9	0.6532	9.7	996.1	0.6527	33.8
14	353.7	0.6531	10.0	1156.4	0.6506	47.5
15	397.4	0.6532	10.4	1338.0	0.6450	84.5

[1]dimension; [2]the CPU time (in second) of a single run. The estimate ($\widehat{\alpha}$) and its standard deviation SD are calculated based on 10 independent runs.

For comparison, parallel tempering was also applied to this example with the following setting: the number of temperature levels $N = d$, the target temperature $T_N = 1$, and the highest temperature $T_1 = d$. The temperature T_1 is so high that the local updating sampler almost did a random walk at that level. The intermediate temperatures were set such that their inverses are equally spaced between $1/T_1$ and $1/T_N$. In the local updating step, each sample was updated iteratively by the MH algorithm for i steps as in SPT. Each run consisted of $2.01e + 6$ iterations for $d = 5, \ldots, 10$ and $5.01e + 6$ iterations for $d = 11, \ldots, 15$. In these runs, the first 10 000 iterations were discarded for the burn-in process, and the others were used for the estimation. The computational results are also summarized in Table 5.4. Figure 5.4 compares the estimated CPU time $T(A, d) = Time(A, d)/72.5 \times (SD(A, d)/8.5e - 4)^2$, which represents the CPU time needed on a computer for algorithm A and dimension d to attain an estimate of α with $SD = 8.5e - 4$. The plot shows that, for this example, SPT can significantly reduce the curse of dimensionality suffered by the Gibbs sampler, but parallel tempering cannot. A linear fitting on the logarithms of $T(\cdot, \cdot)$ and d shows that $T(SPT, d) \sim d^{1.76}$ and $T(PT, d) \sim d^{6.50}$, where PT represents parallel tempering. Later, SPT was applied to simulate from $\pi_{100}(x)$. With 13 730 seconds on the same computer, SPT got one estimate of α with $SD = 2.3e - 3$. Different temperature ladders were tried for parallel tempering, for example, $m \propto \sqrt{d}$, but the resulting CPU time scale (against dimensions) is about the same as that reported above.

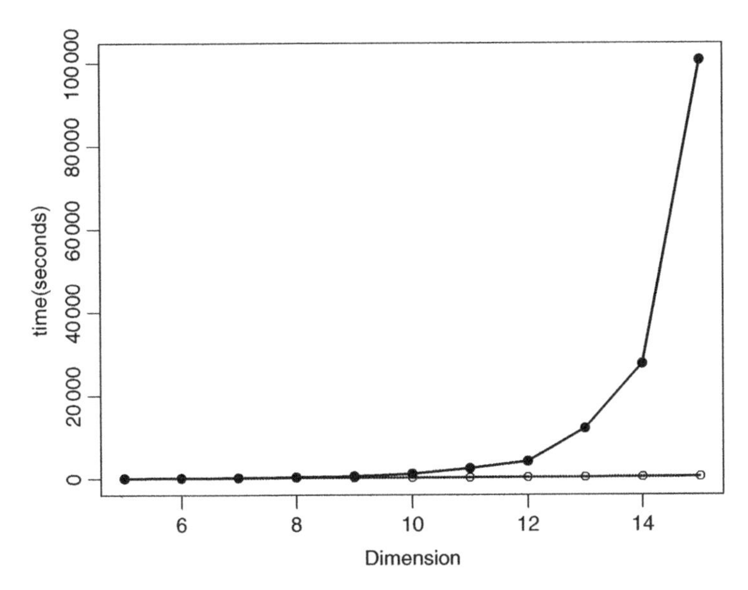

Figure 5.4 Estimated running times $T(SPT, d)$ (solid line) and $T(PT, d)$ (dotted line) for $d = 5, 6, \ldots, 15$ (Liang, 2003).

In addition to the witch's hat distribution, Liang (2003) applies the SPT sampler to an Ising model of size 128^2, for which a build-up ladder is also constructed using the technique of marginalization. A great success has been obtained by SPT for mixing the two opposite magnetization states of the Ising model, while it is known that parallel tempering fails to mix the two magnetization states in a single run, even with many temperature levels, due to the divergence of the specific heat when the temperature is near the critical point.

5.6.4 Discussion

Theoretically, SPT implements the distribution decomposition

$$f(\tilde{x}_1, \tilde{x}_2, \cdots, \tilde{x}_d) = f(\tilde{x}_1)f(\tilde{x}_2|\tilde{x}_1)\cdots f(\tilde{x}_i|\tilde{x}_1, \cdots, \tilde{x}_{i-1})\cdots f(\tilde{x}_d|\tilde{x}_1, \cdots, \tilde{x}_{d-1})$$

in sampling. It avoids directly sampling in the high dimensional space, and thus avoids the curse of dimensionality. The efficiency of SPT can be argued, based on the Rao-Blackwellization procedure (see, e.g., Liu, 2001). Suppose we are interested in estimating one integral $I = E_f h(x)$ with respect to a distribution $f(x)$. The simple sampling method is to first draw independent samples $x^{(1)}, \cdots, x^{(m)}$ from $f(x)$, and then estimate I by

$$\hat{I} = \frac{1}{m}\{h(x^{(1)} + \cdots + h(x^{(m)})\}.$$

If x can be decomposed into two parts $(\tilde{x}_1, \tilde{x}_2)$ and the conditional expectation $E[h(x)|\tilde{x}_2]$ can be carried out analytically, then I can be estimated alternatively by a mixture estimator

$$\tilde{I} = \frac{1}{m}\{E[h(x)|\tilde{x}_2^{(1)}] + \cdots + E[h(x)|\tilde{x}_2^{(m)}]\}.$$

It is easy to see that both $\hat{I}$ and $\tilde{I}$ are unbiased, but $\tilde{I}$ has a smaller variance because of the simple facts

$$E_f h(x) = E_f[E(h(x)|\tilde{x}_2)],$$

and

$$\text{var}\{h(x)\} = \text{var}\{E[h(x)|\tilde{x}_2]\} + E\{\text{var}[h(x)|\tilde{x}_2]\}.$$

The latter equation implies that

$$\text{var}(\hat{I}) = \frac{1}{m}\text{var}\{h(x)\} \geq \frac{1}{m}\text{var}\{E[h(x)|\tilde{x}_2]\} = \text{var}(\tilde{I}).$$

SPT implements a sequential Monte Carlo integration for $E[h(x)|\tilde{x}_d]$ along the build-up ladder, and is thus more efficient than the sampler which tries to sample from $f(\tilde{x}_1, \ldots, \tilde{x}_d)$ directly.

5.7 Equi-Energy Sampler

Let $f(x) \propto \exp\{-H(x)\}$, $x \in \mathcal{X}$, denote the target distribution, where $H(x)$ is called the energy function. Let $E_0 < E_1 < \cdots < E_N < \infty$ denote a ladder of energy levels, and let $T_0 < T_1 < \cdots < T_N < \infty$ denote a ladder of temperatures, where $E_0 = -\infty$ and $T_0 = 1$. Based on the two ladders, the equi-energy sampler (Kou *et al.*, 2006) defines the trial distributions of the population by

$$f_i(x) \propto \exp\{-\max(H(x), E_i)/T_i\}, \quad i = 0, 1, 2 \ldots, N. \tag{5.18}$$

Thus, $f_0(x)$ corresponds to the target distribution.

Define $D_i = \{x : H(x) \in [E_i, E_{i+1})\}$ for $i = 0, \ldots, N$, where $E_{N+1} = \infty$. Thus, $D_0, D_1, \ldots, D_N$ form a partition of the sample space. Each D_i is called an energy ring associated with the energy ladder. Let $I(x)$ denote the index of the energy ring that x belong to; that is, $I(x) = k$ if and only if $H(x) \in [E_k, E_{k+1})$. The equi-energy sampler consists of $N+1$ stages:

1. The equi-energy sampler begins with a MH chain $\{X_N\}$, which targets the highest order distribution $f_N(x)$. After a burn-in period of B steps, the sampler starts to group its samples into energy rings $\widehat{D}_i^{(N)}$, $i = 0, \ldots, N$, where the sample $X_{N,j}$ is grouped into $\widehat{D}_i^{(N)}$ if $I(X_{N,j}) = i$.

2. After a period of M steps, the equi-energy sampler starts to construct a parallel MH chain $\{X_{N-1}\}$, which targets the second highest order distribution $f_{N-1}(x)$, while keeping the chain $\{X_N\}$ running and updating. The chain $\{X_{N-1}\}$ is updated with two types of moves, local MH move and equi-energy jump, with respective probabilities $1 - p_{ee}$ and p_{ee}.

   - Local MH move: Update the chain $\{X_{N-1}\}$ with a single MH transition.
   - Equi-energy jump: Let $x_{N-1,t}$ denote the current state of the chain $\{X_{N-1}\}$, and let $k = I(x_{N-1,t})$. Choose y uniformly from the energy ring $hD_k^{(N)}$, and accept y as the next state of the chain $\{X_{N-1}\}$ with probability

$$\min\left\{1, \frac{f_{N-1}(y)f_N(x_{N-1,t})}{f_{N-1}(x_{N-1,t})f_N(y)}\right\}.$$

   Otherwise, set $x_{N-1,t+1} = x_{N-1,t}$.

   After a burn-in period of B steps, the equi-energy sampler starts to group its samples into different energy rings $\widehat{D}_i^{(N-1)}$, $i = 0, 1, \ldots, N$.

3. After a period of another M steps, the equi-energy sampler starts to construct another parallel MH chain $\{X_{N-2}\}$, which targets the third highest order distribution $f_{N-2}(x)$, while keeping the chains $\{X_N\}$ and $\{X_{N-1}\}$ running and updating. As for the chain $\{X_{N-1}\}$, the chain $\{X_{N-2}\}$ is updated with the local MH move and the equi-energy jump with respective probabilities $1 - p_{ee}$ and p_{ee}.

   ......

N+1. Repeat the above procedure until the level 0 has been reached. Simulation at level 0 results in the energy rings $\widehat{D}_i^{(0)}$, which deposit samples for the target distribution $f(x)$.

Under the assumptions (i) the highest order chain $\{X_N\}$ is irreducible and aperiodic, (ii) for $i = 0, 1, \ldots, N-1$, the MH transition kernel K_i of $\{X_i\}$ connects adjacent energy rings in the sense that for any j, there exist sets $A_1 \subset D_j$, $A_2 \subset D_j$, $B_1 \subset D_{j-1}$ and $B_2 \subset D_{j+1}$ with positive measure such that the transition probabilities $K_i(A_1, B_1) > 0$ and $K_i(A_2, B_2) > 0$, and (iii) the energy ring probabilities $p_{ij} = P_{f_i}(X \in D_j) > 0$ for all i and j, Kou *et al.*, (2006) showed that each chain $\{X_i\}$ is ergodic with f_i as its invariant distribution.

It is clear that the equi-energy sampler stems from parallel tempering, but is different from parallel tempering in two respects. The first difference is on the exchange operation, which is called the equi-energy jump in the equi-energy sampler: in parallel tempering, the current state of a lower order Markov chain is proposed to be exchanged with the current state of a higher order Markov chain; in the equi-energy sampler, the current state of a lower order Markov chain is proposed to be replaced by a past, energy-similar state of a higher order Markov chain. Since the two states have similar energy values, the equi-energy jump usually has a high acceptance rate, and this increases the interaction between different chains. The second difference is on distribution tempering: parallel tempering gradually tempers the target distribution; the equi-energy sampler tempers a sequence of low energy truncated distributions. It is apparent that simulation of the low energy truncated distribution reduces the chance of getting trapped in local energy minima. As demonstrated by Kou *et al.* (2006), the equi-energy sampler is more efficient than parallel tempering. For example, for the multimodal example studied in Section 5.5.4, the equi-energy sampler can sample all the modes within a reasonable CPU time, while parallel tempering fails to do so.

The equi-energy sampler contains several free parameters, the energy ladder, the temperature ladder, the equi-energy jump probability p_{ee}, and the proposal distributions used at different levels. Kou *et al.* (2006) suggest setting the energy levels by a geometric progression, and then to set the temperature levels accordingly such that $(E_{i+1} - E_i)/T_i \approx c$, where c denotes a constant. They find that the algorithm often works well with $c \in [1, 5]$. In addition, they find that the probability p_{ee} is not necessarily very large, and a value between 0.05 and 0.3 often works well for their examples. Since different distributions are simulated at different levels, different proposals should be used at different levels such that a moderate acceptance rate for the local MH moves can be achieved at each level.

5.8 Applications

Population-based MCMC methods have been used in various sampling problems. In this section, we consider three applications of the EMC algorithm, including Bayesian curve fitting, protein folding simulations, and nonlinear time series forecasting.

5.8.1 Bayesian Curve Fitting

The problem. Consider a nonparametric regression model, where observations (z_i, y_i), $i = 1, \cdots, n$ satisfy

$$y_i = g(z_i) + \epsilon_i, \quad i = 1, \cdots, n, \tag{5.19}$$

where $\epsilon_i \sim N(0, \sigma^2)$, $a \leq z_1 \leq \cdots \leq z_n \leq b$, $g \in C_2^m[a, b]$, and m denotes the order of continuity of g. The least squares spline is to approximate the unknown function $g(\cdot)$ using a regression polynomial function of the form

$$S(z) = \sum_{j=1}^{m} \alpha_j z^{j-1} + \sum_{j=1}^{p} \beta_j (z - t_j)_+^{m-1}, \tag{5.20}$$

for $z \in [a, b]$, where $d_+ = \max(0, d)$, t_j's ($j = 1, \cdots, p$) denote the knot locations, p is the number of knots, and α's and β's are the regression coefficients.

The curve fitting problem has been traditionally tackled by the kernel smoothing approaches, see, e.g., Kohn and Ansley (1987), Müller and Stadtmüller (1987), Friedman and Silverman (1989), and Hastie and Tibshirani (1990). Recently, Denison *et al.* (1998) proposed a Bayesian approach implemented with a hybrid sampler, and Liang *et al.* (2001) proposed a fully Bayesian approach implemented with the EMC algorithm, to tackle this problem.

The Model and Posterior. Liang *et al.* (2001) considered a modified least square spline,

$$S'(z) = \sum_{l=1}^{m} \beta_{l,0}(z - t_0)^{l-1} + \sum_{j=1}^{p} \sum_{l=m_0}^{m} \beta_{l,j}(z - t_j)_+^{l-1}, \tag{5.21}$$

where $t_0 = z_1$, and all possible knots are assumed to be on n regular points on $[a, b]$. The modification reduces the continuity constraints on (5.20), and hence, increases the flexibility of the spline (5.21). Replacing the function $g(x)$ in (5.19) by $S'(x)$, then

$$y_i = \sum_{l=1}^{m} \beta_{l,0}(z_i - t_0)^{l-1} + \sum_{j=1}^{p} \sum_{l=m_0}^{m} \beta_{l,j}(z_i - t_j)_+^{l-1} + \epsilon_i, \tag{5.22}$$

where $\epsilon_i \sim N(0, \sigma^2)$ for $i = 1, \cdots, n$. In the matrix-vector form, Equation (5.22) can be written as

$$\boldsymbol{Y} = \boldsymbol{Z}_p \boldsymbol{\beta} + \boldsymbol{\epsilon}, \tag{5.23}$$

where $\boldsymbol{Y}$ is an n-vector of observations, $\boldsymbol{\epsilon} \sim N(0, \sigma^2 I)$, $\boldsymbol{Z}_p = [\boldsymbol{1}, (\boldsymbol{z} - t_0)_+^1, \cdots, (\boldsymbol{z} - t_0)_+^{m-1}, (\boldsymbol{z} - t_1)_+^{m_0}, \cdots, (\boldsymbol{z} - t_p)_+^{m-1}]$, and $\boldsymbol{\beta} = (\beta_{1,0}, \cdots, \beta_{m-1,0}, \beta_{m_0,1}, \cdots, \beta_{m-1,p})$. The $\boldsymbol{\beta}$ includes $m + (m - m_0 + 1)p$ individual parameters, which may be larger than the number of observations.

Let ξ denote a n-binary vector, for which the elements of ones indicate the locations of the knots. Assuming the automatic priors for $\boldsymbol{\beta}$ and σ^2 (see Liang *et al.*, 2001 for details), and integrating out $\boldsymbol{\beta}$ and σ^2, the log-posterior of ξ can be expressed as

$$\begin{aligned} \log P(\xi^{(p)}|\boldsymbol{Y}) &= p \log\left(\frac{\mu}{1-\mu}\right) + \frac{n - wp - m}{2} \log 2 + \log \Gamma\left(\frac{n - wp - m}{2}\right) \\ &\quad - \frac{n - wp - m}{2} \log[\boldsymbol{Y}'\boldsymbol{Y} - \boldsymbol{Y}'\boldsymbol{X}_p(\boldsymbol{X}_p'\boldsymbol{X}_p)^{-1}\boldsymbol{X}_p'\boldsymbol{Y}], \end{aligned} \tag{5.24}$$

where $\xi^{(p)}$ denotes a spline model with p knots, $w = m - m_0 + 1$ denotes the number of extra terms needed to add after adding one more knot to the regression, and μ denotes a hyperparameter to be determined by the user. The value of μ reflects the user's prior knowledge of the smoothness of the underling function $g(\cdot)$. If one believes that it is very smooth, μ should be set to a small value. Otherwise, it should be larger. Some other priors can also be used here, e.g., the priors used in Raftery *et al.* (1997), Fernández *et al.* (2001), and Liang (2002a). Those priors also lead to a closed form of $P(\xi^{(p)}|\boldsymbol{Y})$.

Following standard Bayesian estimation theory, for a given set of samples $\xi_1, \ldots, \xi_M$ from the posterior (5.24), $g(z)$ can be estimated by

$$\widehat{g}(z) = \frac{1}{M}\sum_{i=1}^{M} \boldsymbol{Z}_i(\boldsymbol{Z}_i'\boldsymbol{Z}_i)^{-1}\boldsymbol{Z}_i'\boldsymbol{Y},$$

where $\boldsymbol{Z}_i$ denotes the design matrix corresponding to the model ξ_i.

EMC Algorithm. To sample from the posterior (5.24) using EMC, Liang *et al.* (2001) develop a new mutation operator, which incorporates the reversible jump moves, namely, the 'birth', 'death' and 'simultaneous' moves, proposed by Green (1995). Let S denote the set of knots included in the current model, let S^c denote the complementary set of S, and let $p = |S|$ denote the cardinality of the set S. In the 'birth' step, a knot is randomly selected from S^c and then proposed to add to the model. In the 'death' step, a knot is randomly selected from S and then proposed to delete from the model. The 'simultaneous' means that the 'birth' and 'death' moves are operated simultaneously. In this step, a knot, say t_c, is randomly selected from S; meanwhile, another knot, say t_c^*, is randomly selected from S^c, then t_c is proposed to be replaced by t_c^*. Let $pr(p, \text{birth})$, $pr(p, \text{death})$ and $pr(p, \text{simultaneous})$ denote the proposal probabilities of the three moves, respectively. Then the corresponding transition probability ratios are as follows. For the 'birth' step,

$$\frac{T(\boldsymbol{x}', \boldsymbol{x}')}{T(\boldsymbol{x}, \boldsymbol{x}')} = \frac{pr(p+1, \text{death})}{pr(p, \text{death})}\frac{k-p}{p+1};$$

for the 'death' step,

$$\frac{T(\boldsymbol{x}', \boldsymbol{x})}{T(\boldsymbol{x}, \boldsymbol{x}')} = \frac{pr(p-1, \text{birth})}{pr(p, \text{death})}\frac{p}{k-p+1},$$

and for the 'simultaneous' step, $T(\boldsymbol{x}', \boldsymbol{x})/T(\boldsymbol{x}, \boldsymbol{x}') = 1$, where $\boldsymbol{x}$ and $\boldsymbol{x}'$ denote, respectively, the current and proposed populations.

For this problem, only the 1-point crossover is used, where the two parental chromosomes are selected according to (5.9). The exchanged operation is standard; only the exchanges between the nearest neighboring levels are allowed.

Numerical Results. EMC was tested on the regression function

$$g(z) = 2\sin(4\pi z) - 6|z - 0.4|^{0.3} - 0.5\text{sign}(0.7 - z), \quad z \in [0, 1].$$

The observations were generated with $\sigma = 0.2$ on a grid of 1000 points equally spaced between 0 and 1. Figure 5.5(a) shows the simulated data and the true regression curve. This function has a narrow spike at 0.4 and a jump at 0.7. The approximation to the spike puts a great challenge on the existing curve fitting approaches. This example has been studied by several authors using regression splines, see, for example, Wang (1995) and Koo (1997).

Liang *et al.*, (2001) applied EMC to this example with $m_0 = 2$ and $m = 3$. Therefore, the resulting estimate $\widehat{g}(z)$ is a continuous quadratic piecewise polynomial. In EMC simulations, the parameters were set as follows: The population size $N = 20$, the highest temperature $t_1 = 5$, the lowest temperature $t_N = 1$, the intermediate temperatures are equally spaced between 5 and 1, and the mutation rate $q_m = 0.5$. Figure 5.5(b) shows the maximum a posteriori estimates of the knot locations and the regression curve obtained in one run of EMC with $\mu = 0.01$. Figure 5.6 (a) and (b) show two Bayesian

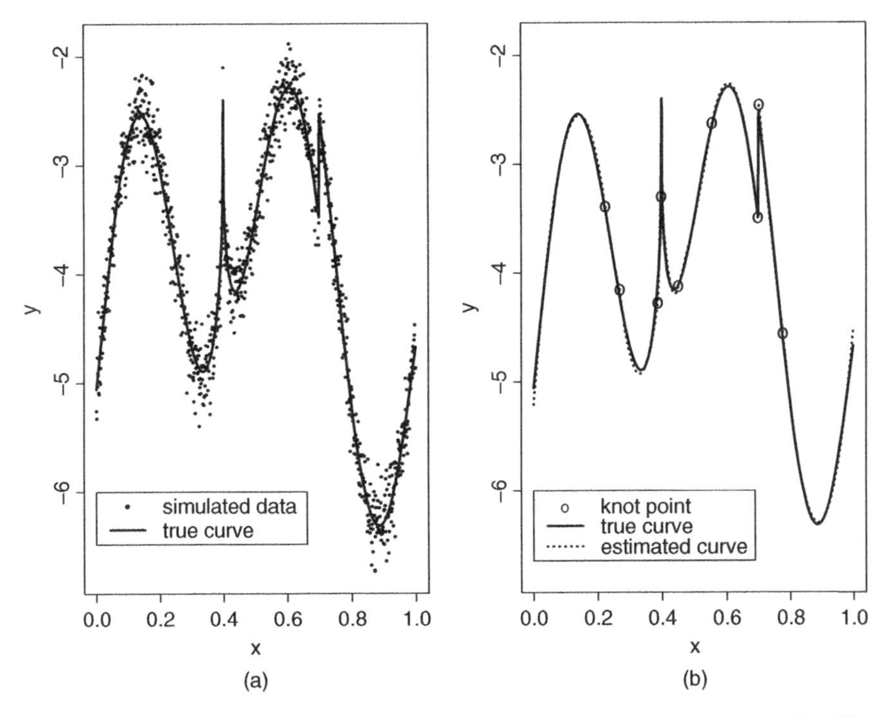

Figure 5.5 (a) The simulated data and the true regression curve. (b) The MAP estimates of the knot points and the regression curve obtained in one run with $\mu = 0.01$ (Liang *et al.*, 2001).

estimates of the regression curve obtained in two runs of EMC with $\mu = 0.01$ and $\mu = 0.015$, respectively. The plots show that the true regression curve can be well approximated using the EMC approach, including the spike and the jump.

For comparison, the hybrid sampler (Denison *et al.*, 1998) was also applied to this example with $m_0 = 2$ and $m = 3$. In the hybrid sampler, p is assumed to follow a priori a truncated Poisson distribution with parameter λ, and σ^2 is assumed to follow a priori the inverted Gamma distribution $IG(0.001, 0.001)$. One iteration of the hybrid sampler consists of two steps:

1. Updating the knots $t_1, \cdots, t_p$;

2. Updating σ^2.

Step 1 can be accomplished by reversible jump MCMC (Green, 1995), and Step 2 can be accomplished by the Gibbs sampler. Given the knots $t_1, \cdots, t_p$, the coefficients $\boldsymbol{\beta}$ are estimated by the standard least square estimates. Figure 5.6 (c) and (d) show two estimates produced by the hybrid sampler in two runs with $\lambda = 3$ and 5, respectively. The CPU time used by each run is

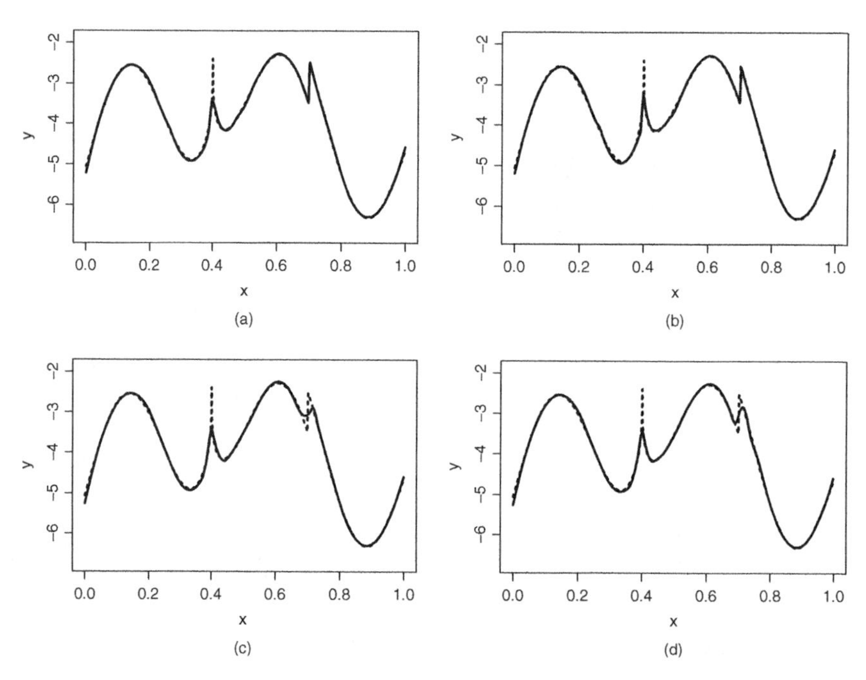

Figure 5.6 Comparison of EMC and the hybrid sampler. The solid line is the estimated curve, and dotted line is the true curve. (a) The automatic Bayesian approach with $\mu = 0.01$. (b) The automatic Bayesian approach with $\mu = 0.015$. (c) The hybrid sampler with $\lambda = 3$. (d) The hybrid sampler with $\lambda = 5$ (Liang *et al.*, 2001).

about the same as that used by EMC. It is easy to see that the jump at 0.7 is less well approximated by the hybrid sampler. The hybrid sampler was also run with $\lambda = 1, 2$ and 10, the results were all similar. This example indicates that EMC is superior to the hybrid sampler for Bayesian curve fitting.

5.8.2 Protein Folding Simulations: 2D HP Model

In recent years, the challenge of prediction of the native structure of a protein from its sequence have attracted a great deal of attention. The difficulty of this problem is that the energy landscape of the system consists of a multitude of local minima separated by high energy barriers. Traditional Monte Carlo and molecular dynamics simulations tend to get trapped in local energy minima, rendering a failure of identification of the native structure and biased estimates of the thermodynamic quantities that are of interest.

The 2D HP Model. In the 2D HP model, a simplification of the real protein model, the protein is composed of only two types of 'amino acids', hydrophobic (H for nonpolar) and hydrophilic (P for polar), and the sequence is folded on a two-dimensional square lattice. At each point, the 'polymer' chain can turn 90° left, right, or continue ahead. Only self-avoiding conformations are valid with energies $\epsilon_{HH} = -1$ and $\epsilon_{HP} = \epsilon_{PP} = 0$ for interactions between non covalently bound neighbors. Interest in this model derives from the fact that, although very simple, it exhibits many of the features of real protein folding (Lau and Dill, 1990; Crippen, 1991); low energy conformations are compact with a hydrophobic core, and the hydrophilic residues are forced to the surface. The 2D HP model has been used by chemists to evaluate new hypotheses of protein structure formation (Sali *et al.*, 1994). Its simplicity permits rigorous analysis of the efficiency of protein folding algorithms. EMC was tested on the 2D HP model by Liang and Wong (2001b), whose approach can be briefly described as follows.

EMC Algorithm. To apply EMC to the 2D HP model, each conformation of a protein is coded by a vector $x = (x^{(1)}, \cdots, x^{(d)})$, where $x^{(i)} \in \{0, 1, 2\}$, and each digit represents a torsion angle: 0, right; 1, continue and 2, left. The energy function of the conformation is denoted by $H(x)$ and the Boltzmann distribution is defined by

$$f(x) \propto \exp\{-H(x)/\tau\},$$

where τ is called the temperature, and it is not necessarily set to 1.

The mutation operators used for the model include a k-point mutation, a three-bead flip, a crankshaft move and a rigid rotation. Let x_m denote the individual selected to undergo a mutation. In the k-point mutation, one randomly choose k positions from x_m, and then replace their values by the ones sampled uniformly from the set $\{0, 1, 2\}$. Here, k is also a random variable,

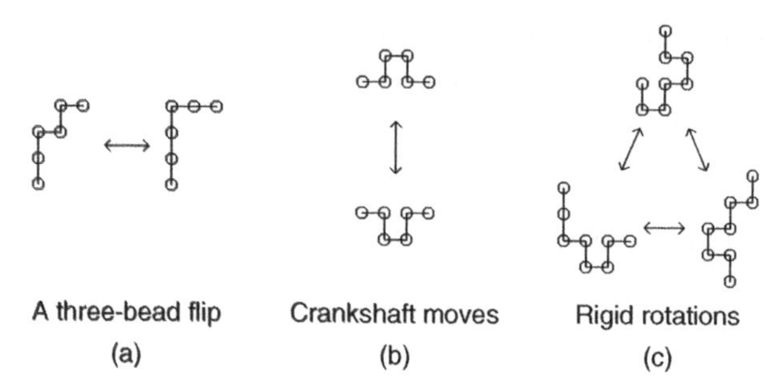

Figure 5.7 Mutation operators used in 2D HP models. The circle denotes a residue, hydrophobic or hydrophilic. (a) A three-bead flip. (b) A crankshaft move. (c) A rigid rotation (Liang and Wong, 2001b).

which is sampled uniformly from 1 to d. When $k = d$, the operator produces a completely new random conformation independent of x_m, and this effectively prevents the population from becoming homogeneous. The other mutation operators are, as depicted by Figure 5.7, identical to the local moves used in Chan and Dill (1993). For example, the two crankshaft structures can be coded by 2002 and 0220, respectively. In the crankshaft move, one first searches for all of the crankshaft structures from x_m and then randomly select one to mutate by reversing (i.e., change 0220 to 2002 and vice versa). The other operators are performed similarly.

For this model, the 1-point and 2-point crossovers were used, where the parental chromosomes were selected according to the probability (5.9). The exchange operation is standard: only the exchanges between the nearest neighboring levels are allowed.

Numerical Results. Liang and Wong (2001b) apply EMC to the sequences given in Unger and Moult (1993), also given in Appendix 5A of this chapter. The results are summarized in Table 5.5. For sequences of length 20, 24 and 25, EMC was run for 5000 iterations with the population size 100, the temperature ladder being equally spaced between 20 and 0.3, and the selection temperature 0.3. For the other sequences, EMC was run for 1000 iterations with the population size 500, the temperature ladder being equally spaced between 20 and 0.3, and the selection temperature 0.5. In all simulations, the mutation rate was set to 0.25. In the mutation step, all individuals of the current population are independently subject to an attempt of mutation, and the proportions of the k-point mutation, three-bead flip, crankshaft move and rigid rotation set to 1/2,1/6,1/6 are 1/6, respectively.

For comparison, Table 5.5 also presents the results from the genetic algorithm and Metropolis Monte Carlo for the sequences. Here, the comparison is based on the number of energy evaluations. This is reasonable, as, in protein

Table 5.5 Comparison of EMC with the genetic algorithm (GA) and Metropolis Monte Carlo (MC). (Liang and Wong, 2001b).

length[a]	ground energy	EMC	GA[b]	MC[c]
20	−9	−9 (9,374)	−9 (30,492)	−9 (292,443)
24	−9	−9 (6,929)	−9 (30,491)	−9 (2,492,221)
25	−8	−8 (7,202)	−8 (20,400)	−8 (2,694,572)
36	−14	−14 (12,447)	−14 (301,339)	−13 (6,557,189)
48	−23	−23 (165,791)	−22 (126,547)	−20 (9,201,755)
50	−21	−21 (74,613)	−21 (592,887)	−21 (15,151,203)
60	−36	−35 (203,729)	−34 (208,781)	−33 (8,262,338)
64	−42	−38 (889,036)	−37 (187,393)	−35 (7,848,952)

For each sequence, the three algorithms were all run 5 times independently, the lowest energy values achieved during the most efficient run were reported in the respective columns together with the number of valid conformations scanned before that value was found. (a) The length denotes the number of residues of the sequence. (b) The results of the genetic algorithm reported in Unger and Moult (1993). The GA was run with the population size 200 for 300 generations. (c) The results of Metropolis Monte Carlo reported in Unger and Moult (1993). Each run of MC consists of 50 000 000 steps.

folding simulations, the dominant factor of CPU cost is energy evaluation, which is performed once for each valid conformation. The comparison indicates that EMC is faster than the genetic algorithm and Metropolis Monte Carlo for locating the putative ground states of proteins. The computational amounts used by EMC were only about 10% to 30% of that used by the genetic algorithm, and 1% to 3% of that used by Metropolis Monte Carlo. For sequences of length 48, 60 and 64, EMC found the same energy states with smaller computational amounts than the genetic algorithm and Metropolis Monte Carlo. With slightly longer CPU times, EMC found some new lower energy states, as reported in Table 5.5.

Note that the pruned-enriched Rosenbluth method (PERM) (Bastolla *et al.*, 1998) generally works well for 2D HP models. A direct comparison of PERM with EMC, Metropolis Monte Carlo and the genetic algorithm is unfair, since the latter three algorithms only perform 'blind' searches over the whole conformation space. In contrast, PERM makes use of more information from the sequence when it folds a protein. PERM may build up its chain from any part of a sequence, for example, from a subsequence of hydrophobic residues. The authors argue for this idea that real proteins have folding nuclei and it should be most efficient to start from such a nucleus. Later, with the use of the secondary structure information, EMC found the putative ground states for the sequence of length 64 and another sequence of length 85, and it also found the putative ground state for the sequence of length 48 with

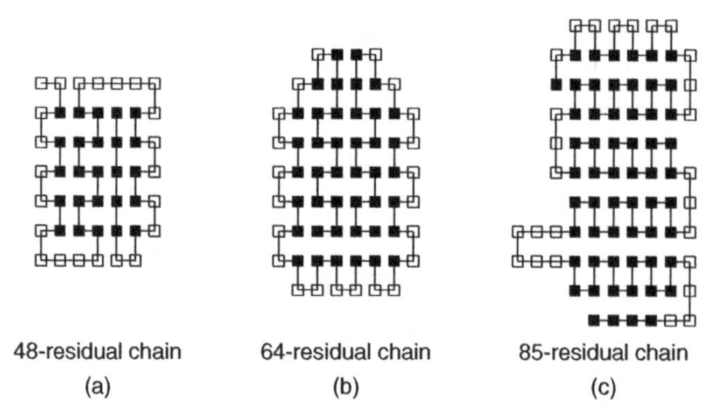

Figure 5.8 Putative ground conformations found by EMC with secondary structure constraints. (a) A putative ground conformation of energy −23 for the 48-mer sequence with the subsequence (residues 17–26) being constrained to the secondary structure. (b) A putative ground conformation of energy −42 for the 64-mer sequence with the subsequence (residues 1–10 and 55–64) being constrained to the secondary structures. (c) A putative ground conformation of energy −52 for the 85-mer sequence with the subsequence (residues 9–18, 27–36, 42–51, 57–66) being constrained to the secondary structures (Liang and Wong, 2001b).

a much smaller number of energy evaluations. Figure 5.8 shows the ground states found by EMC for these sequences.

5.8.3 Bayesian Neural Networks for Nonlinear Time Series Forecasting

Let y_t denote a univariate time series modeled by

$$y_t = f(\boldsymbol{z}_t) + \epsilon_t, \quad t = 1, 2, \ldots, n,$$

where $f(\cdot)$ is an unknown function, $\boldsymbol{z}_t = (y_{t-1}, \ldots, y_{t-p})$ is a vector of lagged values of y_t, and $\{\epsilon_t\}$ is iid noise with mean 0 and unknown finite variance σ^2. In the context of time series, ϵ_t is often called an innovation or disturbance, and p is called the order of the autoregressive model. Determining the function $f(\cdot)$ has been one of central topics in statistics for a long time.

Bayesian Neural Network Models. Bayesian neural networks (BNNs) have been applied to nonlinear time series analysis by Neal (1996), Penny and Roberts (2000), Liang and Wong (2001a), and Liang (2005a), among others. In Liang (2005a), an innovative BNN model is proposed, for which each connection is associated with an indicator function indicating the effectiveness of the connection. Liang's model can be written as

$$\widehat{f}(\boldsymbol{z}_t) = \alpha_0 I_{\alpha_0} + \sum_{i=1}^{p} y_{t-i}\alpha_i I_{\alpha_i} + \sum_{j=1}^{M} \beta_j I_{\beta_j} \psi\left(\gamma_{j0} I_{\gamma_{j0}} + \sum_{i=1}^{p} y_{t-i}\gamma_{ji} I_{\gamma_{ji}}\right), \tag{5.25}$$

where I_ζ is the indicator function associated with the connection ζ, $\psi(z) = \tanh(z)$ is called the activation function, and M is the maximum number of hidden units allowed by the model. The choice $\psi(z) = \tanh(z)$ ensures that the output of a hidden unit is 0 if all the connections to the hidden unit from input units have been eliminated, and thus the hidden unit can be removed from the network without any effect on the network outputs. This is not the case for the sigmoid function, which will return a constant of 0.5 if the input is zero; hence, extra work is needed to make the constant be absorbed by the bias term if one wants to remove the hidden unit from the network. Let Λ be a vector consisting of all indicators in (5.25), which specifies the structure of the network. Liang's model is different from other BNN models in two respects. First, it allows for an automatic selection of input variables. Second, its structure is usually sparse and its performance depends less on initial specifications of the input pattern and the number of hidden units. These two features mean Liang's model often has better prediction performance than other BNN models.

Let $\boldsymbol{\alpha} = (\alpha_0, \alpha_1, \cdots, \alpha_p)$, $\boldsymbol{\beta} = (\beta_1, \cdots, \beta_M)$, $\boldsymbol{\gamma}_j = (\gamma_{j0}, \cdots, \gamma_{jp})$, $\boldsymbol{\gamma} = (\boldsymbol{\gamma}_1, \cdots, \boldsymbol{\gamma}_M)$, and $\boldsymbol{\theta} = (\boldsymbol{\alpha}, \boldsymbol{\beta}, \boldsymbol{\gamma}, \sigma^2)$. Then the model (5.25) can be completely specified by the tuple $(\boldsymbol{\theta}, \Lambda)$. In the following, we will denote by $(\boldsymbol{\theta}, \Lambda)$ a BNN model, and denote by $\widehat{g}(\boldsymbol{z}_t, \boldsymbol{\theta}, \Lambda)$ a BNN estimator of $g(\boldsymbol{z}_t)$. To conduct a Bayesian analysis for (5.25), the following prior distributions are specified for the model and parameters: $\alpha_i \sim N(0, \sigma_\alpha^2)$ for $i = 0, \cdots, p$, $\beta_j \sim N(0, \sigma_\beta^2)$ for $j = 1, \cdots, M$, $\gamma_{ji} \sim N(0, \sigma_\gamma^2)$ for $j = 1, \cdots, M$ and $i = 0, \cdots, p$, $\sigma^2 \sim IG(\nu_1, \nu_2)$, where σ_α^2, σ_β^2, σ_γ^2 and λ are hyper-parameters to be specified by users. The total number of effective connections is $m = \sum_{i=0}^{p} I_{\alpha_i} + \sum_{j=1}^{M} I_{\beta_j}\delta(\sum_{i=0}^{p} I_{\gamma_{ji}}) + \sum_{j=1}^{M}\sum_{i=0}^{p} I_{\beta_j} I_{\gamma_{ji}}$, where $\delta(z)$ is 1 if $z > 0$ and 0 otherwise. The model Λ is subject to a prior probability specified by a truncated Poisson with rate λ; that is,

$$P(\Lambda) = \begin{cases} \frac{1}{Z}\frac{\lambda^m}{m!}, & m = 3, 4, \ldots, U, \\ 0, & \text{otherwise}, \end{cases}$$

where m is the number of effective connections in Λ, and $U = (M+1)(p+1) + M$ is the number of connections of the full model for which all $I_\zeta = 1$, and $Z = \sum_{\Lambda \in \Omega} \lambda^m/m!$. Here, Ω denotes the set of all possible models with $3 \leq m \leq U$. It is reasonable to set the minimum number of m to be 3, as neural networks are usually used for complex problems, and 3 has been small enough as the size of a limiting neural network. Furthermore, we assume that these prior distributions are independent a priori, and the innovation ϵ_t follows a normal distribution $N(0, \sigma^2)$. Given the above specifications, the log-posterior (up to an additive constant) of the model $(\boldsymbol{\theta}, \Lambda)$ is

$$\log P(\boldsymbol{\theta}, \Lambda|\mathcal{D}) = \text{Constant} - \left(\frac{n}{2} + \nu_1 + 1\right) \log \sigma^2 - \frac{\nu_2}{\sigma^2} - \frac{1}{2\sigma^2}\sum_{t=1}^{n}(y_t - \hat{f}(\boldsymbol{x}_t))^2$$
$$- \frac{1}{2}\sum_{i=0}^{p} I_{\alpha_i}\left(\log \sigma_\alpha^2 + \frac{\alpha_i^2}{\sigma_\alpha^2}\right) - \frac{1}{2}\sum_{j=1}^{M} I_{\beta_j}\delta\left(\sum_{i=0}^{p} I_{\gamma_{ji}}\right)\left(\log \sigma_\beta^2 + \frac{\beta_j^2}{\sigma_\beta^2}\right)$$
$$- \frac{1}{2}\sum_{j=1}^{M}\sum_{i=0}^{p} I_{\beta_j} I_{\gamma_{ji}}\left(\log \sigma_\gamma^2 + \frac{\gamma_{ji}^2}{\sigma_\gamma^2}\right) - \frac{m}{2}\log(2\pi) + m\log\lambda - \log(m!). \tag{5.26}$$

For data preparation and hyperparameter settings, Liang (2005a) made the following suggestions: To accommodate different scales of input and output variables, all input and output variables should be normalized before feeding to the neural network. For example, in Liang (2005a), all data were normalized by $(y_t - \bar{y})/S_y$, where $\bar{y}$ and S_y denote the mean and standard deviation of the training data, respectively. Since a neural network with highly varied connection weights has usually a poor generalization performance, σ_α^2, σ_β^2, and σ_γ^2 should be set to some moderate values to penalize high variability of the connection weights. For example, Liang (2005a) set $\sigma_\alpha^2 = \sigma_\beta^2 = \sigma_\gamma^2 = 5$ for all examples. This setting is expected to work for other problems. A non-informative prior can be put on σ^2, for example, setting $\nu_1 = \nu_2 = 0.05$ or 0.01. The values of M and λ, which work together controlling the network size, can be set through a cross-validation procedure, or tuned according to the suggestion of Weigend *et al.* (1990) that the number of connections of a neural network should be about one tenth of the number of training patterns.

Time Series Forecasting with BNN Models. Suppose that a series of samples, $(\boldsymbol{\theta}_1, \Lambda_1)$, $\cdots$, $(\boldsymbol{\theta}_{\mathcal{N}}, \Lambda_{\mathcal{N}})$, have been drawn from the posterior (5.26). Let $\widehat{y}_{t+1}$ denote the one step ahead forecast. One good choice of $\widehat{y}_{t+1}$ is

$$\widehat{y}_{t+1} = \frac{1}{\mathcal{N}}\sum_{i=1}^{\mathcal{N}} \widehat{f}(\boldsymbol{z}_{t+1}, \boldsymbol{\theta}_i, \Lambda_i), \tag{5.27}$$

which is unbiased and consistent by the standard theory of MCMC (Smith and Roberts, 1993). However, for the multi-step case, it is not easy to obtain an unbiased forecast for neural networks or any other nonlinear models. For the multi-step ahead forecast, Liang (2005a) propose the following procedure, which is unbiased and calculable with posterior samples.

1. For each sample $(\boldsymbol{\theta}_i, \Lambda_i)$, forecast y_{t+l} recursively for $l = 1, \ldots, h$ and repeat the process $\mathcal{M}$ times; that is, setting

$$\begin{aligned}
\widehat{y}_{t+1}^{(i,j)} &= \widehat{f}(y_{t-p+1}, \ldots, y_t, \boldsymbol{\theta}_i, \Lambda_i), \\
\widehat{y}_{t+2}^{(i,j)} &= \widehat{f}(y_{t-p+2}, \ldots, y_t, \widehat{y}_{t+1}^{(i,j)} + e_{t+1}^{(i,j)}, \boldsymbol{\theta}_i, \Lambda_i), \\
&\vdots \\
\widehat{y}_{t+h}^{(i,j)} &= \widehat{f}(y_{t-p+h}, \ldots, y_t, \widehat{y}_{t+1}^{(i,j)} + e_{t+1}^{(i,j)}, \ldots, \widehat{y}_{t+h-1}^{(i,j)} + e_{t+h-1}^{(i,j)}, \boldsymbol{\theta}_i, \Lambda_i).
\end{aligned}$$

where $(e_{t+1}^{(i,j)}, \ldots, e_{t+h-1}^{(i,j)})$, $i = 1, \ldots, \mathcal{N}, j = 1, \ldots, \mathcal{M}$, are future disturbances drawn from $N(0, \widehat{\sigma}_i^2)$, and $\widehat{\sigma}_i^2$ is an element of $\boldsymbol{\theta}_i$ and is itself an unbiased estimate of σ^2.

2. Average $\widehat{y}_{t+h}^{(i,j)}$, $i = 1, \ldots, \mathcal{N}$, $j = 1, \ldots, \mathcal{M}$ to get the forecast

$$\widehat{y}_{t+h}^{un} = \frac{1}{\mathcal{M}\mathcal{N}} \sum_{i=1}^{\mathcal{N}} \sum_{j=1}^{\mathcal{M}} \widehat{y}_{t+h}^{(i,j)}.$$

For a large value of $\mathcal{N}$, a reasonable choice of $\mathcal{M}$ is $\mathcal{M} = 1$ as used in this subsection.

Although $\hat{y}_{t+h}^{un}$ is unbiased, it has often a large variance due to the extra randomness introduced by simulating future disturbances. Liang (2005a) proposes an ad hoc forecast for y_{t+l} by setting the future disturbances to zero. That is,

1. For each sample $(\boldsymbol{\theta}_i, \Lambda_i)$, forecast y_{t+l} recursively for $l = 1, \ldots, h$ by the formula

$$\begin{aligned}
\widehat{y}_{t+1}^{(i)} &= \widehat{f}(y_{t-p+1}, \ldots, y_t, \boldsymbol{\theta}_i, \Lambda_i), \\
\widehat{y}_{t+2}^{(i)} &= \widehat{f}(y_{t-p+2}, \ldots, y_t, \widehat{y}_{t+1}^{(i)}, \boldsymbol{\theta}_i, \Lambda_i), \\
&\vdots \\
\widehat{y}_{t+h}^{(i)} &= \widehat{f}(y_{t-p+h}, \ldots, y_t, \widehat{y}_{t+1}^{(i)}, \ldots, \widehat{y}_{t+h-1}^{(i)} \boldsymbol{\theta}_i, \Lambda_i),
\end{aligned}$$

2. Average $\widehat{y}_{t+h}^{(i)}$, $i = 1, \ldots, \mathcal{N}$ to get the forecast

$$\widehat{y}_{t+h}^{ad} = \frac{1}{\mathcal{N}} \sum_{i=1}^{\mathcal{N}} \widehat{y}_{t+h}^{(i)}.$$

Although the forecast $\widehat{y}_{t+h}^{ad}$ is biased, it often has a smaller mean squared prediction error (MSPE) than $\widehat{y}_{t+h}^{un}$. The h-step MSPE for a general forecast

$\widehat{y}_{t+h}$ is defined as

$$\text{MSPE}_h = \sum_{T=t}^{n-h} [y_{T+h} - \widehat{y}_{T+h}]^2/(n-h-t+1),$$

which will be used to evaluate various forecasts in this subsection.

Wolfer Sunspot Numbers. The data set consists of annual sunspot numbers for the years 1700–1955 (Waldmeirer, 1961). It has been used by many authors to illustrate various time series models, for example, the ARIMA model (Box and Jenkins, 1970), SETAR model (Tong and Lim, 1980; Tong, 1990), bilinear model (Gabr and Subba Rao, 1981), and neural network model (Park *et al.*, 1996). Liang (2005a) tested the BNN model (5.25) and the EMC algorithm on this data.

As in Tong and Lim (1980) and Gabr and Subba Rao (1981), Liang (2005a) uses the first 221 observations for model building and the next 35 observations for forecasting. To determine the training pattern, Liang (2005a) suggests

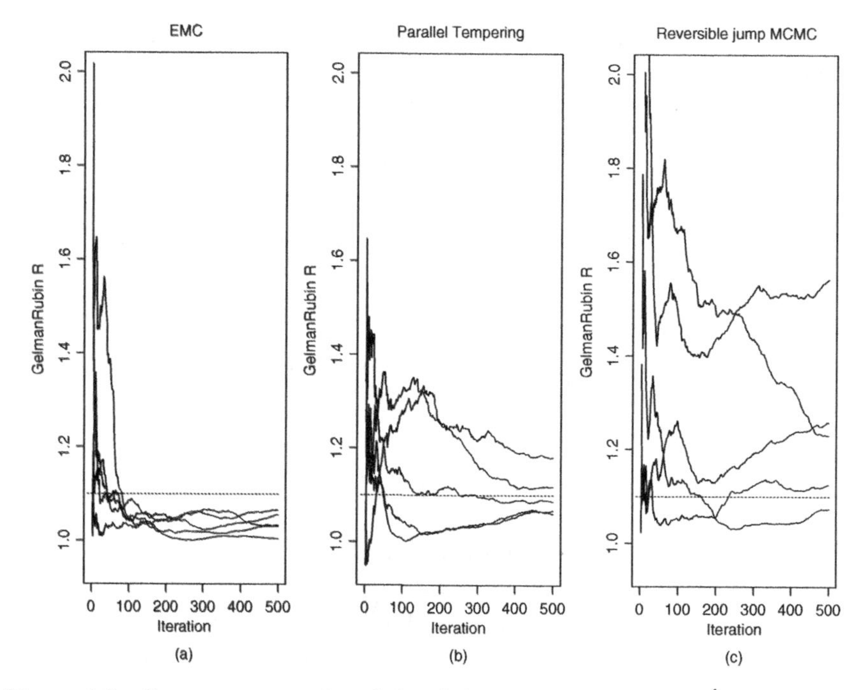

Figure 5.9 Convergence paths of the Gelman-Rubin statistic $\hat{R}$, where each path is computed with 5 independent runs. (a) EMC; (b) parallel tempering; (c) reversible jump MCMC (Liang, 2005a).

Table 5.6 Comparison of BNN and other time series models for the sunspot example (Recompiled from Liang, 2005a).

model	AR(9)	SETAR	bilinear	BNN^{ad}	BNN^{un}	TNN
MSE	199.27	153.71	124.33	$124.74_{1.30}$	$124.74_{1.30}$	$162.94_{1.13}$
size	10	19	11	$27.18_{0.26}$	$27.18_{0.26}$	76
MSPE_1	190.89	148.21	123.77	$142.03_{3.84}$	$142.87_{3.87}$	$171.99_{0.66}$
MSPE_2	414.83	383.90	337.54	$347.85_{9.03}$	$346.44_{8.81}$	$407.71_{1.08}$
MSPE_3	652.21	675.59	569.79	$509.60_{13.57}$	$513.99_{12.27}$	$607.18_{2.27}$
MSPE_4	725.85	773.51	659.05	$482.21_{13.04}$	$521.47_{10.40}$	$615.52_{4.37}$
MSPE_5	771.04	784.27	718.87	$470.35_{14.43}$	$561.94_{12.58}$	$617.24_{6.03}$
MSPE_6	–	–	–	$468.18_{13.88}$	$577.79_{16.71}$	$578.15_{6.92}$

The 'MSE' denotes the mean squared fitting error, 'size' denotes the number of parameters included in the model, 'MSPE_h' denotes the mean squared h-step ahead prediction error for $h = 1, \ldots, 6$. In the BNN^{ad}, BNN^{un}, TNN columns, the number and its subscript denote the averaged MSPE (over 10 independent runs) and the standard deviation of the average, respectively. BNN^{ad}: BNN models with the ad hoc predictor; BNN^{un}: BNN models with the unbiased predictor; TNN: traditional neural network models; The results of AR, SETAR, and bilinear models are taken from Gabr and Subba Rao (1981).

a partial autocorrelation function (PACF)-based approach. Based on that approach, $(y_{t-9}, \ldots, y_{t-1})$ were chosen as the input pattern for this data.

Liang (2005a) tested the efficiency of EMC for BNN training via comparisons with parallel tempering (Geyer, 1991) and reversible jump MCMC (Green, 1995). The three algorithms were all run for this data with the same CPU time, and Figure 5.9 was produced, where the output frequencies of the algorithms have been adjusted such that each produced the same number of samples in each run. Figure 5.9 shows 10 convergence paths of the Gelman-Rubin statistic $\hat{R}$ for each algorithm, where each path was computed based on 5 independent runs. It indicates that EMC can converge faster than parallel tempering and reversible jump MCMC.

Table 5.6 summarizes the prediction results obtained by EMC with 10 independent runs. For comparison, the prediction results from the AR, SETAR, bilinear, and traditional neural network (TNN) models are also shown in Table 5.6. BNN^{ad} is only inferior to the bilinear model for the one- and two-step ahead forecasts. As the forecast horizon extends, it outperforms all other models. This result is consistent with the finding of Hills *et al.* (1996) and Kang (1991) that neural networks generally perform better in the latter period of the forecast horizon. BNN^{un} performs equally well as BNN^{ad} for the short term forecasts ($h \leq 3$). As the forecast horizon extends, future disturbances cause its performance to deteriorate.

Exercises

5.1 Implement the CGMC, EMC and parallel tempering algorithms for the following 5-dimensional mixture normal distribution

$$\frac{1}{3}N_5(\mathbf{0}, I_5) + \frac{2}{3}N_5(\mathbf{5}, I_5),$$

where $\mathbf{0} = (0, \ldots, 0)'$, $\mathbf{5} = (5, \ldots, 5)'$, and I_5 denotes a 5×5 identity matrix.

5.2 Use both EMC and the equi-energy sampler to simulate the multimodal distribution studied in Section 5.5.4 and compare their efficiency.

5.3 Implement the equi-energy sampler for a population constructed as in parallel tempering, and discuss why it is important to use low energy truncated distributions for the population construction.

5.4 Discuss how to make mutation and crossover operations when using EMC to simulate from the posterior (5.26) of the BNN model.

5.5 Consider how to apply sequential parallel tempering to simulate an Ising model at a subcritical temperature:

(a) Discuss how to design the build-up ladder for the Ising model.

(b) Discuss how to make the extrapolation and projection operations along the ladder constructed in part (a).

5.6 Use the parallel tempering to simulate an Ising model at a subcritical temperature, and discuss why it tends to fail in mixing the two ground states of the model when the size of the model is large.

5.7 Show that the CGMC, EMC, and equi-energy sampler are all proper; that is, they can be used to generate correct samples from a target distribution.

5.8 Discuss how to construct the build-up ladder for the traveling salesman problem, and how to use EMC to simulate from the Boltzmann distribution (see Section 6.1.3) defined for the traveling salesman problem.

Appendix 5A: Protein Sequences for 2D HP Models

(20) HPHPPHHPHPPHPHHPPHPH;

(24) HHPPHPPHPPHPPHPPHPPHPPHH;

(25) PPHPPHHPPPPHHPPPPHHPPPPHH;

(36) PPPHHPPHHPPPPPHHHHHHHPPHHPPPPHHPPHPP;

(48) PPHPPHHPPHHPPPPPHHHHHHHHHHPPPPPPHHPPHHPPHPPHHHHH;

(50) HHPHPHPHPHHHHPHPPPHPPPHPPPPHPPPHPPPHPHHHHPHPHPHPHH;

(60) PPHHHPHHHHHHHHPPPHHHHHHHHHHPHPPPHHHHHHHHHHHHPPPPHHHHHHP
HHPHP;

(64) HHHHHHHHHHHHPHPHPPHHPPHHPPHPPHHPPHHPPHPPHHPPHHPPHPHPHHH
HHHHHHHHH;

(85) HHHHPPPPHHHHHHHHHHHHPPPPPPHHHHHHHHHHHHPPPHHHHHHHHHHHHPP
PHHHHHHHHHHHH HPPPHPPHHPPHHPPHPH.

Chapter 6

Dynamic Weighting

The Metropolis-Hastings algorithm (Metropolis *et al.*, 1953; Hastings, 1970) has a stringent requirement for the detailed balance condition. To move across an energy barrier, the expected waiting time is roughly exponential to the energy difference. Hence, the algorithm suffers from a *waiting time dilemma*: either to wait forever in a deep local energy minimum or to have an incorrect equilibrium distribution, in simulations from a complex system for which the energy landscape is rugged.

Wong and Liang (1997) proposed a way out of the waiting time dilemma, which can be described loosely as follows: If necessary, the system may make a transition against a steep probability barrier without a proportionally long waiting time. To account for the bias introduced thereby, an importance weight is computed and recorded along with the sampled values. This transition rule does not satisfy the detailed balance condition any more, but it satisfies what is called *invariance with respect to importance weights* (IWIW). At equilibrium, Monte Carlo approximations to integrals are obtained by the importance-weighted average of the sampled values, rather than the simple average as in the Metropolis-Hastings algorithm.

6.1 Dynamic Weighting

In this section, we describe the IWIW principle and the dynamic weighting algorithm proposed by Wong and Liang (1997).

6.1.1 The IWIW Principle

In dynamic weighting, the state of the Markov chain is augmented by an importance weight to (x, w), where the weight w carries the information of

Advanced Markov Chain Monte Carlo Methods: Learning from Past Samples
Faming Liang, Chuanhai Liu and Raymond J. Carroll

the past samples and can help the system escape from local-traps. Let (x_t, w_t) denote the current state of the Markov chain, a dynamic weighting transition involves the following steps:

1. Draw y from a proposal function $T(x_t, y)$.

2. Compute the dynamic ratio
$$r_d = w_t \frac{f(y)T(y, x_t)}{f(x_t)T(x_t, y)}.$$

3. Let θ_t be a nonnegative number, which can be set as a function of (x_t, w_t). With probability $a = r_d/(\theta_t + r_d)$, set $x_{t+1} = y$ and $w_{t+1} = r_d/a$; otherwise set $x_{t+1} = x_t$ and $w_{t+1} = w_t/(1-a)$.

This transition is called the R-type move in Wong and Liang (1997). It does not satisfy the detailed balance condition, but is *invariant with respect to the importance weight* (IWIW); that is, if

$$\int w_t g(x_t, w_t) dw_t \propto f(x_t) \tag{6.1}$$

holds, then after one step of transition,

$$\int w_{t+1} g(x_{t+1}, w_{t+1}) dw_{t+1} \propto f(x_{t+1}) \tag{6.2}$$

also holds, where $g(x, w)$ denotes the joint density of (x, w). The IWIW property can be shown as follows. Let $x = x_t$, $w = w_t$, $x' = x_{t+1}$ and $w' = w_{t+1}$. Then

$$\begin{aligned}
&\int_0^\infty w' g(x', w') dw' \\
&= \int_{\mathcal{X}} \int_0^\infty [\theta_t + r_d(x, x', w)] g(x, w) T(x, x') \frac{r_d(x, x', w)}{\theta_t + r_d(x, x', w)} dw dx \\
&\quad + \int_{\mathcal{X}} \int_0^\infty \frac{w[\theta_t + r_d(x', z, w)]}{\theta_t} g(x', w) T(x', z) \frac{\theta_t}{\theta_t + r_d(x', z, w)} dw dz \\
&= \int_{\mathcal{X}} \int_0^\infty w g(x, w) \frac{f(x')T(x', x)}{f(x)} dw dx + \int_{\mathcal{X}} \int_0^\infty w g(x', w) T(x', z) dw dz \\
&\propto f(x') \int_{\mathcal{X}} T(x', x) dx + f(x') \\
&= 2f(x').
\end{aligned}$$

Hence, given a sequence of dynamic weighting samples $(x_1, w_1), (x_2, w_2), \ldots, (x_n, w_n)$, the weighted average [of a state function $h(x)$ over the sample]

$$\hat{\mu} = \sum_{i=1}^n w_i h(x_i) / \sum_{i=1}^n w_i \tag{6.3}$$

will converge to $E_f h(x)$, the expectation of $h(x)$ with respect to the target distribution $f(x)$.

The merit of dynamic weighting is as follows: If one trial is rejected, then the dynamic weight will be self-adjusted to a larger value by dividing the rejection probability of that trial, rendering a smaller total rejection probability in the next trial. Using importance weights provides a means for dynamic weighting to make transitions that are not allowed by the standard MH rule, and thus can traverse the energy landscape of the system more freely. But this advantage comes at a price: the importance weights have an infinite expectation, and the estimate (6.3) is of high variability and converges to the true values very slowly, seemingly at a rate of $\log(n)$ (Liu *et al.*, 2001). In short, the infinite waiting time in the standard MH process now manifests itself as an infinite weight quantity in the dynamic weighting process.

To achieve a stable estimate of $\mu = E_f h(x)$, the operations of stratification and trimming on the importance weights are usually performed before computing the weighted estimate of μ. First, the samples are stratified according to the value of the function $h(x)$. The strata are of roughly the same size and within each stratum the variation of $h(x)$ is small. The highest $k\%$ (usually $k = 1$) weights of each stratum are then truncated to the $(100-k)\%$ percentile of the weights within that stratum. In our experience, these operations induce negligible bias in the weighted estimate but can reduce its variance substantially. Note that the trimmed weights depend on the function of interest.

It is interesting to point out that the usual MH transition can be regarded as a special type of IWIW transition: If we apply a MH transition to x and leave w unchanged, then the result satisfies IWIW. This can be shown as follows:

$$
\begin{aligned}
&\int w' g(x', w') dw' = \int w g(x, w) K(x \to x') dw dx \\
&\propto \int f(x) K(x \to x') dx = \int f(x') K(x' \to x) dx \\
&= f(x'),
\end{aligned}
$$

where $K(\cdot \to \cdot)$ denotes a MH transition kernel with $f(x)$ as its invariant distribution. Therefore, correctly weighted distributions will remain when dynamic weighting transitions and MH transitions are alternated in the same run of the Markov chain. This observation leads directly to the tempering dynamic weighting algorithm (Liang and Wong, 1999).

6.1.2 Tempering Dynamic Weighting Algorithm

As discussed in Section 4.2, simulated tempering (Marinari and Parisi, 1992) often suffers from difficulty in transition between different temperature levels. To alleviate this difficulty, one has to employ many temperature levels in the augmented system, which adversely affect the efficiency of the algorithm.

Alternatively, this difficulty can be alleviated by the dynamic weighting rule as prescribed by Liang and Wong (1999).

The tempering dynamic weighting (TDW) algorithm (Liang and Wong, 1999) is essentially the same as the simulated tempering algorithm except that a dynamic weight is now associated with the configuration (x, i) and the dynamic weighting rule is used to guide the transitions between adjacent temperature levels. Let $f_i(x)$ denote the trial distribution at level i, $i = 1, \ldots, N$. Let $0 < \alpha < 1$ be specified in advance and let (x_t, i_t, w_t) denote the current state of the Markov chain. One iteration of the TDW algorithm consists of the following steps:

Tempering Dynamic Weighting Algorithm

1. Draw U from the uniform distribution $U[0, 1]$.

2. If $U \leq \alpha$, set $i_{t+1} = i_t$ and $w_{t+1} = w_t$ and simulate x_{t+1} from $f_{i_t}(x)$ via one or several MH updates.

3. If $U > \alpha$, set $x_{t+1} = x_t$ and propose a level transition, $i_t \to i'$, from a transition function $q(i_t, i')$. Conduct a dynamic weighting transition to update (i_t, w_t):

   - Compute the dynamic weighting ratio

   $$r_d = w_t \frac{c_i f_{i'}(x_t) q(i', i_t)}{c_{i'} f_i(x_t) q(i_t, i')},$$

   where c_i denotes the pseudo-normalizing constant of $f_i(x)$.

   - Accept the transition with probability $a = r_d/(\theta_t + r_d)$, where θ_t can be chosen as a function of (i_t, w_t). If it is accepted, set $i_{t+1} = i'$ and $w_{t+1} = r_d/a$; otherwise, set $i_{t+1} = i_t$ and $w_{t+1} = w_t/(1 - a)$.

Ising Model Simulation at Sub-Critical Temperature. Consider a 2-D Ising model with the Boltzmann density

$$f(\boldsymbol{x}) = \frac{1}{Z(K)} \exp\left\{ K \sum_{i \sim j} x_i x_j \right\}, \tag{6.4}$$

where the spins $x_i = \pm 1$, $i \sim j$ denotes the nearest neighbors on the lattice, $Z(K)$ is the partition function, and K is the inverse temperature. When the temperature is at or below the critical point ($K = 0.4407$), the model has two oppositely magnetized states separated by a very steep energy barrier. Because of its symmetry, the Ising model is more amenable to theoretical analysis. However, for a sampling algorithm that does not rely on the symmetry of the model, such as simulated tempering, parallel tempering and dynamic weighting, this is very hard, especially when the size of the model is large. In the literature, the Ising model has long served as a benchmark example for testing efficiency of new Monte Carlo algorithms.

Liang and Wong (1999) performed TDW simulations on the lattices of size 32^2, 64^2 and 128^2 using 6, 11, and 21 temperature levels (with the values of K being equally spaced between 0.4 and 0.5), respectively. At the same temperature level, the Gibbs sampler (Geman and Geman, 1984) is used to generate new configurations; meanwhile, the weights were left unchanged. The dynamic weighting rule is only used to govern transitions between levels. After each sweep of Gibbs updates, it is randomly proposed to move to an adjacent temperature level with equal probability. The parameter θ_t is set to 1 if $w_t < 10^6$ and 0 otherwise. Five independent runs were performed for the model. In each run, the simulation continues until 10 000 configurations are obtained at the final temperature level. For the model of size 128^2, the average number of sweeps in each run is 776 547.

For the Ising model, one quantity of interest is the spontaneous magnetization, which is defined by $M = \sum_i x_i/d^2$, where d is the linear size of the lattice. Figure 6.1 plots the spontaneous magnetization obtained at the level $K = 0.5$ in a run for the 128^2 lattice. Clearly, the TDW algorithm has succeeded in crossing the very steep barrier separating the two ground states and the system is able to traverse freely between the two energy wells.

Figure 6.2 plots the expectation of $|M|$, the absolute spontaneous magnetization, at various values of K for the lattices of size 32^2, 64^2 and 128^2. The smooth curve is the cerebrated infinite lattice result found by Onsager (1949), and proved by Yang (1952). The expected value of $|M|$ was calculated

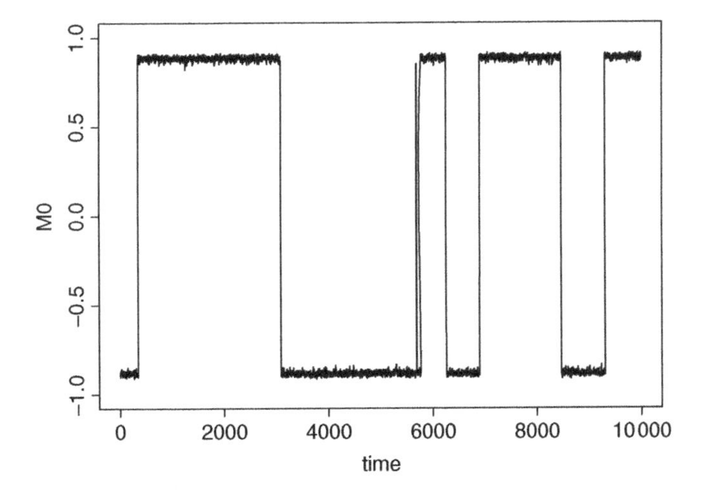

Figure 6.1 Spontaneous magnetization against iterations for a lattice of size 128^2 at the level $K = 0.5$ (Liang and Wong, 1999).

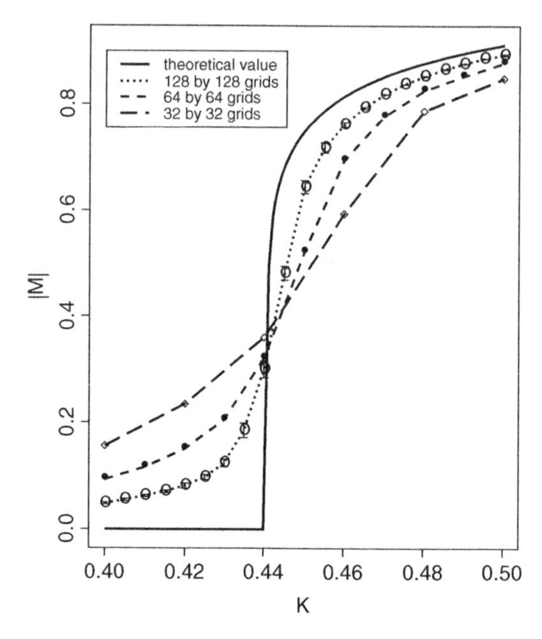

Figure 6.2 Expectation of the absolute spontaneous magnetization against the inverse temperature K for the lattices of size 32^2, 64^2 and 128^2. The points are averages over 5 independent runs. For clarity, error bars are plotted only for the 128^2 lattice. The smooth curve corresponds to the theoretical result of infinite lattice (Liang and Wong, 1999).

by the method of stratified truncation in the following procedure: For each run, the initial 500 samples were discarded for the burn-in process, the remaining samples were then stratified into 10 strata according to the value of $|M|$ with the $(j \times 10)$ percentiles as strata boundaries, $j = 1, 2, \cdots, 9$, and the highest 1% weights of each stratum were truncated to the 99% percentile of the weights of that stratum.

Figure 6.2 shows that the phase transition phenomenon of the 2-D Ising model has been captured by the simulations with increasing accuracy as the lattice size grows. The critical point can be estimated quite well from the crossing of the curves for the 64^2 and 128^2 cases. We note that a major strength of the TDW algorithm is that it can yield in a single run accurate estimates over the entire temperature range extending well below the critical point.

For comparison, Liang and Wong (1999) also applied simulated tempering to the Ising model on the same range $[0.4, 0.5]$ of K. With 51 equally spaced levels, simulated tempering performs well for the 32^2 lattice. For the 64^2 and 128^2 lattices, simulated tempering was run until 50 000 configurations generated at the final level, but failed to sample the two energy wells in the same runs. The problem remains even with more temperature levels. Liang and Wong (1999) reported that they tried 101 levels for the 64^2 lattice, and 101 and 201 levels for the 128^2 lattice, but failed to sample two energy wells in a single run.

In another comparison, Liang and Wong (1999) applied parallel tempering (Geyer, 1991) to the same Ising models. With 51 equally spaced temperature levels, parallel tempering works well for the 32^2 and 64^2 lattices. But for the 128^2 lattice, it failed to sample both energy wells in a single run even with 1 000 000 Gibbs sweeps and 201 temperature levels. The reason for this poor behavior of simulated tempering and parallel tempering is that the Boltzmann distribution (6.4) varies drastically with K. To satisfy the algorithms' requirement that successive tempering distributions should have considerable overlap, a huge number of temperature levels may be needed. See Section 4.2 for discussion on this issue. Since the traversal time of a temperature ladder grows at least as the square of the number of temperature levels, this may lead to an extremely long ergodicity time for a large lattice.

6.1.3 Dynamic Weighting in Optimization

In this section, we use the traveling salesman problem (TSP) (Reinelt, 1994) to illustrate the use of dynamic weighting in optimization. Let n be the number of cities in an area and let d_{ij} denote the distance between city i and city j. The TSP is to find a permutation x of the cities such that the tour length

$$H(x) = \sum_{i=1}^{n-1} d_{x(i),x(i+1)} + d_{x(n),x(1)}, \tag{6.5}$$

is minimized. It is known that the TSP is a NP-complete problem.

To apply dynamic weighting to the TSP, Wong and Liang (1997) first created a sequential build-up order of the cities. Let V denote the set of cities, let A denote the set of cities that have been ordered, and let $A^c = V \setminus A$ denote the set of cities not yet ordered. Then the cities can be ordered as follows:

- Randomly select a city from V.
- Repeat steps (1) and (2) until A^c is empty:

    1. set $k = \arg\max_{i \in A^c} \min_{j \in A} d_{ij}$;
    2. set $A = A \cup \{k\}$ and $A^c = A^c \setminus \{k\}$.

This procedure ensures that each time the city added into A is the one having the maximum separation from the set of already ordered cities.

Then a sequence of TSPs of increasing complexity were considered. At the lowest level the TSP only includes first m_0 (e.g., 10 or 15) cities in the build-up order. At the next lowest level they added a block of the next m (e.g. $m = 5$)

cities to get a slightly larger TSP. The block size m depends on the number of levels one wants to employ. In this way a ladder of complexity consisting of TSPs can be built up on increasing subsets of the cities. Typically 15–25 levels are used, and usually the size s of the TSP at the highest complexity level is still much smaller than the size n of the original TSP. Wong and Liang (1997) suggest choosing s between $0.2n$ to $0.4n$, as such a s-city problem can be a good approximation to the n-city problem, in the sense that a good tour for the former can be an outline of good complete tours for the latter.

Given the complexity ladder of cities, the minimum tour search consists of two steps. First, the s-city tours are sampled using dynamic weighting from the Boltzmann distribution

$$f(x) \propto \exp\left(-H(x)\right), \tag{6.6}$$

where $H(x)$ is as specified in (6.5). For each s-city tour generated by dynamic weighting, the branch and bound algorithm is used to insert the remaining cities, one by one according to the build-up order, by searching over a series of small (15 cities) path spaces. In the dynamic weighting step, to add a city to a tour, the 15 (say) nearest neighbors on the tour to the city are considered and a new sub-tour through these 16 cities is sampled. To add a block of m cities to the tour, the cities are added one by one as described, and an acceptance/rejection decision is made after all m cities are added with the importance weight being updated accordingly. Deletion of cities from a tour follows a similar procedure.

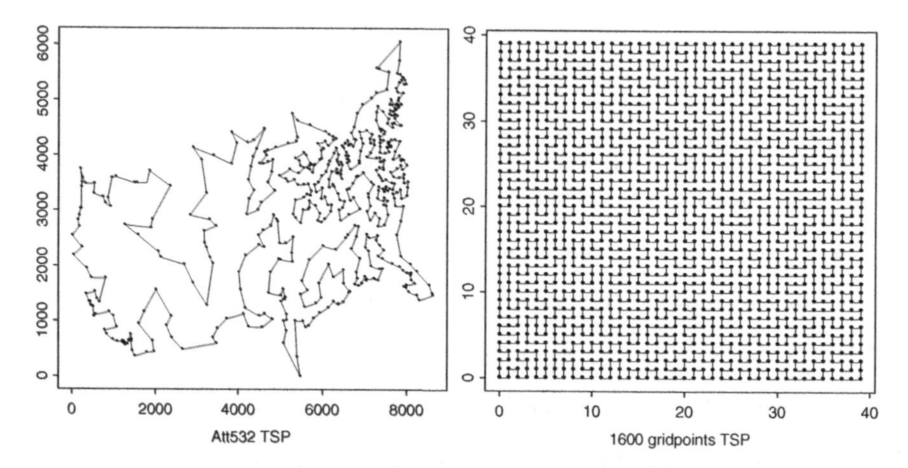

Figure 6.3 (a) The best tour found by dynamic weighting for a 532-city TSP. The tour length is 27 744 compared to the exact minimum of 27 686. (b) One of the exactly optimal tour found by dynamic weighting for a 1600-city TSP (Wong and Liang, 1997).

Figure 6.3 shows the best tours found by dynamic weighting for two large TSPs: att532 and grid1600 (Wong and Liang, 1997). The computation time was 4 h and 10 h respectively on an Ultrasparc I. For att532, the tour produced by dynamic weighting has an excess of 0.2% over the exact optimum. This is much better than tours from other heuristic search methods, except for a special version of the genetic algorithm, for which the powerful Lin-Kernighan move (Lin and Kernighan, 1973) was used for mutation. The tour is the exact optimum for the grid1600 problem, which has not been solved by other heuristic search methods.

6.2 Dynamically Weighted Importance Sampling

In this section we describe a population extension of dynamic weighting – dynamically weighted importance sampling (Liang, 2002b), which has the dynamic importance weights controlled to a desirable range while maintaining the IWIW property of resulting samples.

6.2.1 The Basic Idea

Because the importance weight in dynamic weighting is of high variability, achieving a stable estimate requires techniques such as stratification and truncation (Liu *et al.*, 2001). However, any attempt to shorten the tails of the weight distribution may lead to a biased estimate. To overcome this difficulty, Liang (2002b) proposed a population version of dynamic weighting, the so-called dynamically weighted importance sampling (DWIS) algorithm, which has the importance weight successfully controlled to a desired range, while keeping the estimate unbiased. In DWIS, the state space of the Markov chain is augmented to a population of size N, denoted by $(\boldsymbol{x}, \boldsymbol{w}) = \{x_1, w_1; \ldots; x_N, w_N\}$. With a slight abuse of notation, (x_i, w_i) is called an individual state of the population. Given the current population $(\boldsymbol{x}_t, \boldsymbol{w}_t)$, one iteration of DWIS (as illustrated by Figure 6.4) involves two steps:

1. *Dynamic weighting*: each individual state of the current population is updated via one dynamic weighting transition step to form a new population.

2. *Population control*: duplicate the samples with large weights and discard the samples with small weights. The bias induced thereby is counter-balanced by giving different weights to the new samples produced.

These two steps ensure that DWIS can move across energy barriers like dynamic weighting, but the weights are well controlled and have a finite expectation, and the resulting estimate can converge much faster than that of dynamic weighting.

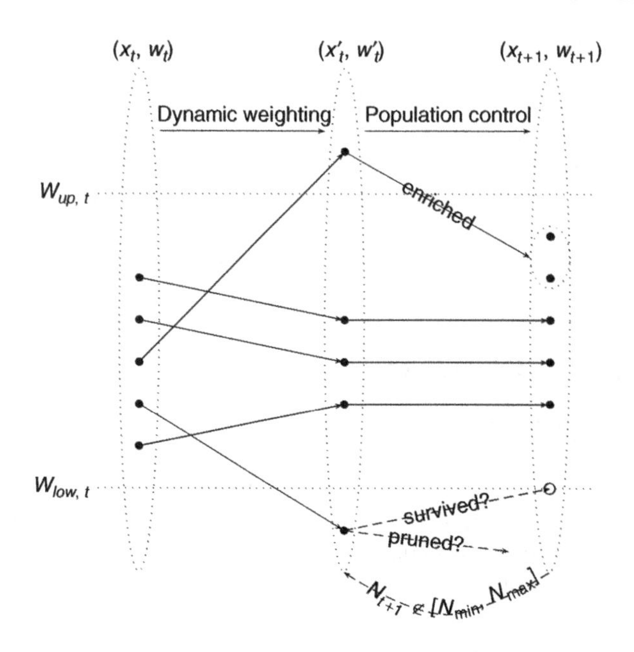

Figure 6.4 A diagram of the DWIS algorithm.

6.2.2 A Theory of DWIS

Let $g_t(x, w)$ denote the joint density of (x, w), an individual state of $(\boldsymbol{x}_t, \boldsymbol{w}_t)$, and let $f(x)$ denote the target distribution we are working with.

Definition 6.2.1 *The distribution $g_t(x, w)$ defined on $\mathcal{X} \times (0, \infty)$ is called correctly weighted with respect to $f(x)$ if the following conditions hold,*

$$\int w g_t(x, w) dw = c_{tx} f(x), \tag{6.7}$$

$$\frac{\int_{\mathcal{A}} c_{tx} f(x) dx}{\int_{\mathcal{X}} c_{tx} f(x) dx} = \int_{\mathcal{A}} f(x) dx, \tag{6.8}$$

where $\mathcal{A}$ is any Borel set, $\mathcal{A} \subseteq \mathcal{X}$.

Definition 6.2.2 *If $g_t(x, w)$ is correctly weighted with respect to $f(x)$, and samples $(x_{t,i}, w_{t,i})$ are simulated from $g_t(x, w)$ for $i = 1, 2, \ldots, n_t$, then $(\boldsymbol{x}_t, \boldsymbol{w}_t) = (x_{t,1}, w_{t,1}; \cdots; x_{t,n_t}, w_{t,n_t})$ is called a correctly weighted population with respect to $f(x)$.*

Let $(\boldsymbol{x}_t, \boldsymbol{w}_t)$ be a correctly weighted population with respect to $f(x)$, and let $y_1, \ldots, y_m$ be distinct states in $\boldsymbol{x}_t$. Generate a random variable/vector Y

such that

$$P\{Y = y_i\} = \frac{\sum_{j=1}^{n_t} w_j I(y_i = x_{t,j})}{\sum_{j=1}^{n_t} w_j}, \quad i = 1, 2, \ldots, m, \tag{6.9}$$

where $I(\cdot)$ is an indicator function. Then Y is approximately distributed as $f(\cdot)$ if n_t is large. This can be summarized as the following theorem:

Theorem 6.2.1 *The distribution of the random vector Y generated in (6.9) converges as $n_t \to \infty$ to $f(\cdot)$, if the importance weight has a finite expectation.*

Proof: Let $(x_{t,i}, w_{t,i})$, $i = 1, \ldots, n_t$ be n_t random samples generated by DWIS with joint density g_t. For any Borel set $\mathcal{A} \subseteq \mathcal{X}$,

$$P\{Y \in \mathcal{A} | x_{t,1}, w_{t,1}; \cdots; x_{t,N_t}, w_{t,N_t}\} = \frac{\sum_{i=1}^{n_t} w_{t,i} I(x_{t,i} \in \mathcal{A})}{\sum_{i=1}^{n_t} w_{t,i}}. \tag{6.10}$$

By the strong law of large numbers, it follows that as $n_t \to \infty$,

$$\sum_{i=1}^{n_t} w_{t,i} I(\boldsymbol{x}_{t,i} \in \mathcal{A})/n_t \to \int_{\mathcal{A}} \int w g_t(x, w) dw dx = \int_{\mathcal{A}} c_{tx} f(x) dx$$

and

$$\sum_{i=1}^{n_t} w_{t,i}/n_t \to \int_{\mathcal{X}} \int w g_t(x, w) dw dx = \int_{\mathcal{X}} c_{tx} f(x) dx.$$

Dividing numerator and denominator of (6.10) by n_t shows that

$$\frac{\sum_{i=1}^{N_t} w_{t,i} I(x_{t,i} \in \mathcal{A})}{\sum_{i=1}^{n_t} w_{t,i}} \to \frac{\int_{\mathcal{A}} c_{tx} f(x) dx}{\int_{\mathcal{X}} c_{tx} f(x) dx} = \int_{\mathcal{A}} f(y) dy,$$

which, by Lebesgue's dominated converge theorem, implies that

$$P\{Y \in \mathcal{A}\} = E[P\{Y \in \mathcal{A} | x_{t,1}, w_{t,1}; \cdots; x_{t,N_t}, w_{t,N_t}\}] \to \int_{\mathcal{A}} f(y) dy$$

and the result is proved. ∎

Let $(\boldsymbol{x}_1, \boldsymbol{w}_1)$, ..., $(\boldsymbol{x}_N, \boldsymbol{w}_N)$ be a series of correctly weighted populations generated by a DWIS algorithm with respect to $f(x)$. Then the quantity $\mu = E_f h(x)$, provided that $h(x)$ is integrable, can be estimated by

$$\widehat{\mu} = \frac{\sum_{t=1}^{N} \sum_{i=1}^{n_t} w_{t,i} h(x_{t,i})}{\sum_{t=1}^{N} \sum_{i=1}^{n_t} w_{t,i}}. \tag{6.11}$$

Let $U_t = \sum_{i=1}^{n_t} w_{t,i} h(x_{t,i})$, $S_t = \sum_{i=1}^{n_t} w_{t,i}$, $S = ES_t$, and $V_t = U_t - \mu S_t$. If the variance of U_t and V_t are both finite, then the standard error of $\widehat{\mu}$

can be calculated using the ratio estimate as in finite population sampling (Ripley, 1987). As shown by Liang (2002b), $\widehat{\mu}$ is consistent and asymptotically normally distributed; that is,

$$\sqrt{N}(\widehat{\mu} - \mu) \rightarrow N(0, \sigma^2), \tag{6.12}$$

where σ^2 is defined as $\text{Var}(V_t)/S^2$.

Definition 6.2.3 *A transition rule for a population* $(\boldsymbol{x}, \boldsymbol{w})$ *is said to be* invariant with respect to the importance weight *(IWIW$_p$) if the joint density of* $(\boldsymbol{x}, \boldsymbol{w})$ *remains correctly weighted whenever the initial joint density is correctly weighted.*

To make this rule distinguishable from the IWIW rule given in (6.1) and (6.2), which is defined for the single Markov chain, we denote it by IWIW$_p$ with the subscript representing for population.

6.2.3 Some IWIW$_p$ Transition Rules

Many transition rules are IWIW$_p$ exactly or approximately. The following are some useful examples. Let $(\boldsymbol{x}_t, \boldsymbol{w}_t)$ denote the current population, let (x, w) denote an individual state of $(\boldsymbol{x}_t, \boldsymbol{w}_t)$, and let (x', w') denote the individual state transmitted from (x, w) in one transition step.

6.2.3.1 A General Dynamic Weighting Sampler

1. Draw y from some transition proposal distribution $T(x, y)$, and compute the dynamic ratio
$$r_d = w\frac{f(y)T(y, x)}{f(x)T(x, y)}.$$

2. Choose $\theta_t = \theta(\boldsymbol{x}_t, \boldsymbol{w}_t) \geq 0$ and draw $U \sim unif(0, 1)$. Update (x, w) as (x', w')
$$(x', w') = \begin{cases} (y, (1 + \delta_t)r_d/a) & \text{if } U \leq a \\ (x, (1 + \delta_t)w/(1 - a)) & \text{otherwise.} \end{cases}$$
where $a = r_d/(r_d + \theta_t)$; and θ_t and δ_t are both functions of $(\boldsymbol{x}_t, \boldsymbol{w}_t)$, but they remain constants for each individual state of the same population.

For simplicity, we consider only the cases where $\theta_t = 0$ or 1, $\delta_t = 0$ or $1/(\alpha + \beta h(W_{up,t})$, where both α and β are positive constants, $W_{up,t}$ is the upper weight control bound of the population $(\boldsymbol{x}_t, \boldsymbol{w}_t)$, and $h(\cdot)$ is an appropriate function defined latter. If $\theta_t \equiv 0$, then the move is essentially a random walk induced by $T(x, y)$ on the space $\mathcal{X}$, herein called the W-type move. If $\delta_t \equiv 1$, then the sampler is reduced to the R-type move discussed in Section 6.1.1.

The following theorem shows that the dynamic weighting sampler is IWIW_p when the initial condition

$$\int_{\mathcal{X}} c_{t_0,x} T(x', x) dx = c_{t_0,x'} \tag{6.13}$$

holds for some population $(\boldsymbol{x}_0, \boldsymbol{w}_0)$.

Theorem 6.2.2 *The general dynamic weighting sampler is $IWIW_p$; that is, if the joint distribution $g_t(x, w)$ for $(\boldsymbol{x}, \boldsymbol{w})$ is correctly weighted with respect to $f(x)$ and (6.13) holds, after one dynamic weighting step, the new joint distribution $g_{t+1}(x', w')$ for $(\boldsymbol{x}', \boldsymbol{w}')$ is also correctly weighted with respect to $f(x)$, and the condition (6.13) also holds.*

Proof: For the case where $\theta > 0$,

$$\begin{aligned}
&\int_0^\infty w' g_{t+1}(x', w') dw' \\
&= \int_{\mathcal{X}} \int_0^\infty (1 + \delta_t)[r_d(x, x', w) + \theta_t] g_t(x, w) T(x, x') \frac{r_d(x, x', w)}{r_d(x, x', w) + \theta_t} dw dx \\
&\quad + \int_{\mathcal{X}} \int_0^\infty (1 + \delta_t) \frac{w[r_d(x', z, w) + \theta_t]}{\theta_t} g_t(x', w) T(x', z) \\
&\quad \times \frac{\theta_t}{r_d(x', z, w) + \theta_t} dw dz \\
&= (1 + \delta_t) \left\{ \int_{\mathcal{X}} \int_0^\infty w g_t(x, w) \frac{f(x') T(x', x)}{f(x)} dw dx \right. \\
&\quad \left. + \int_{\mathcal{X}} \int_0^\infty w g_t(x', w) T(x', z) dw dz \right\} \\
&= (1 + \delta_t) \left\{ f(x') \int_{\mathcal{X}} c_{t,x} T(x', x) dx + c_{t,x'} f(x') \right\} \\
&= 2(1 + \delta_t) c_{t,x'} f(x').
\end{aligned}$$

For the case where $\theta_t = 0$, only the first terms remain on the right sides of the foregoing equations. Thus

$$\int_0^\infty w' g_{t+1}(x', w') dw' = (1 + \delta_t) c_{t,x'} f(x').$$

By defining $c_{t+1,x'} = 2(1 + \delta_t) c_{t,x'}$ for the case where $\theta_t > 0$ and $c_{t+1,x'} = (1 + \delta_t) c_{t,x'}$ for the case where $\theta_t = 0$, it is easy to see that the condition (6.13) still holds for the new population. Hence, $g_{t+1}(x', w')$ is still correctly weighted with respect to $f(x)$. ■

In practice, the initial condition (6.13) can be easily satisfied. For example, if $g_0(x,w) = g(x)$, a trial distribution for drawing from $f(x)$ used in conventional importance sampling, and $w = f(x)/g(x)$, then $\int wg_0(x,w)dw = c_0$, the normalizing constant ratio of g and f, independent of x. Thus, the initial condition (6.13) is satisfied because $\int c_0 T(x',x)dx = c_0$ always holds.

To avoid an extremely large weight caused by a nearly zero divisor, we make the following assumption for the MH ratio: There exists a constant r_0 such that for any pair $(x,x') \in \mathcal{X} \times \mathcal{X}$,

$$r_0 \leq \frac{f(x')}{f(x)} \frac{T(x',x)}{T(x,x')} \leq \frac{1}{r_0}. \tag{6.14}$$

In practice, this condition can be easily satisfied by restricting $\mathcal{X}$ to a compact set. This is sensible for many statistical problems. For proposal distributions, one may choose them to be symmetric, which implies $T(x',x)/T(x,x') = 1$.

6.2.3.2 Adaptive Pruned-Enriched Population Control Scheme (APEPCS)

Given the current population $(\boldsymbol{\theta}_t, \boldsymbol{w}_t)$, adjusting the values of the importance weights and the population size to suitable ranges involves the following scheme. Let $(x_{t,i}, w_{t,i})$ be the ith individual state of the population, let n_t and n'_t denote the current and new population sizes, let $W_{low,t}$ and $W_{up,t}$ denote the lower and upper weight control bounds, let $n_{\min}$ and $n_{\max}$ denote the minimum and maximum population size allowed by the user, and let n_{low} and n_{up} denote the lower and upper reference bounds of the population size.

1. (Initialization) Initialize the parameters $W_{low,t}$ and $W_{up,t}$ by

$$W_{low,t} = \sum_{i=1}^{n_t} w_{t,i}/n_{up}, \quad W_{up,t} = \sum_{i=1}^{n_t} w_{t,i}/n_{low}.$$

   Set $n'_t = 0$ and $\lambda > 1$. Do steps 2–4 for $i = 1, 2, \cdots, n_t$.

2. (Pruned) If $w_{t,i} < W_{low,t}$, prune the state with probability $q = 1 - w_{t,i}/W_{low,t}$. If it is pruned, drop $(x_{t,i}, w_{t,i})$ from $(\boldsymbol{x}_t, \boldsymbol{w}_t)$; otherwise, update $(x_{t,i}, w_{t,i})$ as $(x_{t,i}, W_{low,t})$ and set $n'_t = n'_t + 1$.

3. (Enriched) If $w_{t,i} > W_{up,t}$, set $d = [w_{t,i}/W_{up,t} + 1]$, $w'_{t,i} = w_{t,i}/d$, replace $(x_{t,i}, w_{t,i})$ by d identical states $(x_{t,i}, w'_{t,i})$, and set $n'_t = n'_t + d$, where $[z]$ denotes the integer part of z.

4. (Unchanged) If $W_{low,t} \leq w_{t,i} \leq W_{up,t}$, keep $(x_{t,i}, w_{t,i})$ unchanged, and set $n'_t = n'_t + 1$.

5. (Checking) If $n'_t > n_{\max}$, set $W_{low,t} \leftarrow \lambda W_{low,t}$, $W_{up,t} \leftarrow \lambda W_{up,t}$ and $n'_t = 0$, do step 2–4 again for $i = 1, 2, \cdots, n_t$. If $n'_t < n_{\min}$, set $W_{low,t} \leftarrow W_{low,t}/\lambda$, $W_{up,t} \leftarrow W_{up,t}/\lambda$ and $n'_t = 0$, do step 2–4 again for $i = 1, 2, \cdots, n_t$. Otherwise, stop.

In this scheme, λ is required to be greater than 1, say 2. In addition, n_{low}, n_{up}, $n_{\min}$ and $n_{\max}$ are required to satisfy the constraint $n_{\min} < n_{low} < n_{up} < n_{\max}$. With APEPCS, the population size is strictly controlled to the range $[n_{min}, n_{max}]$, and the weights are adjusted to the range $[W_{low,t}, W_{up,t}]$. Therefore, the APEPCS avoids the possible overflow or extinction of a population in simulations.

The APEPCS is closely related to the resampling technique that is often used in the context of importance sampling, see, for example, Liu and Chen (1998). It can also be viewed as a generalization of the pruned-enriched Rosenbluth method (PERM) (Grassberger, 1997) and the rejection controlled sequential importance sampler (RCSIS) (Liu *et al.*, 1998). In PERM, $W_{low,t}$ and $W_{up,t}$ are pre-fixed. For each individual state with $w_i < W_{low,t}$, it is pruned with a fixed probability of 1/2, and set to $w'_i = 2w_i$ if it is not pruned. In RCSIS, no enrichment step is performed, and $W_{low,t}$ is also a pre-fixed number.

Theorem 6.2.3 shows that the APEPCS is IWIW_p.

Theorem 6.2.3 *The APEPCS is $IWIW_p$; that is, if the joint distribution $g_t(x, w)$ for $(\boldsymbol{x}_t, \boldsymbol{w}_t)$ is correctly weighted with respect to $f(x)$, then after one run of the scheme, the new joint distribution $g_{t+1}(x', w')$ for $(\boldsymbol{x}_{t+1}, \boldsymbol{w}_{t+1})$is also correctly with respect to $f(x)$.*

Proof: In the population control step, only the weights w's are possibly modified. Thus,

$$\begin{aligned}
\int_0^\infty w' g_{t+1}(x', w') dw' &= \int_0^{W_{low,t}} 0 g_t(x', w) \left(1 - \frac{w}{W_{low,t}}\right) dw \\
&+ \int_0^{W_{low,t}} W_{low,t} g_t(x', w) \frac{w}{W_{low,t}} dw \\
&+ \int_{W_{low,t}}^{W_{up,t}} w g_t(x', w) dw + \int_{W_{up,t}}^\infty \left(\sum_{i=1}^d \frac{w}{d}\right) g_t(x', w) dw \\
&= \int_0^\infty w g_t(x', w) dw = c_{t,x'} f(x').
\end{aligned}$$

Hence, $g_{t+1}(x', w')$ is still correctly weighted with respect to $f(x)$ by setting $c_{t+1,x'} = c_{t,x'}$. ∎

6.2.4 Two DWIS Schemes

Since both the dynamic weighting sampler and the population control scheme are IWIW_p, they can be used together in any fraction, while leaving the IWIW_p property unchanged. Based on this observation, we compose the following two DWIS schemes, both of which alter between the dynamic weighting step and the population control step. Let λ be a number greater than 1. Let W_c denote a dynamic weighting move switching parameter, which switches the value of θ_t between 0 and 1 depending on the value of $W_{up,t-1}$.

Dynamically Weighted Importance Sampling (Scheme-$\boldsymbol{R}$)

- *Dynamic weighting*: Apply the R-type move to the population $(\boldsymbol{x}_{t-1}, \boldsymbol{w}_{t-1})$. If $W_{up,t-1} \leq W_c$, then set $\theta_t = 1$. Otherwise, set $\theta_t = 0$. The new population is denoted by $(\boldsymbol{x}'_t, \boldsymbol{w}'_t)$.

- *Population Control*: Apply APEPCS to $(\boldsymbol{x}'_t, \boldsymbol{w}'_t)$. The new population is denoted by $(\boldsymbol{x}_t, \boldsymbol{w}_t)$.

Dynamically Weighted Importance Sampling (Scheme-$\boldsymbol{W}$)

- *Preweight control*: If $n_{t-1} < n_{low}$, then set $W_{low,t} = W_{low,t-1}/\lambda$ and $W_{up,t} = W_{up,t-1}/\lambda$. If $n_{t-1} > n_{up}$, then set $W_{low,t} = \lambda W_{low,t-1}$ and $W_{up,t} = \lambda W_{up,t-1}$. Otherwise, set $W_{low,t} = W_{low,t-1}$ and $W_{up,t} = W_{up,t-1}$.

- *Dynamic weighting*: Apply the W-type move to the population $(\boldsymbol{x}_{t-1}, \boldsymbol{w}_{t-1})$ with $\delta_t = 1/(\alpha + \beta W_{up,t}^{1+\epsilon})$ for some $\epsilon > 0$. The new population is denoted by $(\boldsymbol{x}'_t, \boldsymbol{w}'_t)$.

- *Population Control*: Apply APEPCS to $(\boldsymbol{x}'_t, \boldsymbol{w}'_t)$. The new population is denoted by $(\boldsymbol{x}_t, \boldsymbol{w}_t)$.

Let $(\boldsymbol{x}_1, \boldsymbol{w}_1), \ldots, (\boldsymbol{x}_N, \boldsymbol{w}_N)$ denote a series of populations generated by a DWIS scheme. Then, according to (6.11), the quantity $\mu = E_f h(x)$ can be estimated by

$$\widehat{\mu} = \frac{\sum_{t=N_0+1}^{N} \sum_{i=1}^{n_t} w_{t,i} h(x_{t,i})}{\sum_{t=N_0}^{N} \sum_{i=1}^{n_t} w_{t,i}}, \tag{6.15}$$

where N_0 denotes the number of burn-in iterations. Following from (6.11) and (6.12), $\widehat{\mu}$ is consistent and asymptotically normally distributed.

6.2.5 Weight Behavior Analysis

To analyze the weight behavior of the DWIS, we first introduce the following lemma.

Lemma 6.2.1 *Let $\pi(x_0)$ denote the marginal equilibrium distribution under transition T and let $\pi(x_0, x_1) = \pi(x_0) \times T(x_0, x_1)$ and $r(x_0, x_1) = \pi(x_1, x_0)/\pi(x_0, x_1)$ be the MH ratio. Then*

$$e_0 = E_\pi \log r(x_0, x_1) \leq 0,$$

where the equality holds when it induces a reversible Markov chain, that is, $\pi(x_0, x_1) = \pi(x_1, x_0)$.

Proof: By Jensen's inequality,

$$e_0 = E_\pi \log \frac{\pi(x_1, x_0)}{\pi(x_0, x_1)} \leq \log E_\pi \frac{\pi(x_1, x_0)}{\pi(x_0, x_1)} = 0,$$

where the equality holds only when $\pi(x_0, x_1) = \pi(x_1, x_0)$. ∎

For simplicity, let (x_t, w_t) denote an individual state of the population $(\boldsymbol{x}_t, \boldsymbol{w}_t)$. When $\delta_t \equiv 0$ and $\theta_t \equiv 0$, the weights of the W-type move and the R-type move evolve as

$$\log w_t = \log w_{t-1} + \log r(x_{t-1}, x_t),$$

which results in

$$\log w_t = \log w_0 + \sum_{s=1}^{t} \log r(x_{s-1}, x_s). \tag{6.16}$$

Following from Lemma 6.2.1 and the ergodicity theorem (under stationarity),

$$\frac{1}{t} \sum_{s=1}^{t} \log r(x_{s-1}, x_s) \to e_0 < 0, \quad a.s. \tag{6.17}$$

as $t \to \infty$. Hence, w_t will go to 0 almost surely as $t \to \infty$.

Scheme-R. When $\theta_t \equiv 1$, the expectation of w_t, conditional on x_{t-1}, x_t and w_{t-1}, can be calculated as

$$\begin{aligned} E[w_t | x_{t-1}, x_t, w_{t-1}] &= (r_d + 1)\frac{r_d}{r_d + 1} + w_{t-1}(r_d + 1)\frac{1}{r_d + 1} = r_d + w_{t-1} \\ &= w_{t-1}[1 + r(x_{t-1}, x_t)]. \end{aligned} \tag{6.18}$$

Since $r(x_{t-1}, x_t) \geq 0$, the weight process $\{w_t\}$ is driven by an inflation drift. Hence, w_t will go to infinity almost surely as t becomes large.

To prevent the weight process from going to 0 or ∞, scheme-R alters the use of $\theta_t = 0$ and $\theta_t = 1$. When $W_{up,t} > W_c$, θ_t is set to 0, so the weight process of scheme-R can be bounded above by

$$\log w_t = \log w_0 + \sum_{s=1}^{t} \log r(x_{s-1}, x_s) - \sum_{s=1}^{t} \log(d_s),$$

provided that $\theta_1 = \cdots = \theta_t = 0$, where d_s is a positive integer as defined in the APEPCS.

Scheme-W To prevent the weight process from going to 0, the factor $(1+\delta_t)$ is multiplied at each iteration, where

$$\delta_t = \frac{1}{\alpha + \beta W_{up,t}^{1+\epsilon}}, \quad \alpha > 0, \quad \beta > 0, \quad \epsilon > 0.$$

For example, it was set $\alpha = \beta = 1$ and $\epsilon = 0.5$ in Liang (2002b). When $W_{up,t}$ is very small, $\delta_t \approx 1/\alpha$, the weight process is driven by an inflation drift (i.e., choosing α such that $\log(1+\alpha^{-1}) + e_0 > 0$) and is prevented from going to 0. When $W_{up,t}$ is very large, $\delta_t \approx 0$, the weight process is driven by the negative drift e_0 and is prevented from going to infinity. The preweight control step tunes the value of δ to help the population control scheme to control the population size and the weight quantity to a suitable range.

In this scheme, the weight evolves as

$$\log w_t = \begin{cases} -\infty & w_t' < W_{low,t} \text{ and pruned}, \\ \log(W_{low,t}) & w_t' < W_{low,t} \text{ and not pruned}, \\ \log w_t' - \log d_t & w_t' \geq W_{low,t}, \end{cases} \tag{6.19}$$

where $w_t' = (1+\delta_t) w_{t-1} r(x_{t-1}, x_t)$, and d_t is a positive integer number such that $\log(W_{low,t}) \leq \log w_t' - \log(d_t) \leq \log(W_{up,t})$. Following from (6.19), $\log w_t$ is bounded above by the process

$$\begin{aligned} \log w_t &= \log w_0 + \sum_{s=1}^{t} \log r(x_{s-1}, x_s) + \sum_{s=1}^{t} \log(1+\delta_s) - \sum_{s=1}^{t} \log(d_s) \\ &\approx \log w_0 + \sum_{s=1}^{t} \log r(x_{s-1}, x_s) + \sum_{s=1}^{t} \delta_s - \sum_{s=1}^{t} \log(d_s) \\ &\leq c + \sum_{s=1}^{t} \log r(x_{s-1}, x_s) - \sum_{s=1}^{t} \log(d_s), \end{aligned} \tag{6.20}$$

where c is a constant. The last inequality follows from the fact that $\sum_{s=1}^{t} \delta_s$ is upper bounded, as we set $\delta_s = 1/[a + bW_{up,s}^{1+\epsilon}]$ with $\epsilon > 0$.

Moments of DWIS Weights In summary, the weight process of the two DWIS schemes can be characterized by the following process:

$$Z_t = \begin{cases} Z_{t-1} + \log r(x_{t-1}, x_t) - \log(d_t), & \text{if } Z_{t-1} > 0, \\ 0, & \text{if } Z_{t-1} < 0, \end{cases} \tag{6.21}$$

where there exists a constant C such that $|Z_t - Z_{t-1}| \leq C$ almost surely. The existence of C follows directly from condition (6.14). Let $T_0 = 0$, $T_i = \min\{t : t > T_{i-1}, Z_t = 0\}$, and $L_i = T_i - T_{i-1}$ for $i \geq 1$. From (6.17) and the fact that

$d_t \geq 1$, it is easy to see that L_i is almost surely finite; that is, there exists an integer M such that $P(L_i < M) = 1$. This implies that for any fixed $\eta > 0$,

$$E\exp(\eta Z_t) \leq \exp(\eta MC), \quad a.s.$$

This leads to the following theorem:

Theorem 6.2.4 *Under the assumption (6.14), the importance weight in DWIS almost surely has finite moments of any order.*

The ratio $\kappa = W_{up,t}/W_{low,t}$ determines the moving ability of DWIS, and it is called the freedom parameter of DWIS by Liang (2002b). In the APEPCS, κ is also equal to the ratio n_{up}/n_{low}. When choosing the value of κ, the efficiency of the resulting sampler should be taken into account. A large κ will ease the difficulty of escaping from local energy minima, but the variability of the weights will also increase accordingly. The efficiency of DWIS may be reduced by an excessively large κ. Our experience shows that a value of κ between 2 and 10 works well for most problems.

6.2.6 A Numerical Example

Suppose we are interested in simulating from a distribution $f \propto (1, 1000, 1, 2000)$ with the transition matrix

$$T = \begin{pmatrix} \frac{1}{2} & \frac{1}{2} & 0 & 0 \\ \frac{2}{3} & 0 & \frac{1}{3} & 0 \\ 0 & \frac{4}{7} & 0 & \frac{3}{7} \\ 0 & 0 & \frac{1}{2} & \frac{1}{2} \end{pmatrix}.$$

This example characterizes the problem of simulation from a multimodal distribution.

For the MH algorithm, once it reaches state 4, it will have a very hard time leaving it. On average it will wait 4667 trials for one successful transition from state 4 to state 3. Figure 6.5 (b) shows that it almost always got stuck at state 4. Figure 6.5 (a) shows that the MH algorithm fails to estimate the mass function by plotting the standardized errors of $\hat{f}_{MH}$ [i.e., $(\sum_{i=1}^{4}(\hat{f}_{MH,i} - f_i)^2/f_i)^{1/2}$] against log-iterations, where $\hat{f}_A$ denotes one estimate of f by algorithm A.

Dynamic weighting (the R-type move) was applied to this example with $\theta = 1$. It started with $(x_0, w_0) = (4, 1)$ and was run for 35 000 iterations. The ability of weight self-adjustment enables it to move from state 4 to mix with the other states. Figure 6.5 (c) shows that even only with 200 ($\approx e^{5.3}$) iterations, dynamic weighting already produces a very good estimate for f in the run. But as the run continues, the estimate is not improved and occasionally is even contaminated by the big jumps of importance weights. An occasional big jump in importance weights is an inherent feature of dynamic weighting.

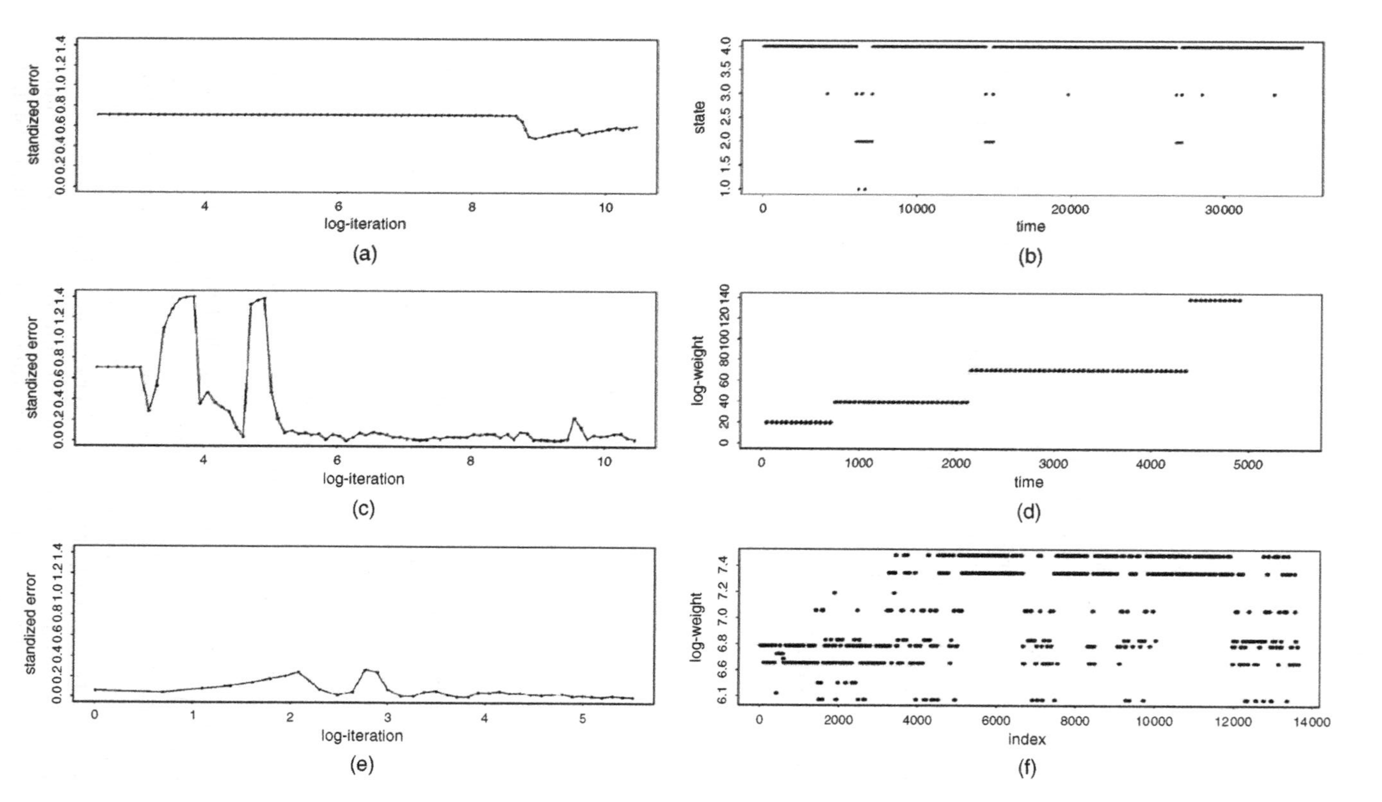

Figure 6.5 Comparison of MH, dynamic weighting (DW) and DWIS for the 4-state example. (a) Time plot of standardized errors of $\hat{f}_{MH}$ against log-iterations. (b) Time plot of state transitions of the MH run. (c) Time plot of standardized errors of $\hat{f}_{DW}$ against log-iterations. (d) Time plot of the log-weights collected at state 4 in the run of dynamic weighting. (e) Time plot of standardized errors of $\hat{f}_{DWIS}$ against log-iterations. (f) Time plot of the log-weights collected at state 4 in the run of DWIS (Liang, 2002b).

It is caused by an occasional rejection of a transition with a large weight and a large dynamic ratio. Figure 6.5 (d) shows the time plot of the weights collected at state 4. It shows that at state 4, the weight increases monotonically, and a big jump happens occasionally. Although the jump happens increasingly rarely, the weight increasing amplitude of each jump becomes larger. If the simulation stops at an unsuitable time (e.g., just after a big weight jump), then the resulting estimate may be severely biased. In this case, the last several weights will dominate all the others, and the effective sample size will be very small. To achieve a stable estimate, some additional techniques, such as the stratified truncation as described in Section 6.1.1, must be used.

DWIS was applied to this example with $N_{min} = 25$, $N_{max} = 200$, $N_{low} = 50$, $N_{up} = 100$, $N_0 = 80$, and $\kappa = 1000$. It was run for 250 iterations, and produced a total of 34 110 weighted samples. Figure 6.5 (e) shows that the DWIS estimate converges very fast to the true mass function, and the estimate improves continuously as the run goes on. Figure 6.5 (f) plots the weights collected at state 4. It shows that in DWIS, the weights have been well controlled and no longer increase monotonically, and no weight jump occurs. This example shows that DWIS can move across energy barriers like dynamic weighting, but the weights are well controlled and the estimates can be improved continuously like the MH algorithm.

6.3 Monte Carlo Dynamically Weighted Importance Sampling

In this section, we describe a Monte Carlo version of dynamically weighted importance sampling (Liang and Cheon, 2009), which can be used to sample from distributions with intractable integrals.

6.3.1 Sampling from Distributions with Intractable Normalizing Constants

Reconsider the problem of sampling from distributions with intractable normalizing constants that has been considered in Sections 4.6–4.8. Suppose we have data D generated from a statistical model with the likelihood function

$$f(D|x) = \frac{p(x, D)}{Z(x)}, \tag{6.22}$$

where, to keep the notation consistent with previous sections, x is used to denote the parameters of the model, and $Z(x)$ is the normalizing constant which depends on x and is not available in closed form. Let $\pi(x)$ denote the prior distribution of x. The posterior distribution of x is then given by

$$\pi(x|D) = \frac{1}{Z(x)} p(x, D)\pi(x). \tag{6.23}$$

Since the closed form of $Z(x)$ is not available, sampling from (6.23) poses a great challenge on the current statistical methods. As discussed in Section 4.6, such a sampling problem often arises in Bayesian spatial statistics. Two famous examples are image segmentation (Hurn *et al.*, 2003) and disease mapping (Green and Richardson, 2002), both involving a hidden Potts model whose normalizing constant is intractable.

It is known that the MH algorithm cannot be applied to simulate from $\pi(x|D)$, because the acceptance probability requires computing the ratio $Z(x)/Z(x')$, where x' denotes the proposed value. As summarized in Section 4.6, the existing methods for the problem can be divided into two categories. The first category is called the likelihood approximation-based methods, which include the maximum pseudo-likelihood method (Besag, 1974), the Monte Carlo maximum likelihood method (Geyer and Thompson, 1992), stochastic approximation Monte Carlo (Liang *et al.*, 2007; Liang, 2007a), and the double MH algorithm (Liang, 2009c), among others. Note that the double MH algorithm does not approximate the likelihood directly, but uses a sample generated in a finite number of MH transitions to substitute for a perfect sample of $f(\cdot)$. The second category is called the auxiliary variable MCMC methods, which include the Møller algorithm (Møller *et al.*, 2006) and the exchange algorithm (Murray *et al.*, 2006). A serious weakness of these two algorithms is that they require perfect sampling (Propp and Wilson, 1996) of D. Unfortunately, perfect sampling is very expensive or impossible for many statistical models whose normalizing constant is intractable.

In Section 6.3.2, we describe a new Monte Carlo method, the so-called Monte Carlo dynamically weighted importance sampling (MCDWIS) method (Liang and Cheon, 2009), for the problem. The advantage of MCDWIS is that it avoids the requirement for perfect sampling, and thus can be applied for a wide range of problems for which perfect sampling is not available or very expensive.

6.3.2 Monte Carlo Dynamically Weighted Importance Sampling

The MCDWIS algorithm is a Monte Carlo version of the DWIS algorithm described in Section 6.2. As in DWIS, the state space of the Markov chain is augmented to a population, a collection of weighted samples $(\boldsymbol{x}, \boldsymbol{w}) = \{x_1, w_1; \ldots; x_n, w_n\}$. Given the current population $(\boldsymbol{x}_t, \boldsymbol{w}_t)$, one iteration of the MCDWIS consists of two steps:

1. *Monte Carlo dynamic weighting (MCDW):* update each individual state of the current population by a MCDW transition.

2. *Population control:* split or replicate the individual states with large weights and discard the individual states with small weights.

The MCDW step allows for the use of Monte Carlo estimates in MCMC simulations. The bias induced thereby is counterbalanced by giving different weights to the new samples produced. Therefore, a Monte Carlo estimate of $Z(x)/Z(x')$ can be incorporated into the simulation, while leaving $\pi(x|D)$ invariant with respect to dynamic importance weights. Note that conventional MCMC algorithms do not allow for the use of Monte Carlo estimates in simulations. Otherwise, the detailed balance condition will be violated.

The population control is the same as that used in DWIS, which has the importance weight and population size controlled to suitable ranges.

6.3.2.1 A MCDW Sampler

Let $(\boldsymbol{x}_t, \boldsymbol{w}_t)$ denote the current population, let (x, w) denote an individual state of the population, and let (x', w') denote the individual state transmitted from (x, w) in one transition step. The MCDW sampler can be described as follows.

1. Draw x^* from some proposal distribution $T(x, x^*)$.

2. Simulate auxiliary samples $D_1, \ldots, D_m$ from $f(D|x^*)$ using a MCMC algorithm, say, the MH algorithm. Estimate the normalizing constant ratio $R_t(x, x^*) = Z(x)/Z(x^*)$ by

$$\widehat{R}_t(x, x^*) = \frac{1}{m} \sum_{i=1}^{m} \frac{p(D_i, x)}{p(D_i, x^*)}, \tag{6.24}$$

   which is also known as the importance sampling (IS) estimator of $R_t(x, x^*)$.

3. Calculate the Monte Carlo dynamic weighting ratio

$$r_d = r_d(x, x^*, w) = w\widehat{R}_t(x, x^*) \frac{p(D, x^*)}{p(D, x)} \frac{T(x^*, x)}{T(x, x^*)}.$$

4. Choose $\theta_t = \theta_t(\boldsymbol{x}_t, \boldsymbol{w}_t) \geq 0$ and draw $U \sim unif(0, 1)$. Update (x, w) as (x', w')

$$(x', w') = \begin{cases} (x^*, r_d/a), & \text{if } U \leq a, \\ (x, w/(1-a)), & \text{otherwise,} \end{cases}$$

   where $a = r_d/(r_d + \theta_t)$; θ_t is a function of $(\boldsymbol{x}_t, \boldsymbol{w}_t)$, but remains a constant for each individual state of the same population.

The sampler is termed 'Monte Carlo' in the sense that $R_t(x, x^*)$ is substituted by its Monte Carlo estimator in simulations. If $R_t(x, x^*)$ is available analytically, then MCDWIS is reduced to DWIS. Regarding this sampler, we have two remarks.

Remark 1 As pointed out by Chen *et al.* (2000), $\widehat{R}_t(x,x^*)$ is an unbiased and consistent estimator of $R_t(x,x^*)$. Following from the central limit theorem, we have

$$\sqrt{m}\left(\widehat{R}_t(x,x^*) - R_t(x,x^*)\right) \to N\left(0,\sigma_t^2\right), \tag{6.25}$$

where σ_t^2 can be expressed as

$$\sigma_t^2 = \text{Var}\left(\frac{p(D_1,x)}{p(D_1,x^*)}\right) + 2\sum_{i=2}^{\infty}\text{Cov}\left(\frac{p(D_1,x)}{p(D_1,x^*)}, \frac{p(D_i,x)}{p(D_i,x^*)}\right).$$

Alternatively, we can write $\widehat{R}_t(x,x^*) = R_t(x,x^*)(1+\epsilon_t)$, where $\epsilon_t \sim N(0, \sigma_t^2/[mR_t^2(x,x^*)])$.

Remark 2 This sampler is designed according to the scheme-R of DWIS. A similar sampler can also be deigned according to the scheme-W of DWIS. As discussed previously, the parameter θ_t can be specified as a function of the population $(\boldsymbol{x}_t, \boldsymbol{w}_t)$. For simplicity, we here concentrate only on the cases where $\theta_t = 0$ or 1.

The following theorem shows that the MCDWIS sampler is IWIW$_p$ when the initial condition

$$\int_{\mathcal{X}} c_{t_0,x} T(x',x)dx = c_{t_0,x'} \tag{6.26}$$

holds for some population $(\boldsymbol{x}_{t_0}, \boldsymbol{w}_{t_0})$.

Theorem 6.3.1 *The Monte Carlo dynamic weighting sampler is IWIW$_p$; that is, if the joint distribution $g_t(x,w)$ for $(\boldsymbol{x}_t, \boldsymbol{w}_t)$ is correctly weighted with respect to $\pi(x|D)$ and (6.26) holds, after one Monte Carlo dynamic weighting step, the new joint density $g_{t+1}(x',w')$ for $(\boldsymbol{x}_{t+1}, \boldsymbol{w}_{t+1})$ is also correctly weighted with respect to $\pi(x|D)$ and (6.26) also holds.*

Proof: For the case $\theta_t > 0$,

$$\begin{aligned}
&\int_0^\infty w' g_{t+1}(x',w')dw' \\
&\quad = \int_{\mathcal{X}}\int_0^\infty\int_{-\infty}^\infty [r_d(x,x',w)+\theta_t] g_t(x,w) T(x,x')\varphi(\epsilon_t) \\
&\qquad \times \frac{r_d(x,x',w)}{r_d(x,x',w)+\theta_t} d\epsilon_t dw dx \\
&\qquad + \int_{\mathcal{X}}\int_0^\infty\int_{-\infty}^\infty \frac{w[r_d(x',z,w)+\theta_t]}{\theta_t} g_t(x',w) T(z|x')\varphi(\epsilon_t) \\
&\qquad \times \frac{\theta_t}{r_d(x',z,w)+\theta_t} d\epsilon_t dw dz
\end{aligned}$$

$$= \int_{\mathcal{X}} \int_0^\infty \int_{-\infty}^{\infty} wR(x, x')(1+\epsilon_t) \frac{p(D, x')\pi(x')}{p(D, x)\pi(x)}$$
$$\times T(x', x) g_t(x, w) \varphi(\epsilon_t) d\epsilon_t dw dx$$
$$+ \int_{\mathcal{X}} \int_0^\infty \int_{-\infty}^{\infty} w g_t(x', w) T(x', z) \varphi(\epsilon_t) d\epsilon_t dw dz$$
$$= \int_{\mathcal{X}} \int_0^\infty wR(x, x') \frac{p(D, x')\pi(x')}{p(D, x)\pi(x)} T(x', x) g_t(x, w) dw dx$$
$$+ \int_{\mathcal{X}} \int_0^\infty w g_t(x', w) T(x', z) dw dz$$
$$= \int_{\mathcal{X}} \frac{\pi(x'|D)}{\pi(x|D)} T(x', x) \left(\int_0^\infty w g_t(x, w) dw \right) dx$$
$$+ \int_{\mathcal{X}} c_{t,x'} \pi(x'|D) T(x', z) dz$$
$$= \pi(x'|D) \int_{\mathcal{X}} c_{t,x} T(x', x) dx + c_{t,x'} \pi(x'|D)$$
$$= \pi(x'|D) \int_{\mathcal{X}} c_{t,x} T(x', x) dx + c_{t,x'} \pi(x'|D)$$
$$= 2c_{t,x'} \pi(x'|D),$$

where $\varphi(\cdot)$ denotes the density of ϵ_t. For the case $\theta_t = 0$, only the term (I) remains. Thus,

$$\int_0^\infty w' g_{t+1}(x', w') dw' = c_{t,x'} \pi(x'|D).$$

By defining $c_{t+1,x'} = 2c_{t,x'}$ for the case $\theta_t > 0$ and $c_{t+1,x'} = c_{t,x'}$ for the case $\theta_t = 0$, it is easy to see that the condition (6.26) still holds for the new population. Hence, $g_{t+1}(x', w')$ is still correctly weighted with respect to $\pi(x|D)$. ■

In practice, the initial condition (6.26) can be easily satisfied. For example, choose $g_0(x, w) = g(x)$ and set $w = \widehat{R} p(D, x)\pi(x)/g(x)$, where $\widehat{R}$ denotes an unbiased estimator of $1/Z(x)$, then $\int\int w g_0(x, w)\varphi(\epsilon) d\epsilon dw = \pi(x|D)$, that is, $c_{t,x} \equiv 1$. The initial condition is thus satisfied due to the identity $\int T(x', x) dx = 1$.

To avoid an extremely large weight caused by a nearly zero divisor, similarly to DWIS, we can make the following assumption for the target distribution: for any pair of $(x, x^*) \in \mathcal{X} \times \mathcal{X}$ and any sample $D \in \mathcal{D}$, there exists a constant r_1 such that

$$r_1 \le \frac{p(D, x)}{p(D, x^*)} \le \frac{1}{r_1}. \tag{6.27}$$

This condition can be easily satisfied by restricting both $\mathcal{X}$ and $\mathcal{D}$ to be compact. This is natural for many spatial problems, such as the autologistic model studied in Section 6.3.4. In addition, for the proposal distribution, we can assume that there exists a constant r_2 such that for any pair $(x, x^*) \in \mathcal{X} \times \mathcal{X}$,

$$r_2 \leq \frac{T(x^*, x)}{T(x, x^*)} \leq \frac{1}{r_2}. \tag{6.28}$$

This condition can be easily satisfied by using a symmetric proposal, which implies $r_2 = 1$. Following from (6.24), (6.27) and (6.28), it is easy to see that there exists a constant r_0 such that for any pair $(x, x^*) \in \mathcal{X} \times \mathcal{X}$,

$$r_0 \leq \widehat{R}_t(x, x^*) \frac{p(D, x^*)}{p(D, x)} \frac{T(x^*, x)}{T(x, x^*)} \leq \frac{1}{r_0}. \tag{6.29}$$

6.3.2.2 Monte Carlo Dynamically Weighted Importance Sampling

Similar to the DWIS, we compose the following MCDWIS algorithm, which alters between the MCDW and population control steps. Since both the MCDW step and the population control step are IWIW$_p$, the MCDWIS algorithm is also IWIW$_p$. Let W_c denote a dynamic weighting move switching parameter, which switches the value of θ_t between 0 and 1 depending on the value of $W_{up,t-1}$. One iteration of the MCDWIS algorithm can be described as follows.

- *MCDW*: Apply the Monte Carlo dynamic weighting move to the population $(\boldsymbol{x}_{t-1}, \boldsymbol{w}_{t-1})$. If $W_{up,t-1} \leq W_c$, then set $\theta_t = 1$. Otherwise, set $\theta_t = 0$. The new population is denoted by $(\boldsymbol{x}'_t, \boldsymbol{w}'_t)$.
- *Population Control*: Apply APEPCS to $(\boldsymbol{x}'_t, \boldsymbol{w}'_t)$. The new population is denoted by $(\boldsymbol{x}_t, \boldsymbol{w}_t)$.

Let $(\boldsymbol{x}_1, \boldsymbol{w}_1), \ldots, (\boldsymbol{x}_N, \boldsymbol{w}_N)$ denote a series of populations generated by MCDWIS. Then, according to (6.11), the quantity $\mu = E_\pi h(x)$ can be estimated in (6.15). As for DWIS, this estimator is consistent and asymptotically normally distributed. Below, we will show that the dynamic importance weights have finite moments of any order.

Weight Behavior Analysis To analyze the weight behavior of the MCDWIS, we introduce the following lemma.

Lemma 6.3.1 *Let $f(D|x) = p(D, x)/Z(x)$ denote the likelihood function of D, let $\pi(x)$ denote the prior distribution of x, and let $T(\cdot, \cdot)$ denote a proposal distribution of x. Define $p(x, x'|D) = p(D, x)\pi(x)T(x, x')$, and $r(x, x') = \widehat{R}(x, x')p(x', x|D)\ /p(x, x'|D)$ to be a Monte Carlo MH ratio, where $\widehat{R}(x, x')$ denotes an unbiased estimator of $Z(x)/Z(x')$. Then*

$$e_0 = E \log r(x, x') \leq 0,$$

where the expectation is taken with respect to the joint density $\varphi(\widehat{R}) \times p(x,x'|D)/Z(x)$.

Proof: By Jensen's inequality,

$$e_0 = E \log \left[\widehat{R}(x,x') \frac{p(x',x|D)}{p(x,x'|D)} \right] \leq \log E \left[\widehat{R}(x,x') \frac{p(x',x|D)}{p(x,x'|D)} \right] = 0,$$

where the equality holds when $p(x',x|D) = p(x,x'|D)$, and $\varphi(\cdot)$ is a Dirac measure with $\varphi(\widehat{R} = R) = 1$ and 0 otherwise. ■

Given this lemma, it is easy to see that, theoretically, MCDWIS shares the same weight behavior with scheme-R of DWIS; that is, the following theorem holds for MCDWIS.

Theorem 6.3.2 *Under the assumptions (6.27) and (6.28), the MCDWIS almost surely has finite moments of any order.*

6.3.3 Bayesian Analysis for Spatial Autologistic Models

Consider again the spatial autologistic model described in Section 4.8.2. Let $D = \{d_i : i \in S\}$ denote the observed binary data, where i is called a spin and S is the set of indices of the spins. Let $|S|$ denote the total number of spins in S, and let $n(i)$ denote the set of neighbors of spin i. The likelihood function of the model can be written as

$$f(D|x) = \frac{1}{Z(x)} \exp \left\{ x_a \sum_{i \in S} d_i + \frac{x_b}{2} \sum_{i \in S} d_i \left(\sum_{j \in n(i)} d_j \right) \right\}, \quad (x_a, x_b) \in \mathcal{X}, \tag{6.30}$$

where $x = (x_a, x_b)$, and the normalizing constant $Z(x)$ is intractable.

To conduct a Bayesian analysis for the model, a uniform prior

$$(x_a, x_b) \in \mathcal{X} = [-1, 1] \times [0, 1],$$

is assumed for the parameters. The MCDWIS can then be applied to simulate from the posterior distribution $f(x|D)$. Since, for this model, $\mathcal{D}$ is finite and $\mathcal{X}$ is compact, the condition (6.27) is satisfied. To ensure the condition (6.28) to be satisfied, a Gaussian random walk proposal is used for updating x. Following from Theorem 6.3.2, the importance weights will have finite moments of any order for this model.

6.3.3.1 US Cancer Mortality Data

Liang and Cheon (2009) tested the MCDWIS on the US cancer mortality data. See Section 4.8.2 for a description of the data. The MCDWIS was run

10 times for this data, and each run consisted of 100 iterations. The CPU time cost by each run was about 5.8 m on a 3.0 GHz personal computer.

In each run, the MCDWIS was initialized with $n_0 = 250$ random samples of x generated in a short run of the double MH algorithm (Liang, 2009), and each sample was assigned an equal weight of 1. The double MH run consisted of 3500 iterations, where the first 1000 iterations were discarded for burn-in, and then 250 samples were collected with an equal time space from the remaining 2500 iterations and were used as the initial samples. At each iteration, the auxiliary variable is generated with a single MH update. As previously mentioned, the double MH algorithm is very fast, but can only sample from the posterior distribution $\pi(x|D)$ approximately, even when the run is long.

After initialization, MCDWIS iterates between the MCDW and population control steps. In the MCDW step, the normalizing constant ratio $R_t(x, x^*)$ was estimated using 50 auxiliary samples, which were generated from 50 cycles of Gibbs updates starting with D; and $T(\cdot|\cdot)$ was a Gaussian random walk proposal $N_2(x, 0.03^2 I_2)$. In the population control step, the parameters were set as follows: $W_c = e^5$, $n_{low} = 200$, $n_{up} = 500$, $n_{\min} = 100$ and $n_{\max} = 1000$.

Figure 6.6 shows the time plots of the population size, θ_t and $W_{up,t}$ produced in a run of MCDWIS. It indicates that after a certain number of burn-in iterations, the population size and the magnitude of the weights can evolve stably. In each run, the first 20 iterations were discarded for the burn-in process, and the samples generated from the remaining iterations were used for inference. Averaging over the ten runs produced the following estimate: $(\widehat{x}_a, \widehat{x}_b) = (-0.3016, 0.1229)$ with the standard error $(1.9 \times 10^{-3}, 8.2 \times 10^{-4})$.

For comparison, Liang and Cheon (2009) also applied the exchange algorithm (Murray *et al.*, 2006), see Section 4.7, to this example. The exchange algorithm was also run 10 times, where each run consisted of 55 000 iterations and took about 6.5 m CPU time on a 3.0 GHz computer. By averaging over ten runs, Liang and Cheon (2009) obtained the estimate $(\widehat{x}_a, \widehat{x}_b) = (-0.3018, 0.1227)$ with the standard error $(2.7 \times 10^{-4}, 1.4 \times 10^{-4})$. In averaging, the first 5000 iterations of each run have been discarded for the burn-in process.

In summary, MCDWIS and the exchange algorithm produced almost identical estimates for this example, although the estimate produced by the former has slightly larger variation. The advantage of MCDWIS can be seen in the next subsection, where for some datasets the exchange algorithm is not feasible while MCDWIS still works well.

6.3.3.2 Simulation Studies

To assess accuracy of the estimates produced by MCDWIS, Liang and Cheon (2009) also applied MCDWIS and the exchange algorithm to the simulated

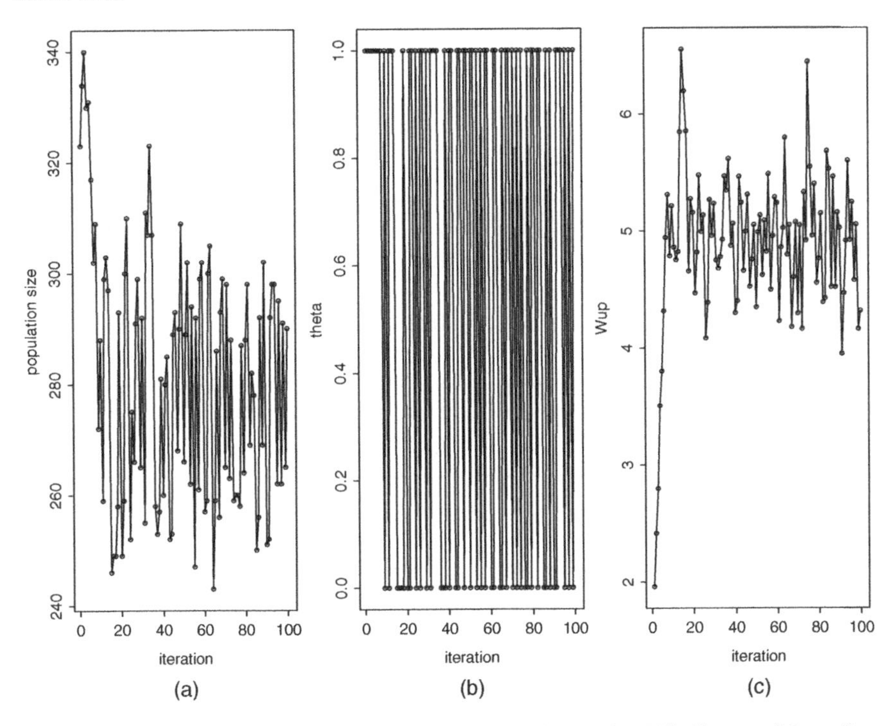

Figure 6.6 Simulation results of the MCDWIS for the US Cancer Mortality example: (a) time plot of population size; (b) time plot of θ_t; and (c) time plot of $\log(W_{up,t})$. The dotted line in plot (c) shows the value of $\log(W_c)$.

U.S. cancer mortality data by Liang (2009c); see Section 4.8.1.1 for a description of the data. Both algorithms were run as for the previous example, except that each run of the exchange algorithm was shortened to 10 500 iterations with the first 500 iterations being discarded for burn-in. The results were summarized in Table 6.1. For a thorough comparison, the maximum pseudo-likelihood estimates (MPLEs) were also included in Table 6.1, which were obtained by Liang (2009c) by maximizing the pseudo-likelihood function

$$\tilde{L}(x|D) = \prod_{i \in D} \left[\frac{e^{x_a d_i + x_b \sum_{j \in n(i)} d_i d_j}}{e^{x_a + x_b \sum_{j \in n(i)} d_j} + e^{-x_a - x_b \sum_{j \in n(i)} \sum d_j}} \right],$$

using the downhill simplex method (Press *et al.*, 1992). Comparing to other deterministic optimization methods, the downhill simplex method avoids the requirement for the gradient information of the objective function, and can thus be easily applied to the constraint optimization problems.

This example shows that MCDWIS can produce almost the same accurate results as the exchange algorithm, and more accurate results than

Table 6.1 Computational results for the simulated US cancer mortality data (recompiled from Liang and Cheon, 2009).

	MCDWIS			exchange algorithm			MPLE	
(x_a, x_b)	$\widehat{x}_a$	$\widehat{x}_b$	CPU[a]	$\widehat{x}_a$	$\widehat{x}_b$	CPU[b]	$\widehat{x}_a$	$\widehat{x}_b$
(0,.1)[e]	−.004[c]	.100	5.8	−.004	.100	1.2	−.004	.102
	.003[d]	.002		.002	.002		.002	.002
(0,.2)[e]	−.002	.202	5.8	−.002	.201	2.8	−.002	.203
	.002	.002		.002	.002		.002	.002
(0,.3)[e]	−.001	.298	5.8	−.001	.298	7.9	−.002	.298
	.001	.002		.001	.002		.002	.002
(0,.4)[e]	.000	.396	5.8	.000	.398	145.8	.002	.401
	.001	.002		.001	.001		.001	.002
(.1,.1)[e]	.103	.099	5.8	.103	.099	1.1	.102	.100
	.003	.002		.003	.002		.003	.002
(.3,.3)[e]	.295	.304	5.8	.301	.301	3.5	.290	.304
	.011	.005		.010	.004		.010	.005
(.5,.5)[f]	.495	.517	5.8	–	–	–	.561	.485
	.035	.012		–	–		.039	.012

[a]The CPU time (in minutes) cost by a single run of the MCDWIS on a 3.0 GHz Dell personal computer. [b]The CPU time (in minutes) cost by a single run of the exchange algorithm on a 3.0 GHz Dell personal computer. [c]The estimate of the corresponding parameter. [d]The standard error of the estimate. [e]The data were simulated using the perfect sampler. [f]The data were simulated using the Gibbs sampler, started with a random configuration and then iterated for 100 000 Gibbs cycles.

MPLE especially when both x_a and x_b are large. It is remarkable that the CPU time cost by the MCDWIS is independent of the setting of x. Whereas, when x_b increases, the CPU time cost by the exchange algorithm increases exponentially.

The novelty of the MCDWIS is that it allows for the use of Monte Carlo estimates in MCMC simulations, while still leaving the target distribution invariant with respect to importance weights. The MCDWIS can also be applied to Bayesian inference for missing data or random effects models, for example, the generalized linear mixed model, which often involves simulations from a posterior distribution with intractable integrals.

6.4 Sequentially Dynamically Weighted Importance Sampling

In this section, we describe a sequential version of dynamically weighted importance sampling, which can be viewed a general form of sequential importance sampling.

Although DWIS has significantly improved the mixing of the MH algorithm, it may still have a hard time in simulating from a system for which the attraction basin of the global minimum energy solution is very narrow. One approach to alleviating this difficulty is to use DWIS in conjunction with a complexity ladder by simulating from a sequence of systems with gradually flattened energy landscapes. This approach is called sequentially dynamically weighted importance sampling (SDWIS).

Suppose that one wants to simulate from a distribution $f(x)$, and a sequence of trial distributions $f_1, \cdots, f_k$ has been constructed with $f_k \equiv f$. For example, $f_{k-1}(x_{k-1})$ can be set as a marginal distribution of $f_k(x_k)$. See Section 5.6.1 for discussions on how to construct such a complexity ladder of distributions.

1.
   - *Sample* Sample $x_{1,1}, \ldots, x_{1,N_1}$ from $f_1(\cdot)$ using a MCMC algorithm, and set $w_{1,i} = 1$ for $i = 1, \ldots, N_1$. The samples form a population $(\boldsymbol{x}_1, \boldsymbol{w}_1) = (x_{1,1}, w_{1,1}; \ldots; x_{1,N_1}, w_{1,N_1})$.
   - *DWIS* Generate $(\boldsymbol{x}'_1, \boldsymbol{w}'_1)$ from $(\boldsymbol{x}_1, \boldsymbol{w}_1)$ using DWIS with $f_1(x)$ working as the target distribution.

2.
   - *Extrapolation* Generate $x_{2,i}$ from $x'_{1,i}$ with the extrapolation operator $T_{12}(x'_{1,i}, x_{2,i})$, and set

$$w_{2,i} = w'_{1,i} \frac{f_2(x_{2,i})}{f_1(x'_{1,i}) T_{12}(x'_{1,i}, x_{2,i})},$$

   for each $i = 1, 2, \cdots, N'_1$.
   - *DWIS* Generate $(\boldsymbol{x}'_2, \boldsymbol{w}'_2)$ from $(\boldsymbol{w}_2, \boldsymbol{w}_2)$ using DWIS with $f_2(x)$ working as the target distribution.
   - *Extrapolation* Generate $x_{k,i}$ from $x'_{k-1,i}$ with the extrapolation operator $T_{k-1,k}(x'_{k-1,i}, x_{k,i})$, and set

$$w_{k,i} = w'_{k-1,i} \frac{f_k(x_{k,i})}{f_{k-1}(x'_{k-1,i}) T_{k-1,k}(x'_{k-1,i}, x_{k,i})},$$

   for $i = 1, 2, \cdots, N'_{k-1}$.
   - *DWIS*: Generate $(\boldsymbol{x}'_k, \boldsymbol{w}'_k)$ from $(\boldsymbol{x}_k, \boldsymbol{w}_k)$ using DWIS with $f_k(x)$ working as the target distribution.

At each step, SDWIS includes two sub-steps: extrapolation and DWIS. In the extrapolation step, the samples from the preceding level are extrapolated to the current level. This step may involve dimension jumping, which can be accommodated by the proposal $T_{i-1,i}(\cdot,\cdot)$.

The validity of SDWIS can be verified as follows. First, it is easy to see that the condition (6.13) is satisfied by the population $(\boldsymbol{x}_1, \boldsymbol{w}_1)$, where $w_{1,i} \equiv 1$ for all i, $g(x, w) = f_1(x)$, and $c_{1,x} = 1$. Hence, the DWIS step is IWIW$_p$ with respect to the corresponding target distribution. To show the validity of SDWIS, it suffices to show that the extrapolation step is also IWIW$_p$. For simplicity, the unnecessary subscripts are suppressed in the remainder of this section. Let (x', w') denote a sample of the population $(\boldsymbol{x}'_{i-1}, \boldsymbol{w}'_{i-1})$, and let (x, w) denote a sample of the population $(\boldsymbol{x}_i, \boldsymbol{w}_i)$. Assume that (x, w) is extrapolated from (x', w'), and the population $(\boldsymbol{x}'_{i-1}, \boldsymbol{w}'_{i-1})$ is correctly weighted with respect to $f_{i-1}(x)$. For the extrapolation step, we have

$$
\begin{aligned}
&\int_0^\infty w g_i(x, w) dw \\
&= \int_0^\infty w' \frac{f_i(x)}{f_{i-1}(x') T_{i-1,i}(x', x)} g_{i-1}(x', w') T_{i-1,i}(x', x) dw' \\
&= \int_0^\infty w' g_{i-1}(x', w') \frac{f_i(x)}{f_{i-1}(x')} dw' \\
&= c_{i-1,x} f_i(x),
\end{aligned}
$$

which implies that the new population $(\boldsymbol{x}_i, \boldsymbol{w}_i)$ is also correctly weighted with respect to $f_i(x)$ by setting $c_{i,x} = c_{i-1,x}$.

The advantage of SDWIS over conventional sequential importance sampling (SIS) algorithms is apparent: the DWIS steps will remove the bad seed samples at the early stages of sampling and force the good seed samples to produce more offspring. More importantly, SDWIS overcomes the sample degeneracy problem suffered by conventional sequential importance sampling or particle filter algorithms by including the DWIS step, which maintains the diversity of the population. Together with a radial basis function network, SDWIS has been successfully applied to the modeling of the sea surface temperatures by Ryu *et al.* (2009): in a numerical example, they *et al.* show that SDWIS can be more efficient than the standard SIS and the partial rejection control SIS (Liu *et al.*, 1998) algorithms, and SDWIS indeed avoids the sample degeneracy problem.

The framework of SDWIS is so general that it has included several other algorithms as special cases. If only the population control scheme is performed at the DWIS step, SDWIS is reduced to the pruned-enriched Rosenbluth method (Grassberger, 1997). If only some MH or Gibbs steps are performed at the DWIS step (with $f_k(x)$ being chosen as a power function of $f(x)$), SDWIS is reduced to annealed importance sampling (Neal, 2001). Note that the MH or Gibbs step will not alter the correctly weightedness of a population, as they are IWIW as shown at the end of Section 6.1.1. With this observation, MacEachern *et al.* (1999) propose improving the performance of sequential importance sampling by mixing in some MCMC steps.

Exercises

6.1 In addition to the one described in Section 6.1.1, Wong and Liang (1997) prescribe the following dynamic weighting transition (type-Q transition): Draw y from a proposal transition function $T(x_t, y)$ and compute the dynamic ratio $r_d(x_t, y, w_t)$. Let $\theta > 0$ be a positive number. With probability $a = \min(1, r_d(x_t, y, w_t)/\theta)$, set $x_{t+1} = y$ and $w_{t+1} = r_d(x_t, y, w_t)/a$; otherwise, set $x_{t+1} = x_t$ and $w_{t+1} = w_t/q$, where q is the conditional probability of rejection of a proposal. If $1/q$ is unknown, its unbiased estimate, constructed based on further independent trials, can be used. Show this transition is IWIW.

6.2 Implement the dynamic weighting algorithm for the distribution $f(x) = (1/7,\ 2/7, 4/7)$ with state space $\{1, 2, 3\}$ and the following proposal transition matrices:

$$(a) \begin{pmatrix} 0 & \frac{1}{2} & \frac{1}{2} \\ \frac{1}{2} & 0 & \frac{1}{2} \\ \frac{1}{2} & \frac{1}{2} & 0 \end{pmatrix}, \qquad (b) \begin{pmatrix} 0 & \frac{1}{3} & \frac{2}{3} \\ \frac{1}{2} & 0 & \frac{1}{2} \\ \frac{4}{7} & \frac{3}{7} & 0 \end{pmatrix}.$$

(a) Estimate $E(X)$ and find the standard error of the estimate.

(b) Compare the standard errors resulted from the two different transition matrices.

6.3 Implement the dynamic weighting and DWIS algorithms for the distribution $f(x) \propto (1, 1000, 1, 2000)$ with state space $\{1, 2, 3, 4\}$ and the following proposal transition matrix:

$$\begin{pmatrix} \frac{1}{2} & \frac{1}{2} & 0 & 0 \\ \frac{2}{3} & 0 & \frac{1}{3} & 0 \\ 0 & \frac{4}{7} & 0 & \frac{3}{7} \\ 0 & 0 & \frac{1}{2} & \frac{1}{2} \end{pmatrix}.$$

Explore the convergence rates of the two algorithms through their respective estimates of $E(X)$.

6.4 Ritter and Tanner (1992) fitted the nonlinear regression model

$$y = \theta_1(1 - \exp(-\theta_2 x)) + \epsilon, \quad \epsilon \sim N(0, \sigma^2),$$

to the biochemical oxygen demand (BOD) data versus time (Source: Marske, 1967).

Table 6.2 Biochemical oxygen demand (BOD) versus time (Source: Marske 1967).

time(days)	BOD(mg/l)
1	8.3
2	10.3
3	19.0
4	16.0
5	15.6
7	19.8

Assume the prior of σ^2 is flat on $(0, \infty)$ and the prior of (θ_1, θ_2) is uniform on the region $[-20, 50] \times [-2, 6]$. Integrating out σ^2 from the posterior yields

$$f(\theta_1, \theta_2 | x, y) \approx [S(\theta_1, \theta_2)]^{-n/2-1},$$

where $S(\theta_1, \theta_2)$ is the sum of squares. Implement DWIS for the posterior density $f(\theta_1, \theta_2|x, y)$, and find a discrete estimate of $f(\theta_1, \theta_2|x, y)$.

6.5 Implement the simulated example described in Section 6.3.3.2, and try to reproduce Table 6.1.

6.6 Discuss how to apply the dynamic weighting algorithm to training a feed-forward neural network.

6.7 The DWIS algorithm works on a population of unequally weighted samples. Discuss how to make crossover operations for DWIS.

Chapter 7

Stochastic Approximation Monte Carlo

Consider sampling from a distribution with the density/mass function

$$f(x) = \frac{1}{Z(\tau)} \exp\{-H(x)/\tau\}, \quad x \in \mathcal{X} \tag{7.1}$$

on a sample space $\mathcal{X}$, where τ is called the temperature, and $H(x)$ the energy function in terms of physics. Conventional MCMC algorithms, such as the Metropolis-Hastings algorithm, would draw each configuration $x \in \mathcal{X}$ with a probability proportional to its Boltzmann weight

$$w_b(x) = \exp(-H(x)/\tau).$$

In statistical physics, this type of simulation is called the *canonical ensemble* simulation, which yields the usually bell-shaped probability distribution of energy $U = H(x)$:

$$P_{CE}(u) = \frac{1}{Z(\tau)} g(u) \exp(-u/\tau),$$

where $g(u)$ is called the *density of states* (or spectral density).

Local updates have probability $\exp(-\Delta u/\tau)$ for the system to cross an energy barrier of Δu. Hence, at low temperatures, the sampler tends to get trapped in one of the local energy minima, rendering the simulation ineffective. In their seminal work on umbrella sampling, Torrie and Valleau (1977) consider two general strategies to alleviate this type of difficulty. One is to use a global updating scheme, and the other is to sample from an ensemble for which each configuration is assigned a weight different from $w_b(\cdot)$ so that the local-trap problem does not exist any more. The population-based algorithms

Advanced Markov Chain Monte Carlo Methods: Learning from Past Samples
Faming Liang, Chuanhai Liu and Raymond J. Carroll

described in Chapter 5 and some of the auxiliary variable-based algorithms described in Chapter 4 (e.g., simulated tempering and clustering algorithms) are all based on the first strategy. For example, in simulated tempering and parallel tempering, the simulation at high temperatures provides a global proposal for simulation at the target temperature. In this chapter, we will describe some algorithms that utilize the second strategy, in particular, the stochastic approximation Monte Carlo (SAMC) (Liang *et al.*, 2007), and related algorithms.

7.1 Multicanonical Monte Carlo

Multicanonical Monte Carlo (Berg and Neuhaus, 1991, 1992) seeks to draw samples in an ensemble where each configuration with energy $u = H(x)$ is assigned a weight

$$w_m(u) \propto \frac{1}{g(u)} = e^{-S(u)},$$

where $S(u) = \log(g(u))$ is called the microcanonical entropy. A simulation with this weight function will yield a uniform distribution of energy:

$$P_m(u) \propto g(u)w_m(u) = \text{constant},$$

and lead to a free random walk in the space of energy. This allows the sampler to escape any energy barriers, and to explore any regions of the sample space even for those with small $g(u)$'s. The samples generated in the simulation will form a flat histogram in the space of energy, hence, the multicanonical algorithm is also called a flat histogram Monte Carlo algorithm (Liang, 2006).

Since $g(u)$'s are generally unknown a priori, Berg and Neuhaus (1992) provided a recursive procedure to learn their values. The procedure initializes the estimate of $g(u)$ through a Monte Carlo simulation with a tempered target distribution $f_T(x) \propto \exp(-H(x)/T)$. For simplicity, suppose that the energy function U only takes values on a finite set $\{u_1, \ldots, u_m\}$. Let $x_1, \ldots, x_N$ denote the MCMC samples drawn from $f_T(x)$, and let $N_T(i) = \#\{x_j : H(x_j) = u_i\}$ denote the number of samples with energy u_i. As $N \to \infty$,

$$N_T(i)/N \approx \frac{1}{Z(T)} g(u_i) e^{-u_i/T}, \quad i = 1, \ldots, m,$$

then the spectral density can be estimated by

$$\widehat{g}(u_i) = \frac{N_T(i)e^{u_i/T}}{\sum_{j=1}^m N_T(j)e^{u_j/T}}, \quad i = 1, \ldots, m.$$

In practice, the temperature T should be sufficiently high such that each value of the energy can be visited with reasonably large frequency. Given the initial spectral density estimate, the multicanonical algorithm iterates between the following two steps:

1. Run a MCMC sampler, say, the MH algorithm, sufficiently long according to the current weighting function
$$w_m^{(t)}(x) \propto \frac{1}{\widehat{g}_t(H(x))}, \tag{7.2}$$
where t indexes the stages of the simulation.

2. Update the spectral density estimate by
$$\log\left(\widehat{g}_{t+1}(u_i)\right) = c + \log\left(\widehat{g}_t(u_i)\right) + \log\left(\widehat{\pi}_t(i) + \alpha_i\right), \quad i = 1, \ldots, m, \tag{7.3}$$
where the constant c is introduced to ensure that $\log(\widehat{g}_{t+1})$ is an estimate of $\log(g)$, and $\widehat{\pi}_t(i)$ is the relative sampling frequency of the energy u_i at stage t, and $\alpha_1, \ldots, \alpha_m$ are small positive constants which serve as 'prior values' to smooth out the estimate $\widehat{g}$.

Since the number of iterations performed in step 1 is sufficiently large, it is reasonable to assume that the simulation has reached equilibrium, and thus
$$\widehat{\pi}_t(i) \propto \frac{g(u_i)}{\widehat{g}_t(u_i)}, \quad i = 1, \ldots, m. \tag{7.4}$$
Substituting (7.4) into (7.3), then
$$\log\left(\widehat{g}_{t+1}(u_i)\right) = c + \log(g(u_i)), \quad i = 1, \ldots, m, \tag{7.5}$$
which implies the validity of the algorithm for estimating $g(u)$ (up to a multicanonical constant). On the other hand, in (7.5), the independence of $\widehat{g}_{t+1}(u)$ on the previous estimate $\widehat{g}_t(u)$ implies that the spectral density estimate can only reach limited accuracy, which is determined by the length of the simulation performed in Step 1. After certain stage, increasing the number of stages will not improve the accuracy of the spectral density estimate.

Given a set of multicanonical samples $x_t^{(1)}, \ldots, x_t^{(N_t)}$ generated at stage t, the quantity $E_f\rho(x)$ (the expectation of $\rho(x)$ with respect to the target distribution $f(x)$) can be estimated using the reweighting technique by
$$\widehat{E}_f^{(t)}\rho(x) = \frac{\sum_{i=1}^{N} \widehat{g}_t(H(x_t^{(i)}))\rho(x_t^{(i)})}{\sum_{i=1}^{N_t} \widehat{g}_t(H(x_t^{(i)}))},$$
which is consistent as the conventional importance sampling estimator. More generally, $E_f\rho(x)$ can be estimated, based on the samples generated in multiple stages, by
$$\widetilde{E}_f\rho(x) = \sum_{k=t_0+1}^{t} \lambda_k \widehat{E}_f^{(k)}\rho(x),$$

where t_0 denotes the number of burn-in stages, and λ_k's are subject to the constraint $\sum_{k=t_0+1}^{t} \lambda_k = 1$ and can be chosen to minimize the total variance of the estimator.

The multicanonical algorithm has been employed by Hansmann and Okamoto (1997) to optimization for a continuous system. They discretize the energy as $-\infty < u_1 < \cdots < u_m < \infty$, and redefine $N(u_i)$ as the number of samples with energy between u_{i-1} and u_i, i.e., $N(u_i) = \#\{x_j : u_{i-1} < H(x_j) \leq u_i\}$. In this case, $g(u_i)$ is no longer the spectral density of the system, but the hyper-volume of the subregion $E_i = \{x : u_{i-1} < H(x) \leq u_i\}$. Multicanonical Monte Carlo has been employed for an increasing number of applications in physics, chemistry, structural biology and other areas. See Berg *et al.* (1999) for details.

7.2 $1/k$-Ensemble Sampling

Similar to multicanonical Monte Carlo, $1/k$-ensemble sampling (Hesselbo and Stinchcombe, 1995) seeks to draw samples in an ensemble where each configuration x with energy $u = H(x)$ is assigned a weight

$$w_{1/k}(u) \propto \frac{1}{k(u)},$$

where $k(u) = \sum_{u' \leq u} g(u')$, that is, the cumulative spectral density function of the distribution. Hence, $1/k$-ensemble sampling will produce the following distribution of energy:

$$P_{1/k}(u) \propto \frac{g(u)}{k(u)} = \frac{d \log k(u)}{du}.$$

Since, in many physical systems, $k(u)$ is a rapidly increasing function of u, $\log k(u) \approx \log g(u)$ for a wide range of u, and the simulation will lead to an approximately random walk in the space of entropy. Recall that $S(u) = \log g(u)$ is called the microcanonical entropy of the system. Comparing to multicanonical Monte Carlo, $1/k$-ensemble sampling is designed to spend more time in exploring low energy regions, hence, it is potentially more suitable for optimization problems. Improvement over multicanonical Monte Carlo has been observed in ergodicity of the simulations for the Ising model and travel salesman problems (Hesselbo and Stinchcombe, 1995).

In practice, $k(u)$ is not known a priori. It can be estimated using the same recursive procedure as described in the previous section. Obviously, with a good estimate of $g(u)$, one would be able to get a good estimate of $k(u)$, and vice versa. Motivated by the Wang-Landau algorithm (Wang and Landau, 2001) described in Section 7.3, Liang (2004) proposes an online procedure for estimation of $k(u)$, which facilitates the implementation of the algorithm. Liang's procedure can be described as follows.

Consider the problem of sampling from the distribution (7.1), where the sample space $\mathcal{X}$ can be discrete or continuous. Let $T(\cdot \to \cdot)$ denote a proposal distribution, and let $\psi(x)$ denote a general non-negative function defined on $\mathcal{X}$ with $\int_{\mathcal{X}} \psi(x)dx < \infty$. For example, for a continuous system one may set $\psi(x) = \exp\{-H(x)/t\}$ with $t \geq \tau$; and for a discrete system one may set $\psi(x) \equiv 1$. Issues on the choice of $\psi(x)$ will be discussed later. Suppose that the sample space has been partitioned according to the energy function into m disjoint subregions: $E_1 = \{x : H(x) \leq u_1\}, E_2 = \{x : u_1 < H(x) \leq u_2\}, \ldots$, $E_{m-1} = \{x : u_{m-2} < H(x) \leq u_{m-1}\}$, and $E_m = \{x : H(x) > u_{m-1}\}$, where $-\infty < u_1 < u_2 < \ldots < u_m < \infty$ are specified by the user. Let $G^{(i)} = \int_{\cup_{j=1}^{i} E_i} \psi(x)dx$ for $i = 1, \ldots, m$, and let $\boldsymbol{G} = (G^{(1)}, \ldots, G^{(m)})$. Since m is finite and the partition $E_1, \ldots, E_m$ is fixed, there must exist a number $\epsilon > 0$ such that

$$\epsilon < \min_{i,j} \frac{G^{(i)}}{G^{(j)}} < \max_{i,j} \frac{G^{(i)}}{G^{(j)}} < \frac{1}{\epsilon}, \tag{7.6}$$

for all $G^{(i)}, G^{(j)} > 0$. With a slight generalization, $1/k$-ensemble sampling is reformulated by Liang (2004) as an algorithm of sampling from the following distribution

$$f_G(x) \propto \sum_{i=1}^{m} \frac{\psi(x)}{G^{(i)}} I(x \in E_i), \tag{7.7}$$

where $I(\cdot)$ is the indicator function.

Since $\boldsymbol{G}$ is unknown in general, the simulation from (7.7) can proceed as follows. Let $\{\gamma_t\}$ denote a positive, non-increasing sequence satisfying

$$(i) \quad \overline{\lim_{t\to\infty}} |\gamma_t^{-1} - \gamma_{t+1}^{-1}| < \infty, \quad (ii) \quad \sum_{t=1}^{\infty} \gamma_t = \infty, \quad (iii) \quad \sum_{t=1}^{\infty} \gamma_t^{\zeta} < \infty, \tag{7.8}$$

for some $\zeta \in (1, 2]$. For example, one may set

$$\gamma_t = \frac{t_0}{\max\{t_0, t\}}, \quad t = 1, 2, \ldots, \tag{7.9}$$

for some value $t_0 > 1$. Let $\widehat{G}_t^{(i)}$ denote the working estimate of $G^{(i)}$ obtained at iteration t. By (7.6), it is reasonable to restrict $\widehat{\boldsymbol{G}}$ to taking values from a compact set. Denote the compact set by Θ, which can be set to a huge set, say, $\Theta = [0, 10^{100}]^m$. As a practical matter, this is equivalent to setting $\Theta = \mathbb{R}^m$. The simulation starts with an initial estimate of $\boldsymbol{G}$, say $\widehat{G}_0^{(i)} = i$ for $i = 1, \ldots, m$, and a random sample $x_0 \in \mathcal{X}$, and then iterates between the following steps:

On-line $1/k$-Ensemble Sampling Algorithm

1. Propose a new configuration x^* in the neighborhood of x_t according to a prespecified proposal distribution $T(\cdot \to \cdot)$.

2. Accept x^* with probability

$$\min\left\{\frac{\widehat{G}_t^{(J(x_t))}}{\widehat{G}_t^{(J(x^*))}}\frac{\psi(x^*)}{\psi(x_t)}\frac{T(x^* \to x_t)}{T(x_t \to x^*)}, 1\right\}, \tag{7.10}$$

   where $J(z)$ denotes the index of the subregion that z belongs to. If it is accepted, set $x_{t+1} = x^*$. Otherwise, set set $x_{t+1} = x_t$.

3. Update the estimates of $G^{(i)}$'s by setting

$$\widehat{G}_{t+\frac{1}{2}}^{(i)} = \widehat{G}_t^{(i)},$$

   for $i = 1, \ldots, J(x_{t+1}) - 1$, and

$$\widehat{G}_{t+\frac{1}{2}}^{(i)} = \widehat{G}_t^{(i)} + \delta_s \widehat{G}_t^{(J(x_{t+1}))},$$

   for $i = J(x_{t+1}), \ldots, m$. If $\widehat{\boldsymbol{G}}_{t+\frac{1}{2}} \in \Theta$, then set $\widehat{\boldsymbol{G}}_{t+1} = \widehat{\boldsymbol{G}}_{t+\frac{1}{2}}$; otherwise, set $\widehat{\boldsymbol{G}}_{t+1} = c^*\widehat{\boldsymbol{G}}_{t+\frac{1}{2}}$, where c^* is chosen such that $c^*\widehat{\boldsymbol{G}}_{t+\frac{1}{2}} \in \Theta$.

This algorithm falls into the category of stochastic approximation algorithms (Robbins and Monro, 1951; Benveniste *et al.*, 1990; Andrieu *et al.*, 2005). Under mild conditions, it can be shown (similar to the proof, presented in Section 7.7.1, for the convergence of SAMC) that as $t \to \infty$,

$$\widehat{G}_t^{(i)} \to c \int_{\cup_{j=1}^{i} E_j} \psi(x)dx, \tag{7.11}$$

for $i = 1, \ldots, m$, where c is an arbitrary constant. Note that f_G is invariant with respect to scale transformations of $\boldsymbol{G}$. To determine the value of c, extra information is needed, such as $G^{(m)}$ being equal to a known number.

For discrete systems, if one sets $\psi(x) \equiv 1$, the algorithm reduces to the original $1/k$-ensemble algorithm. For continuous systems, a general choice for $\psi(x)$ is $\psi(x) = \exp\{-H(x)/T\}$. In this case, $G^{(i)}$ corresponds to the partition function of a truncated distribution whose unnormalized density function is given by $\exp\{-H(x)/T\} I_{x \in \cup_{j=1}^{i} E_j}$. In the limit $T \to \infty$, the algorithm will work as for a discrete system.

7.3 The Wang-Landau Algorithm

Like multicanonical Monte Carlo, the Wang-Landau algorithm (Wang and Landau, 2001) seeks to draw samples in an ensemble where each configuration with energy u is assigned a weight

$$w_m(u) \propto \frac{1}{g(u)},$$

where $g(u)$ is the spectral density. The difference between the two algorithms is on their learning procedures for the spectral density.

The Wang-Landau algorithm can be regarded as an innovative implementation of multicanonical Monte Carlo, but it goes beyond that. In multicanonical Monte Carlo, the sampler tends to be blocked by the edge of the already visited area, and it takes a long time to traverse an area because of the general feature of 'random walk'. The WL algorithm succeeds in removing these problems by timely penalizing moving to and staying at the energy that have been visited many times.

Suppose that the sample space $\mathcal{X}$ is finite and the energy function $H(x)$ takes values on a finite set $\{u_1, \ldots, u_m\}$. The simulation of the Wang-Landau algorithm consists of several stages. In the first stage, it starts with an initial setting of $\widehat{g}(u_1), \ldots, \widehat{g}(u_m)$, say $\widehat{g}(u_1) = \ldots = \widehat{g}(u_m) = 1$, and a random sample x_0 drawn from $\mathcal{X}$, and then iterates between the following steps:

Wang-Landau Algorithm

1. Simulate a sample x by a single Metropolis update which admits the invariant distribution $\widehat{f}(x) \propto 1/\widehat{g}(H(x))$.

2. Set $\widehat{g}(u_i) \leftarrow \widehat{g}(u_i)\delta^{I(H(x)=u_i)}$ for $i = 1, \ldots, m$, where δ is a modification factor greater than 1 and $I(\cdot)$ is the indicator function.

The algorithm iterates till a flat histogram has been produced in the space of energy. A histogram is usually considered to be flat if the sampling frequency of each u_i is not less than 80% of the average sampling frequency. Once this condition is satisfied, the estimates $\widehat{g}(u_i)$'s and the current sample x are passed on to the next stage as initial values, the modification factor is reduced to a smaller value according to a specified scheme, say, $\delta \leftarrow \sqrt{\delta}$, and the sampler collector is resumed. The next stage simulation is then started, continuing until the new histogram is flat again. The process is repeated until δ is very close to 1, say, $\log(\delta) < 10^{-8}$.

Liang (2005b) generalized the Wang-Landau algorithm to continuum systems. The generalization is mainly in three aspects, namely the sample space, the working function and the estimate updating scheme. Suppose that the sample space $\mathcal{X}$ is continuous and has been partitioned according to a chosen parameterization of x, say, the energy function $H(x)$, into m disjoint subregions: $E_1 = \{x : H(x) \le u_1\}, E_2 = \{x : u_1 < H(x) \le u_2\}, \ldots, E_{m-1} = \{x : u_{m-2} < H(x) \le u_{m-1}\}$, and $E_m = \{x : H(x) > u_{m-1}\}$, where $-\infty < u_1 < u_2 < \ldots < u_{m-1} < \infty$ are specified by the user. Let $\psi(x)$ be a non-negative function defined on the sample space with $0 < \int_{\mathcal{X}} \psi(x)dx < \infty$, which is also called the working function of the generalized Wang-Landau algorithm.

In practice, one often sets $\psi(x) = \exp(-H(x)/\tau)$. Let $\boldsymbol{g} = (g^{(1)}, \ldots, g^{(m)})$ be the weight associated with the subregions. One iteration of the generalized Wang-Landau algorithm consists of the following steps:

Generalized Wang-Landau Algorithm

1. Simulate a sample x by a number, denoted by κ, of MH steps which admits
$$\widehat{f}(x) \propto \sum_{i=1}^{m} \frac{\psi(x)}{\widehat{g}^{(i)}} I(x \in E_i), \tag{7.12}$$
as the invariant distribution.

2. Set $\widehat{g}^{(k)} \leftarrow \widehat{g}^{(k)} + \delta \varrho^{k-J(x)} \widehat{g}^{(J(x))}$ for $k = J(x), \ldots, m$, where $J(x)$ is the index of the subregion that x belongs to and $\varrho > 0$ is a parameter which controls the sampling frequency for each of the subregions.

Liang (2005b) noted that, as the number of stages becomes large,
$$\begin{aligned} \widehat{g}^{(1)} &\approx g^{(1)} = c \int_{E_1} \psi(x)dx, \\ \widehat{g}^{(i)} &\approx g^{(i)} = c \int_{E_i} \psi(x)dx + \rho g^{(i-1)}, \quad i = 2, \ldots, m, \end{aligned}$$
where c is an unknown constant; and the visiting frequency to each subregion will be approximately proportional to $\int_{E_i} \psi(x)dx/g^{(i)}$ for $i = 1, \ldots, m$. The generalization from the spectral density to the integral $\int_{E_i} \psi(x)dx$ is of great interest to statisticians, which leads to a wide range of applications of the algorithm in statistics, such as the model selection and some other Bayesian computational problems, as discussed in Liang (2005b).

If κ is small, say, $\kappa = 1$ as adopted in the Wang-Landau algorithm, there is no rigorous theory to support the convergence of $\widehat{g}_i$'s. In fact, some deficiencies of the Wang-Landau algorithm have been observed in simulations. Yan and de Pablo (2003) note that the estimates of $g(u_i)$ can only reach a limited statistical accuracy which will not be improved with further iterations, and the large number of configurations generated towards the end of the simulation make only a small contribution to the estimates. The deficiency of the Wang-Landau algorithm is caused by the choice of the modification factor δ. This can be explained as follows. Let n_s be the number of iterations performed in stage s, and let δ_s be the modification factor used in stage s. Without loss of generality, let $n_s \equiv n$, which is so large that a flat histogram of energy can be produced at each stage. Let $\delta_s = \sqrt{\delta_{s-1}}$ as suggested by Wang and Landau (2001). Then the tail sum $n \sum_{s=s_0+1}^{\infty} \log \delta_s < \infty$ for any value of s_0. Since the tail sum represents the total correction to the current estimate made in the iterations that follow, the numerous configurations generated toward the end of the simulation make only a small contribution to the estimate. To overcome this deficiency, Liang *et al.* (2007) suggest guiding the choice of δ by the stochastic approximation algorithm (Robbins and Monro, 1951), which ensures that the estimate of $\boldsymbol{g}$ can be improved continuously as the simulation goes on. The resulting algorithm is called the stochastic approximation Monte Carlo (SAMC) algorithm.

7.4 Stochastic Approximation Monte Carlo

The Algorithm. Consider the problem of sampling from the distribution (7.1). Suppose, as in the generalized Wang-Landau algorithm (Liang, 2005b), the sample space $\mathcal{X}$ has been partitioned according to a function $\lambda(x)$ into m subregions, $E_1 = \{x : \lambda(x) \leq u_1\}, E_2 = \{x : u_1 < \lambda(x) \leq u_2\}, \ldots, E_{m-1} = \{x : u_{m-2} < \lambda(x) \leq u_{m-1}\}, E_m = \{x : \lambda(x) \geq u_{m-1}\}$, where $-\infty < u_1 < \cdots < u_{m-1} < \infty$. Here $\lambda(\cdot)$ can be any function of x, such as a component of x, the energy function $H(x)$, etc. Let $\psi(x)$ be a non-negative function with $0 < \int_{\mathcal{X}} \psi(x)dx < \infty$, which is called the working function of SAMC. In practice, one often sets $\psi(x) = \exp(-H(x)/\tau)$. Let $g_i = \int_{E_i} \psi(x)dx$ for $i = 1, \ldots, m$, and $\boldsymbol{g} = (g_1, \ldots, g_m)$. The subregion E_i is called an empty subregion if $g_i = 0$. An inappropriate specification of the cut-off points u_i's may result in some empty subregions. Technically, SAMC allows for the existence of empty subregions in simulations. To present the idea clearly, we temporarily assume that all subregions are nonempty; that is, assuming $g_i > 0$ for all $i = 1, \ldots, m$. SAMC seeks to sample from the distribution

$$f_g(x) \propto \sum_{i=1}^{m} \frac{\pi_i \psi(x)}{g_i} I(x \in E_i), \tag{7.13}$$

where π_i's are prespecified frequency values such that $\pi_i > 0$ for all i and $\sum_{i=1}^{m} \pi_i = 1$. It is easy to see, if $g_1, \ldots, g_m$ are known, sampling from $f_g(x)$ will result in a random walk in the space of the subregions, with each subregion being visited with a frequency proportional to π_i. The distribution $\boldsymbol{\pi} = (\pi_1, \ldots, \pi_m)$ is called the desired sampling distribution of the subregions.

Since $\boldsymbol{g}$ is unknown, Liang *et al.* (2007) estimate $\boldsymbol{g}$ under the framework of the stochastic approximation algorithm (Robbins and Monro, 1951), which ensures consistency of the estimation of $\boldsymbol{g}$. Let $\theta_t^{(i)}$ denote the estimate of $\log(g_i/\pi_i)$ obtained at iteration t, and let $\theta_t = (\theta_t^{(1)}, \ldots, \theta_t^{(m)})$. Since m is finite, the partition $E_1, \ldots, E_m$ is fixed, and $0 < \int_{\mathcal{X}} \psi(x) < \infty$, there must exist a number $\epsilon > 0$ such that

$$\epsilon < \min_i \log\left(\frac{g_i}{\pi_i}\right) < \max_i \log\left(\frac{g_i}{\pi_i}\right) < \frac{1}{\epsilon},$$

which implies that θ_t can be restricted to taking values on a compact set. Henceforth, this compact set will be denoted by Θ. In practice, Θ can be set to a huge set, say, $\Theta = [-10^{100}, 10^{100}]^m$. As a practical matter, this is equivalent to set $\Theta = \mathbb{R}^m$. Otherwise, if one assumes $\Theta = \mathbb{R}^m$, a varying truncation version of this algorithm can be considered, as in Liang (2009e). For both mathematical and practical simplicity, Θ is restricted to be compact in this chapter.

Let $\{\gamma_t\}$ denote the gain factor sequence, which satisfies the condition (A_1):

(A_1) The sequence $\{\gamma_t\}$ is positive and nonincreasing, and satisfies the conditions:

$$(i)\quad \overline{\lim_{t\to\infty}} |\gamma_t^{-1} - \gamma_{t+1}^{-1}| < \infty, \quad (ii)\quad \sum_{t=1}^{\infty} \gamma_t = \infty, \quad (iii)\quad \sum_{t=1}^{\infty} \gamma_t^{\eta} < \infty, \tag{7.14}$$

for some $\eta \in (1, 2]$.

In practice, one often sets

$$\gamma_t = \frac{t_0}{\max\{t_0, t^{\xi}\}}, \quad t = 0, 1, 2, \ldots, \tag{7.15}$$

for some prespecified values $t_0 > 1$ and $\frac{1}{2} < \xi \leq 1$. A large value of t_0 will allow the sampler to reach all subregions very quickly even for a large system. Let $J(x)$ denote the index of the subregion that the sample x belongs to. SAMC starts with a random sample x_0 generated in the space $\mathcal{X}$ and an initial estimate $\theta_0 = (\theta_0^{(1)}, \ldots, \theta_0^{(m)}) = (0, \ldots, 0)$, then iterates between the following steps:

The SAMC Algorithm

(a) *Sampling* Simulate a sample x_{t+1} by a single MH update which admits the following distribution as its invariant distribution:

$$f_{\theta_t}(x) \propto \sum_{i=1}^{m} \frac{\psi(x)}{\exp(\theta_t^{(i)})} I(x \in E_i). \tag{7.16}$$

(a.1) Generate y in the sample space $\mathcal{X}$ according to a proposal distribution $q(x_t, y)$.

(a.2) Calculate the ratio

$$r = e^{\theta_t^{(J(x_t))} - \theta_t^{(J(y))}} \frac{\psi(y) q(y, x_t)}{\psi(x_t) q(x_t, y)}.$$

(a.3) Accept the proposal with probability $\min(1, r)$. If it is accepted, set $x_{t+1} = y$; otherwise, set $x_{t+1} = x_t$.

(b) *Weight updating* For $i = 1, \ldots, m$, set

$$\theta_{t+\frac{1}{2}}^{(i)} = \theta_t^{(i)} + \gamma_{t+1} \left(I_{\{x_{t+1} \in E_i\}} - \pi_i \right). \tag{7.17}$$

If $\theta_{t+\frac{1}{2}} \in \Theta$, set $\theta_{t+1} = \theta_{t+\frac{1}{2}}$; otherwise, set $\theta_{t+1} = \theta_{t+\frac{1}{2}} + \boldsymbol{c}^*$, where $\boldsymbol{c}^* = (c^*, \ldots, c^*)$ can be any constant vector satisfying the condition $\theta_{t+\frac{1}{2}} + \boldsymbol{c}^* \in \Theta$.

- *Remark 1* In the weight updating step, $\theta_{t+\frac{1}{2}}$ is adjusted by adding a constant vector $\boldsymbol{c}^*$ when $\theta_{t+\frac{1}{2}} \notin \Theta$. The validity of this adjustment is simply due to the fact that $f_{\theta_t}(x)$ is invariant with respect to a location shift of θ_t.

  The compactness constraint on θ_t should only apply to the components of θ for which the corresponding subregions are unempty. In practice, one can place an indicator on each subregion, indicating whether or not the subregion has been visited or is known to be unempty. The compactness check for $\theta_{t+\frac{1}{2}}$ should be done only for the components for which the corresponding subregions have been visited or are known to be unempty.

- *Remark 2* The explanation for the condition (A_1) can be found in advanced books on stochastic approximation, see, for example, Nevel'son and Has'minskiĭ (1973). The condition $\sum_{t=1}^{\infty} \gamma_t = \infty$ is necessary for the convergence of θ_t. Otherwise, it follows from Step (b) that, assuming the adjustment of $\theta_{t+\frac{1}{2}}$ did not occur,

$$\sum_{t=0}^{\infty} |\theta_{t+1}^{(i)} - \theta_t^{(i)}| \leq \sum_{t=0}^{\infty} \gamma_{t+1} |I_{\{x^{(t+1)} \in E_i\}} - \pi_i| \leq \sum_{t=0}^{\infty} \gamma_{t+1} < \infty.$$

  Thus, θ_t cannot reach $\log(\boldsymbol{g}/\boldsymbol{\pi})$ if, for example, the initial point θ_0 is sufficiently far away from $\log(\boldsymbol{g}/\boldsymbol{\pi})$. On the other hand, γ_t cannot be too large. An overly large γ_t will prevent convergence. It turns out that the third condition in (7.14) asymptotically damps the effect of random errors introduced by new samples. When it holds, we have $\gamma_{t+1}|I_{\{x_{t+1} \in E_i\}} - \pi_i| \leq \gamma_{t+1} \to 0$ as $t \to \infty$.

- *Remark 3* A striking feature of the SAMC algorithm is that it possesses a self-adjusting mechanism: if a proposed move is rejected at an iteration, then the weight of the subregion that the current sample belongs to will be adjusted to a larger value, and the total rejection probability of the next iteration will be reduced. This mechanism enables the algorithm to escape from local energy minima very quickly. The SAMC algorithm represents a significant advance for simulations of complex systems for which the energy landscape is rugged.

Convergence. To provide a rigorous theory for the algorithm, we need to assume that the Markov transition kernels used in Step (a) satisfy the drift condition given in Section 7.7.1. This assumption is classical in the literature of Markov chain, which implies the existence of a stationary distribution $f_\theta(x)$

for any $\theta \in \Theta$ and uniform ergodicity of the Markov chain. However, to check the drift condition, it is usually difficult. For example, for the random walk MH kernel, the distribution $f(x)$, or equivalently, the distribution $f_\theta(x)$ worked on at each iteration, needs to satisfy the following conditions as prescribed by Andrieu *et al.* (2005):

(i) $f(x)$ is bounded, bounded away from zero on every compact set, and continuously differentiable (a.e.).

(ii) $f(x)$ is superexponential, that is,

$$\lim_{|x|\to+\infty} \left\langle \frac{x}{|x|}, \nabla \log f(x) \right\rangle = -\infty.$$

(iii) The contours $\partial A(x) = \{y : f(y) = f(x)\}$ are asymptotically regular, that is,

$$\lim_{|x|\to+\infty} \sup \left\langle \frac{x}{|x|}, \frac{\nabla f(x)}{|\nabla f(x)|} \right\rangle < 0.$$

These conditions control the tail behavior of $f(x)$. Since they are usually difficult to verify, for mathematical simplicity, one may assume that $\mathcal{X}$ is compact. For example, one may restrict $\mathcal{X}$ to the set $\{x : f(x) \geq \epsilon_0\}$, where ϵ_0 is a sufficiently small number, say, $\epsilon_0 = \sup_{x\in\mathcal{X}} f(x)/10^{100}$, where $\sup_{x\in\mathcal{X}} f(x)$ can be estimated through a pilot run. This, as a practical matter, is equivalent to imposing no constraints on $\mathcal{X}$. To ease verification of the drift condition, one may assume further that the proposal distribution $q(x,y)$ satisfies the following local positive condition:

(A_2) For every $x \in \mathcal{X}$, there exist $\epsilon_1 > 0$ and $\epsilon_2 > 0$ such that

$$\|x - y\| \leq \epsilon_1 \Longrightarrow q(x,y) \geq \epsilon_2, \tag{7.18}$$

where $\|x - y\|$ denotes a certain distance measure between x and y.

This is a natural condition in studying the convergence of the MH algorithm (see, e.g., Roberts and Tweedie, 1996, and Jarner and Hansen, 2000). In practice, this kind of proposal can be easily designed for both discrete and continuum systems. For a continuum system, $q(x,y)$ can be set to the random walk Gaussian proposal $y \sim N(x, \sigma^2 I)$ with σ^2 being calibrated to have a desired acceptance rate. For a discrete system, $q(x,y)$ can be set to a discrete distribution defined on a neighborhood of x, assuming that the states have been ordered in a certain way.

Under the assumptions (A_1), (A_2) and the compactness of $\mathcal{X}$, Liang *et al.* (2007) prove the following theorem regarding the convergence of SAMC, where (A_2) and the compactness of $\mathcal{X}$ directly leads to the holding of the drift condition.

Theorem 7.4.1 *Assume (A_1) and the drift condition (B_2) (given in Section 7.7.1) hold. Then, as $t \to \infty$,*

$$\theta_t^{(i)} \to \theta_*^{(i)} = \begin{cases} C + \log(\int_{E_i} \psi(x)dx) - \log(\pi_i + \nu), & \text{if } E_i \neq \emptyset, \\ -\infty, & \text{if } E_i = \emptyset, \end{cases} \tag{7.19}$$

where C is an arbitrary constant, $\nu = \sum_{j \in \{i: E_i = \emptyset\}} \pi_j/(m - m_0)$, and m_0 is the number of empty subregions.

The proof of this theorem can be found in Section 7.7.1, where the convergence for a general stochastic approximation MCMC (SAMCMC) algorithm is proved. As discussed in Section 7.7.1.4, SAMC can be viewed as a special instance of the general SAMCMC algorithm. Since $f_{\theta_t}(x)$ is invariant with respect to a location shift of θ_t, the unknown constant C in (7.19) can not be determined by the samples drawn from $f_{\theta_t}(x)$. To determine the value of C, extra information is needed; for example, $\sum_{i=1}^m e^{\theta_t^{(i)}}$ is equal to a known number. Let $\widehat{\pi}_i^{(t)}$ denote the realized sampling frequency of the subregion E_i by iteration t. As $t \to \infty, \widehat{\pi}_i^{(t)}$ converges to $\pi_i + \nu$ if $E_i \neq \emptyset$ and 0 otherwise. In this sense, SAMC can be viewed as a dynamic stratified sampling method. Note that for a nonempty subregion, its sampling frequency is independent of its probability $\int_{E_i} f(x)dx$. This implies that SAMC is capable of exploring the whole sample space, even for the regions with tiny probabilities. Potentially, SAMC can be used to sample rare events from a large sample space.

Convergence Rate. Theorem 7.4.2 concerns the convergence rate of θ_t, which gives a L^2 upper bound for the mean squared error of θ_t.

Theorem 7.4.2 *Assume the gain factor sequence is chosen in (7.15) and the drift condition (B_2) (given in Section 7.7.1) holds. Then there exists a constant λ such that*

$$E\|\theta_t - \theta_*\|^2 \leq \lambda \gamma_t,$$

where $\theta_* = (\theta_*^{(1)}, \ldots, \theta_*^{(m)})$ is as specified in (7.19).

The proof of this theorem can be found in Section 7.7.2.

Monte Carlo Integration. In addition to estimating the normalizing constants g_i's, SAMC can be conveniently used for Monte Carlo integration, estimating the expectation $E_f \rho(x) = \int_{\mathcal{X}} \rho(x) f(x)$ for an integrable function $\rho(x)$. Let $(x_1, \theta_1), \ldots, (x_n, \theta_n)$ denote the samples generated by SAMC during the first n iterations. Let $y_1, \ldots, y_{n'}$ denote the distinct samples among $x_1, \ldots, x_n$. Generate a random variable/vector Y such that

$$P(Y = y_i) = \frac{\sum_{t=1}^n \exp\{\theta_t^{(J(x_t))}\} I(x_t = y_i)}{\sum_{t=1}^n \exp\{\theta_t^{(J(x_t))}\}}, \quad i = 1, \ldots, n', \tag{7.20}$$

where $I(\cdot)$ is the indicator function. Under the assumptions $(A_1), (A_2)$ and the compactness of $\mathcal{X}$, Liang (2009b) showed that Y is asymptotically distributed as $f(\cdot)$.

Theorem 7.4.3 *Assume* (A_1) *and the drift condition* (B_2) *(given in Section 7.7.1) hold. For a set of samples generated by SAMC, the random variable/vector* Y *generated in (7.20) is asymptotically distributed as* $f(\cdot)$.

The proof of this theorem can be found in Section 7.7.3. It implies that for an integrable function $\rho(x), E_f\rho(x)$ can be estimated by

$$\widehat{E_f\rho(x)} = \frac{\sum_{t=1}^{n} \exp\{\theta_t^{(J(x_t))}\} h(x_t)}{\sum_{t=1}^{n} \exp\{\theta_t^{(J(x_t))}\}}. \tag{7.21}$$

As $n \to \infty, \widehat{E_f\rho(x)} \to E_f\rho(x)$ for the same reason that the usual importance sampling estimate converges (Geweke, 1989).

Some Implementation Issues. For an effective implementation of SAMC, several issues need to be considered.

- *Sample space partition* This can be done according to our goal and the complexity of the given problem. For example, if we aim to minimize the energy function, the sample space can be partitioned according to the energy function. The maximum energy difference in each subregion should be bounded by a reasonable number, say, 2, which ensures that the local MH moves within the same subregion have a reasonable acceptance rate. Note that within the same subregion, sampling from the working density (7.16) is reduced to sampling from $\psi(x)$. If our goal is model selection, then the sample space can be partitioned according to the index of models, as illustrated in Section 7.6.1.2.

- *The desired sampling distribution* If our goal is to estimate $\boldsymbol{g}$, then we may set the desired distribution to be uniform. However, if our goal is optimization, then we may set the desired sampling distribution biased to low energy regions. As illustrated by Hesselbo and Stinchcombe (1995) and Liang (2005b), biasing sampling to low energy regions often improves the ergodicity of the simulation.

- *The choice of the gain factor sequence and the number of iterations* To estimate $\boldsymbol{g}, \gamma_t$ should be very close to 0 at the end of simulations. Otherwise, the resulting estimates will have a large variation. Under the setting of (7.15), the speed of γ_t going to zero is controlled by ξ and t_0. In practice, one often fixes ξ to 1 and choose t_0 according to the complexity of the problem. The more complex the problem, the larger the value of t_0 one should choose. A large t_0 will force the sampler to reach all subregions quickly, even in the presence of multiple local energy minima.

The appropriateness of the choices of the gain factor sequence and the number of iterations can be determined by checking the convergence of multiple runs (starting with different points) through **examining the variation** of $\widehat{\boldsymbol{g}}$ or $\widehat{\boldsymbol{\pi}}$, where $\widehat{\boldsymbol{g}}$ and $\widehat{\boldsymbol{\pi}}$ denote, respectively, the estimates of $\boldsymbol{g}$ and $\boldsymbol{\pi}$ obtained at the end of a run. The idea of monitoring convergence of MCMC simulations using multiple runs has been discussed early on by Gelman and Rubin (1992) and Geyer (1992). A rough examination for $\widehat{\boldsymbol{g}}$ is to see visually whether or not the $\widehat{\boldsymbol{g}}$ vectors produced in the multiple runs follow the same pattern. Existence of different patterns implies that the gain factor is still large at the end of the runs or some parts of the sample space are not visited in all runs. The examination for $\widehat{\boldsymbol{g}}$ can also be done by a statistical test under the assumption of multivariate normality. Refer to Jobson (1992) – for the testing methods for multivariate outliers.

To examine the variation of $\widehat{\boldsymbol{\pi}}$, Liang *et al.* (2007) defined the statistic $\epsilon_f(E_i)$, which measures the deviation of $\widehat{\pi}_i$, the realized sampling frequency of subregion E_i in a run, from its theoretical value. The statistic is defined as

$$\epsilon_f(E_i) = \begin{cases} \frac{\widehat{\pi}_i - (\pi_i + \widehat{\nu})}{\pi_i + \widehat{\nu}} \times 100\%, & \text{if } E_i \text{ is visited}, \\ 0, & \text{otherwise}, \end{cases} \tag{7.22}$$

for $i = 1, \ldots, m$, where $\widehat{\nu} = \sum_{j \notin \mathcal{S}} \pi_j / |\mathcal{S}|$ and $\mathcal{S}$ denotes the set of subregions that have been visited during the simulation. Here, $\widehat{\nu}$ works as an estimate of ν in (7.19). It is said that $\{\epsilon_f(E_i)\}$, output from all runs and for all subregions, matches well if the following two conditions are satisfied: (i) There does not exist such a subregion which is visited in some runs but not in others; and (ii) $\max_{i=1}^{m} |\epsilon_f(E_i)|$ is less than a threshold value, say, 10%, for all runs. A group of $\{\epsilon_f(E_i)\}$ which does not match well implies that some parts of the sample space are not visited in all runs, t_0 is too small (the self-adjusting ability is thus weak), or the number of iterations is too small.

In practice, to have a reliable diagnostic for the convergence, we may check both $\widehat{\boldsymbol{g}}$ and $\widehat{\boldsymbol{\pi}}$. In the case where a failure of multiple-run convergence is detected, SAMC should be re-run with more iterations or a larger value of t_0. Determining the value of t_0 and the number of iterations is a trial-and-error process.

Two Illustrative Examples. In the following two examples, we illustrate the performance of SAMC. Example 7.1 shows that SAMC can produce a consistent estimate of $\boldsymbol{g}$. Example 7.2 shows that SAMC can result in a 'free' random walk in the space of the subregions, and thus is essentially immune to the local-trap problem suffered by conventional MCMC algorithms.

■ **Example 7.1 Estimation Consistency**

The distribution consists of 10 states with the unnormalized mass function $P(x)$ given in Table 7.1. It has two modes which are well separated by low mass states.

Table 7.1 Unnormalized mass function of the 10-state distribution.

x	1	2	3	4	5	6	7	8	9	10
$P(x)$	1	100	2	1	3	3	1	200	2	1

Using this example, Liang *et al.* (2007) demonstrate estimation consistency of SAMC. In their simulations, the sample space is partitioned according to the mass function into the following five subregions: $E_1 = \{8\}, E_2 = \{2\}$, $E_3 = \{5,6\}, E_4 = \{3,9\}$ and $E_5 = \{1,4,7,10\}$. When $\psi(x) = 1$, the true value of $\boldsymbol{g}$ is $(1,1,2,2,4)$, which counts the number of states in the respective subregions. The proposal used in the MH step is a stochastic matrix, each row of which is generated independently from the Dirichlet distribution $Dir(1,\ldots,1)$, the uniform in the 9-simplex space. The desired sampling distribution is uniform; that is, $\pi_1 = \ldots = \pi_5 = 1/5$. The sequence $\{\gamma_t\}$ is as given in (7.15) with $t_0 = 10$. SAMC was run 100 times, each independent run consisting of 5×10^5 iterations. The estimation error of $\boldsymbol{g}$ is measured by the function $\epsilon_e(t) = \sqrt{\sum_{E_i \neq \emptyset}(\widehat{g}_i^{(t)} - g_i)^2/g_i}$ at 10 equally spaced time points $t = 5 \times 10^4, \ldots, 5 \times 10^5$, where $\widehat{g}_i^{(t)}$ denotes the estimate of g_i obtained at iteration t. Figure 7.1(a) shows the curve of $\epsilon_e(t)$ obtained by averaging over the 100 runs. The statistic $\epsilon_f(E_i)$ is calculated at time $t = 10^5$ for each run. The results show that they match well. Figure 7.1(b) shows the box-plots of $\epsilon_f(E_i)$'s of the 100 runs. The deviations are <3%. This indicates that SAMC has achieved the desired sampling distribution and the choice of t_0 and the number of iterations are appropriate. Other choices of t_0, including $t_0 = 20$ and 30, give similar results.

For comparison, Liang *et al.* (2007) apply the Wang-Landau algorithm to this example with the proposal distribution being the same as that used in SAMC and the gain factor being set as in Wang and Landau (2001). The gain factor starts with $\delta_0 = 2.718$ and then decreases in the scheme $\delta_{s+1} \to \sqrt{\delta_s}$. At stage s, n_s iterations are performed, where n_s is a large constant, which ensures a flat histogram can be produced at each stage. The values of n_s tried include $n_s = 1000$, 2500, 5000 and 10 000. Figure 7.1(a) shows the curves of $\epsilon_e(t)$ for each choice of n_s, where each curve is obtained by averaging over 100 independent runs.

The comparison shows that SAMC produced more accurate estimates for $\boldsymbol{g}$ and converged much faster than WL. More importantly, the SAMC estimates can be improved continuously as the simulation goes on, while the WL estimates can only reach a certain accuracy depending on the value of n_s.

Liang (2009b) re-uses this example to illustrate that SAMC can work as a dynamic importance sampling algorithm. For this purpose, SAMC is re-run with the foregoing setting except for $\psi(x) = P(x)$. In each run, the first 10^4 iterations are discarded for the burn-in process and the samples generated in

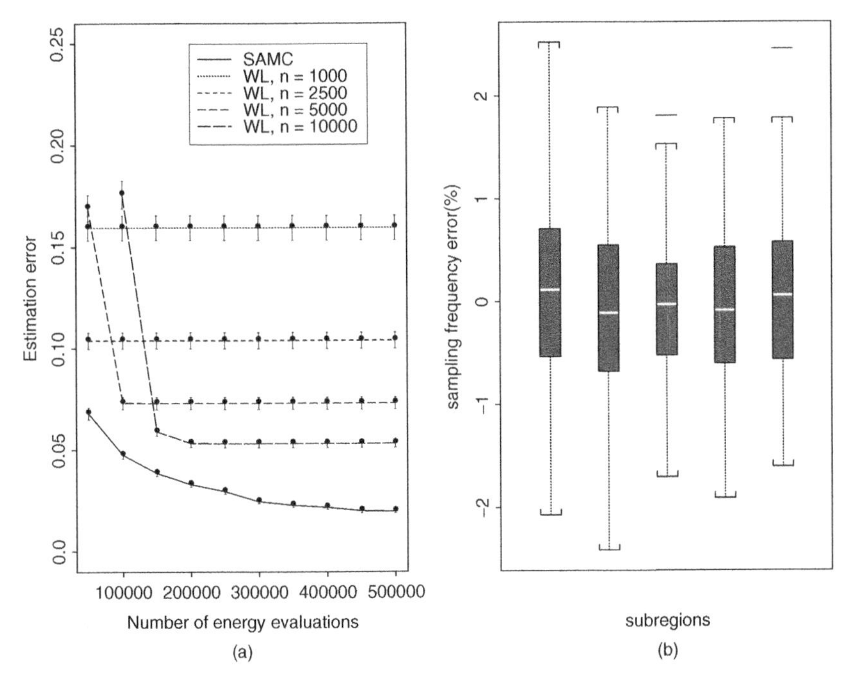

Figure 7.1 Comparison of the WL and SAMC algorithms. (a) Average $\epsilon_e(t)$ curves obtained by SAMC and WL. The vertical bars show the ±one-standard-deviation of the average of the estimates. (b) Box-plots of $\{\epsilon_f(E_i)\}$ obtained in 100 runs of SAMC (Liang *et al.*, 2007).

Table 7.2 Comparison of SAMC and MH for the 10-state example (Liang, 2009b).

algorithm	bias ($\times 10^{-3}$)	standard error ($\times 10^{-3}$)	CPU time (seconds)
SAMC	−0.528	1.513	0.38
MH	−3.685	4.634	0.20

The Bias and Standard Error (of the Bias) were calculated based on 100 independent runs, and the CPU times were measured on a 2.8 GHz computer for a single run.

the remaining iterations are used for estimation. Table 7.2 summarizes the estimates of $E_f(X)$ obtained by SAMC.

For comparison, Liang (2009b) also applies the MH algorithm to this example, with the same transition proposal matrix. The algorithm is run 100 times independently, each run consisting of 5.1×10^5 iterations; the samples generated in the last 5×10^5 iterations are used for estimation. The numerical results in Table 7.2 indicate that for this example, SAMC is significantly better than MH in terms of standard errors. After accounting for the CPU cost, SAMC can still make about fourfold improvement over MH. For more complex problems, such as the phylogeny estimation problem considered in Cheon and Liang (2008), SAMC and MH will cost about the same CPU time for the same number of iterations; in this case, the CPU time cost by each algorithm is dominated by the part used for energy evaluation, and the part used for weight updating in SAMC can be ignored.

The reason why SAMC outperforms MH can be explained by Figure 7.2 (a) &(b). For this example, MH mixes very slowly due to the presence of two separated modes, whereas SAMC can still mix very fast due to its self-adjusting mechanism. Figure 7.2(c) shows the evolution of the log-weights of SAMC samples. It indicates that the log-weights can evolve very stably. In SAMC simulations, θ_t was updated in (7.17), where the term $-\gamma_{t+1}\boldsymbol{\pi}$ helps to stabilize the magnitude of the importance weights by keeping the sum of all components of θ_t unchanged over iterations.

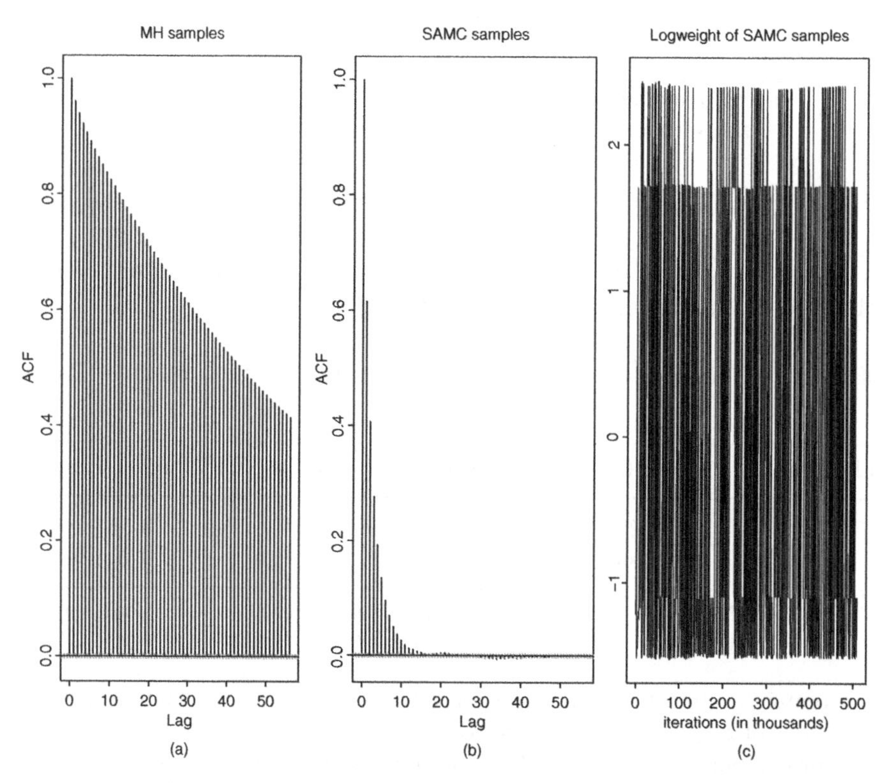

Figure 7.2 Computational results for the 10-state example: (a) Autocorrelation plot of the MH samples; (b) Autocorrelation plot of the SAMC samples; (c) Log-weights of the SAMC samples (Liang, 2009b).

■ **Example 7.2 Multimodal Sampling**

This problem is to sample from a multimodal distribution defined by $f(\boldsymbol{x}) \propto \exp\{-H(\boldsymbol{x})\}$, where $\boldsymbol{x} = (x_1, x_2) \in [1.1, 1.1]^2$ and

$$H(\boldsymbol{x}) = -\left\{x_1 \sin(20x_2) + x_2 \sin(20x_1)\right\}^2 \cosh\left\{\sin(10x_1)x_1\right\} \\ -\left\{x_1 \cos(10x_2) - x_2 \sin(10x_1)\right\}^2 \cosh\left\{\cos(20x_2)x_2\right\}.$$

Liang *et al.* (2007) use this example to illustrate the performance of SAMC in multimodal sampling; it has been used widely in the literature to demonstrate the multimodal sampling problem (see e.g., Robert and Casella, 2004). Figure 7.3 (a) shows that $H(\boldsymbol{x})$ has a multitude of local energy minima separated by high energy barriers. In applying SAMC to this example, Liang *et al.* (2007) partitioned the sample space into 41 subregions with an equal energy bandwidth: $E_1 = \{\boldsymbol{x} : H(\boldsymbol{x}) \leq -8.0\}, E_2 = \{\boldsymbol{x} : -8.0 < H(\boldsymbol{x}) \leq -7.8\}, \ldots,$ and $E_{41} = \{\boldsymbol{x} : -0.2 < H(\boldsymbol{x}) \leq 0\}$, and set the other parameters as follows: $\psi(\boldsymbol{x}) = \exp\{-H(\boldsymbol{x})\}, t_0 = 200, \pi_1 = \cdots = \pi_{41} = 1/41$, and a random walk proposal $q(\boldsymbol{x}_t, \cdot) = N_2(\boldsymbol{x}_t, 0.25^2 I_2)$. SAMC was run for 20 000 iterations, and 2000 samples were collected at equally spaced time points. Figure 7.3 (b) shows the evolving path of the 2000 samples. For comparison, MH was applied to simulate from the distribution $f_{st}(\boldsymbol{x}) \propto \exp\{-H(\boldsymbol{x})/5\}$. MH was run for 20 000 iterations with the same proposal $N_2(\boldsymbol{x}_t, 0.25^2 I_2)$, and 2000 samples were collected at equally spaced time points. Figure 7.3(c) shows the evolving path of the 2000 samples, which characterizes the performance of simulated/parallel tempering at high temperatures.

Under the foregoing setting, SAMC samples almost uniformly in the space of energy (i.e., the energy bandwidth of each subregion is small, and the sample distribution closely matches the contour plot of $H(\boldsymbol{x})$), whereas simulated tempering tends to sample uniformly in the sample space $\mathcal{X}$ when the temperature is high. Because one does not know a priori where the high-energy and low energy regions are and how much the ratio of their 'volumes' is, one cannot control the simulation time spent on low and high energy regions in simulated tempering. However, one can control almost exactly, up to the constant ν in (7.19), the simulation time spent on low energy and high energy regions in SAMC by choosing the desired sampling distribution $\boldsymbol{\pi}$. SAMC can go to high energy regions, but it spends only limited time over there to help the system to escape from local energy minima, and spends other time exploring low energy regions. This smart simulation time distribution scheme makes SAMC potentially more efficient than simulated tempering in optimization. Liang (2005b) reported a neural network training example which shows that the generalized Wang-Landau algorithm is more efficient than simulated tempering in locating global energy minima.

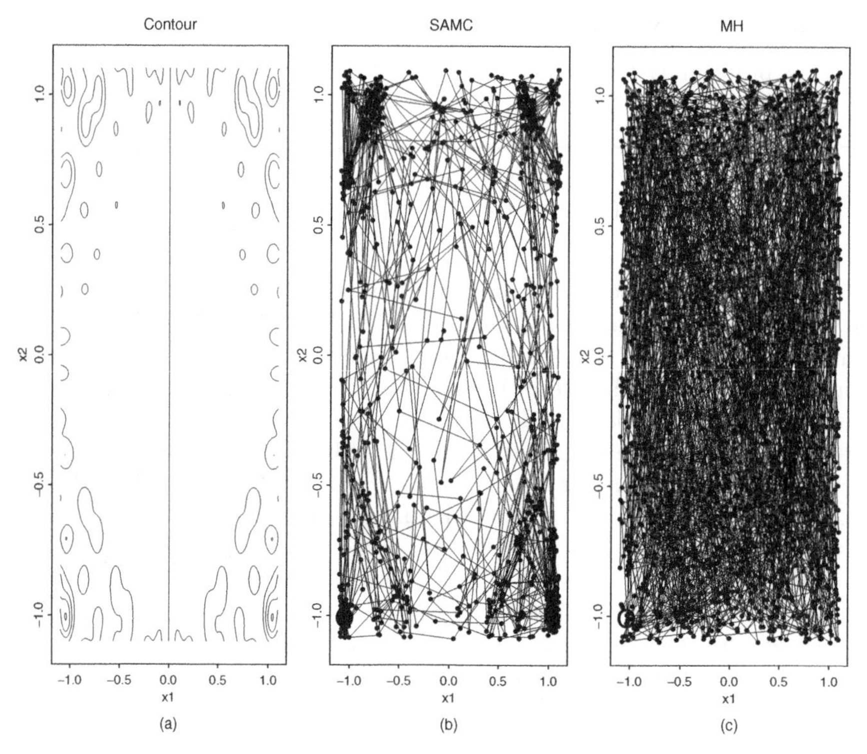

Figure 7.3 (a) Contour of $H(\boldsymbol{x})$. (b) Sample path of SAMC. (c) Sample path of MH at the temperature $T = 5$ (Liang *et al.*, 2007).

7.5 Applications of Stochastic Approximation Monte Carlo

In this section, we present several applications of SAMC, which all take advantage of the self-adjusting mechanism of SAMC. For the p-value evaluation problem (Section 7.5.1), the self-adjusting mechanism means the sampler can traverse the permutation space very quickly; for problems of phylogeny inference (Section 7.5.2) and Bayesian network learning (Section 7.5.3), the self-adjusting mechanism enables the sampler to escape from local-trap problems, and leads to correct inference for the quantities of interest.

7.5.1 Efficient p-Value Evaluation for Resampling-Based Tests

P-value evaluation is a crucial ingredient of hypothesis testing. Most often, the p-value can be evaluated according to the asymptotic distribution of the test statistic when the sample size is sufficiently large. However, in many applications, the asymptotic distribution is unavailable, or unreliable due to insufficient sample size, and a resampling-based procedure, such as permutation or bootstrap (Good, 2005), has to be used for evaluating p-values.

The resampling-based approach can be very time consuming. For example, in the current genome-wide association study (GWAS) (see, e.g., Pearson and Manolio, 2008), about half a million to a million genetic markers are tested for their association with the outcome of interest. To maintain the family-wide false positive error rate at an acceptance level, an individual test has to have a very small p-value to be declared globally significant. For instance, when testing half a million genetic markers, a commonly accepted threshold for global significance is about 10^{-7}. Thus, if the resampling-based approach is used, it generally requires more than 10^9 permutations to get a reliable estimate for a p-value at the level of 10^{-7}. This is highly challenging to current computing power.

To alleviate this difficulty, importance sampling algorithms have been proposed for some specific tests, such as the scan statistic used in genetic linkage studies (Anquist and Hossjer, 2004; Shi *et al.*, 2007) and the genetic association test based on contingency table analysis (Kimmel and Shamir, 2006). Although these algorithms work well in these specific contexts, they are not general enough to be adopted for other applications. Yu and Liang (2009) propose evaluating the test p-value using SAMC. Their approach is very general, which can be easily used for any resampling-based test. The numerical results indicate that SAMC can achieve 100 to 1000 times as efficient a result as the standard resampling-based procedure, when evaluating a small p-value.

7.5.1.1 *P*-value Evaluation Through SAMC

Let X denote the observed data, and let $\lambda(X)$ denote the test statistic of interest to us. Let $\mathcal{X}$ denote the permutation space of X, which is composed of all possible combinations of the elements of X under the null hypothesis that we are testing for. Clearly, $\mathcal{X}$ is finite. To use SAMC to evaluate the p-value, we first partition the permutation space $\mathcal{X}$ according to the test statistic $\lambda(x)$ into m subregions: $E_1 = \{x : \lambda(x) \leq \lambda_1, x \in \mathcal{X}\}, E_2 = \{x : \lambda_1 < \lambda(x) \leq \lambda_2, x \in \mathcal{X}\}, \ldots, E_{m-1} = \{x : \lambda_{m-2} < \lambda(x) \leq \lambda_{m-1}, x \in \mathcal{X}\}, E_m = \{x : \lambda(x) > \lambda_{m-1}, x \in \mathcal{X}\}$, where $-\infty < \lambda_1 < \cdots < \lambda_{m-1} < \infty$.

Since $\mathcal{X}$ is finite, if conditions (A_1) and (A_2) hold, then Theorem 7.4.1 follows. By Theorem 7.4.1, if we set $\psi(x) \propto 1$ for all $x \in \mathcal{X}$, then $\int_{E_i} \psi(x)dx$ counts the number of permutations belonging to the subregion E_i, and

$$\widehat{P}_t(\lambda(X) > \lambda_k) = \frac{\sum_{i=k+1}^{m} \exp\{\theta_t^{(i)}\}(\pi_i + \widehat{\nu}_t)}{\sum_{j=1}^{m} \exp\{\theta_t^{(j)}\}(\pi_j + \widehat{\nu}_t)}, \tag{7.23}$$

will converge to the tail probability $P(\lambda(X) > \lambda_k)$ when t becomes large, where $\widehat{\nu}_t = \sum_{j \notin \mathcal{S}_t} \pi_j / |\mathcal{S}_t|$, and $\mathcal{S}_t$ denotes the set of the subregions that have been visited by iteration t. For a non-cut point, the tail probability can be obtained by linear interpolation. For example, $\lambda_k < \lambda^* < \lambda_{k+1}$, then

$$\widehat{P}_t(\lambda(X) > \lambda^*) = \frac{\lambda^* - \lambda_k}{\lambda_{k+1} - \lambda_k} \widehat{P}_t(\lambda(X) > \lambda_{k+1}) + \frac{\lambda_{k+1} - \lambda^*}{\lambda_{k+1} - \lambda_k} \widehat{P}_t(\lambda(X) > \lambda_k), \tag{7.24}$$

will form a good estimate of $P(\lambda(X) > \lambda^*)$ at the end of the run.

For the p-value evaluation problem, SAMC generally prefers a fine partition: the proposal distribution $q(\cdot,\cdot)$ is usually local, and so a fine partition will reduce the chance that both the current and proposed samples fall into the same subregion, thus helping the self-adjusting mechanism of the system to transverse the permutation space quickly. In practice, $\lambda_1, \ldots, \lambda_m$ are often chosen to be an equal difference sequence, with the difference ranging from $\sigma/100$ to $\sigma/50$, where σ denotes the standard deviation of $\lambda(X)$. The σ can be estimated by drawing a small number of permutations from $\mathcal{X}$. The value of λ_1 and λ_{m-1}, which specify the range of interest for the statistic $\lambda(X)$, can be determined by a pilot run.

7.5.1.2 An Illustrative Example

Suppose that there are two groups of observations, $x_{1,1}, \ldots, x_{1,n_1} \sim N(\mu_1, \sigma^2)$ and $x_{2,1}, \ldots, x_{2,n_2} \sim N(\mu_2, \sigma^2)$. To test the null hypotheses $H_0 : \mu_1 = \mu_2$ versus $H_1 : \mu_1 \neq \mu_2$, the test statistic

$$\lambda(\boldsymbol{x}) = \frac{\bar{x}_{1\cdot} - \bar{x}_{2\cdot}}{S_{x_1,x_2}\sqrt{\frac{1}{n_1} + \frac{1}{n_2}}},$$

is usually used, where $\bar{x}_{1\cdot} = \sum_{i=1}^{n_1} x_{1i}/n_1, \bar{x}_{2\cdot} = \sum_{i=1}^{n_2} x_{2i}/n_2$, and

$$S_{x_1,x_2} = \sqrt{[(n_1 - 1)S_{x_1}^2 + (n_2 - 1)S_{x_2}^2]/(n_1 + n_2 + 1)}$$

is the pooled estimate of σ^2. It is known that $\lambda(\boldsymbol{x})$ follows a t-distribution with the degree of freedom $n_1 + n_2 - 2$. To simulate the data, we set $n_1 = n_2 = 1000, \mu_1 = \mu_2 = 0$, and $\sigma_1 = \sigma_2 = 1$. To illustrate the performance of SAMC, we estimate the empirical distribution of $|\lambda(\boldsymbol{x})|$ based on the simulated observations.

In SAMC simulations, the gain factor was chosen in (7.15) with $\xi = 1$ and $t_0 = 1000$, the total number of iterations was set to $N = 5 \times 10^6$, and λ_1 and λ_{m-1} were fixed to 0 and 6. Different values of $m = 120$, 300 and 600 were tried. For each value of $m, \lambda_2, \ldots, \lambda_{m-2}$ were chosen to be equally spaced between λ_1 and λ_{m-1}. Let $\boldsymbol{x}_t = (\boldsymbol{x}_t^{(1)}, \boldsymbol{x}_t^{(2)})$ denote a permutation of the two groups of observations obtained at iteration t. The new permutation $\boldsymbol{y}$ is obtained by exchanging s pairs of observations from the first and second groups, where each pair is drawn at random, with replacement, from the space $\{1, \ldots, n_1\} \times \{1, \ldots, n_2\}$. Therefore, the local positive condition (A_2)

Table 7.3 Mean absolute relative errors (%) of tail probability estimates produced by SAMC (Yu and Liang, 2009).

	exchange rate			
partition	2%	5%	10%	20%
$m = 120$	2.364 (0.285)	1.625 (0.189)	2.377 (0.224)	2.419 (0.193)
$m = 300$	1.393 (0.128)	1.398 (0.157)	1.077 (0.112)	1.773 (0.168)
$m = 600$	1.353 (0.212)	0.983 (0.128)	1.045 (0.114)	1.584 (0.150)

The numbers in the parentheses represent the standard deviations of the estimates.

is satisfied. For this example, we tried different values of $s = 20, 50, 100$ and 200, which correspond to $2\%, 5\%, 10\%$ and 20% of observations, respectively.

For each cross-setting $(m, k) \in \{120, 300, 600\} \times \{20, 50, 100\}$, SAMC was run 20 times. Let $\widehat{P}(\lambda(X) > \lambda_k)$ denotes the estimate of $P(\lambda(X) > \lambda_k)$. Table 7.3 reports the mean absolute relative error (MARE) of the estimates produced by SAMC, where the MARE is defined by

$$\text{MARE} = \frac{1}{m-1} \sum_{k=1}^{m-1} \frac{|\widehat{P}(\lambda(X) > \lambda_k) - P(\lambda(X) > \lambda_k)|}{P(\lambda(X) > \lambda_k)} \times 100\%. \tag{7.25}$$

The results indicate that SAMC can produce very accurate estimates, with about 1% of the relative error, for the empirical distribution of $\lambda(X)$, when the partition is fine (with $300 \sim 600$ intervals over a range of 6 standard deviations), and the exchange rate is around 10%.

The results also indicate that SAMC prefers a fine partition of $\mathcal{X}$ and a local update of $\boldsymbol{x}_t$ (i.e., a low exchange rate). As previously explained, a fine partition enhances the ability of the self-adjusting mechanism of SAMC to transverse quickly over the permutation space. However, preferring a local update of $\boldsymbol{x}_t$ is unusual, as the MCMC algorithm usually prefers a global update. This can be explained as follows. Suppose that $\boldsymbol{x}_t$ is an extreme permutation with a large value of $|\lambda(\boldsymbol{x}_t)|$. If a high exchange rate was applied to $\boldsymbol{x}_t$, then $\boldsymbol{y}$ will be almost independent of $\boldsymbol{x}_t$ and the chance of getting an extremer value of $|\lambda(\boldsymbol{y})|$ will be very low. However, if a low exchange rate was applied to $\boldsymbol{x}_t$, then $|\lambda(\boldsymbol{y})|$ would have a value around $|\lambda(\boldsymbol{x}_t)|$, and the chance of having $|\lambda(\boldsymbol{y})| > |\lambda(\boldsymbol{x}_t)|$ will be relatively higher. Hence, a local updating of $\boldsymbol{x}_t$ enhances the ability of SAMC to explore the tails of the distribution of $\lambda(X)$. Based on their experience, Yu and Liang (2009) recommended an updating rate between 5% and 10% for such types of problem.

For comparison, the traditional permutation method was also applied to this example. It was run 20 times, and each run consisted of 5×10^6 permutations. The average MARE of the 20 estimates is 21.32% with standard deviation 1.757, which is much worse than that produced by SAMC. Figure 7.4

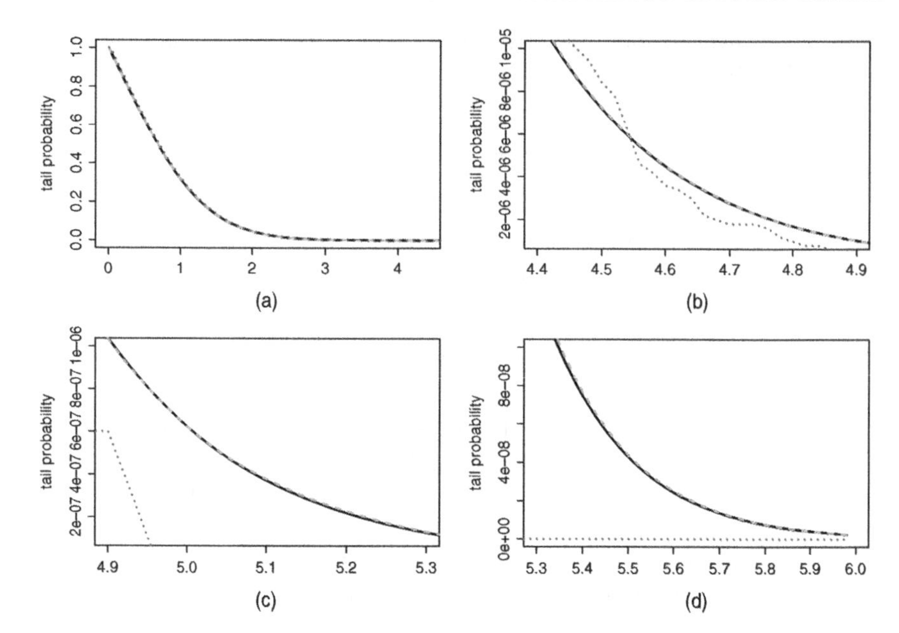

Figure 7.4 Comparison of the true p-value (solid curve), the estimate produced by SAMC (dotted curve), and the estimate produced by the permutation method (dashed read curve) at 300 equally spaced points. The estimates for the true p-values (a) greater than 10^{-5}, (b) between 10^{-6} and 10^{-5}, (c) between 10^{-7} and 10^{-6}; and (d) between 10^{-9} and 10^{-7} (Yu and Liang, 2009).

shows the tail probability estimates obtained in a single run with $m = 300$. The permutation method works well for the region with the true p-value $>10^{-5}$, and fails for the region with the true p-value $<10^{-6}$. To get an estimate with comparable accuracy with the SAMC estimate, the permutation method may need about 10^{11} permutations.

Figure 7.5 shows the progressive plots of the highest test statistic values sampled by SAMC and the permutation method. It indicates that SAMC can sample extreme test statistic values very fast. Almost in all runs, SAMC can sample some test statistic with the true p-value $<10^{-9}$ with only about 10^4 iterations. However, with a total of $10^8 (= 5\,000\,000 \times 20)$ permutations, the permutation method never sampled a test statistic with the true p-value $<10^{-8}$.

7.5.2 Bayesian Phylogeny Inference

Phylogeny inference is one of fundamental topics in molecular evolution. The traditional methods select a single 'best' tree, either by the neighbor joining (NJ) method (Saitou and Nei, 1987) or according to some optimality criterion, such as minimum evolution (Kidd and Sgaramella-Zonta, 1971; Rzhetsky

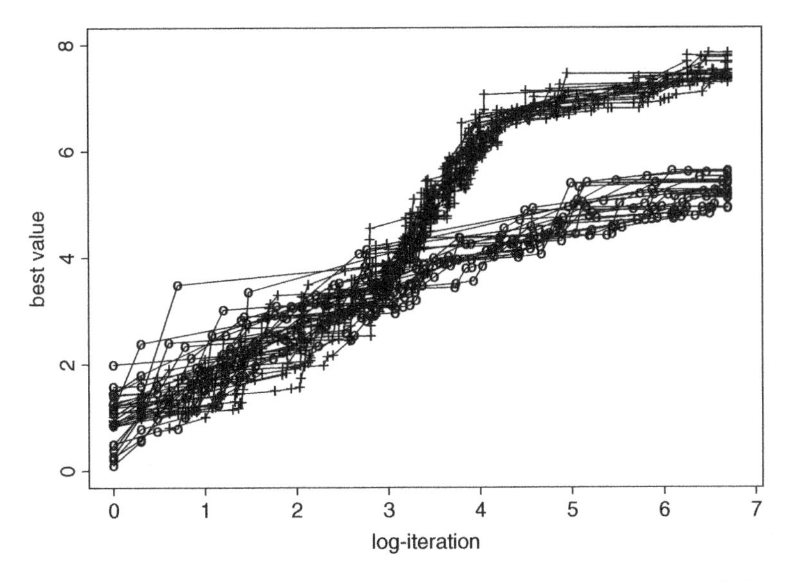

Figure 7.5 Progressive plots of the highest test statistic values sampled by the SAMC ('+') and the permutation methods ('o') for the simulation example. The x-axis plots the logarithm (with base 10) of iteration numbers (Yu and Liang, 2009).

and Nei, 1992), maximum parsimony (Fitch, 1971; Maddison, 1991), and maximum likelihood (Felsenstein, 1981, 1993; Salter and Pearl, 2001). Although the traditional methods work well for many problems, they do not produce valid inferences beyond point estimates.

To account for uncertainty of phylogeny estimation, Bayesian methods have been widely used in recent literature. Significant work include Rannala and Yang (1996), Mau and Newton (1997), Mau *et al.* (1999), Larget and Simon (1999), Newton *et al.* (1999), and Li *et al.* (2000), where the MH algorithm (Metropolis *et al.*, 1953; Hastings, 1970) is employed to simulate from a posterior distribution defined on a parameter space that includes tree topologies as well as branch lengths and the parameters of the sequence evolutionary model. However, the MH algorithm tends to get trapped in local energy minima, rendering ineffective inference for the phylogeny. A number of authors have employed advanced MCMC algorithms to try to resolve this difficulty. For example: Huelsenbeck and Ronquist (2001) and Altekar *et al.* (2004) employ parallel tempering; (Geyer, 1991) and Feng *et al.* (2003) employ the multiple-try Metropolis algorithm (Liu *et al.*, 2000); Cheon and Liang (2008) apply a sequential Monte Carlo algorithm, focusing on maximum a posteriori (MAP) trees.

Cheon and Liang (2009) apply SAMC, and compare SAMC with the popular Bayesian phylogeny software, BAMBE (Larget and Simon, 1999), and

MrBayes (Huelsenbeck and Ronquist, 2001). On simulated and real datasets, the numerical results indicate that SAMC outperforms BAMBE and MrBayes. Of the three methods, SAMC produces the consensus trees that most resemble true trees, and parameter estimates with the smallest mean square errors, but cost the least CPU time.

7.5.2.1 Bayesian Phylogeny Analysis

A phylogenetic tree can be represented by a rooted binary tree, each node with descendants representing the most recent common ancestor of the descendants, and the root representing the most common ancestor of all the entities at the leaves of the tree. In general, a phylogenetic tree of n leaves has $n-2$ internal nodes (excluding the root node) and $2n-2$ branches. The length of branch represents the distance between two end node sequences, and it is often calculated from a model of substitution of residues over the course of evolution.

Suppose that we are interested in conducting a phylogeny analysis of n nucleotide sequences (taxa). The problem for analyzing protein sequences is similar. The nucleotide sequences can be arranged as a n by N matrix, where N is the common number of sites or the common length of the sequences. For a demonstration purpose, we assume that that evolution among sites is independent conditional on the given genealogy, although this assumption is probably violated by most coding sequence datasets. As explained by Galtier *et al.* (2005), the sites in a protein (or an RNA sequence) interact to determine the selected 3-dimensional structure of the protein, so the evolutionary process of interacting sites are not independent.

Under the independence assumption, several evolutionary models have been proposed for nucleotides, such as the one-parameter model (Jukes and Cantor, 1969), two-parameter model (Kimura, 1980), Felsenstein model (Felsenstein, 1981), and HKY85 model (Hasegawa *et al.*, 1985); the most flexible, is the HKY model, which possesses a general stationary distribution of nucleotides, and allows for different rates of transitional and transversional events. The transition probability matrix of the HKY85 model is given by

$$\begin{aligned} Q_{j|i}(h) \\ = \begin{cases} \pi_j + \pi_j\left(\frac{1}{\lambda_j}-1\right)e^{-\alpha h} + \left(\frac{\lambda_j-\pi_j}{\lambda_j}\right)e^{-\alpha\gamma_j h}, & \text{if } i=j, \\ \pi_j + \pi_j\left(\frac{1}{\lambda_j}-1\right)e^{-\alpha h} - \left(\frac{\pi_j}{\lambda_j}\right)e^{-\alpha\gamma_j h}, & \text{if } i\neq j \text{ (transition)}, \\ \pi_j(1-e^{-\alpha h}), & \text{if } i\neq j \text{ (transversion)}, \end{cases} \end{aligned} \tag{7.26}$$

where h denotes the evolution time or the branch length of the phylogenetic tree, α denotes the evolutionary rate, $\lambda_j = \pi_A + \pi_G$ if base j is a purine (A or G) and $\pi_C + \pi_T$ if base j is a pyrimidine (C or T), $\gamma_j = 1 + (\kappa - 1)\lambda_j$, and κ is a parameter responsible for distinguishing between transitions and transversions. The stationary probabilities of the four nucleotides are π_A, π_C, π_G, and π_T, respectively.

Let $\omega = (\boldsymbol{\tau},\ \boldsymbol{h},\ \boldsymbol{\phi})$ denote a phylogenetic tree, where $\boldsymbol{\tau}$, $\boldsymbol{h}$ and $\boldsymbol{\phi}$ denote the tree topology, branch lengths, and model parameters, respectively. The likelihood can be calculated using the pruning method developed by Felsenstein (Felsenstein, 1981). The pruning method produces a collection of partial likelihoods of subtrees, starting from the leaves and working recursively to the root for each site. Let $\mathcal{S} = \{A, C, G, T\}$ denote the set of nucleotides. For site k of a leaf e, define $L_e^k(i) = 1$ if state i matches the base found in the sequence and 0 otherwise, where i indexes the elements of $\mathcal{S}$. At site k of an internal node v, the conditional probability of descendant data given state i is

$$L_v^k(i) = \left(\sum_{j \in \mathcal{S}} L_u^k(j) Q_{j|i}(h_{vu})\right) \times \left(\sum_{j \in \mathcal{S}} L_w^k(j) Q_{j|i}(h_{vw})\right), \quad i \in \mathcal{S},$$

where u and w denote the two children nodes of v, and h_{ab} denotes the length of the branch ended with the nodes a and b. The likelihood can then be written as

$$L(\omega|D) = \prod_{k=1}^{N} \sum_{i \in \mathcal{S}} \pi_0(i) L_r^k(i), \tag{7.27}$$

where D denotes the observed sequences of n taxa, r denotes the root node, and π_0 is the initial probability distribution assigned to the ancestral root sequence. In simulations, π_0 is often set to the observed frequency of nucleotides of the given sequences.

Following Mau *et al.* (1999) and Larget and Simon (1999), we placed a uniform prior on ω. The resulting posterior distribution is

$$f(\omega|D) \propto L(\omega|D), \tag{7.28}$$

for which the MAP tree coincides with the maximum likelihood tree.

In applying SAMC to sample from the posterior (7.28), Cheon and Liang (2009) adopt the local moves used in Larget and Simon (1999) for updating phylogenetic trees. The moves consist of three types of updates: the update for model parameters; the update for branch lengths; and the update for tree topology. Lee $\rho(\omega)$ denote a quantity of interest for phylogeny analysis, such as the presence/absence of a branch or an evolutionary parameter. Following from the theory of SAMC, $E_f\rho(\omega)$, the expectation of $r(\omega)$ with respect to the posterior (7.28), can be estimated by

$$\widehat{E_f\rho(\omega)} = \frac{\sum_{t=n_0+1}^{n} \rho(\omega_t) \exp\{\theta_t^{(J(\omega_t))}\}}{\sum_{t=n_0+1}^{n} \exp\{\theta_t^{(J(\omega_t))}\}}, \tag{7.29}$$

where $(\omega_{n_0+1}, \theta_{n_0+1}^{(J(\omega_{n_0+1}))}), \ldots, (\omega_n, \theta_n^{(J(\omega_n))})$ denote the samples generated by SAMC, and n_0 denotes the number of burn-in iterations.

7.5.2.2 Simulation Examples

In this study, 100 nucleotide datasets were generated according to a given tree of 30 taxa (shown in Figure 7.6(a)), a given root sequence (shown in Table 7.4), and a HKY85 model with parameters $\kappa = 2, \alpha = 1$, and $\pi_A = \pi_G = \pi_C = \pi_T = 0.25$. The length of each sequence is 300. SAMC, BAMBE and MrBayes were applied to each of the 100 datasets. For each dataset, each algorithm was run for 2.2×10^5 iterations. In all simulations, α was restricted to be 1, as the evolutionary rate of each site of the nucleotide sequences was modeled equally (see Cheon and Liang (2009) for details of the simulations).

Figure 7.6 compares the consensus trees produced by the three methods for one dataset. Other results are summarized in Table 7.5, where MAST (maximum agreement subtree) provides a measurement for the similarity between the true and consensus trees constructed by respective methods. The MAST scores were calculated using the software `TreeAnalyzer`, developed by Dong and Kraemer (2004), based on the tree comparison algorithms of Farach *et al.* (1995) and Goddard *et al.* (1994). Note that `TreeAnalyzer` has normalized its output by 100; that is, the MAST score of two identical trees

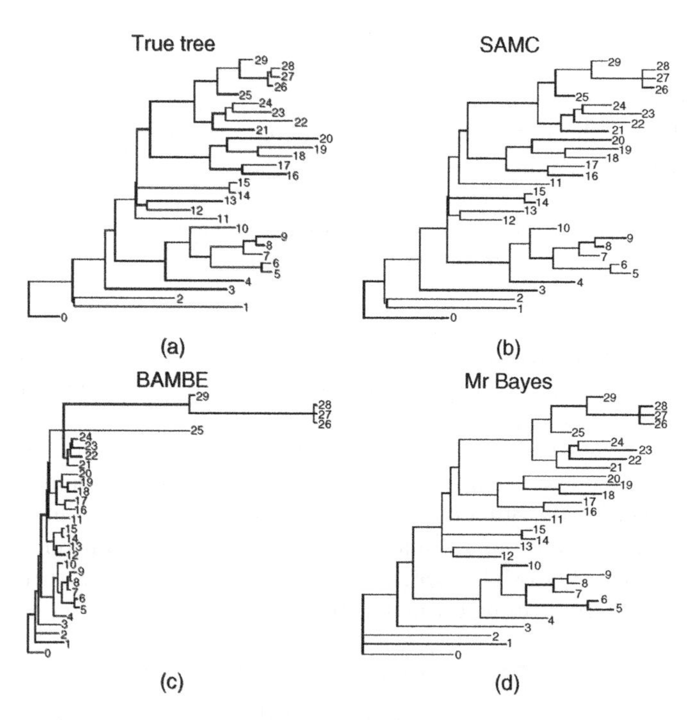

Figure 7.6 Comparison of consensus trees produced by SAMC, BAMBE and MrBayes for a simulated 30-taxa dataset (Cheon and Liang, 2009).

Table 7.4 The root sequence for the simulated example (Cheon and Liang, 2009).

AACAAAGCCACAATTATTAATACTCTTGCTACATCCTGAGCAAAAGCCCC
CGCCCTCACAGCTCTCACCCCCCTTATTCTTCTTTCACTAGGGGGCCTCC
CCCCTCTCACGGGCTTTATACCAAAATGACTGATTCTTCAAGAACTAACC
AAACAAGGCCTTGCCCCCACCGCAACCCTAGCAGCCCTCTCAGCACTCC
TTAGCCTCTATTTCTACCTGCGCCTCTCCTACACAATAACCCTCACTATTT
CCCCCAACAGCCTTCTAGGTACCACCCCCTGACGTTTGCCTTCTACCCAA

Table 7.5 Bayesian phylogeny inference for the simulated data (Cheon and Liang, 2009).

method	CPU	AveL	κ	π_A	π_G	π_C	π_T	MAST
SAMC	6.2	9.74	1.998	0.277	0.178	0.294	0.250	78.12
		(3.28)	(.001)	(.002)	(.001)	(.001)	(.001)	(0.33)
MrBayes	18.7	5.15	2.154	0.274	0.181	0.295	0.252	75.18
		(3.17)	(.032)	(.002)	(.001)	(.001)	(.001)	(0.35)
BAMBE	6.4	0	1.514	0.271	0.174	0.303	0.252	72.41
		(4.52)	(.016)	(.002)	(.001)	(.002)	(.002)	(0.31)

CPU: CPU time (in minutes) cost by a single run of the algorithm on an Intel Pentium III computer. AveL: the difference of the averaged log-likelihood values produced by the corresponding method and BAMBE. MAST: Maximum Agreement Subtree. Each entry is calculated by averaging over 100 datasets, and the number in the parentheses represent the standard deviation of the corresponding estimate.

is 100. Table 7.5 shows that the consensus trees produced by SAMC tend to resemble true trees more than those produced by MrBayes and BAMBE. In addition to MAST scores, SAMC also produces more accurate estimates of κ than MrBayes and BAMBE. MrBayes employs a parallel tempering algorithm for simulations, where multiple Markov chains are run in parallel at different temperatures; it therefore costs more CPU time than single chain methods BAMBE and SAMC, for the same number of iterations. Cheon and Liang (2009) also compare SAMC with BAMBE and MrBayes on the sequences generated for the trees with 20 taxa and 40 taxa. The results are similar.

7.5.3 Bayesian Network Learning

The use of graphs to represent statistical models has become popular in recent years. In particular, researchers have directed interest in Bayesian networks and applications of such models to biological data (see, e.g., Friedman *et al.*, 2000 and Ellis and Wong 2008). The Bayesian network illustrated by

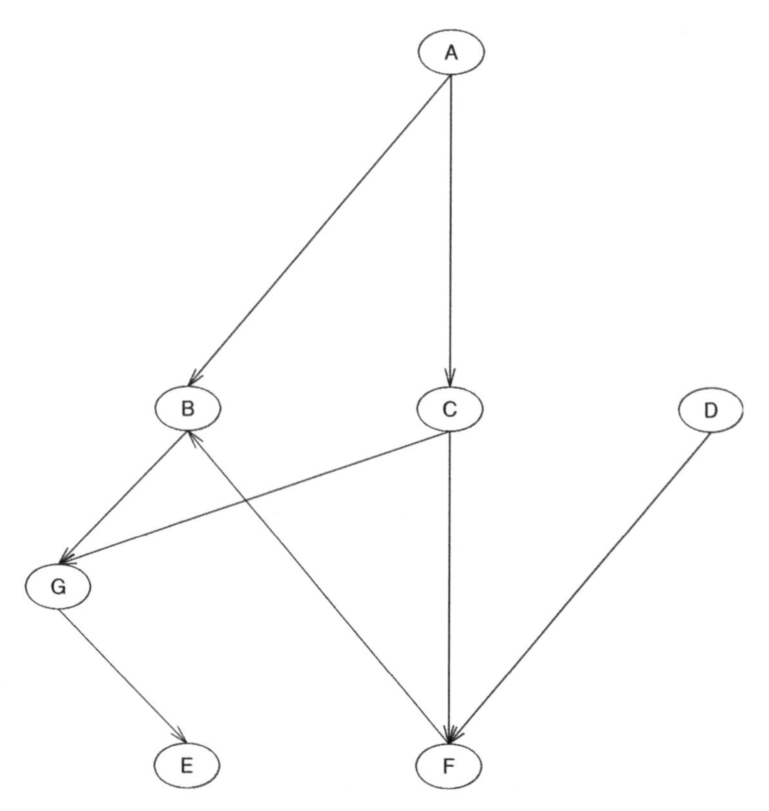

Figure 7.7 An example of Bayesian networks (Liang and Zhang, 2009).

Figure 7.7, is a directed acyclic graph (DAG) for which the nodes represent the variables in the domain and the edges correspond to direct probabilistic dependencies between them. As indicated by many applications, the Bayesian network is a powerful knowledge representation and reasoning tool under the conditions of uncertainty typical of real-life applications (see, e.g., Taroni *et al.*, 2006).

Bayesian networks are used in a variety of approaches, including: the conditional independence test-based approach (Spirtes *et al.*, 1993; Wermuth and Lauritzen, 1983; de Campos and Huete, 2000); the optimization-based approach (Herskovits and Cooper, 1990; Lam and Bacchus, 1994; Heckerman *et al.*, 1995; Chickering, 1996); and the Bayesian approach (Madigan and Raftery, 1994; Madigan and York, 1995; Giudici and Green, 1999). Due to its advantage in uncertainty analysis, the Bayesian approach is pursued increasingly by researchers and practitioners.

For the Bayesian approach, the posterior distribution of the Bayesian network is usually simulated using the MH algorithm, which is prone to get stuck in local minima on the energy landscape. To overcome the local-trap problem suffered by the MH algorithm, Friedman and Koller (2003) introduce a two-stage algorithm: they use the MH algorithm to sample a temporal order of the nodes, and then sample a network structure compatible with the given node order. As discussed in Friedman and Koller (2003), for any Bayesian networks, there exists a temporal order of the nodes such that for any two nodes X and Y, if there is an edge from X and Y, then X must precede Y. For example, for the Bayesian network shown in Figure 7.7a temporal order is ACDFBGE. The two-stage algorithm improves mixing over the space of network structures. However, as pointed out by Ellis and Wong (2008), the structures sampled by it does not follow the correct posterior distribution, because the temporal order does not induce a partition of network structures space. A network may be compatible with more than one temporal order. For example, the network in Figure 7.7 is compatible with both the orders ACDFBGE and ADCFBGE.

Liang and Zhang (2009) propose using SAMC to learn Bayesian networks: the network features can be inferred by dynamically weighted averaging of the samples generated in the learning process. The numerical results indicate that SAMC can mix much faster over the space of Bayesian networks than the MH simulation-based approach.

7.5.3.1 Bayesian Networks

A Bayesian network model can be defined as a pair $B = (\mathcal{G}, \boldsymbol{\rho})$, where $\mathcal{G} = (\mathcal{V}, \mathcal{E})$ is a directed acyclic graph that represents the structure of the network, $\mathcal{V}$ represents the set of nodes, $\mathcal{E}$ represents the set of edges, and $\boldsymbol{\rho}$ is a vector of conditional probabilities. For a node $V \in \mathcal{V}$, a parent of V is a node from which there is a directed link to V. The set of parents of V is denoted by $pa(V)$. For illustration purpose, we consider here only the discrete case where V is a categorical variable taking values in a finite set $\{v_1, \ldots, v_{r_i}\}$. There are $q_i = \prod_{V_j \in pa(V_i)} r_j$ possible values for the joint state of the parents of V_i. Each element of $\boldsymbol{\rho}$ represents a conditional probability. For example, ρ_{ijk} is the probability of variable V_i in state j conditioned on that $pa(V_i)$ is in state k. Naturally, $\boldsymbol{\rho}$ is restricted by the constraints $\rho_{ijk} \geq 0$ and $\sum_{j=1}^{r_i} \rho_{ijk} = 1$. The joint distribution of the variables $\boldsymbol{V} = \{V_1, \ldots, V_d\}$ can be specified by the decomposition

$$P(\boldsymbol{V}) = \prod_{i=1}^{d} P\big(V_i | pa(V_i)\big). \tag{7.30}$$

Let $\mathcal{D} = \{\boldsymbol{V}_1, \ldots, \boldsymbol{V}_N\}$ denote a set of independently and identically distributed samples drawn from (7.30). Let n_{ijk} denote the number of samples for which V_i is in state j and $pa(V_i)$ is in state k. Then, the count $(n_{i1k}, \ldots, n_{ir_ik})$

follows a multinomial distribution; that is,

$$(n_{i1k}, \dots, n_{ir_ik}) \sim \text{Multinomial}\left(\sum_{j=1}^{r_i} n_{ijk}, \boldsymbol{\rho}_{ik}\right), \tag{7.31}$$

where $\boldsymbol{\rho}_{ik} = (\rho_{i1k}, \dots, \rho_{ir_ik})$. Thus, the likelihood function of the Bayesian network model can be written as

$$P(\mathcal{D}|\mathcal{G}, \boldsymbol{\rho}) = \prod_{i=1}^{d} \prod_{k=1}^{q_i} \binom{\sum_{j=1}^{r_i} n_{ijk}}{n_{i1k}, \dots, n_{ir_ik}} \rho_{i1k}^{n_{i1k}} \cdots \rho_{ir_ik}^{n_{ir_ik}}. \tag{7.32}$$

Since a network with a large number of edges is often less interpretable and there is a risk of over-fitting, it is important to use priors over the network space that encourage sparsity. For this reason, we let $\mathcal{G}$ be subject to the prior

$$P(\mathcal{G}|\beta) \propto \left(\frac{\beta}{1-\beta}\right)^{\sum_{i=1}^{d} |pa(V_i)|}, \tag{7.33}$$

where $0 < \beta < 1$ is a user-specified parameter. For example, $\beta = 0.1$ may be a reasonable choice for β. The parameters $\boldsymbol{\rho}$ is subject to a product Dirichlet distribution

$$P(\boldsymbol{\rho}|\mathcal{G}) = \prod_{i=1}^{d} \prod_{k=1}^{q_i} \frac{\Gamma(\sum_{j=1}^{q_i} \alpha_{ijk})}{\Gamma(\alpha_{i1k}) \cdots \Gamma(\alpha_{ir_ik})} \rho_{i1k}^{\alpha_{i1k}-1} \cdots \rho_{ir_ik}^{\alpha_{ir_ik}-1}, \tag{7.34}$$

where $\alpha_{ijk} = 1/(r_i q_i)$, as suggested by Ellis and Wong (2008). Combining with the likelihood function and the prior distributions and integrating out $\boldsymbol{\rho}$, we get the posterior distribution:

$$P(\mathcal{G}|\mathcal{D}) \propto \prod_{i=1}^{d} \left(\frac{\beta}{1-\beta}\right)^{|pa(V_i)|} \prod_{k=1}^{q_i} \frac{\Gamma(\sum_{j=1}^{r_i} \alpha_{ijk})}{\Gamma(\sum_{j=1}^{r_i} (\alpha_{ijk} + n_{ijk}))} \prod_{j=1}^{r_i} \frac{\Gamma(\alpha_{ijk} + n_{ijk})}{\Gamma(\alpha_{ijk})}, \tag{7.35}$$

which contains all the network structure information provided by the data.

Note that Bayesian networks are conceptually different from causal Bayesian networks. In a causal Bayesian network, each edge can be interpreted as a direct causal relation between a parent node and a child node, relative to the other nodes in the network (Pearl, 1988). The formulation of Bayesian networks, as described above, is not sufficient for causal inference. To learn a causal Bayesian network, one needs a dataset obtained through experimental interventions. In general, one cannot learn a causal Bayesian network from observational data alone (see Cooper and Yoo, 1999, and Ellis and Wong, 2008, for more discussions on this issue).

7.5.3.2 Learning Bayesian Networks using SAMC

Let $\mathcal{G}$ denote a feasible Bayesian network for the data $\mathcal{D}$. At each iteration of SAMC, the sampling step can be performed as follows.

(a) Randomly choose between the following possible changes to the current network $\mathcal{G}_t$ producing $\mathcal{G}'$.

  (a.1) *Temporal order change.* Swap the order of two neighboring models. If there is an edge between them, reverse its direction.

  (a.2) *Skeletal change.* Add (or delete) an edge between a pair of randomly selected nodes.

  (a.3) *Double skeletal change.* Randomly choose two different pairs of nodes, and add (or delete) edges between each pair of the nodes.

(b) Calculate the ratio

$$r = e^{\theta_t^{(J(\mathcal{G}'))} - \theta_t^{(J(\mathcal{G}_t))}} \frac{\psi(\mathcal{G}')}{\psi(\mathcal{G}_t)} \frac{T(\mathcal{G}' \to \mathcal{G}_t)}{T(\mathcal{G}_t \to \mathcal{G}')},$$

where $\psi(\mathcal{G})$ is defined as the right hand side of 7.35, and the ratio of the proposal probabilities $T(\mathcal{G}' \to \mathcal{G}_t)/T(\mathcal{G}_t \to \mathcal{G}') = 1$ for all of the three types of the changes. Accept the new network structure $\mathcal{G}'$ with probability $\min(1, r)$. If it is accepted, set $\mathcal{G}_{t+1} = \mathcal{G}'$; otherwise, set $\mathcal{G}_{t+1} = \mathcal{G}_t$.

Clearly, the proposal used above satisfies the condition (A_2). We note that a similar proposal has been used by Wallace and Korb (1999) in a Metropolis sampling process for Bayesian networks, and that the double changes are not necessary for the algorithm to work, but are included to help accelerate the sampling process.

Following the ergodicity theory of SAMC, any quantity of interest, such as the presence/absence of an edge or a future observation, can be estimated in (7.21) given a set of SAMC samples $(\mathcal{G}_{n_0+1}, \theta_{n_0+1}^{(J(\mathcal{G}_{n_0+1}))}), \ldots, (\mathcal{G}_n, \theta_n^{(J(\mathcal{G}_n))})$ generated from the posterior distribution, where n_0 denotes the number of burn-in iterations.

7.5.3.3 SPECT Heart Data

This dataset, which is available at machine learning repository `http://archive.ics.uci.edu/ml`, describes diagnosing cardiac single proton emission computed tomography (SPECT) images. The patients are classified into two categories: normal and abnormal. The database of 267 SPECT image sets (patients) was processed to extract features that summarize the original SPECT images. As a result, 22 binary feature patterns were created for each patient. The SPECT dataset has been used by a few researchers, including Cios *et al.* (1997) and Kurgan *et al.* (2001), to demonstrate their machine

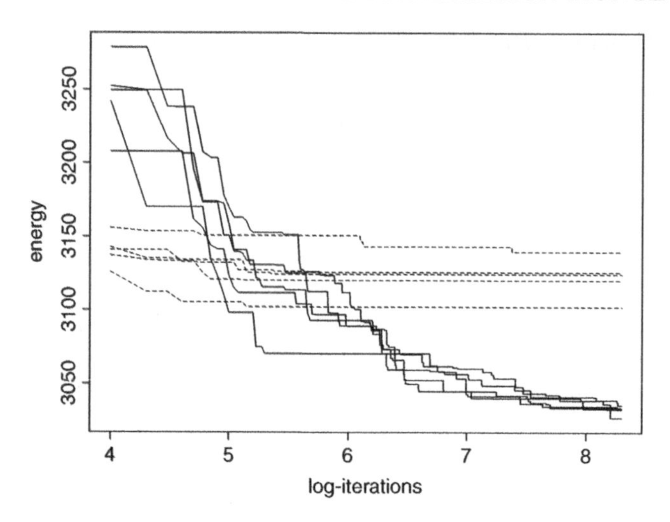

Figure 7.8 Progression paths of minimum energy values produced in five runs of SAMC (solid lines) and five runs of MH (dashed lines) for the SPECT data (Liang and Zhang, 2009).

learning algorithms. Liang and Zhang (2009) built a Bayesian network using SAMC for the features and the overall classification of the patients.

In applying SAMC to this dataset, the sample space was partitioned into 2001 subregions with an equal energy bandwidth, $E_1 = \{x : U(x) \leq 2000\}, E_2 = \{x : 2000 < U(x) \leq 2001\}, \ldots, E_{2000} = \{x : 3998 < U(x) \leq 3999\}$, and $E_{2001} = \{x : U(x) > 3999\}$. SAMC was run 5 times, and each run consisted of 2×10^8 iterations. The overall acceptance rate of the SAMC moves is about 0.13. For comparison, MH was also run 5 times with the same proposal, and each run consisted of 2.0×10^8 iterations. The overall acceptance rate of the MH moves is only about 0.006. Figure 7.8 compares the progression paths of minimum energy values produced by SAMC and MH in the respective runs. SAMC outperforms MH for this example; the minimum energy value produced by SAMC in any of the five runs is much lower than that produced by MH in all of the five runs.

Figure 7.9 (left panel) shows the putative MAP Bayesian network learned by SAMC over the five runs, where the node 23 corresponds to the overall classification of the patients. The plot indicates that conditional on the features 17 and 21, the classification of the patients is independent of other features. Figure 7.9 (right panel) shows the consensus network for which each edge presents in the posterior network samples with a probability higher than 0.5. For example, the edge from 17 to 23 has a probability of 0.67 presenting in the posterior network samples, and the edge from 21 to 23 has a probability of 0.91. Note that the consensus network may contain several subnetworks, which are disconnected from each other.

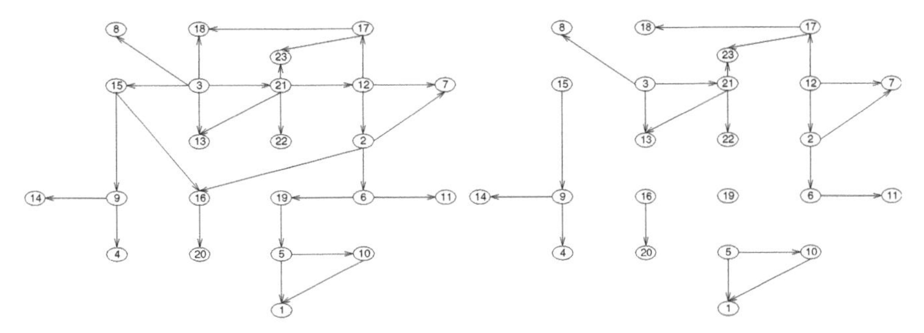

Figure 7.9 SPECT Data: (left) The putative MAP Bayesian network learned by SAMC, and (right) the consensus Bayesian network learned by SAMC (Liang and Zhang, 2009).

7.6 Variants of Stochastic Approximation Monte Carlo

In this section, we present several variants of SAMC, which improve the performance of SAMC for certain types of problems.

7.6.1 Smoothing SAMC for Model Selection Problems

Suppose that one is working with a distribution as specified in (7.1) and that the sample space $\mathcal{X}$ has been partitioned into m disjoint subregions $E_1, \ldots, E_m$ according to a function $\lambda(x)$. As previously discussed, the superiority of SAMC in sample space exploration is due to its self-adjusting mechanism. If a subregion, say E_i, is visited at iteration t, θ_t will be updated accordingly such that E_i will have a smaller probability to be revisited at iteration $t+1$. However, this mechanism has not yet reached its maximum efficiency because it does not differentiate neighboring and non-neighboring subregions of E_i when updating θ_t. We note that for many problems, the subregions $E_1, \ldots, E_m$ form a sequence of naturally ordered categories with $g_1, \ldots, g_m$ changing smoothly along the index of subregions, where $g_i = \int_{E_i} \psi(x)dx$ for $i = 1, \ldots, m$, and $\psi(x)$ is a non-negative function with $\int_{\mathcal{X}} \psi(x)dx < \infty$. For example, for model selection problems, $\mathcal{X}$ can be partitioned according to the index of models, the subregions can be ordered according to the number of parameters contained in each model, and the neighboring subregions often contain similar probability values. Intuitively, the sample x_t may contain some information on its neighboring subregions, so the visiting to its neighboring subregions should also be penalized to some extent at the next iteration. Consequently, this improves the ergodicity of the simulation. Henceforth, we will call a partition with $g_1, \ldots, g_m$ changing smoothly a smooth partition, or say the sample space is partitioned smoothly. In this subsection, we assume that the sample space has been partitioned smoothly.

Liang (2009a) shows that the efficiency of SAMC can be improved by including at each iteration a smoothing step, which distributes the information contained in each sample to its neighboring subregions. The new algorithm is called smoothing-SAMC, or SSAMC.

7.6.1.1 Smoothing-SAMC Algorithm

SSAMC is different from SAMC in two respects. First, the gain factor sequence used in SSAMC is a little more restrictive than that used in SAMC. In SSAMC, the gain factor sequence is required to satisfy the following condition:

(B_1) The sequence $\{\gamma_t\}$ is positive and nonincreasing, and satisfies the conditions:

$$(i) \quad \overline{\lim_{t\to\infty}} |\gamma_t^{-1} - \gamma_{t+1}^{-1}| < \infty, \quad (ii) \quad \sum_{t=1}^{\infty} \gamma_t = \infty, \quad (iii) \quad \sum_{t=1}^{\infty} \gamma_t^{\zeta} < \infty, \tag{7.36}$$

for any $\zeta > 1$.

The trade-off with the condition (A_1), given in Section 7.4, is that a higher order noise term can be included in updating θ_t as prescribed in (7.41). For example, $\{\gamma_t\}$ can be set as

$$\gamma_t = \frac{t_0}{\max\{t_0, t\}}, \quad t = 1, 2, \ldots, \tag{7.37}$$

where t_0 is a pre specified number. More discussions on condition (B_1) can be found at the end of this subsection. There we can see that the condition (B_1) can be relaxed a bit according to the setting of other parameters of this algorithm.

Second, SSAMC allows multiple samples to be generated at each iteration, and employs a smoothed estimate of $p_t^{(i)}$ in updating θ_t, where $p_t^{(i)} = \int_{E_i} f_{\theta_t}(x)dx$ is the limiting probability that a sample is drawn from E_i at iteration t, and $f_{\theta_t}(x)$ is as defined in (7.16). Let $x_t^{(1)}, \ldots, x_t^{(\kappa)}$ denote the samples generated by a MH kernel with the invariant distribution $f_{\theta_t}(x)$. Since κ is usually a small number, say, 10 to 20, the samples form a sparse frequency vector $\boldsymbol{e}_t = (e_t^{(1)}, \ldots, e_t^{(m)})$ with $e_t^{(i)} = \sum_{l=1}^{\kappa} I(x_t^{(l)} \in E_i)$. Because the law of large numbers does not apply here, $\boldsymbol{e}_t/\kappa$ is not a good estimator of $\boldsymbol{p}_t = (p_t^{(1)}, \ldots, p_t^{(m)})$. As suggested by many authors, including Burman (1987), Hall and Titterington (1987), Dong and Simonoff (1994), Fan *et al.* (1995), and Aerts *et al.* (1997), the frequency estimate can be improved by a smoothing method. Since the partition has been assumed to be smooth, information in nearby subregions can be borrowed to help produce more accurate estimates of $\boldsymbol{p}_t$.

Liang (2009a) smoothes the frequency estimator $\boldsymbol{e}_t/\kappa$ using the Nadaraya-Watson kernel method. Then

$$\widehat{p}_t^{(i)} = \frac{\sum_{j=1}^{m} W\left(\frac{\Lambda(i-j)}{mh_t}\right) \frac{e_t^{(j)}}{\kappa}}{\sum_{j=1}^{m} W\left(\frac{\Lambda(i-j)}{mh_t}\right)}, \tag{7.38}$$

where $W(z)$ is a kernel function with bandwidth h_t, and Λ is a rough estimate of the range of $\lambda(x), x \in \mathcal{X}$. Here, $W(z)$ is chosen to have a bounded support; that is, there exists a constant C such that $W(z) = 0$ if $|z| > C$. With this condition, it is easy to show $\widehat{p}_t^{(i)} - e_t^{(i)}/\kappa = O(h_t)$. There are many choices for $W(z)$, for example, an Epanechnikov kernel or a double-truncated Gaussian kernel. The former is standard, and the latter can be written as

$$W(z) = \begin{cases} \exp(-z^2/2), & \text{if } |z| < C, \\ 0, & \text{otherwise.} \end{cases} \tag{7.39}$$

The bandwidth h_t is chosen as a power function of γ_t; that is, $h_t = a\gamma_t^b$ for $a > 0$ and $b \in (0, 1]$, where b specifies the decay rate of the smoothing adaptation in SSAMC. For a small value of b, the adaptation can decay very slowly. In Liang (2009a), $W(z)$ was set to the double-truncated Gaussian kernel with $C = 3$ and

$$h_t = \min\left\{\gamma_t^b, \frac{\text{range}\{\lambda(x_t^{(1)}), \ldots, \lambda(x_t^{(\kappa)})\}}{2(1 + \log_2(\kappa))}\right\}, \tag{7.40}$$

where $b = 1/2$, and the second term in $\min\{\cdot, \cdot\}$ is the default bandwidth used in conventional density estimation procedures, see, e.g., S-PLUS 5.0 (Venables and Ripley, 1999). Clearly, $h_t = O(\gamma_t^b)$ when t becomes large.

In summary, one iteration of SSAMC consists of the following steps:

SSAMC Algorithm (Steps)

1. *Sampling*. Simulate samples $x_{t+1}^{(1)}, \ldots, x_{t+1}^{(\kappa)}$ using the MH algorithm from the distribution $f_{\theta_t}(x)$ as defined in (7.16). The simulation should be done in an iterative manner; that is, generating $x_{t+1}^{(i+1)}$ with a proposal $q(x_{t+1}^{(i)}, \cdot)$, where where $x_{t+1}^{(0)} = x_t^{(\kappa)}$.

2. *Smoothing*. Calculate $\widehat{\boldsymbol{p}}_t = (\widehat{p}_t^{(1)}, \ldots, \widehat{p}_t^{(m)})$ in (7.38).

3. *Weight updating*. Set

$$\theta_{t+\frac{1}{2}} = \theta_t + \gamma_{t+1}(\widehat{\boldsymbol{p}}_t - \boldsymbol{\pi}). \tag{7.41}$$

If $\theta_{t+\frac{1}{2}} \in \Theta$, set $\theta_{t+1} = \theta_{t+\frac{1}{2}}$; otherwise, set $\theta_{t+1} = \theta_{t+\frac{1}{2}} + \mathbf{c}^*$, where $\mathbf{c}^* = (c^*, \ldots, c^*)$ can be any vector which satisfies the condition $\theta_{t+\frac{1}{2}} + \mathbf{c}^* \in \Theta$.

As in SAMC, Θ can be restricted to a compact set. To ensure the transition kernel to satisfy the drift condition, we restrict $\mathcal{X}$ to be either finite (for a discrete system) or compact (for a continuum system), and choose the proposal distribution to satisfy the condition (A_2). As previously discussed, when $\mathcal{X}$ is continuous, $\mathcal{X}$ can be restricted to the region $\{x : \psi(x) \geq \psi_{\min}\}$, where $\psi_{\min}$ is sufficiently small such that the region $\{x : \psi(x) < \psi_{\min}\}$ is not of interest. Under these assumptions, Liang (2009a) establishes the following theorem concerning the convergence of SSAMC, whose proof can be found in Section 7.7.1.3.

Theorem 7.6.1 *Assume* (B_1) *and the drift condition* (B_2) *(given in Section 7.7.1) hold. Then,*

$$\theta_i^{(t)} \to \begin{cases} C + \log(\int_{E_i} \psi(x)dx) - \log(\pi_i + \nu), & \text{if } E_i \neq \emptyset, \\ -\infty. & \text{if } E_i = \emptyset, \end{cases} \tag{7.42}$$

as $t \to \infty$, *where* C *is an arbitrary constant,* $\nu = \sum_{j \in \{i : E_i = \emptyset\}} \pi_j/(m - m_0)$, *and* m_0 *is the number of empty subregions.*

Regarding implementation of SSAMC, most issues previously discussed for SAMC, including those on sample space partitioning, convergence diagnostic, and parameter setting (for $\boldsymbol{\pi}, t_0$ and the total number of iterations), are applicable to SSAMC. Below we add three more issues specific to SSAMC.

- *The choice of smoothing estimators.* Theoretically, any smoothing estimator, which satisfies the condition $\widehat{p}_t^{(i)} - e_t^{(i)}/\kappa = O(h^\tau)$ for some $\tau > 0$, can be used in SSAMC. Other than the Nadaraya-Watson kernel estimator, the local log-likelihood estimator (Tibshirani and Hastie, 1987) and the local polynomial estimator (Aerts *et al.*, 1997) can also be used in SSAMC (see Simonoff, 1998, for a comprehensive review for smoothing estimators).

  When no smoothing operator is used, that is, replacing (7.41) by

$$\theta_{t+\frac{1}{2}} = \theta_t + \gamma_{t+1}(\mathbf{e}_t/\kappa - \boldsymbol{\pi}), \tag{7.43}$$

  SSAMC is reduced to a multiple-sample version of SAMC. Henceforth, we will call this version of SAMC multiple-SAMC or MSAMC.

- *The choice of* κ. Since the convergence of SSAMC is determined by three parameters κ, t_0 and N, where N denotes the total number of iterations, the value of κ should be determined together with the values of t_0 and N. In practice, κ is usually set to a number less than 20. Since the gain factor is kept at a constant at each iteration, a run with a large κ has to end at a large value of γ_t, provided that the total running time is fixed. Note that the estimate of $\boldsymbol{g}$ produced in a run ended at a large value of γ_t will be highly variable. In our experience, SSAMC can benefit from the smoothing operation even when κ is as small as 5.

- *The condition* (B_1). When proving Theorem 7.6.1, an important step is to identify the order τ of the perturbation term

$$\widehat{\boldsymbol{p}}_t - \boldsymbol{\pi} - H(\theta_t, \boldsymbol{x}_{t+1}) = O(\gamma_{t+1}^{\tau}),$$

where $\boldsymbol{x}_{t+1} = (x_{t+1}^{(1)}, \ldots, x_{t+1}^{(\kappa)})$ and $H(\theta_t, \boldsymbol{x}_{t+1}) = \boldsymbol{e}_t/\kappa - \boldsymbol{\pi}$. The choice of the gain factor sequence $\{\gamma_t\}$ depends very much on the value of τ. If the value of τ is known, (B_1)-(iv) can be relaxed to $\sum_{t=1}^{\infty} \gamma_t^{\zeta} < \infty$ for some $\zeta \in [1+\tau, 2]$. For MSAMC, this condition can be further relaxed to $\sum_{t=1}^{\infty} \gamma_t^2 < \infty$, as the corresponding perturbation term is 0.

7.6.1.2 Bayesian Model Selection

Liang *et al.* (2007) and Martinez *et al.* (2009) apply SAMC to Bayesian model selection problems and compare it to the reversible jump MCMC (RJMCMC) algorithm (Green, 1995). They conclude that SAMC outperforms RJMCMC when the model space is complex, for example, when the model space contains several modes which are well separated from each other, or some tiny probability models, which are of interest to us. However, when the model space is simple, that is, it contains only a single mode and neighboring models have comparable posterior probabilities, SAMC may not work better than RJMCMC, as the self-adjusting mechanism of SAMC is no longer crucial for mixing the models. Liang (2009a) shows that for Bayesian model selection problems, SSAMC can make significant improvement over SAMC and that it can work better than RJMCMC, even when the model space is simple. This is illustrated below by the change-point identification problem that we considered in Section 3.3.2.

Given the posterior distribution (3.20), the marginal posterior distribution $P(\mathcal{X}_k|Z)$ can be estimated using SSAMC. Without loss of generality, we restrict our considerations to the models with $k_{\min} \le k \le k_{\max}$. Let $E_k = \mathcal{X}_k$ and $\psi(\cdot) \propto P(\boldsymbol{\vartheta}^{(k)}|Z)$. It follows from 7.42 that $\widehat{g}_t^{(i)}/\widehat{g}_t^{(j)} = \exp\{\theta_t^{(i)} - \theta_t^{(j)}\}$ forms a consistent estimator of the ratio $P(\mathcal{X}_i|Z)/P(\mathcal{X}_j|Z)$ when $\boldsymbol{\pi}$ is set to be uniform and t becomes large. The sampling step can be performed as in RJMCMC, including the 'birth', 'death' and 'simultaneous' moves. For the 'birth' move, the acceptance probability is

$$\min\left\{1, \frac{e^{\theta_t^{(k)}}}{e^{\theta_t^{(k+1)}}} \frac{P(\boldsymbol{\vartheta}_*^{(k+1)}|X)}{P(\boldsymbol{\vartheta}_t^{(k,l)}|X)} \frac{q_{k+1,k}}{q_{k,k+1}} \frac{c_{u+1} - c_u - 1}{1}\right\}; \tag{7.44}$$

for the 'death' move, the acceptance probability is

$$\min\left\{1, \frac{e^{\theta_t^{(k)}}}{e^{\theta_t^{(k-1)}}} \frac{P(\boldsymbol{\vartheta}_*^{(k-1)}|X)}{P(\boldsymbol{\vartheta}_t^{(k,l)}|X)} \frac{q_{k-1,k}}{q_{k,k-1}} \frac{1}{c_{u+1}-c_{u-1}-1}\right\}; \tag{7.45}$$

and for the 'simultaneous' move, the acceptance probability is

$$\min\left\{1, \frac{P(\boldsymbol{\vartheta}_*^{(k)}|X)}{P(\boldsymbol{\vartheta}_t^{(k,l)}|X)}\right\}, \tag{7.46}$$

since, for which, the proposal is symmetric in the sense $T(\boldsymbol{\vartheta}_t^{(k,l)} \to \boldsymbol{\vartheta}_*^{(k)}) = T(\boldsymbol{\vartheta}_*^{(k)} \to \boldsymbol{\vartheta}_t^{(k,l)}) = 1/(c_{u+1} - c_{u-1} - 2)$.

Liang (2009a) applies SSAMC to the simulated example given in Section 3.3.2 with the parameters being set as follows: $\alpha = \beta = 0.05, \lambda = 1, k_{\min} = 7$, and $k_{\max} = 14$. In general, the values of $k_{\min}$ and $k_{\max}$ can be determined with a short pilot run of SSAMC. SSAMC was run 20 times independently with $\kappa = 20, t_0 = 5, N = 10^5, \Lambda = k_{\max} - k_{\min} + 1, m = 8$, and $\pi_1 = \cdots = \pi_m = \frac{1}{m}$. The results are summarized in Table 7.6.

For comparison, Liang (2009a) also applies SAMC and RJMCMC to the same example with 20 runs. SAMC employs the same transition proposals and the same parameter setting as SSAMC except for $t_0 = 100$ and $N = 2 \times 10^6$. RJMCMC employs the same transition proposals as those used by SSAMC and SAMC, and performs 2×10^6 iterations in each run. Therefore, SSAMC, SAMC and RJMCMC perform the same number of energy evaluations in each run. The CPU times cost by a single run of them are 28.5 seconds, 25.5 seconds, and 23.9 seconds, respectively, on a 2.8 GHz computer. Table 7.6 shows that

Table 7.6 Estimated posterior probability for the change-point example (recompiled from Liang, 2009a).

	SSAMC		SAMC		MSAMC		RJMCMC	
k	P(%)	SD	P(%)	SD	P(%)	SD	P(%)	SD
7	0.101	0.002	0.094	0.003	0.098	0.002	0.091	0.005
8	55.467	0.247	55.393	0.611	55.081	0.351	55.573	0.345
9	33.374	0.166	33.373	0.357	33.380	0.223	33.212	0.205
10	9.298	0.103	9.365	0.279	9.590	0.135	9.354	0.144
11	1.566	0.029	1.579	0.069	1.646	0.030	1.569	0.040
12	0.177	0.004	0.180	0.010	0.187	0.004	0.185	0.010
13	0.016	0.001	0.015	0.001	0.017	0.000	0.017	0.001
14	0.002	0.000	0.001	0.000	0.002	0.000	0.001	0.000

P(%): the posterior probability $P(\mathcal{X}_k|Z)$ (normalized to 100%). SD: standard deviation of the estimates.

SSAMC works best among the three algorithms. As known by many people, RJMCMC is a general MH algorithm, and it is really hard to find another Monte Carlo algorithm to beat it for such a simple single-modal problem. However, SSAMC does. SSAMC is different from RJMCMC in two respects. First, like SAMC, SSAMC has the capability to self-adjust its acceptance rate. This capability enables it to overcome any difficulty in dimension jumping and to explore the entire model space very quickly. Second, SSAMC has the capability to make use of nearby model information to improve estimation. However, this can hardly be done in RJMCMC due to the stringent requirement for its Markovian property. These two capabilities make SSAMC potentially more efficient than RJMCMC for general Bayesian model selection problems.

It is worth pointing out that for this example, although the overall performance of SAMC is worse than that of RJMCMC, SAMC tends to work better than RJMCMC for the low probability model spaces, for example, those with 7 and 14 change points. This is due to the fact that SAMC samples equally from each model space, while RJMCMC samples from each model space proportionally to its probability. MSAMC have also been applied to this example with the same setting as that used by SSAMC. Its results, reported in Table 7.6, indicate that averaging over multiple samples can improve the convergence of SAMC, but a further smoothing operation on the frequency estimator is also important.

7.6.1.3 Discussion

SSAMC provides a general framework on how to improve efficiency of Monte Carlo simulations by incorporating some sophisticated nonparametric techniques. For illustrative purposes, the Nadaraya-Watson kernel estimator is employed above. Advanced smoothing techniques, such as the local log-likelihood estimator, should work better in general. For SAMC, allowing multiple samples to be generated at each iteration is important, as it provides us much freedom to incorporate data-mining techniques into simulations.

Liang *et al.* (2007) discuss potential applications of SAMC to problems for which the sample space is jointly partitioned according to two functions. The applicability of SSAMC to these problems is apparent: for joint partitions, the subregions can usually be ordered as a contingency table, and the smoothing estimator used above can thus be easily applied there.

7.6.2 Continuous SAMC for Marginal Density Estimation

A common computational problem in statistics is estimation of marginal densities. Let x denote a d-dimensional random vector, and let $f(x)$ denote its density function, which is known up to a normalizing constant; that is,

$$f(x) = \frac{1}{C}\psi(x), \quad x \in \mathcal{X}, \tag{7.47}$$

where $\mathcal{X}$ is the sample space, C is the unknown normalizing constant, and $\psi(x)$ is fully known or at least calculable for any points in $\mathcal{X}$. Let $y = \lambda(x)$ denote an arbitrary function which maps $\mathcal{X}$ to a lower dimensional space $\mathcal{Y}$ with dimension $d_\lambda < d$. Our task is to estimate the marginal density $g(y) = \int_{\{x:y=\lambda(x)\}} f(x)dx$ for $y \in \mathcal{Y}$.

When $g(y)$ is not available analytically, one must turn to approximations. To this end, approximate samples are generated from $f(x)$ via a MCMC sampler, and $g(y)$ can then be estimated using the kernel density estimation method (see, e.g., Wand and Jones, 1995). The kernel density estimation method allows for dependent samples (Hart and Vieu, 1990; Yu, 1993; Hall *et al.*, 1995). When iid samples are available, other nonparametric density estimation methods, such as local likelihood (Loader, 1999), smoothing spline (Gu, 1993; Gu and Qiu, 1993), and logspline (Kooperberg and Stone, 1991; Kooperberg, 1998), can also be applied to estimate the marginal density.

The problem has also been tackled by some authors from a different angle. For example, Chen (1994) proposed an importance sampling-based parametric method for the case where y is a subvector of x. Chen's estimator also allows for dependent samples. The major shortcoming of Chen's methods is its strong dependence on the knowledge of the analytical form of the inverse transformation $x = \lambda^{-1}(y)$. Other parametric methods, such as those by Gelfand *et al.* (1992) and Verdinelli and Wasserman (1995), suffer from similar handicaps.

Liang (2007a) proposes a continuous version of SAMC for marginal density estimation. Henceforth, the new algorithm will be abbreviated as the continuous SAMC algorithm or CSAMC. CSAMC abandons the use of sample space partitioning and incorporates the technique of kernel density estimation into simulations. CSAMC is very general. It works for any transformation $\lambda(x)$ regardless of availability of the analytical form of the inverse transformation. Like SAMC, CSAMC has the capability to self-adjust the acceptance probability of local moves. This mechanism enables it to escape from local energy minima to sample relevant parts of the sample space very quickly.

7.6.2.1 The CSAMC Algorithm

In what follows, CSAMC is described for the case where y is a bivariate vector. Let $g(y)$ be evaluated at the grid points of a lattice. Without loss of generality, we assume that the endpoints of the lattice form a rectangle denoted by $\mathcal{Y}$. In practice, the rectangle can be chosen such that its complementary space is of little interest to us. For example, in Bayesian statistics, the parameter space is often unbounded, and $\mathcal{Y}$ can then be set to a rectangle which covers the high posterior density region. A rough high posterior density region can usually be identified based on a preliminary analysis of the data. Alternatively, the high posterior density rectangle can be identified by a trial-and-error process, starting with a small rectangle and increasing it gradually until the resulting estimates converges.

Let the grid points be denoted by $\{z_{ij} : i = 1, \ldots, L_1, j = 1, \ldots, L_2\}$, where $z_{ij} = (z_i^{(1)}, z_j^{(2)})$. Let $d_1 = z_2^{(1)} - z_1^{(1)}$ and $d_2 = z_2^{(2)} - z_1^{(2)}$ be the horizontal and vertical neighboring distances of the lattice, respectively. Let $g_{ij} = g(z_{ij})$ be the true marginal density value at the point z_{ij}, and let $\widehat{g}_{ij}^{(t)}$ be the working estimate of g_{ij} obtained at iteration t. For any point $\tilde{y} = (\tilde{y}_1, \tilde{y}_2) \in \mathcal{Y}$, the density value can then be approximated by bilinear interpolation as follows. If $z_i^{(1)} < \tilde{y}_1 < z_{i+1}^{(1)}$ and $z_j^{(2)} < \tilde{y}_2 < z_{j+1}^{(2)}$ define i and j, then

$$\widehat{g}^{(t)}(\tilde{y}) = (1-u)(1-v)\widehat{g}_{ij}^{(t)} + u(1-v)\widehat{g}_{i+1,j}^{(t)} + (1-u)v\widehat{g}_{i,j+1}^{(t)} + uv\widehat{g}_{i+1,j+1}^{(t)}, \tag{7.48}$$

where $u = (\tilde{y}_1 - z_i^{(1)})/(z_{i+1}^{(1)} - z_i^{(1)})$ and $v = (\tilde{y}_2 - z_j^{(2)})/(z_{j+1}^{(2)} - z_j^{(2)})$.

Similarly to the desired sampling distribution used in SAMC, $\pi(y)$ is specified as the desired sampling distribution on the marginal space $\mathcal{Y}$. Theoretically, $\pi(y)$ can be any distribution defined on $\mathcal{Y}$. In practice, $\pi(y)$ is often set to be uniform over $\mathcal{Y}$. Let $\{\gamma_t : t = 1, 2, \ldots\}$ denote a sequence of gain factors as used in SAMC, and let $\boldsymbol{H}_t = \text{diag}(h_{t1}^2, h_{t2}^2)$ denote the bandwidth matrix used in the density estimation procedure at iteration t. The sequence $\{h_{ti} : t = 1, 2, \ldots\}$ is positive and non-increasing, and converges to 0 as $t \to \infty$. In Liang (2007a), it takes the form

$$h_{ti} = \min\left\{\gamma_t^b, \frac{\text{range}(\tilde{y}_i)}{2\,(1 + \log_2(M))}\right\}, \quad i = 1, 2, \tag{7.49}$$

where $b \in (\frac{1}{2}, 1]$, and the second term in $\min\{\cdot,\cdot\}$ is the default bandwidth used in conventional density estimation procedures (see, e.g., the procedure *density(·)* in S-PLUS 5.0, Venables and Ripley, 1999). With the above notations, one iteration of CSAMC consists of the following steps:

CSAMC Algorithm (Steps)

(a) *Sampling*: Draw samples $x_{t+1}^{(k)}, k = 1, \ldots, \kappa$, from the working density

$$\widehat{f}_t(x) \propto \frac{\psi(x)}{\widehat{g}^{(t)}(\lambda(x))}, \tag{7.50}$$

via a MCMC sampler, e.g., the MH algorithm or the Gibbs sampler, where $\psi(x)$ is as defined in (7.47), and $\widehat{g}^{(t)}(\lambda(x))$ is as defined in (7.48).

(b) *Estimate updating*:

(b.1) Estimate the density of the transformed samples $y_{t+1}^{(1)} = \lambda(x_{t+1}^{(1)})$, $\ldots, y_{t+1}^{(\kappa)} = \lambda(x_{t+1}^{(\kappa)})$ using the kernel method. Evaluate the density at the grid points,

$$\zeta_u^{(t+1)}(z_{ij}) = \frac{1}{\kappa}\sum_{k=1}^{\kappa} |\boldsymbol{H}_t|^{-\frac{1}{2}} K\left((\boldsymbol{H}_t)^{-\frac{1}{2}}(z_{ij} - y_{t+1}^{(k)})\right), \tag{7.51}$$

for $i = 1, \ldots, L_1$ and $j = 1, \ldots, L_2$, where $K(y)$ is a bivariate kernel density function.

(b.2) Normalize $\zeta_u^{(t+1)}(z_{ij})$ on the grid points by setting

$$\zeta^{(t+1)}(z_{ij}) = \frac{\zeta_u^{(t+1)}(z_{ij})}{\sum_{i'=1}^{L_1}\sum_{j'=1}^{L_2}\zeta_u^{(t+1)}(z_{i'j'})},$$
$$i = 1, \ldots, L_1, \quad j = 1, \ldots, L_2. \tag{7.52}$$

(b.3) Update the working estimate $\widehat{g}_{ij}^{(t)}$ in the following manner,

$$\log \widehat{g}_{ij}^{(t+1)} = \log \widehat{g}_{ij}^{(t)} + \gamma_{t+1}\left(\zeta^{(t+1)}(z_{ij}) - \pi'(z_{ij})\right), \tag{7.53}$$

for $i = 1, \ldots, L_1$ and $j = 1, \ldots, L_2$, where $\pi'(z_{ij}) = \pi(z_{ij})/[\sum_{i'=1}^{L_1}\sum_{j'=1}^{L_2}\pi(z_{i'j'})]$.

(c) *Lattice refinement*. Refine the lattice by increasing the values of L_1, L_2 or both, if $\max\left\{\frac{d_1}{h_{t1}}, \frac{d_2}{h_{t2}}\right\} > \upsilon$, where υ is a threshold value.

Since $\boldsymbol{x}_{t+1}^{(1)}, \ldots, \boldsymbol{x}_{t+1}^{(\kappa)}$ are generated from (7.50), it follows from Wand and Jones (1995) that

$$E\zeta_u^{(t+1)}(\boldsymbol{z}_{ij}) = \Psi_t(\boldsymbol{z}_{ij}) + \frac{1}{2}\mu_2(K)\mathrm{tr}\left\{\boldsymbol{H}_t\mathcal{H}_t(\boldsymbol{z}_{ij})\right\} + o\left\{\mathrm{tr}(\boldsymbol{H}_t)\right\}, \tag{7.54}$$

where $\Psi_t(\boldsymbol{z})$ is a density function proportional to $g(\boldsymbol{z})/\widehat{g}^{(t)}(\boldsymbol{z}), \mu_2(K) = \int z^2K(z)dz$, and $\mathcal{H}_t$ is the Hessian matrix of $\Psi_t(\boldsymbol{z})$. Thus, $\zeta_u^{(t)}(\boldsymbol{z})$ forms an (asymptotically) unbiased estimator of $\Psi_t(\boldsymbol{z})$ as $h_{t\cdot}$ goes to 0.

As in SAMC, Θ can be restricted to a compact set by assuming that all the grid points are in the support set of $f(\boldsymbol{x})$. By assuming that the transition kernel satisfies the drift condition, we have the following theorem, whose proof follows from (7.54) and the proof of Theorem 7.6.1.

Theorem 7.6.2 *Assume* (B_1) *and the drift condition* (B_2) *(given in Section 7.7.1) hold. Let* $\{h_{ti} : t = 1, 2, \ldots\}$ *be a sequence as specified in (7.49). Then*

$$P\left\{\lim_{t\to\infty} \log \widehat{g}^{(t)}(z_{ij}) = c + \log g(z_{ij}) - \log \pi(z_{ij})\right\} = 1, \tag{7.55}$$

for $i = 1, \ldots, L_1$ *and* $j = 1, \ldots, L_2$, *where* c *is an arbitrary constant, which can be determined by imposing an additional constraint on* $\widehat{g}(z_{ij})$*'s.*

Like SSAMC, CSAMC tends to converge faster than SAMC. This is because CSAMC employs the technique of kernel density estimation, and thus estimates can be updated in large blocks at an early stage of the simulation. In CSAMC, the bandwidth plays a similar role as the gain factor. It is meant to enhance the ability of the sampler in sample space exploration. The difference is that the bandwidth controls the moving step size of the sampler, while the gain factor controls the moving mobility of the sampler. CSAMC is equipped with both the bandwidth and the gain factor.

CSAMC is different from other density estimators in its dynamic nature. In CSAMC, the estimates are updated iteratively, each updating is based on a small number of MCMC samples simulated from the working density, and the updating manner turns gradually from global to local due to the gradual decrease of the kernel bandwidth as the simulation proceeds. Hence, the rules developed for the choice of bandwidth matrix for conventional kernel density estimators may not work for CSAMC. Liang (2007) suggests associating the choice of bandwidth matrix with the choice of $\boldsymbol{\pi}$. For example, if $\boldsymbol{\pi}$ is set to be uniform on $\mathcal{Y}$, then the samples $y_t^{(1)}, \ldots, y_t^{(\kappa)}$ tend to be uniformly distributed on $\mathcal{Y}$ when t becomes large, and this suggests the use of a diagonal bandwidth matrix with the diagonal elements accounting for the variability (range) of the samples in directions of respective coordinates. Other bandwidth matrices, such as the full bandwidth matrix, are not recommended here due to the uniformity of the samples. When unequally spaced grid points are used in evaluation of the density, the bandwidth h_{ti} may be allowed to vary with the position of the grid point. However, data-driven approaches, for example, the nearest neighbor and balloon approaches (Loftsgaarden and Quesenberry, 1965; Terrell and Scott, 1992), may not be appropriate for CSAMC, because the convergence of bandwidth is out of our control in these approaches.

The kernel method requires one to evaluate the kernel density at all grid points for any given sample $y_t^{(k)}$. For fast computation, one only needs to evaluate the kernel density at the grid points in a neighborhood of $y_t^{(k)}$, for example, the grid points within a cycle of radius $\max\{4h_{t1}, 4h_{t2}\}$ and centered at $y_t^{(k)}$. This will save a lot of CPU time when the lattice size is large, while keeping the estimation less affected by the kernel density truncation.

7.6.2.2 An Illustrative Example

Consider the mixture Gaussian distribution,

$$f(\boldsymbol{x}) = \frac{1}{3}N_3(-\boldsymbol{\mu}_1, \Sigma_1) + \frac{2}{3}N_3(\boldsymbol{\mu}_2, \Sigma_2), \tag{7.56}$$

where $\boldsymbol{x} = (x_1, x_2, x_3), \boldsymbol{\mu}_1 = (-5, -5, -5), \boldsymbol{\mu}_2 = (10, 25, 1)$,

$$\Sigma_1 = \begin{pmatrix} 4 & 5 & 0 \\ 5 & 64 & 0 \\ 0 & 0 & 1 \end{pmatrix}, \quad \text{and} \quad \Sigma_2 = \begin{pmatrix} 1/4 & 0 & 0 \\ 0 & 1/4 & 0 \\ 0 & 0 & 1/4 \end{pmatrix}.$$

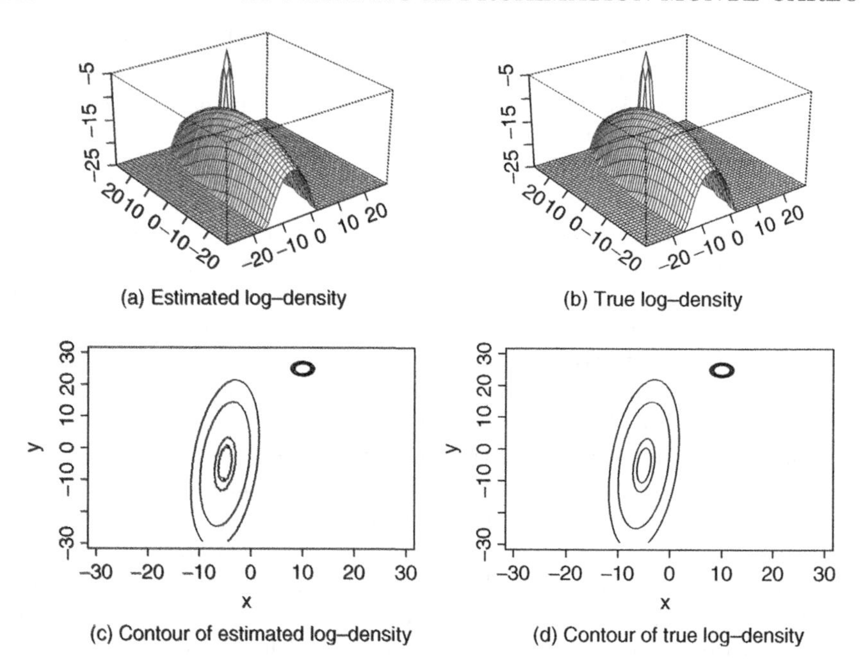

Figure 7.10 Computational results of CCMC for the mixture Gaussian distribution. (a) The log-marginal density estimate; (b) the true log-marginal density; (c) the contour plot of the log-marginal density estimate; (d) the contour plot of the true log-marginal density. The circles in the two contour plots corresponds to the 95%, 90%, 50%, and 10% percentiles of the log-density, respectively (Liang, 2007a).

Suppose that we want to estimate the marginal density of $\boldsymbol{y} = (x_1, x_2)$. This example mimics a multimodal posterior distribution in Bayesian inference, where the two modes are well separated.

In applying CSAMC to this example, $\mathcal{Y}$ is restricted to the square $[-30.5, 30.5] \times [-30.5, 30.5]$, and the marginal density is evaluated on a square lattice 245×245. At each iteration, the working density was simulated using the MH algorithm with the random walk proposal $N_3(\boldsymbol{x}_t, 5^3 I)$, and the sample size κ was set to 10. With 5×10^6 iterations, CSAMC produced a log-marginal density estimate as shown in Figure 7.10(a). Liang (2007a) also applies CSAMC with great success to estimate the normalizing constant function for a spatial autologistic model:

7.6.3 Annealing SAMC for Global Optimization

During the past several decades, simulated annealing (Kirkpatrick *et al.*, 1983) and the genetic algorithm (Holland, 1975; Goldberg, 1989) have been applied successfully by many authors to highly complex optimization problems in different fields of science and engineering. In spite of their successes, both algorithms suffer from some difficulties in convergence to the global optima.

Suppose that one is interested in minimizing a function $H(x)$ over a given space $\mathcal{X}$. Throughout this subsection, $H(x)$ is called the energy function in terms of physics. Simulated annealing works by simulating a sequence of distributions specified by

$$f_k(x) = \frac{1}{Z_k} \exp\{-H(x)/T_k\}, \quad x \in \mathcal{X}, \quad k = 1, 2, \ldots,$$

where T_k is called the temperature, and $T_1 > T_2 > \cdots$ forms a decreasing ladder. As discussed in Section 4.1, the simulation will get stuck at a local minimum unless the temperature decreases at a rate of logarithm. However, the CPU time required by this cooling schedule can be too long to be affordable for challenging problems.

The genetic algorithm tries to solve the minimization problem by mimicking the natural evolutionary process. A population of candidate solutions (also known as individuals), generated at random, are tested and evaluated for their energy values (also known as fitness values); the best of them are then bred through mutation and crossover operations; the process is repeated over many generations, until an individual of satisfactory performance is found. Through the crossover operation, information distributed across the population is effectively used in the minimization process. Schmitt (2001) shows that, under certain conditions, the genetic algorithm can converge asymptotically to the global minima at a logarithmic rate in analogy to simulated annealing.

As previously discussed, a remarkable feature of SAMC is that is self-adjusting mechanism makes it immune to the local-trap problem. We now consider applications of SAMC on optimization. Two modified versions of SAMC, annealing SAMC (ASAMC) (Liang, 2007b) and annealing evolutionary SAMC (AESAMC) (Liang, 2009d), are discussed. The ASAMC algorithm works in the same spirit as simulated annealing but with the sample space instead of temperature shrinking with iterations. The AESAMC algorithm represents a further improvement of annealing SAMC by incorporating some crossover operators originally used by the genetic algorithm into the search process. Under mild conditions, both ASAMC and AESAMC can converge weakly toward a neighboring set of global minima in the space of energy.

7.6.3.1 Annealing SAMC

Like conventional MCMC algorithms, SAMC is able to find the global energy minima if the run is long enough. However, due to the broadness of the sample space, the process may be slow even when sampling has been biased to low energy subregions. To accelerate the search process, Liang (2007b) proposed to shrink the sample space over iterations.

Suppose that the subregions $E_1, \ldots, E_m$ have been arranged in ascending order by energy; that is, if $i < j$, then $H(x) < H(y)$ for any $x \in E_i$ and $y \in E_j$. Let $\varpi(u)$ denote the index of the subregion that a sample x with energy u belongs to. For example, if $x \in E_j$, then $\varpi(H(x)) = j$. Let $\mathcal{X}_t$ denote the sample space at iteration t. Annealing SAMC initiates its search in the entire sample space $\mathcal{X}_0 = \bigcup_{i=1}^m E_i$, and then iteratively searches in the set

$$\mathcal{X}_t = \bigcup_{i=1}^{\varpi(u_t^*+\aleph)} E_i, \quad t = 1, 2, \ldots, \tag{7.57}$$

where u_t^* denotes the best function value obtained by iteration t, and $\aleph > 0$ is a user specified parameter which determines the broadness of the sample space at each iteration. Since the sample space shrinks iteration by iteration, the algorithm is called annealing SAMC. Let Θ_t denote the state space of θ_t. In summary, ASAMC consists of the following steps:

Annealing SAMC Algorithm

1. *Initialization*. Partition the sample space $\mathcal{X}$ into m disjoint subregions $E_1, \ldots, E_m$ according to the objective function $H(x)$; specify a desired sampling distribution $\boldsymbol{\pi}$; initialize x_0 by a sample randomly drawn from the sample space $\mathcal{X}, \theta_0 = (\theta_0^{(1)}, \ldots, \theta_0^{(m)}) = (0, 0, \ldots, 0), \aleph$, and $\mathcal{X}_0 = \bigcup_{i=1}^m E_i$; and set the iteration number $t = 0$.

2. *Sampling*. Draw sample x_{t+1} by a single or few MH moves which admit the following distribution as the invariant distribution,

$$f_{\theta_t}(x) \propto \sum_{i=1}^{\varpi(u_t^*+\aleph)} \frac{\psi(x)}{\exp\{\theta_t^{(i)}\}} I(x \in E_i), \tag{7.58}$$

   where $I(x \in E_i)$ is the indicator function, $\psi(x) = \exp\{-H(x)/\tau\}$, and τ is a user-specified parameter.

3. *Working weight updating*. Update the log-weight θ_t as follows:

$$\theta_{t+\frac{1}{2}} = \theta_t^{(i)} + \gamma_{t+1}\big[I(x_{t+1} \in E_i) - \pi_i\big], \quad i = 1, \ldots, \varpi(u_t^* + \aleph),$$

   where the gain factor sequence $\{\gamma_t\}$ is subject to the condition (A_1). If $\theta_{t+\frac{1}{2}} \in \Theta$, set $\theta_{t+1} = \theta_{t+\frac{1}{2}}$; otherwise, set $\theta_{t+1} = \theta_{t+\frac{1}{2}} + \mathbf{c}^*$, where $\mathbf{c}^* = (c^*, \ldots, c^*)$ and c^* is chosen such that $\theta_{t+\frac{1}{2}} + \mathbf{c}^* \in \Theta$.

4. *Termination*. Check the termination condition, e.g., a fixed number of iterations has been reached. Otherwise, set $t \to t + 1$ and go to step 2.

Liang (2007b) shows that if the gain factor sequence satisfies condition (A_1) and the proposal distribution satisfies the minorisation condition, that is,

$$\sup_{\theta \in \Theta} \sup_{x,y \in \mathcal{X}} \frac{f_\theta(y)}{q(x,y)} < \infty, \tag{7.59}$$

then ASAMC can converge weakly toward a neighboring set of the global minima of $H(x)$ in the space of energy, where $f_\theta(y)$ is defined by

$$f_\theta(y) \propto \sum_{i=1}^{m} \frac{\psi(y)}{\exp(\theta^{(i)})} I(y \in E_i).$$

More precisely, we have the following theorem:

Theorem 7.6.3 *Assume (A_1) and (7.59) hold. Then, as $t \to \infty$, x_t generated by ASAMC converges in distribution to a random variable with the density function*

$$f_{\theta_*}(x) \propto \sum_{i=1}^{\varpi(u^*+\aleph)} \frac{\pi_i' \psi(x)}{\int_{E_i} \psi(x)dx} I(x \in E_i), \tag{7.60}$$

where $\pi_i' = \pi_i + (1 - \sum_{j \in \mathcal{S}} \pi_j)/|\mathcal{S}|$, $\mathcal{S} = \{k : E_k \neq \emptyset, k = 1, \ldots, \varpi(u^* + \aleph)\}$, u^* is the global minimum value of $H(x)$, and $|\mathcal{S}|$ is the cardinality of $\mathcal{S}$.

The proof of this theorem can be found in Section 7.7.1.4. In practice, the proposals satisfying (7.59) can be easily designed for both discrete and continuum systems. For example, if $\mathcal{X}$ is compact, a sufficient design for the minorisation condition is to choose $q(x,y) > \epsilon_0$ for all $x, y \in \mathcal{X}$, where ϵ_0 is any positive number.

Learning Neural Networks for a Two-Spiral Problem. Over the past several decades, feed-forward neural networks, otherwise known as multiple-layer perceptrons (MLPs), have achieved increased popularity among scientists, engineers, and other professionals as tools for knowledge representation. Given a group of connection weights $\boldsymbol{x} = (\alpha, \beta, \gamma)$, the MLP approximator can be written as

$$\hat{f}(\boldsymbol{z}_k|\boldsymbol{x}) = \varphi_o\left(\alpha_0 + \sum_{j=1}^{p} \gamma_j z_{kj} + \sum_{i=1}^{M} \alpha_i \varphi_h\big(\beta_{i0} + \sum_{j=1}^{p} \beta_{ij} z_{kj}\big)\right), \tag{7.61}$$

where M is the number of hidden units, p is the number of input units, $\boldsymbol{z}_k = (z_{k1}, \ldots, z_{kp})$ is the kth input pattern, and α_i, β_{ij} and γ_j are the weights on the connections from the ith hidden unit to the output unit, from the jth input unit to the ith hidden unit, and from the jth input unit to the output unit, respectively. The connections from input units to the output unit are also called the shortcut connections. In (7.61), the bias unit is treated as a special input unit with a constant input, say 1. The functions $\varphi_h(\cdot)$ and $\varphi_o(\cdot)$ are called the activation functions of the hidden units and the output unit, respectively. Popular choices of $\varphi_h(\cdot)$ include the sigmoid function and the hyperbolic tangent function. The former is defined as $\varphi_h(z) = 1/(1+e^{-z})$ and the latter $\varphi_h(z) = \tanh(z)$. The choice of $\varphi_o(\cdot)$ is problem-dependent. For regression problems, $\varphi_o(\cdot)$ is usually set to the linear function $\varphi_o(z) = z$; and for classification problems, $\varphi_o(\cdot)$ is usually set to the sigmoid function. The problem of MLP training is to minimize the objective function

$$H(\boldsymbol{x}) = \sum_{k=1}^{N}\left(y_k - \widehat{f}(\boldsymbol{z}_k|\boldsymbol{x})\right)^2 + \lambda\left[\sum_{i=0}^{M}\alpha_i^2 + \sum_{i=1}^{M}\sum_{j=0}^{p}\beta_{ij}^2 + +\sum_{j=1}^{p}\gamma_j^2\right], \quad (7.62)$$

by choosing appropriate connection weights, where y_k denotes the target output corresponding to the input pattern $\boldsymbol{z}_k$, the second term is the regularization term, and λ is the regularization parameter. The regularization term is often chosen as the sum of squares of the connection weights, which stabilizes the generalization performance of the MLP. Henceforth, $H(\boldsymbol{x})$ will be called the energy function of the MLP.

As known by many researchers, the energy landscape of the MLP is rugged. The gradient-based training algorithms, such as back-propagation (Rumelhart *et al.*, 1986) and the BFGS algorithm (Broyden, 1970; Fletcher, 1970; Goldfarb, 1970; Shanno, 1970), tend to converge to a local energy minimum near the starting point. Consequently, the information contained in the training data may not be learned sufficiently. To avoid the local-trap problem, simulated annealing (SA) (Kirkpatrick *et al.*, 1983) has been employed to train neural networks by many authors, including Amato *et al.* (1991), and Owen and Abunawass (1993), among others, who show that for complex learning tasks, SA has a better chance to converge to a global energy minimum than have gradient-based algorithms.

Liang (2007b) compares efficiency of ASAMC, SAMC and simulated annealing (SA) in training MLPs using a two-spiral problem. As depicted by Figure 7.11, the two-spiral problem is to learn a mapping that distinguishes between points on two intertwined spirals. To have a better calibration of the efficiency of these algorithms, Liang (2007b) drops the regularization term in the energy function (7.62) such that it a global minimum value known as 0. The regularization term is usually included when one is concerned about generalization errors. To compare with the results published in the literature, both $\varphi_h(z)$ and $\varphi_o(z)$ are set to the sigmoid function.

It is known that the two-spiral problem can be solved using MLPs with multiple hidden layers: Lang and Witbrock (1989) use a 2-5-5-5-1 MLP

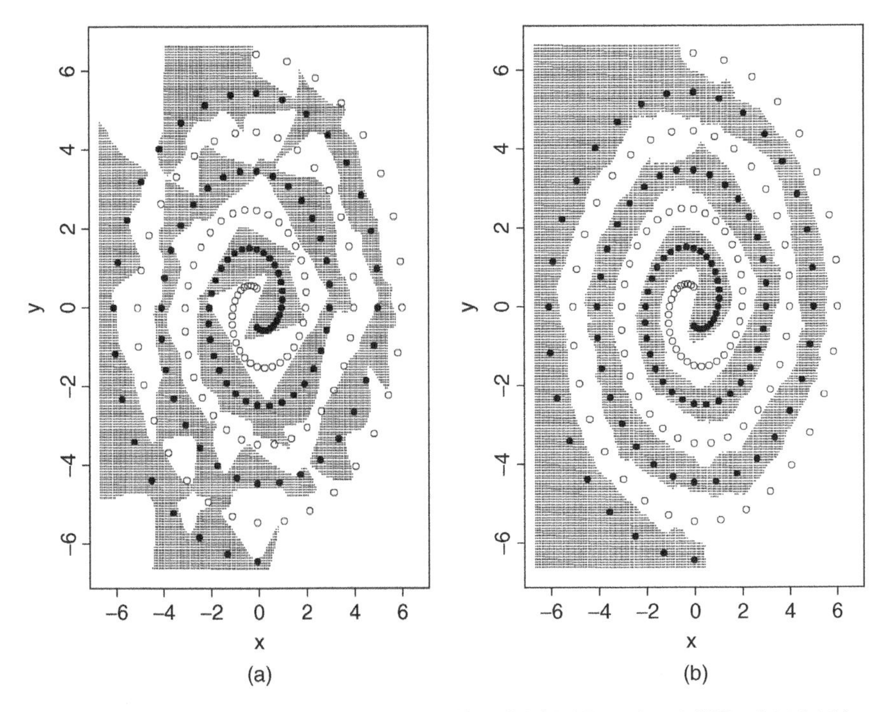

Figure 7.11 Classification maps learned by ASAMC with a MLP of 30 hidden units. The black, and white points show the training data for two different spirals. (a) Classification map learned in one run. (b) Classification map averaged over 20 runs. This figure demonstrates the success of ASAMC in minimization of complex functions (Liang, 2007b).

with shortcut connections (138 trainable connection weights); Fahlman and Lebiere (1990) use cascade-correlation networks with 12–19 hidden units, the smallest network having 114 connections; and Wong and Liang (1997) use a 2-14-4-1 MLP without shortcut connections. It is generally believed that this problem is so very difficult for the standard one-hidden-layer MLP, because it requires the MLP to learn a highly nonlinear separation of the input space. Baum and Lang (1991) report that a solution can be found using a 2-50-1 back-propagation MLP, but only the MLP has been initialized with queries.

For the two-spiral problem, Liang (2007b) trains a 2-30-1 MLP without shortcut connections, which consists of a total of 121 connections. ASAMC was run 20 times independently, and each run consisted of a maximum of 10^7 iterations. The simulation stopped early if a solution with $H(\boldsymbol{x}) < 0.2$ was found. Figure 7.11(a) shows the classification map learned in one run, which indicates that a MLP with 30 hidden units is able to separate the two spirals. Figure 7.11(b) shows the average classification map over the 20 runs by the

Table 7.7 Comparison of ASAMC, SAMC, SA and BFGS for the two-spiral example (Liang, 2007b).

algorithm	mean	SD	min	max	prop	iteration($\times 10^6$)	time
ASAMC	0.620	0.191	0.187	3.23	15	7.07	94m
SAMC	2.727	0.208	1.092	4.089	0	10.0	132m
SA-1	17.485	0.706	9.02	22.06	0	10.0	123m
SA-2	6.433	0.450	3.03	11.02	0	10.0	123m
BFGS	15.50	0.899	10.00	24.00	0	–	3s

Let z_i denote the minimum energy value obtained in the ith run for $i = 1, \ldots, 20$, 'Mean' is the average of z_i, 'SD' is the standard deviation of 'mean', 'min' $= \min_{i=1}^{20} z_i$, 'max' $= \max_{i=1}^{20} z_i$, 'Prop' $= \#\{i : z_i \leq 0.21\}$, 'Iteration' is the average number of iterations performed in each run, and 'Time' is the average CPU time cost by each run. SA-1 employs the linear cooling schedule. SA-2 employs the geometric cooling schedule.

ensemble averaging approach (Perrone, 1993). Each solution in the ensemble was weighted equally. Ensemble averaging smoothes the classification boundary and improves the generalization performance of the MLP.

For comparison, SAMC, SA and BFGS were applied to this example. Both SAMC and SA were run 20 times, and each run consisted of 10^7 iterations. BFGS was also run 20 times, but it converged within 1000 iterations in each run. The numerical results are summarized in Table 7.7, which indicate that ASAMC has made a dramatic improvement over SAMC, SA and BFGS for this example. ASAMC found perfect solutions in 15 out of 20 runs with an average of 7.07×10^6 iterations, while SAMC, SA and BFGS failed to find perfect solutions in all runs. Liang (2007b) reports that the MLP structure considered above is not minimal for this problem; ASAMC can find perfect solutions to this problem using a MLP with 27 hidden units (109 connections). See Liang (2008b) for more discussion on the use of SAMC-based algorithms for MLP training.

7.6.3.2 Annealing Evolutionary Stochastic Approximation Monte Carlo

Like the genetic algorithm, AESAMC works on a population of samples. Let $\boldsymbol{x} = (x_1, \ldots, x_n)$ denote the population, where n is the population size, and $x_i = (x_{i1}, \ldots, x_{id})$ is a d-vector and is called an individual or chromosome in terms of genetic algorithms. Clearly, the minimum of $H(x)$ can be obtained by minimizing the function $\boldsymbol{H}(\boldsymbol{x}) = \sum_{i=1}^{n} H(x_i)$. An unnormalized Boltzmann density can be defined for the population as follows,

$$\psi(\boldsymbol{x}) = \exp\{-\boldsymbol{H}(\boldsymbol{x})/\tau\}, \quad \boldsymbol{x} \in \mathcal{X}^n, \tag{7.63}$$

where $\mathcal{X}^n = \mathcal{X} \times \cdots \times \mathcal{X}$ is a product sample space. The sample space can be partitioned according to the function $\boldsymbol{H}(\boldsymbol{x})$ into m subregions: $\boldsymbol{E}_1 = \{x : \boldsymbol{H}(\boldsymbol{x}) \leq u_1\}, \boldsymbol{E}_2 = \{x : u_1 < \boldsymbol{H}(\boldsymbol{x}) \leq u_2\}, \ldots, \boldsymbol{E}_{m-1} = \{x : u_{m-2} < \boldsymbol{H}(\boldsymbol{x}) \leq u_{m-1}\}$, and $\boldsymbol{E}_m = \{x : \boldsymbol{H}(\boldsymbol{x}) > u_{m-1}\}$, where $u_1 < u_2 < \ldots < u_{m-1}$ are $m-1$ known real numbers. Note that the sample space is not necessarily partitioned according to the function $\boldsymbol{H}(\boldsymbol{x})$. For example, it can also be partitioned according to $\lambda(\boldsymbol{x}) = \min\{H(x_1), \ldots, H(x_n)\}$. The population can then evolve under the framework of ASAMC with an appropriate proposal distribution for the MH moves. At iteration t, the MH moves admit the following distribution as the invariant distribution,

$$\boldsymbol{f}_{\theta_t}(\boldsymbol{x}) \propto \sum_{i=1}^{\varpi(\boldsymbol{u}_t^* + \aleph)} \frac{\psi(\boldsymbol{x})}{\exp\{\theta_t^{(i)}\}} I(\boldsymbol{x} \in \boldsymbol{E}_i), \quad \boldsymbol{x} \in \mathcal{X}_t^n, \tag{7.64}$$

where $\boldsymbol{u}_t^*$ denotes the best value of $\boldsymbol{H}(\boldsymbol{x})$ obtained by iteration t. As previously discussed, $\{\theta_t\}$ can be kept in a compact space in simulations due to the use of a fixed sample space partition.

Since, in AESAMC, the state of the MH chain has been augmented to a population, the crossover operators used in the genetic algorithm can be employed to accelerate the evolution of the population. However, to satisfy the Markov chain reversibility condition, these operators need to be modified appropriately. This can be done in a similar way to that described in Section 5.5 for evolutionary Monte Carlo (Liang and Wong, 2000, 2001a). See Liang (2009d) for the details. As demonstrated by Liang and Wong (2000, 2001a), Goswami and Liu (2007), and Jasra *et al.* (2007), incorporating genetic-type moves into Markov chain Monte Carlo can often improve ergodicity of the simulation. Under mild conditions, Liang (2009d) shows that AESAMC can converge weakly toward a neighboring set of the global minima of $H(x)$ in the space of energy; that is, Theorem 7.6.3 also holds for AESAMC.

Although AESAMC is proposed as an optimization technique, it can also be used as a dynamic importance sampling algorithm as SAMC by keeping the sample space unshrunken over iterations. AESAMC has provided a general framework on how to incorporate crossover operations into dynamically weighted MCMC simulations, for example, dynamic weighting (Wong and Liang, 1997; Liu *et al.*, 2001; Liang, 2002b) and population Monte Carlo (Cappé *et al.*, 2004). This framework is potentially more useful than the MCMC framework provided by evolutionary Monte Carlo. Under the MCMC framework, the crossover operation has often a low acceptance rate, where the MH rule will typically reject an unbalanced pair of offspring for which one has a high density value and the other low. In AESAMC, this difficulty has been much alleviated due to its self-adjusting mechanism.

Multimodal Optimization Problems. Liang (2009d) compares AESAMC with SA, ASAMC and other metaheuristics, including the genetic algorithm (Genocop III) (Michalewicz and Nazhiyath, 1995), scatter search (Laguna and Martí, 2005), directed tabu search (DTS) (Hedar and Fukushima, 2006), and continuous GRASP (C-GRASP) (Hirsch *et al.*, 2006; Hirsch *et al.*, 2007), using a set of benchmark multimodal test functions whose global minima are known. These test functions are given in the Appendix 7A, and are also available in Laguna and Martí (2005), Hedar and Fukushima (2006) and Hirsch *et al.* (2006).

Genocop III is an implementation of the genetic algorithm that is customized for solving nonlinear optimization problems with continuous and bounded variables. The scatter search algorithm is an evolutionary algorithm that, unlike the genetic algorithm, operates on a small population of solutions and makes only limited use of randomization as a proxy for diversification when searching for a globally optimal solution. The DTS algorithm

Table 7.8 Average optimality gap values over the 40 test functions (Liang, 2009d).

Algorithm	5000[a]	10000[a]	20000[a]	50000[a]
Genocop III[b]	636.37	399.52	320.84	313.34
Scatter Search[b]	4.96	3.60	3.52	3.46
C-GRASP[c]	6.20	4.73	3.92	3.02
DTS[d]	4.22	1.80	1.70	1.29
ASAMC[e] ($\aleph = 10$)	4.11(0.11)	2.55(0.08)	1.76(0.08)	1.05(0.06)
ASAMC[e] ($\aleph = 5$)	3.42(0.09)	2.36(0.07)	1.51(0.06)	0.94(0.03)
ASAMC[e] ($\aleph = 1$)	3.03(0.11)	2.02(0.09)	1.39(0.07)	0.95(0.06)
SA[e] ($t_{high} = 20$)	3.58(0.11)	2.59(0.08)	1.91(0.06)	1.16(0.04)
SA[e] ($t_{high} = 5$)	3.05(0.11)	1.80(0.09)	1.17(0.06)	0.71(0.03)
SA[e] ($t_{high} = 2$)	2.99(0.12)	1.89(0.09)	1.12(0.06)	0.69(0.03)
SA[e] ($t_{high} = 1$)	2.36(0.11)	1.55(0.08)	1.06(0.06)	0.67(0.03)
SA[e] ($t_{high} = 0.5$)	2.45(0.11)	1.39(0.07)	1.06(0.07)	0.75(0.06)
AESAMC[f] ($\aleph = 10$)	1.59(0.08)	0.82(0.02)	0.68(0.01)	0.50(0.01)
AESAMC[f] ($\aleph = 1$)	1.93(0.27)	0.79(0.01)	0.66(0.02)	0.49(0.01)

[a]The number of function evaluations. [b]The results are from Laguna and Martí (2005). [c]The results are from Hirsch *et al.* (2006). [d]The results are from Hedar and Fukushima (2006). [e]The number in the parentheses denotes the standard deviation of the average, and it is calculated as $\sqrt{\sum_{i=1}^{40} s_i^2/n_i/40}$ with s_i^2 being the variance of the best function values produced by the algorithm in n_i runs ($n_i = 500$ for the test functions R_{10} and R_{20}, and 5 for all others). [f]The number in the parentheses denotes the standard deviation of the average. It is calculated as for SA and ASAMC, but with $n_i = 5$ for all test functions.

is a hybrid of the tabu search algorithm (Glover and Laguna, 1997) and a direct search algorithm. The role of direct search is to stabilize the search, especially in the vicinity of a local optimum; that is, to generate neighborhood trial moves instead of using completely blind random search. The search strategy employed by the DTS algorithm is an adaptive pattern search. The C-GRASP algorithm is a multistart searching algorithm for continuous optimization problems subject to box constraints, where the starting solution for local improvement is constructed in a greedy randomized fashion. The numerical results reported in Hedar and Fukushima (2006) and Hirsch *et al.* (2006) indicate that the DTS, C-GRASP, and scatter search algorithms represent very advanced metaheuristics in the current optimization community.

All of the AESAMC, ASAMC and SA algorithms fall into the class of stochastic optimization algorithms. To account for the variability of their results, each algorithm was run 5 times for each test function, and the average of the best function values produced in each run were reported as its output. Table 7.8 reports the average optimality gap over the 40 test functions, where the optimality gap for a test function is defined as the absolute difference between the global minimum value of the function and the output value by the algorithm. The comparison indicates that AESAMC significantly outperforms Genocop III, scatter search, C-GRASP, DTS, ASAMC and SA for these test functions. It is remarkable that the average optimality gaps produced by AESAMC with 20 000 function evaluations have been comparable with or better than those produced by any other algorithms (given in Table 7.8) with 50 000 function evaluations.

Figure 7.12 compares the optimality gaps produced by AESAMC, ASAMC, and SA for each of the 40 test functions. It indicates that AESAMC produced smaller optimality gap values than ASAMC and SA for almost all test functions, especially for those of high dimension.

7.7 Theory of Stochastic Approximation Monte Carlo

7.7.1 Convergence

This section is organized as follows. Section 7.7.1.1 describes a general stochastic approximation MCMC (SAMCMC) algorithm, which includes the SAMC, SSAMC, CSAMC, ASAMC and AESAMC as special instances, and gives the conditions for its convergence. Section 7.7.1.2 establishes the convergence of the general SAMCMC algorithm. Section 7.7.1.3 applies the convergence results for the general SAMCMC algorithm and obtain the convergence of SSAMC. Section 7.7.1.4 discusses the convergence of SAMC, CSAMC, ASAMC and AESAMC.

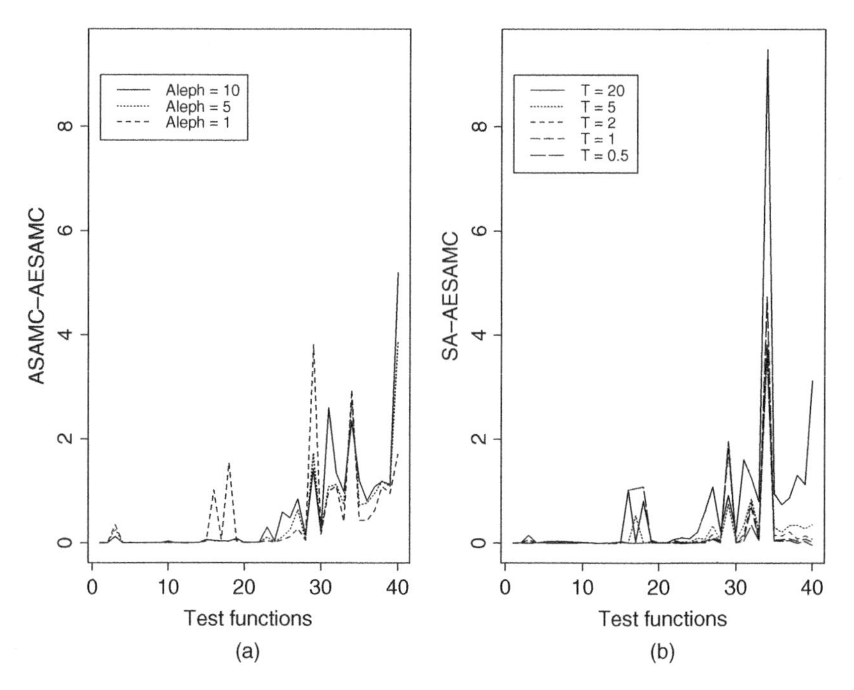

Figure 7.12 Comparison of optimality gap values produced by AESAMC (with $\aleph = 1$), ASAMC and SA with 50 000 function evaluations for each of the 40 test functions. (a) Difference of the optimality gap values produced by ASAMC and AESAMC (with $\aleph = 1$). (b) Difference of the optimality gap values produced by SA and AESAMC (with $\aleph = 1$).

7.7.1.1 Convergence of a general SAMCMC Algorithm

Consider the following stochastic approximation MCMC algorithm:

1. Draw samples $x_{t+1}^{(1)}, \ldots, x_{t+1}^{(\kappa)}$ through a MH kernel which admits $f_{\theta_t}(x)$ as its invariant distribution.

2. Set

$$\theta^* = \theta_t + \gamma_{t+1} H(\theta_t, \boldsymbol{x}_{t+1}) + \gamma_{t+1}^{1+\tau} \eta(\boldsymbol{x}_{t+1}), \tag{7.65}$$

   where $\boldsymbol{x}_{t+1} = (x_{t+1}^{(1)}, \ldots, x_{t+1}^{(\kappa)}), \gamma_{t+1}$ is the gain factor, $\tau > 0$, and $\eta(\boldsymbol{x}_{t+1})$ is bounded, i.e., there exists a constant Δ such that $\|\eta(\boldsymbol{x}_{t+1})\| \leq \Delta$ for all $t \geq 0$. If $\theta^* \in \Theta$, set $\theta_{t+1} = \theta^*$; otherwise, set $\theta_{t+1} = \theta^* + \boldsymbol{c}^*$, where $\boldsymbol{c}^* = (c^*, \ldots, c^*)$ can be any vector satisfying the condition $\theta^* + \boldsymbol{c}^* \in \Theta$.

Here, we assume that Θ is compact. As previously explained, Θ can be naturally restricted to a compact set for SAMC and all its variants described in

Section 7.6. Otherwise, a varying truncation version of the SAMCMC algorithm can be considered, as in Liang (2009e). Let $h(\theta) = \int_{\mathcal{X}^k} H(\theta, \boldsymbol{x}) f_\theta(d\boldsymbol{x})$, and $\xi_{t+1} = H(\theta_t, \boldsymbol{x}_{t+1}) - h(\theta_t) + \gamma_{t+1}^{\tau} \eta(\boldsymbol{x}_{t+1})$, which is called the observation noise. Then, 7.65 can be expressed in a more general form as

$$\theta_{t+1} = \theta_t + \gamma_{t+1} h(\theta_t) + \gamma_{t+1} \xi_{t+1}. \tag{7.66}$$

The convergence of the general SAMCMC algorithm can be established under the following conditions.

Conditions on the Step-Sizes

(B_1) The gain factor sequence $\{\gamma_t\}_{t=0}^{\infty}$ is non-increasing, positive and satisfies the condition 7.36; that is,

$$(i) \quad \overline{\lim_{t \to \infty}} |\gamma_t^{-1} - \gamma_{t+1}^{-1}| < \infty, \quad (ii) \quad \sum_{t=1}^{\infty} \gamma_t = \infty, \quad (iii) \quad \sum_{t=1}^{\infty} \gamma_t^{\zeta} < \infty,$$

for any $\zeta > 1$.

Drift Conditions on the Transition Kernel P_θ. We first give some definitions on general drift and continuity conditions, and then give the specific drift and continuity conditions for SAMCMC on a product sample space.

Assume that a transition kernel P is ψ-irreducible (following the standard notation of Meyn and Tweedie, 1993), aperiodic, and has a stationary distribution on a sample space denoted by $\mathcal{X}$. A set $C \subset \mathcal{X}$ is said to be small if there exists a probability measure ν on $\mathcal{X}$, a positive integer l and $\delta > 0$ such that

$$P_\theta^l(x, A) \geq \delta \nu(A), \quad \forall x \in C, \quad \forall A \in \mathcal{B}_{\mathcal{X}},$$

where $\mathcal{B}_{\mathcal{X}}$ is the Borel set of $\mathcal{X}$. A function $V : \mathcal{X} \to [1, \infty)$ is said to be a drift function outside C if there exist constants $\lambda < 1$ and b such that

$$P_\theta V(x) \leq \lambda V(x) + bI(x \in C), \quad \forall x \in \mathcal{X},$$

where $P_\theta V(x) = \int_{\mathcal{X}} P_\theta(x, y) V(y) dy$. For $g : \mathcal{X} \to \mathbb{R}^d$, define the norm

$$\|g\|_V = \sup_{x \in \mathcal{X}} \frac{|g(x)|}{V(x)},$$

and define the set $\mathcal{L}_V = \{g : \mathcal{X} \to \mathbb{R}^d, \|g\|_V < \infty\}$.

For the general SAMCMC algorithm, the drift and continuity conditions can be specified as follows. Let $\boldsymbol{P}_\theta$ be the joint transition kernel for generating the samples $\boldsymbol{x} = (x^{(1)}, \ldots, x^{(\kappa)})$ at each iteration by ignoring the subscript t, let $\mathcal{X}^\kappa = \mathcal{X} \times \cdots \times \mathcal{X}$ be the product sample space, let $\boldsymbol{A} = A_1 \times \cdots \times A_\kappa$ be a measurable rectangle in $\mathcal{X}^\kappa$ for which $A_i \in \mathcal{B}_{\mathcal{X}}$ for $i = 1, \ldots, \kappa$, and let $\mathcal{B}_{\mathcal{X}^\kappa} = \mathcal{B}_{\mathcal{X}} \times \cdots \times \mathcal{B}_{\mathcal{X}}$ be the σ-algebra generated by measurable rectangles.

(B_2) The transition kernel $\boldsymbol{P}_\theta$ is ψ-irreducible and aperiodic for any $\theta \in \Theta$. There exist a function $V : \mathcal{X}^\kappa \to [1, \infty)$ and constants $\alpha \geq 2$ and $\beta \in (0, 1]$ such that,

(i) For any $\theta \in \Theta$, there exist a set $\boldsymbol{C} \subset \mathcal{X}^\kappa$, an integer l, constants $0 < \lambda < 1, b, \varsigma, \delta > 0$ and a probability measure ν such that:

- $\boldsymbol{P}_\theta^l V^\alpha(\boldsymbol{x}) \leq \lambda V^\alpha(\boldsymbol{x}) + bI(\boldsymbol{x} \in \boldsymbol{C}), \quad \forall \boldsymbol{x} \in \mathcal{X}^\kappa.$ (7.67)
- $\boldsymbol{P}_\theta V^\alpha(\boldsymbol{x}) \leq \varsigma V^\alpha(\boldsymbol{x}), \quad \forall \boldsymbol{x} \in \mathcal{X}^\kappa.$ (7.68)
- $\boldsymbol{P}_\theta^l(\boldsymbol{x}, \boldsymbol{A}) \geq \delta\nu(\boldsymbol{A}), \quad \forall \boldsymbol{x} \in \boldsymbol{C}, \quad \forall \boldsymbol{A} \in \mathcal{B}_{\mathcal{X}^\kappa}.$ (7.69)

(ii) There exists a constant c_1 such that for all $\boldsymbol{x} \in \mathcal{X}^\kappa$ and $\theta, \theta' \in \Theta$,

- $\|H(\theta, \boldsymbol{x})\|_V \leq c_1.$ (7.70)
- $\|H(\theta, \boldsymbol{x}) - H(\theta', \boldsymbol{x})\|_V \leq c_1 \|\theta - \theta'\|^\beta.$ (7.71)

(iii) There exists a constant c_2 such that for all $\theta, \theta' \in \Theta$,

- $\|\boldsymbol{P}_\theta g - \boldsymbol{P}_{\theta'} g\|_V \leq c_2 \|g\|_V |\theta - \theta'|^\beta, \quad \forall g \in \mathcal{L}_V.$ (7.72)
- $\|\boldsymbol{P}_\theta g - \boldsymbol{P}_{\theta'} g\|_{V^\alpha} \leq c_2 \|g\|_{V^\alpha} |\theta - \theta'|^\beta, \quad \forall g \in \mathcal{L}_{V^\alpha}.$ (7.73)

Assumption (B_2)-(i) is classical in literature of Markov chain. It implies the existence of a stationary distribution $f_\theta(\boldsymbol{x})$ for any $\theta \in \Theta$ and V^α-uniform ergodicity (Andrieu *et al.*, 2005). Assumption (B_2)-(ii) gives conditions on the bound of $H(\theta, x)$. For general SAMCMC algorithms, this is a critical condition, which directly leads to boundedness of the observation noise. For SAMC and its variants, the drift function can be simply set as $V(x) = 1$, as $H(\theta, x)$ is a bounded function.

Lyapunov condition on $h(\theta)$ Let $\mathcal{L} = \{\theta \in \Theta : h(\theta) = \mathbf{0}\}$.

(B_3) The mean field function $h : \Theta \to \mathbb{R}^d$ is continuous, and there exists a continuously differentiable function $v : \Theta \to [0, \infty)$ such that $\dot{v}(\theta) = \nabla^T v(\theta) h(\theta) < 0, \forall \theta \in \mathcal{L}^c$ and $\sup_{\theta \in Q} \dot{v}(\theta) < 0$ for any compact set $Q \subset \mathcal{L}^c$.

This condition assumes the existence of a global Lyapunov function v for the mean field h. If h is a gradient field, that is, $h = -\nabla J$ for some lower bounded real-valued and differentiable function $J(\theta)$, then v can be set to J, provided that J is continuously differentiable. This is typical for stochastic optimization problems, for example, machine learning (Tadić, 1997), where a continuously differentiable objective function $J(\theta)$ is minimized. For SAMC and its variants, the Lyapunov function can be chosen accordingly as shown below.

A main convergence result. Let $\mathcal{P}_{\boldsymbol{x}_0,\theta_0}$ denote the probability measure of the Markov chain $\{(\boldsymbol{x}_t, \theta_t)\}$, started in $(\boldsymbol{x}_0, \theta_0)$, and implicitly defined by the sequences $\{\gamma_t\}$. Also define $D(\boldsymbol{z}, A) = \inf_{\boldsymbol{z}' \in A} \|\boldsymbol{z} - \boldsymbol{z}'\|$.

Theorem 7.7.1 *Assume Θ is compact, $(B_1), (B_2)$ and (B_3) hold, and $\sup_{\boldsymbol{x} \in \mathcal{X}^\kappa} V(\boldsymbol{x}) < \infty$. Let the sequence $\{\theta_n\}$ be defined by 7.65. Then for all $(\boldsymbol{x}_0, \theta_0) \in \mathcal{X}^\kappa \times \Theta$,*

$$\lim_{t \to \infty} D(\theta_t, \mathcal{L}) = 0, \qquad \mathcal{P}_{\boldsymbol{x}_0,\theta_0} - a.e.$$

In this theorem, Θ is assumed to be compact and the drift function $V(x)$ is assumed to be uniformly bounded. We note that the former can be relaxed to $\mathbb{R}^d$ and the latter can be weakened to $\sup_{x \in \mathcal{X}_0^\kappa} V(x) < \infty$ for a subset $\mathcal{X}_0 \subset \mathcal{X}$, if the SAMCMC algorithm adopts a varying truncation scheme as in Liang (2009e). Also, if the condition is weakened to $\sum_{t=1}^\infty \gamma_t^{1+\zeta} V^2(\boldsymbol{x}_t) < \infty$, where ζ is defined in (B_1'), the convergence results presented in this section still hold.

7.7.1.2 Proof of Theorem 7.7.1

To prove this theorem, we first introduce some lemmas. Lemma 7.7.1 is a partial restatement of Proposition 6.1 of Andrieu *et al.* (2005).

Lemma 7.7.1 *Assume Θ is compact and the drift condition (B_2) holds. Then the following results hold.*

(C_1) *For any $\theta \in \Theta$, the Markov kernel $\boldsymbol{P}_\theta$ has a single stationary distribution $\boldsymbol{f}_\theta$. In addition $H : \Theta \times \mathcal{X}^\kappa$ is measurable for all $\theta \in \Theta, h(\theta) = \int_{\mathcal{X}^\kappa} H(\theta, \boldsymbol{x}) \boldsymbol{f}_\theta(d\boldsymbol{x}) < \infty$.*

(C_2) *For any $\theta \in \Theta$, the Poisson equation $u(\theta, \boldsymbol{x}) - \boldsymbol{P}_\theta u(\theta, \boldsymbol{x}) = H(\theta, \boldsymbol{x}) - h(\theta)$ has a solution $u(\theta, \boldsymbol{x})$, where $\boldsymbol{P}_\theta u(\theta, \boldsymbol{x}) = \int_{\mathcal{X}^\kappa} u(\theta, \boldsymbol{x}') \boldsymbol{P}_\theta(\boldsymbol{x}, \boldsymbol{x}') d\boldsymbol{x}'$. There exist a function $V : \mathcal{X}^\kappa \to [1, \infty)$ such that the set $\{\boldsymbol{x} \in \mathcal{X}^\kappa : V(\boldsymbol{x}) < \infty\} \neq \emptyset$, and for any constant $\beta \in (0, 1]$,*

$$\begin{aligned}
&\text{(i)} \quad \sup_{\theta \in \Theta} \|H(\theta, \boldsymbol{x})\|_V < \infty, \\
&\text{(ii)} \quad \sup_{\theta \in \Theta} (\|u(\theta, \boldsymbol{x})\|_V + \|P_\theta u(\theta, \boldsymbol{x})\|_V) < \infty, \\
&\text{(iii)} \quad \sup_{(\theta,\theta') \in \Theta \times \Theta} \|\theta - \theta'\|^{-\beta} \Big(\|u(\theta, \boldsymbol{x}) - u(\theta', \boldsymbol{x})\|_V \\
&\qquad\qquad + \|P_\theta u(\theta, \boldsymbol{x}) - P_{\theta'} u(\theta', \boldsymbol{x})\|_V \Big) < \infty.
\end{aligned} \tag{7.74}$$

Tadić studied the convergence for a stochastic approximation MCMC algorithm under different conditions from those given in Andrieu *et al.* (2005); the following lemma corresponds to Theorem 4.1 and Lemma 2.2 of Tadić (1997). The proof we give below is similar to Tadić's except for some necessary changes for including the higher order noise term in ξ_t.

Lemma 7.7.2 *Assume Θ is compact, (B_1) and (B_2) hold, and $\sup_{\boldsymbol{x}\in\mathcal{X}^\kappa} V(\boldsymbol{x}) < \infty$. Then the following results hold:*

(D_1) *There exist $\mathbb{R}^d$-valued random processes $\{\epsilon_t\}_{t\geq 0}, \{\epsilon'_t\}_{t\geq 0}$ and $\{\epsilon''_t\}_{t\geq 0}$ defined on a probability space $(\Omega, \mathcal{F}, \mathcal{P})$ such that*

$$\gamma_{t+1}\xi_{t+1} = \epsilon_{t+1} + \epsilon'_{t+1} + \epsilon''_{t+1} - \epsilon''_t, \quad t \geq 0. \tag{7.75}$$

(D_2) *The series $\sum_{t=0}^{\infty} \|\epsilon'_t\|, \sum_{t=0}^{\infty} \|\epsilon''_t\|^2$ and $\sum_{t=0}^{\infty} \|\epsilon_{t+1}\|^2$ all converge a.s. and*

$$E(\epsilon_{t+1}|\mathcal{F}_t) = 0, \quad a.s., \quad n \geq 0, \tag{7.76}$$

where $\{\mathcal{F}_t\}_{t\geq 0}$ is a family of σ-algebras of $\mathcal{F}$ satisfying $\sigma\{\theta_0\} \subseteq \mathcal{F}_0$ and $\sigma\{\epsilon_t, \epsilon'_t, \epsilon''_t\} \subseteq \mathcal{F}_t \subseteq \mathcal{F}_{t+1}, t \geq 0$.

(D_3) *Let $R_t = R'_t + R''_t, t \geq 1$, where $R'_t = \gamma_{t+1}\nabla^T v(\theta_t)\xi_{t+1}$, and*

$$R''_{t+1} = \int_0^1 \left[\nabla v(\theta_t + s(\theta_{t+1} - \theta_t)) - \nabla v(\theta_t)\right]^T (\theta_{t+1} - \theta_t) ds.$$

Then $\sum_{t=1}^{\infty} \gamma_t\xi_t$ and $\sum_{t=1}^{\infty} R_t$ converge a.s.

Proof: The results $(D_1), (D_2)$ and (D_3) are proved as follows:

- (D_1) The condition (C_2) implies that there exists a constant $c_1 \in \mathbb{R}^+$ such that

  $$\|\theta_{t+1} - \theta_t\| = \|\gamma_{t+1}H(\theta_t, \boldsymbol{x}_{t+1}) + \gamma_{t+1}^{1+\tau}\eta(\boldsymbol{x}_{t+1})\| \leq c_1\gamma_{t+1}[V(\boldsymbol{x}_{t+1}) + \Delta].$$

  The condition (B_1) yields $\gamma_{t+1}/\gamma_t = O(1)$ and $|\gamma_{t+1} - \gamma_t| = O(\gamma_t\gamma_{t+1})$ for $t \to \infty$. Consequently, there exists a constant $c_2 \in \mathbb{R}^+$ such that

  $$\gamma_{t+1} \leq c_2\gamma_t, \quad |\gamma_{t+1} - \gamma_t| \leq c_2\gamma_t^2, \quad t \geq 0.$$

  Let $\epsilon_0 = \epsilon'_0 = 0$, and

  $$\begin{aligned}\epsilon_{t+1} &= \gamma_{t+1}\left[u(\theta_t, \boldsymbol{x}_{t+1}) - \boldsymbol{P}_{\theta_t}u(\theta_t, \boldsymbol{x}_t)\right],\\ \epsilon'_{t+1} &= \gamma_{t+1}\left[\boldsymbol{P}_{\theta_{t+1}}u(\theta_{t+1}, \boldsymbol{x}_{t+1}) - \boldsymbol{P}_{\theta_t}u(\theta_t, \boldsymbol{x}_{t+1})\right]\\ &\quad + (\gamma_{t+2} - \gamma_{t+1})\boldsymbol{P}_{\theta_{t+1}}u(\theta_{t+1}, \boldsymbol{x}_{t+1}) + \gamma_{t+1}^{1+\tau}\eta(\boldsymbol{x}_{t+1}),\\ \epsilon''_t &= -\gamma_{t+1}\boldsymbol{P}_{\theta_t}u(\theta_t, \boldsymbol{x}_t).\end{aligned}$$

  It is easy to verify that (7.75) is satisfied.

- (D_2) Since $\sigma(\theta_t) \subseteq \mathcal{F}_t$, we have

  $$E(u(\theta_t, \boldsymbol{x}_{t+1})|\mathcal{F}_t) = \boldsymbol{P}_{\theta_t}u(\theta_t, \boldsymbol{x}_t),$$

which concludes (7.76). The condition (C_2) implies that there exist constants $c_3, c_4, c_5, c_6, c_7, c_8 \in \mathbb{R}^+$ and $\tau' = \min(\beta, \tau) > 0$ such that

$$
\begin{aligned}
\|\epsilon_{t+1}\|^2 &\leq 2c_3\gamma_{t+1}^2 V^2(\boldsymbol{x}_t), \\
\|\epsilon'_{t+1}\| &\leq c_4\gamma_{t+1}V(\boldsymbol{x}_{t+1})\|\theta_{t+1} - \theta_t\|^\beta + c_5\gamma_{t+1}^2 V(\boldsymbol{x}_{t+1}) + c_6\gamma_{t+1}^{1+\tau}\Delta \\
&\leq c_7\gamma_{t+1}^{1+\tau'}\left[V(\boldsymbol{x}_{t+1}) + \Delta)\right], \\
\|\epsilon''_{t+1}\|^2 &\leq c_8\gamma_{t+1}^2 V^2(\boldsymbol{x}_{t+1}),
\end{aligned}
$$

It follows from condition (B_1) and the condition $\sup_{\boldsymbol{x}} V(\boldsymbol{x}) < \infty$ that the series $\sum_{t=0}^{\infty} \|\epsilon_{t+1}\|^2, \sum_{t=0}^{\infty} \|\epsilon'_t\|$, and $\sum_{t=0}^{\infty} \|\epsilon''_t\|^2$ all converge.

- (D_3) Let $M = \sup_{\theta\in\Theta} \max\{\|h(\theta)\|, \|\nabla v(\theta)\|\}$. By (C_2)-(i), we have $\sup_{\theta\in\Theta} \|h(\theta)\| < \infty$. By the compactness of $\Theta, \sup_{\theta\in\Theta} \|\nabla v(\theta)\| < \infty$, where $v(\theta)$ is defined in (B_3). Therefore, we have $M < \infty$. Let L be the Lipschitz constant of $\nabla v(\cdot)$. Since $\sigma\{\theta_t\} \subset \mathcal{F}_t, E(\nabla^T v(\theta_t)\epsilon_{t+1}|\mathcal{F}_t) = 0$ by (D_2). In addition,

$$
\sum_{t=0}^{\infty} E\big(|\nabla^T v(\theta_t)\epsilon_{t+1}|^2\big) \leq M^2 \sum_{t=0}^{\infty} E\big(\|\epsilon_{t+1}\|^2\big) < \infty.
$$

It follows from the martingale convergence theorem (Hall and Heyde, 1980; Theorem 2.15) that both $\sum_{t=0}^{\infty} \epsilon_{t+1}$ and $\sum_{t=0}^{\infty} \nabla^T v(\theta_t)\epsilon_{t+1}$ converge almost surely. Since

$$
\begin{aligned}
&\sum_{t=0}^{\infty} |\nabla^T v(\theta_t)\epsilon'_{t+1}| \leq M \sum_{t=1}^{\infty} \|\epsilon'_t\|, \\
&\sum_{t=1}^{\infty} \gamma_t^2 \|\xi_t\|^2 \leq 4\sum_{t=1}^{\infty} \|\epsilon_t\|^2 + 4\sum_{t=1}^{\infty} \|\epsilon'_t\|^2 + 8\sum_{t=0}^{\infty} \|\epsilon''_t\|^2,
\end{aligned}
$$

it follows from (D_2) that both $\sum_{t=0}^{\infty} |\nabla^T v(\theta_t)\epsilon'_{t+1}|$ and $\sum_{t=1}^{\infty} \gamma_t^2 \|\xi_t\|^2$ converge. In addition,

$$
\begin{aligned}
\|R''_{t+1}\| &\leq L\|\theta_{t+1} - \theta_t\|^2 = L\|\gamma_{t+1}h(\theta_t) + \gamma_{t+1}\xi_{t+1}\|^2 \\
&\leq 2L\big(M^2\gamma_{t+1}^2 + \gamma_{t+1}^2\|\xi_{t+1}\|^2\big),
\end{aligned}
$$

and

$$
\left|(\nabla v(\theta_{t+1}) - \nabla v(\theta_t))^T \epsilon''_{t+1}\right| \leq L\|\theta_{t+1} - \theta_t\|\|\epsilon''_{t+1}\|,
$$

for all $t \geq 0$. Consequently,

$$
\sum_{t=1}^{\infty} |R''_t| \leq 2LM^2 \sum_{t=1}^{\infty} \gamma_t^2 + 2L\sum_{t=1}^{\infty} \gamma_t^2\|\xi_t\|^2 < \infty,
$$

and

$$\sum_{t=0}^{\infty} \left| (v(\theta_{t+1}) - v(\theta_t))^T \epsilon''_{t+1} \right|$$
$$\leq \left(2L^2 M^2 \sum_{t=1}^{\infty} \gamma_t^2 + 2L^2 \sum_{t=1}^{\infty} \gamma_t^2 \|\xi_t\|^2 \right)^{1/2} \left(\sum_{t=1}^{\infty} \|\epsilon''_t\|^2 \right)^{1/2} < \infty.$$

Since,

$$\sum_{t=1}^{n} \gamma_t \xi_t = \sum_{t=1}^{n} \epsilon_t + \sum_{t=1}^{n} \epsilon'_t + \epsilon''_n - \epsilon''_0,$$
$$\sum_{t=0}^{n} R'_{t+1} = \sum_{t=0}^{n} \nabla^T v(\theta_t) \epsilon_{t+1} + \sum_{t=0}^{n} \nabla^T v(\theta_t) \epsilon'_{t+1} + \nabla^T v(\theta_{n+1}) \epsilon''_{n+1}$$
$$- \sum_{t=0}^{n} (\nabla v(\theta_{t+1} - \nabla v(\theta_t))^T \epsilon''_{t+1} - \nabla^T v(\theta_0) \epsilon''_0,$$

it is obvious that $\sum_{t=1}^{\infty} \gamma_t \xi_t$ and $\sum_{t=1}^{\infty} R_t$ converge almost surely.

The proof for Lemma 7.7.2 is completed. ■

Theorem 7.7.1 can be proved in a similar way to Theorem 2.2 of Tadić (1997). To make the book self-contained, we rewrite the proof below.

Proof: Let $M = \sup_{\theta \in \Theta} \max\{\|h(\theta)\|, |v(\theta)|\}$ and $\mathcal{V}_\varepsilon = \{\theta : v(\theta) \leq \varepsilon\}$. Following the compactness of Θ and the condition $\sup_{\boldsymbol{x}} V(\boldsymbol{x}) < \infty$, we have $M < \infty$. Applying Taylor's expansion formula (Folland, 1990), we have

$$v(\theta_{t+1}) = v(\theta_t) + \gamma_{n+1} \dot{v}(\theta_{t+1}) + R_{t+1}, \quad t \geq 0,$$

which implies that

$$\sum_{i=0}^{t} \gamma_{i+1} \dot{v}(\theta_i) = v(\theta_{t+1}) - v(\theta_0) - \sum_{i=0}^{t} R_{i+1} \geq -2M - \sum_{i=0}^{t} R_{i+1}.$$

Since $\sum_{i=0}^{t} R_{i+1}$ converges (owing to Lemma 7.7.2), $\sum_{i=0}^{t} \gamma_{i+1} \dot{v}(\theta_i)$ also converges. Furthermore,

$$v(\theta_t) = v(\theta_0) + \sum_{i=0}^{t-1} \gamma_{i+1} \dot{v}(\theta_i) + \sum_{i=0}^{t-1} R_{i+1}, \quad t \geq 0,$$

$\{v(\theta_t)\}_{t \geq 0}$ also converges. On the other hand, conditions (B_1) and (B_2) imply $\underline{\lim}_{t \to \infty} d(\theta_t, \mathcal{L}) = 0$. Otherwise, there exists $\varepsilon > 0$ and n_0 such that $d(\theta_t, \mathcal{L}) \geq \varepsilon, t \geq n_0$; as $\sum_{t=1}^{\infty} \gamma_t = \infty$ and $p = \sup\{\dot{v}(\theta) : \theta \in \mathcal{V}_\varepsilon^c\} < 0$, it is obtained that $\sum_{t=n_0}^{\infty} \gamma_{t+1} \dot{v}(\theta_t) \leq p \sum_{t=1}^{\infty} \gamma_{t+1} = -\infty$.

Suppose that $\overline{\lim}_{t\to\infty} d(\theta_t, \mathcal{L}) > 0$. Then, there exists $\varepsilon > 0$ such that $\overline{\lim}_{t\to\infty} d(\theta_t, \mathcal{L}) \geq 2\varepsilon$. Let $t_0 = \inf\{t \geq 0 : d(\theta_t, \mathcal{L}) \geq 2\varepsilon\}$, while $t'_k = \inf\{t \geq t_k : d(\theta_t, \mathcal{L}) \leq \varepsilon\}$ and $t_{k+1} = \inf\{t \geq t'_k : d(\theta_t, \mathcal{L}) \geq 2\varepsilon\}, k \geq 0$. Obviously, $t_k < t_{k'} < t_{k+1}, k \geq 0$, and

$$d(\theta_{t_k}, \mathcal{L}) \geq 2\varepsilon,\ d(\theta_{t'_k}, \mathcal{L}) \leq \varepsilon,\ \text{and}\ d(\theta_t, \mathcal{L}) \geq \varepsilon,\ t_k \leq t < t'_k,\ k \geq 0.$$

Let $q = \sup\{\dot{v}(\theta) : \theta \in \mathcal{V}^c_\varepsilon\}$. Then

$$q \sum_{k=0}^{\infty} \sum_{i=t_k}^{t'_k-1} \gamma_{i+1} \geq \sum_{k=0}^{\infty} \sum_{i=t_k}^{t'_k-1} \gamma_{i+1}\dot{v}(\theta_i) \geq \sum_{t=0}^{\infty} \gamma_{t+1}\dot{v}(\theta_t) > -\infty.$$

Therefore, $\sum_{k=0}^{\infty} \sum_{i=t_k}^{t'_k-1} \gamma_{i+1} < \infty$, and consequently, $\lim_{k\to\infty} \sum_{i=t_k}^{t'_k-1} \gamma_{i+1} = 0$. Since $\sum_{t=1}^{\infty} \gamma_t \xi_t$ converges (owing to Lemma 7.7.2), we have

$$\varepsilon \leq \|\theta_{t'_k} - \theta_{t_k}\| \leq M \sum_{i=t_k}^{t'_k-1} \gamma_{i+1} + \left\| \sum_{i=t_k}^{t'_k-1} \gamma_{i+1}\xi_{i+1} \right\| \longrightarrow 0,$$

as $k \to \infty$. This contradicts with our assumption $\varepsilon > 0$. Hence, $\overline{\lim}_{t\to\infty} d(\theta_t, \mathcal{L}) > 0$ does not hold. Therefore, $\lim_{t\to\infty} d(\theta_t, \mathcal{L}) = 0$ almost surely. ■

7.7.1.3 Convergence of smoothing SAMC

Consider the smoothing SAMC algorithm described in Section 7.6.1. Without loss of generality, we assume that all subregions are unempty. For the empty subregions, the convergence (7.42) is trivial. Thus, Θ can be naturally restricted to a compact set in the proof.

Let $\boldsymbol{e}_t = (e_t^{(1)}, \ldots, e_t^{(m)})$, where $e_t^{(i)} = \sum_{j=1}^{\kappa} I(x_t^{(j)} \in E_i)$. Since the kernel used in (7.38) has a bounded support, $\widehat{p}_t^{(i)} - e_t^{(i)}/\kappa$ can be rewritten as

$$\widehat{p}_t^{(i)} - e_t^{(i)}/\kappa = \frac{\sum_{l=\max\{1,i-k_0\}}^{\min\{m,i+k_0\}} W\left(\frac{\Lambda l}{mh_t}\right)\left(\frac{e_t^{(i+l)}}{\kappa} - \frac{e_t^{(i)}}{\kappa}\right)}{\sum_{l=\max\{1,i-k_0\}}^{\min\{m,i+k_0\}} W\left(\frac{\Lambda l}{mh_t}\right)}, \tag{7.77}$$

where $k_0 = \left[\frac{Cmh_t}{\Lambda}\right]$, and $[z]$ denotes the maximum integer less than z. By noting the inequality $-1 \leq \frac{e_{tj}}{\kappa} - \frac{e_{ti}}{\kappa} \leq 1$, we have $|\widehat{p}_t^{(i)} - e_t^{(i)}/\kappa| \leq 2k_0$, which is true even when $k_0 = 0$. Thus, there exists a bounded function $-2Cm/\Lambda \leq \eta_i^*(\boldsymbol{e}_t) \leq 2Cm/\Lambda$ such that

$$\widehat{p}_t^{(i)} - e_t^{(i)}/\kappa = h_t \eta_i^*(\boldsymbol{e}_t). \tag{7.78}$$

Since h_t is chosen in 7.40 as a power function of γ_t, SSAMC falls into the class of stochastic approximation MCMC algorithms described in Section 7.7.1.2 by letting $\eta(\boldsymbol{x}_t) = (\eta_1^*(\boldsymbol{e}_t), \ldots, \eta_m^*(\boldsymbol{e}_t))$, and Theorem 7.6.1 can be proved by verifying that SSAMC satisfies the conditions (B_1) to (B_3):

- *Condition* (B_1). This condition can be satisfied by choosing an appropriate gain factor sequence, such as the one specified in (7.37).

- *Condition* (B_2). Let $\boldsymbol{x}_{t+1} = (x_{t+1}^{(1)}, \ldots, x_{t+1}^{(\kappa)})$, which can be regarded as a sample produced by a Markov chain on the product space $\mathcal{X}^\kappa = \mathcal{X} \times \cdots \times \mathcal{X}$ with the kernel

$$\boldsymbol{P}_{\theta_t}(\boldsymbol{x}, \boldsymbol{y}) = P_{\theta_t}(x^{(\kappa)}, y^{(1)}) P_{\theta_t}(y^{(1)}, y^{(2)}) \cdots P_{\theta_t}(y^{(\kappa-1)}, y^{(\kappa)}),$$

where $P_{\theta_t}(x, y)$ denotes the one-step MH kernel, and $x^{(\kappa)}$ denotes the last sample generated at the previous iteration. To simplify notations, in what follows we will drop the subscript t, denoting $\boldsymbol{x}_t$ by $\boldsymbol{x}$ and $\theta_t = (\theta_t^{(1)}, \ldots, \theta_t^{(m)})$ by $\theta = (\theta^{(1)}, \ldots, \theta^{(m)})$.

Roberts and Tweedie (1996; Theorem 2.2) show that if the target distribution is bounded away from 0 and ∞ on every compact set of its support $\mathcal{X}$, then the MH chain with a proposal distribution satisfying the local positive condition is irreducible and aperiodic, and every nonempty compact set is small. Following from this result, $P_\theta(x, y)$ is irreducible and aperiodic, and thus $\boldsymbol{P}_\theta(\boldsymbol{x}, \boldsymbol{y})$ is irreducible and aperiodic.

Since $\mathcal{X}$ is compact, Roberts and Tweedie's result implies that $\mathcal{X}$ is a small set and the minorisation condition holds on $\mathcal{X}$ for the kernel $P_\theta(x, y)$; i.e., there exists an integer l', a constant δ, and a probability measure $\nu'(\cdot)$ such that

$$P_\theta^{l'}(x, A) \geq \delta \nu'(A), \quad \forall x \in \mathcal{X}, \ \forall A \in \mathcal{B}_{\mathcal{X}}.$$

Following from Rosenthal (1995; Lemma 7), we have

$$\boldsymbol{P}_\theta^{l}(\boldsymbol{x}, \boldsymbol{A}) \geq \delta \nu(\boldsymbol{A}), \quad \forall \boldsymbol{x} \in \mathcal{X}^\kappa, \ \forall \boldsymbol{A} \in \mathcal{B}_{\mathcal{X}^\kappa},$$

by setting $l = \min\{n : n \times \kappa \geq l', n = 1, 2, 3, \ldots\}$ and defining the measure $\nu(\cdot)$ as follows: Marginally on the first coordinate, $\nu(\cdot)$ agrees with $\nu'(\cdot)$; conditionally on the first coordinate, $\nu(\cdot)$ is defined by

$$\nu(x^{(2)}, \ldots, x^{(\kappa)} | x^{(1)}) = \mathcal{W}(x^{(2)}, \ldots, x^{(\kappa)} | x^{(1)}), \tag{7.79}$$

where $\mathcal{W}(x^{(2)}, \ldots, x^{(\kappa)} | x^{(1)})$ is the conditional distribution of the Markov chain samples generated by the kernel $\boldsymbol{P}_\theta$. Conditional on $x_t^{(1)}$, the samples $x_t^{(2)}, \ldots, x_t^{(\kappa)}$ are generated independently of all previous samples $\boldsymbol{x}_{t-1}, \ldots, \boldsymbol{x}_1$. Hence, $\mathcal{W}(x^{(2)}, \ldots, x^{(\kappa)} | x^{(1)})$ exists. This verifies condition 7.69 by setting $\boldsymbol{C} = \mathcal{X}^\kappa$. Thus, for any $\theta \in \Theta$ the following conditions hold

$$\begin{aligned} \boldsymbol{P}_\theta^l V^\alpha(\boldsymbol{x}) &\leq \lambda V^\alpha(\boldsymbol{x}) + bI(\boldsymbol{x} \in \boldsymbol{C}), \quad \forall \boldsymbol{x} \in \mathcal{X}^\kappa, \\ \boldsymbol{P}_\theta V^\alpha(\boldsymbol{x}) &\leq \varsigma V^\alpha(\boldsymbol{x}), \quad \forall \boldsymbol{x} \in \mathcal{X}^\kappa, \end{aligned} \tag{7.80}$$

by choosing $V(\boldsymbol{x}) = 1, 0 < \lambda < 1, b = 1 - \lambda, \varsigma > 1$, and $\alpha \geq 2$. These conclude that $(B_2$-i) is satisfied.

Let $H^{(i)}(\theta, \boldsymbol{x})$ be the ith component of the vector $H(\theta, \boldsymbol{x}) = (\mathbf{e}/\kappa - \boldsymbol{\pi})$. By construction, $|H^{(i)}(\theta, \boldsymbol{x})| = |e^{(i)}/\kappa - \pi_i| < 1$ for all $\boldsymbol{x} \in \mathcal{X}^\kappa$ and $i = 1, \ldots, m$. Therefore, there exists a constant $c_1 = \sqrt{m}$ such that for any $\theta \in \Theta$ and all $\boldsymbol{x} \in \mathcal{X}^\kappa$,

$$\|H(\theta, \boldsymbol{x})\| \leq c_1. \tag{7.81}$$

Also, $H(\theta, \boldsymbol{x})$ does not depend on θ for a given sample $\boldsymbol{x}$. Hence, $H(\theta, \boldsymbol{x}) - H(\theta', \boldsymbol{x}) = 0$ for all $(\theta, \theta') \in \Theta \times \Theta$, and thus

$$\|H(\theta, \boldsymbol{x}) - H(\theta', \boldsymbol{x})\| \leq c_1 \|\theta - \theta'\|, \tag{7.82}$$

for all $(\theta, \theta') \in \Theta \times \Theta$. Equations (7.81) and (7.82) imply that (B_2-ii) is satisfied by choosing $\beta = 1$ and $V(\boldsymbol{x}) = 1$.

Let $s_\theta(x, y) = q(x, y) \min\{1, r_\theta(x, y)\}$, where $r_\theta(x, y) = [f_\theta(y) q(y, x)]/[f_\theta(x) q(x, y)]$. Thus,

$$\left| \frac{\partial s_\theta(x, y)}{\partial \theta^{(i)}} \right| = \Big| - q(x, y) I(r_\theta(x, y) < 1) I(J(x) = i \text{ or } J(y) = i) \\ I(J(x) \neq J(y)) r_\theta(x, y) \Big| \leq q(x, y),$$

where $I(\cdot)$ is the indicator function, and $J(x)$ denotes the index of the subregion where x belongs to. The mean-value theorem implies that there exists a constant c_2 such that

$$|s_\theta(x, y) - s_{\theta'}(x, y)| \leq q(x, y) c_2 \|\theta - \theta'\|, \tag{7.83}$$

which implies that

$$\sup_x \int_{\mathcal{X}} |s_\theta(x, y) - s_{\theta'}(x, y)| \, dy \leq c_2 \|\theta - \theta'\|. \tag{7.84}$$

Since the MH kernel can be expressed in the form

$$P_\theta(x, dy) = s_\theta(x, dy) + I(x \in dy)[1 - \int_{\mathcal{X}} s_\theta(x, z) dz],$$

for any measurable set $A \subset \mathcal{X}$,

$$\begin{aligned} &|P_\theta(x, A) - P_{\theta'}(x, A)| \\ &= \left| \int_A [s_\theta(x, y) - s_{\theta'}(x, y)] dy + I(x \in A) \int_{\mathcal{X}} [s_{\theta'}(x, z) - s_\theta(x, z)] dz \right| \\ &\leq \int_{\mathcal{X}} |s_\theta(x, y) - s_{\theta'}(x, y)| dy + I(x \in A) \int_{\mathcal{X}} |s_{\theta'}(x, z) - s_\theta(x, z)| dz \\ &\leq 2 \int_{\mathcal{X}} |s_\theta(x, y) - s_{\theta'}(x, y)| dy \\ &\leq 2 c_2 \|\theta - \theta'\|. \end{aligned} \tag{7.85}$$

Since $\boldsymbol{P}_\theta(\boldsymbol{x}, \boldsymbol{A})$ can be expressed in the following form,

$$\boldsymbol{P}_\theta(\boldsymbol{x}, \boldsymbol{A}) = \int_{A_1} \cdots \int_{A_\kappa} P_\theta(x^{(\kappa)}, y^{(1)}) P_\theta(y^{(1)}, y^{(2)}) \cdots P_\theta(y^{(\kappa-1)}, y^{(\kappa)}) \, dy^{(1)} \cdots dy^{(\kappa)},$$

(7.85) implies that there exists a constant c_3 such that

$$\begin{aligned}
\left|\boldsymbol{P}_\theta(\boldsymbol{x}, \boldsymbol{A}) - \boldsymbol{P}_{\theta'}(\boldsymbol{x}, \boldsymbol{A})\right| &= \Bigg|\int_{A_1} \cdots \int_{A_\kappa} \big[P_\theta(x^{(\kappa)}, y^{(1)}) P_\theta(y^{(1)}, y^{(2)}) \\
&\quad \cdots P_\theta(y^{(\kappa-1)}, y^{(\kappa)}) - P_{\theta'}(x^{(\kappa)}, y^{(1)}) P_{\theta'}(y^{(1)}, y^{(2)}) \cdots \\
&\quad P_{\theta'}(y^{(\kappa-1)}, y^{(\kappa)})\big] dy^{(1)} \cdots dy^{(\kappa)}\Bigg| \\
&\leq \int_{A_1} \int_{\mathcal{X}} \cdots \int_{\mathcal{X}} \left|P_\theta(x^{(\kappa)}, y^{(1)}) - P_{\theta'}(x^{(\kappa)}, y^{(1)})\right| P_\theta(y^{(1)}, y^{(2)}) \cdots \\
&\quad P_\theta(y^{(\kappa-1)}, y^{(\kappa)}) dy^{(1)} \cdots dy^{(\kappa)} + \int_{\mathcal{X}} \int_{A_2} \int_{\mathcal{X}} \cdots \int_{\mathcal{X}} P_{\theta'}(x^{(\kappa)}, y^{(1)}) \\
&\quad \left|P_\theta(y^{(1)}, y^{(2)}) - P_{\theta'}(y^{(1)}, y^{(2)})\right| P_\theta(y^{(2)}, y^{(3)}) \cdots P_\theta(y^{(\kappa-1)}, y^{(\kappa)}) dy^{(1)} \\
&\quad \cdots dy^{(\kappa)} + \cdots + \int_{\mathcal{X}} \cdots \int_{\mathcal{X}} \int_{A_\kappa} P_{\theta'}(x^{(\kappa)}, y^{(1)}) \cdots P_{\theta'}(y^{(\kappa-2)}, y^{(\kappa-1)}) \\
&\quad \left|P_\theta(y^{(\kappa-1)}, y^{(\kappa)}) - P_{\theta'}(y^{(\kappa-1)}, y^{(\kappa)})\right| dy^{(1)} \cdots dy^{(\kappa)} \\
&\leq c_3 \|\theta - \theta'\|,
\end{aligned}$$

which implies that (7.72) is satisfied. For any function $g \in \mathcal{L}_V$,

$$\begin{aligned}
\|\boldsymbol{P}_\theta g - \boldsymbol{P}_{\theta'} g\|_V &= \left\| \int (\boldsymbol{P}_\theta(\boldsymbol{x}, d\boldsymbol{y}) - \boldsymbol{P}_{\theta'}(\boldsymbol{x}, d\boldsymbol{y})) g(\boldsymbol{y}) \right\|_V \\
&= \Bigg\| \int_{\mathcal{X}_+^\kappa} (\boldsymbol{P}_\theta(\boldsymbol{x}, d\boldsymbol{y}) - \boldsymbol{P}_{\theta'}(\boldsymbol{x}, d\boldsymbol{y})) g(\boldsymbol{y}) \\
&\quad + \int_{\mathcal{X}_-^\kappa} (\boldsymbol{P}_\theta(\boldsymbol{x}, d\boldsymbol{y}) - \boldsymbol{P}_{\theta'}(\boldsymbol{x}, d\boldsymbol{y})) g(\boldsymbol{y}) \Bigg\|_V \\
&\leq \|g\|_V \left\{ |\boldsymbol{P}_\theta(\boldsymbol{x}, \mathcal{X}_+^\kappa) - \boldsymbol{P}_{\theta'}(\boldsymbol{x}, \mathcal{X}_+^\kappa)| + |\boldsymbol{P}_\theta(\boldsymbol{x}, \mathcal{X}_-^\kappa) - \boldsymbol{P}_{\theta'}(\boldsymbol{x}, \mathcal{X}_-^\kappa)| \right\} \\
&\leq 4 c_2 \|g\|_V |\theta - \theta'| \qquad \text{(following from (7.85))}
\end{aligned}$$

where $\mathcal{X}_+^\kappa = \{\boldsymbol{y} : \boldsymbol{y} \in \mathcal{X}^\kappa, \boldsymbol{P}_\theta(\boldsymbol{x}, d\boldsymbol{y}) - \boldsymbol{P}_{\theta'}(\boldsymbol{x}, \boldsymbol{y}) > 0\}$ and $\mathcal{X}_-^\kappa = \mathcal{X}^\kappa \backslash \mathcal{X}_+^\kappa$. Therefore, condition ($B_2$-iii) is satisfied by choosing $V(\boldsymbol{x}) = 1$ and $\beta = 1$.

- *Condition* (B_3). Since the kernel $P_\theta(x, \cdot)$ admits $f_\theta(x)$ as its invariant distribution, for any fixed θ,

$$E(e^{(i)}/\kappa - \pi_i) = \frac{\int_{E_i} \psi(x) dx / e^{\theta^{(i)}}}{\sum_{k=1}^m [\int_{E_k} \psi(x) dx / e^{\theta^{(k)}}]} - \pi_i = \frac{S_i}{S} - \pi_i, \; i = 1, \ldots, m, \tag{7.86}$$

where $S_i = \int_{E_i} \psi(x)dx/e^{\theta^{(i)}}$ and $S = \sum_{k=1}^m S_k$. Thus, we have

$$h(\theta) = \int_{\mathcal{X}} H(\theta, \boldsymbol{x})\boldsymbol{f}(d\boldsymbol{x}) = \left(\frac{S_1}{S} - \pi_1, \ldots, \frac{S_m}{S} - \pi_m\right)^T.$$

It follows from (7.86) that $h(\theta)$ is a continuous function of θ. Let $v(\theta) = \frac{1}{2}\sum_{k=1}^m (\frac{S_k}{S} - \pi_k)^2$. As shown below, $v(\theta)$ has continuous partial derivatives of the first order.

Solving the system of equations formed by (7.86), we have

$$\mathcal{L} = \left\{(\theta_1, \ldots, \theta_m) : \theta_i = \mathrm{C} + \log\left(\int_{E_i} \psi(\boldsymbol{x})d\boldsymbol{x}\right) - \log(\pi_i), i = 1, \ldots, m\right\},$$

where $\mathrm{C} = \log(S)$ can be determined by imposing a constraint on S. For example, setting $S = 1$ leads to that $c = 0$. It is obvious that $\mathcal{L}$ is nonempty and $v(\theta) = 0$ for every $\theta \in \mathcal{L}$.

To verify the conditions related to $\dot{v}(\theta)$, we have the following calculations:

$$\begin{aligned}
\frac{\partial S}{\partial \theta^{(i)}} &= \frac{\partial S_i}{\partial \theta^{(i)}} = -S_i, \\
\frac{\partial S_i}{\partial \theta^{(j)}} &= \frac{\partial S_j}{\partial \theta^{(i)}} = 0, \\
\frac{\partial (\frac{S_i}{S})}{\partial \theta^{(i)}} &= -\frac{S_i}{S}(1 - \frac{S_i}{S}), \\
\frac{\partial (\frac{S_i}{S})}{\partial \theta^{(j)}} &= \frac{\partial (\frac{S_j}{S})}{\partial \theta^{(i)}} = \frac{S_i S_j}{S^2},
\end{aligned} \tag{7.87}$$

for $i, j = 1, \ldots, m$ and $i \neq j$.

$$\begin{aligned}
\frac{\partial v(\theta)}{\partial \theta^{(i)}} &= \frac{1}{2}\sum_{k=1}^m \frac{\partial (\frac{S_k}{S} - \pi_k)^2}{\partial \theta^{(i)}} \\
&= \sum_{j \neq i} \left(\frac{S_j}{S} - \pi_j\right)\frac{S_i S_j}{S^2} - \left(\frac{S_i}{S} - \pi_i\right)\frac{S_i}{S}\left(1 - \frac{S_i}{S}\right) \\
&= \sum_{j=1}^m \left(\frac{S_j}{S} - \pi_j\right)\frac{S_i S_j}{S^2} - \left(\frac{S_i}{S} - \pi_i\right)\frac{S_i}{S} \\
&= \mu_{\eta^*}\frac{S_i}{S} - \left(\frac{S_i}{S} - \pi_i\right)\frac{S_i}{S},
\end{aligned} \tag{7.88}$$

for $i = 1, \ldots, m$, where $\mu_{\eta^*} = \sum_{j=1}^m (\frac{S_j}{S} - \pi_j)\frac{S_j}{S}$. Thus, we have

$$\begin{aligned}
\dot{v}(\theta) &= \mu_{\eta^*}\sum_{i=1}^m \left(\frac{S_i}{S} - \pi_i\right)\frac{S_i}{S} - \sum_{i=1}^m \left(\frac{S_i}{S} - \pi_i\right)^2 \frac{S_i}{S} \\
&= -\left\{\sum_{i=1}^m \left(\frac{S_i}{S} - \pi_i\right)^2 \frac{S_i}{S} - \mu_{\eta^*}^2\right\} \\
&= -\sigma_{\eta^*}^2 \leq 0,
\end{aligned} \tag{7.89}$$

Table 7.9 Distribution of η^*

state (η^*)	$\frac{S_1}{S} - \pi_1$	$\cdots$	$\frac{S_m}{S} - \pi_m$
Prob.	$\frac{S_1}{S}$	$\cdots$	$\frac{S_m}{S}$

where $\sigma^2_{\eta^*}$ denotes the variance of the discrete distribution defined in Table 7.9,

If $\theta \in \mathcal{L}, \dot{v}(\theta) = 0$; otherwise, $\dot{v}(\theta) < 0$. Therefore, $\sup_{\theta \in Q} \dot{v}(\theta) < 0$ for any compact set $Q \subset \mathcal{L}^c$.

7.7.1.4 Discussion

We now discuss briefly the convergence of SAMC and ASAMC. The proof for the convergence of CSAMC follows directly from that of SSAMC, and the proof for the convergence of AESAMC follows directly from that of ASAMC.

SAMC. It is easy to see that SAMC is a special instance of the general SAMCMC algorithm described in Section 7.7.1.1 with $\kappa = 1$ and $\eta(\cdot) = 0$. Because $\eta(\cdot) = 0$, the condition (B_1) can be relaxed to the condition (A_1), while not changing the validity of Lemma 7.7.2 and Theorem 7.7.1. The conditions (B_2) and (B_3) can be verified as in Section 7.7.1.3. This concludes the proof for Theorem 7.4.1.

ASAMC. ASAMC can also be viewed as a special instance of the general SAMCMC algorithm described in Section 7.7.1.1 with $\kappa = 1$ and $\eta(\cdot) = 0$. Note that the space shrinkage made at each iteration does not affect the convergence theory of the algorithm. The proof for Theorem 7.6.3 can be done on the full sample space $\mathcal{X}$. Comparing to SAMC, ASAMC changes the condition imposed on the proposal distribution, from the local positive condition (A_2) to the minorisation condition 7.59. This change affects only the verification of condition (B_2)-(i) when proving Theorem 7.6.3. The other conditions can still be verified as in Section 7.7.1.3. Below we verify the condition (B_2)-(i) with the minorization condition.

For the MH kernel, we have (as in Section 7.7.1.3, we omit the subscripts of θ_t and x_t in the following proof),

$$
\begin{aligned}
P_\theta(x, A) &= \int_A s_\theta(x, y)dy + I(x \in A)\left(1 - \int_{\mathcal{X}} s_\theta(x, z)dz\right) \\
&\geq \int_A q(x, y) \min\left\{1, \frac{f_\theta(y)q(y, x)}{f_\theta(x)q(x, y)}\right\} dy = \int_A \min\left\{q(x, y), \frac{f_\theta(y)q(y, x)}{f_\theta(x)}\right\} dy \\
&\geq \int_A \min\left\{q(x, y), \frac{f_\theta(y)}{\omega^*}\right\} dy \qquad \text{(by the minorisation condition)} \\
&= \int_A \frac{f_\theta(y)}{\omega^*} dy \geq \frac{\psi(A)}{\omega^* \int_{\mathcal{X}} \psi(x)dx} = \frac{\psi^*(A)}{\omega^*},
\end{aligned}
$$

where $s(x, y) = q(x, y) \min\{1, [f(y)q(y, x)]/[f(x)q(x, y)]\}$, the minorization constant $\omega^* = \sup_{\theta\in\Theta} \sup_{x,y\in\mathcal{X}} f_\theta(x)/q(y, x)$, and $\psi^*(\cdot)$ denotes a normalized measure of $\psi(\cdot)$. Suppose the constraint $\sum_{i=1}^m \int_{E_k} \psi(x)dx/e^{\theta^{(k)}} = 1$ has been imposed on $\theta, e^{\theta^{(k)}}$ is then bounded above by $\int_{\mathcal{X}} \psi(x)dx$. Therefore, the condition

$$\inf_{\theta\in\Theta} P_\theta^l(x, A) \geq \delta\nu(A), \quad \forall \boldsymbol{x} \in \mathcal{X},\ \forall A \in \mathcal{B} \tag{7.90}$$

is satisfied by choosing $\delta = \frac{1}{\omega^*}, l = 1$, and $\nu(\cdot) = \psi^*(\cdot)$. Hence, $C = \mathcal{X}$ is a small set. This further implies

$$P_\theta^l V^\alpha(x) \leq \lambda V^\alpha(x) + bI(x \in C), \quad \forall x \in \mathcal{X}, \tag{7.91}$$

by choosing $V(x) = 1, l = 1, 0 < \lambda < 1, b = 1 - \lambda$, and $\alpha \in [2, \infty)$. Since $V(x) = 1$, the condition

$$P_\theta V^\alpha(x) \leq \varsigma V^\alpha(x) \tag{7.92}$$

holds by choosing $\varsigma \geq 1$. Following from (7.90), (7.91) and (7.92), (B_2-i) is satisfied. As aforementioned, other conditions can be verified as in Section 7.7.1.3. Thus, as $t \to \infty, \theta_t \to \theta_*$ almost surely, and $f_{\theta_t}(x) \to f_{\theta_*}(x)$ almost surely.

Lemma 7.7.3 *(Billingsley, 1986, p.218) Suppose that $F_t(A) = \int_A f_t(x)dx$ and $F(A) = \int_A f(x)dx$ for densities $f_t(x)$ and $f(x)$ defined on $\mathcal{X}$. If $f_t(x)$ converges to $f(x)$ almost surely, then $F_t(A) \longrightarrow F(A)$ as $t \to \infty$ uniformly for any $A \in \mathcal{B}(\mathcal{X})$, where $\mathcal{B}(\mathcal{X})$ denotes the Borel set of the space $\mathcal{X}$.*

Let $F_{\theta_t}(x)$ and $F_{\theta_*}(x)$ denote the cumulative distribution functions corresponding to $f_{\theta_t}(x)$ and $f_{\theta_*}(x)$, respectively. Following from Lemma 7.7.3, $F_{\theta_t}(x)$ converges to $F_{\theta_*}(x)$ as $t \to \infty$; that is, the weak convergence of x_t.

7.7.2 Convergence Rate

7.7.2.1 A General Result

Consider the general SAMCMC algorithm described in Section 7.7.1.1, for which $\{\theta_t\}$ is updated through

$$\theta_{t+1} = \theta_t + \gamma_{t+1} H(\theta_t, \boldsymbol{x}_{t+1}) + \gamma_{t+1}^2 \eta(\boldsymbol{x}_{t+1}), \tag{7.93}$$

where γ_{t+1} is the gain factor and $\eta(\boldsymbol{x}_{t+1})$ is a bounded function. To assess the convergence rate of the algorithm, we need the following additional condition:

(B_4) The mean field function $h(\theta)$ is measurable and locally bounded. There exist a constant $\delta > 0$ and θ_* such that for all $\theta \in \Theta$,

$$(\theta - \theta_*)^T h(\theta) \leq -\delta\|\theta - \theta_*\|^2. \tag{7.94}$$

In a similar way to Benveniste *et al.* (1990; Section 1.10.2), we prove the following theorem, which gives a L^2 upper bound for the approximation error of θ_t.

Theorem 7.7.2 *Assume Θ is compact, the conditions $(B_1), (B_2), (B_3)$, and (B_4) hold, $\sup_{\boldsymbol{x}\in\mathcal{X}^*} V(\boldsymbol{x}) < \infty$, and the gain factor sequence is chosen in the form*

$$\gamma_t = \frac{T_0}{\max\{T_0, t^{\xi}\}}, \tag{7.95}$$

where $1/2 < \xi \leq 1$, and T_0 is a constant. Let the sequence $\{\theta_n\}$ be defined by (7.93). There exists a constant λ such that

$$E\|\theta_t - \theta_*\|^2 \leq \lambda\gamma_t.$$

Proof: Writing $\epsilon_t = \theta_t - \theta_*$, and following from the Poisson equation

$$H(\theta, x) = h(\theta) + u(\theta, x) - P_\theta u(\theta, x),$$

we have

$$\begin{aligned}\|\epsilon_{t+1}\|^2 =& \|\epsilon_t\|^2 + 2\gamma_{t+1}\epsilon_t^T h(\theta_t) + 2\gamma_{t+1}\epsilon_t^T\left[u(\theta_t, x_{t+1}) - P_{\theta_t}u(\theta_t, x_{t+1})\right] \\ &+ 2\gamma_{t+1}^2\epsilon_t^T\eta(\theta_t, x_{n+1}) + \gamma_{t+1}^2\|H(\theta_t, x_{t+1}) + \gamma_{t+1}\eta(\theta_t, x_{t+1})\|^2.\end{aligned}$$

Then, decomposing $u(\theta_t, x_{t+1}) - P_{\theta_t}u(\theta_t, x_{t+1})$ as follows:

$$\begin{aligned}&u(\theta_t, x_{t+1}) - P_{\theta_t}u(\theta_t, x_{t+1}) \\ &= u(\theta_t, x_{t+1}) - P_{\theta_t}u(\theta_t, x_t) + P_{\theta_{t-1}}u(\theta_{t-1}, x_t) \\ &- P_{\theta_t}u(\theta_t, x_{t+1}) + P_{\theta_t}u(\theta_t, x_t) - P_{\theta_{t-1}}u(\theta_{t-1}, x_t).\end{aligned}$$

Note that

$$E\left\{\gamma_{t+1}\epsilon_t\left[u(\theta_t, x_{t+1}) - P_{\theta_t}u(\theta_t, x_t)\right]\right\} = 0, \tag{7.96}$$

and that, by (7.74) in Lemma 7.7.1,

$$\gamma_{t+1}\epsilon_t\|P_{\theta_t}u(\theta_t, x_t) - P_{\theta_{t-1}}u(\theta_{t-1}, x_t)\| = O(\gamma_{t+1}^2), \tag{7.97}$$

and

$$\epsilon_t\left[P_{\theta_{t-1}}u(\theta_{t-1}, x_t) - P_{\theta_t}u(\theta_t, x_{t+1})\right] = z_t - z_{t+1} + (\epsilon_{t+1} - \epsilon_t)P_{\theta_t}u(\theta_t, x_{t+1}), \tag{7.98}$$

where $z_t = \epsilon_t P_{\theta_{t-1}}u(\theta_{t-1}, x_t)$, and

$$\|(\epsilon_{t+1} - \epsilon_t)P_{\theta_t}u(\theta_t, x_{t+1})\| = O(\gamma_{t+1}). \tag{7.99}$$

Thus, from (7.96)–(7.99) and (7.94), we deduce that

$$E\|\epsilon_{t+1}\|^2 \leq (1 - 2\delta\gamma_{t+1})E\|\epsilon_t\|^2 + C_1\gamma_{t+1}^2 + 2\gamma_{t+1}E(z_t - z_{t+1}), \tag{7.100}$$

for some constant C_1. Note that, by (7.74) in Lemma 7.7.1, z_t is bounded; that is, there exists a constant C_2 such that $E\|z_t\| < C_2$ for all $t > 0$. ∎

Lemma 7.7.4 *Suppose t_0 is such that $1 - 2\delta\gamma_{t+1} \geq 0$ for all $t \geq t_0$ and*

$$\frac{1}{\gamma_{t+1}} - \frac{1}{\gamma_t} < 2\delta. \tag{7.101}$$

Consider for $t > t_0$ the finite sequence A_k^t for $k = t_0, \ldots, t$,

$$A_k^t = \begin{cases} 2\gamma_k \prod_{j=k}^{t-1}(1 - 2\delta\gamma_{j+1}) & \text{if } k \leq t-1, \\ 2\gamma_t & \text{if } k = t. \end{cases}$$

Then the sequence A_k^t is increasing.

Proof: If $k + 1 < t$, then

$$A_{k+1}^t - A_k^t = 2\left[\prod_{j=k+1}^{t-1}(1 - 2\delta\gamma_{j+1})\right](\gamma_{k+1} - \gamma_k + 2\delta\gamma_k\gamma_{k+1}).$$

If $k + 1 = t$, then

$$A_t^t - A_{t-1}^t = 2(\gamma_t - \gamma_{t-1} + 2\delta\gamma_{t-1}\gamma_t).$$

Thus, $A_{k+1}^t - A_k^t \geq 0$ for all $t_0 \leq k < t$. ■

Lemma 7.7.5 *Let $\{u_t\}_{t \geq t_0}$ be a sequence of real numbers such that for all $t \geq t_0$*

$$u_{t+1} \geq u_t(1 - 2\delta\gamma_{t+1}) + C_1\gamma_{t+1}^2 \tag{7.102}$$

with additionally

$$E\|\epsilon_{t_0}\|^2 \leq u_{t_0}. \tag{7.103}$$

Then for all $t \geq t_0 + 1$,

$$E\|\epsilon_t\|^2 \leq u_t + \sum_{k=t_0+1}^{t} A_k^t(z_{k-1} - z_k). \tag{7.104}$$

Proof: If (7.104) is true, then following (7.100) and (7.102),

$$\begin{aligned} E\|\epsilon_{t+1}\|^2 &\leq (1 - 2\delta\gamma_{t+1})\left[u_t + \sum_{k=t_0+1}^{t} A_k^t(z_{k-1} - z_k)\right] \\ &\quad + 2\gamma_{t+1}(z_t - z_{t+1}) + C_1\gamma_{t+1}^2 \\ &= (1 - 2\delta\gamma_{t+1})u_t + C_1\gamma_{t+1}^2 + (1 - 2\delta\gamma_{t+1})\sum_{k=t_0+1}^{t} A_k^t(z_{k-1} - z_k) \\ &\quad + 2\gamma_{t+1}(z_t - z_{t+1}) \\ &\leq u_{t+1} + \sum_{k=t_0+1}^{t+1} A_k^{t+1}(z_{k-1} - z_k), \end{aligned}$$

which completes the proof of the lemma by induction. ■

Proof of Theorem 7.7.2 (continued) Take $T > t_0$, where t_0 is as defined in Lemma 7.7.4, and choose λ such that

$$E\|\epsilon_T\|^2 \leq \lambda\gamma_T.$$

For $t \geq T$, we have

$$\sum_{k=T+1}^{t} A_k^t(z_{k-1} - z_k) = \sum_{k=T+1}^{t-1} (A_{k+1}^t - A_k^t)z_k - 2\gamma_t z_t + A_{T+1}^t z_T.$$

Following (7.104) and the result $E\|z_t\| < \infty$, for any sequence $\{u_t\}_{t \geq T}$ satisfying (7.102) and (7.103), we have

$$E\|e_t\|^2 \leq u_t + C_3\gamma_t, \tag{7.105}$$

where C_3 is a suitable constant. We note that the sequence $u_t = \lambda\gamma_t$ with γ_t being specified in (7.95) satisfies the conditions (7.102) and (7.101) when t becomes large. This completes the proof of this theorem.

7.7.2.2 Asymptotic Normality

Under some conditions slightly different from those given in Theorem 7.7.2, Benveniste *et al.* (1990) established the asymptotic normality of θ_t, that is, $(\theta_t - \theta_*)/\sqrt{\gamma_t}$ converges wealy toward a normal random variable, for a type of stochastic approximation MCMC algorithms. Extension of their results to SAMC is easy, if not straightforward. Interested readers can continue to work on this problem.

7.7.2.3 Proof of Theorem 7.4.2

Theorem 7.4.2 concerns the convergence rate of the SAMC algorithm. To prove this theorem, it suffices to verify that the conditions (B_1)–(B_4) hold for SAMC. Since the conditions (B_1)–(B_3) have been verified in Section 7.7.1.3, only (B_4) needs to be verified here. To verify (B_4), we first show that $h(\theta)$ has bounded second derivatives. Continuing the calculation in (7.87), we have

$$\frac{\partial^2(\frac{S_i}{S})}{\partial(\theta^{(i)})^2} = \frac{S_i}{S}\left(1 - \frac{S_i}{S}\right)\left(1 - \frac{2S_i}{S}\right), \quad \frac{\partial^2(\frac{S_i}{S})}{\partial\theta^{(j)}\partial\theta^{(i)}} = -\frac{S_iS_j}{S^2}\left(1 - \frac{2S_i}{S}\right), \tag{7.106}$$

where S and S_i are as defined in Section 7.7.1.3. This implies that the second derivative of $h(\theta)$ is uniformly bounded by noting the inequality $0 < \frac{S_i}{S} < 1$.

Table 7.10 Distribution of Z.

State (Z)	z_1	$\cdots$	z_m
Prob.	$\frac{S_1}{S}$	$\cdots$	$\frac{S_m}{S}$

Let $F = \partial h(\theta)/\partial\theta$. From (7.87) and (7.106), we have

$$F = \begin{pmatrix} -\frac{S_1}{S}(1-\frac{S_1}{S}) & \frac{S_1 S_2}{S^2} & \cdots & \frac{S_1 S_m}{S^2} \\ \frac{S_2 S_1}{S^2} & -\frac{S_2}{S}(1-\frac{S_2}{S}) & \cdots & \frac{S_2 S_m}{S^2} \\ \vdots & \ddots & \vdots & \vdots \\ \frac{S_m S_1}{S^2} & \cdots & \cdots & -\frac{S_m}{S}(1-\frac{S_m}{S}) \end{pmatrix}.$$

Thus, for any nonzero vector $\boldsymbol{z} = (z_1, \ldots, z_m)^T$,

$$\boldsymbol{z}^T F \boldsymbol{z} = -\left[\sum_{i=1}^{m} z_i^2 \frac{S_i}{S} - \left(\sum_{i=1}^{m-1} z_i \frac{S_i}{S}\right)^2\right] = -\mathrm{Var}(Z) < 0, \tag{7.107}$$

where $\mathrm{Var}(Z)$ denotes the variance of the discrete distribution defined in Table 7.10.

Thus, the matrix F is negative definite. Applying Taylor expansion to $h(\theta)$ at a point θ_*, we have

$$(\theta - \theta_*)^T h(\theta) \leq -\delta \|\theta - \theta_*\|^2,$$

for some value $\delta > 0$. Therefore, (B_4) is satisfied by SAMC.

Clearly, Theorem 7.4.2 also holds for ASAMC and AESAMC. For SSAMC and CSAMC, if the bandwidth is chosen to be of order $O(\gamma_t)$, Theorem 7.4.2 also holds for them.

7.7.3 Ergodicity and its IWIW Property

Liang (2009b) studies the ergodicity of SAMC, showing that the samples generated by SAMC can be used in evaluation of $E_f h(x)$ through a weighted average estimator. To show this property, we introduce the following lemmas.

Let $\{Z_t, t \geq 0\}$ be a nonhomogeneous Markov chain with the finite state space $S = \{1, 2, \ldots, k\}$, the initial distribution

$$\big(P(1), P(2), \ldots, P(k)\big), \quad P(i) > 0, \; i \in S, \tag{7.108}$$

and the transition matrix

$$P_t = \big(P_t(j|i)\big), \quad P_t(j|i) > 0, \quad i, j \in S, \; t \geq 1, \tag{7.109}$$

where $P_t(j|i) = P(Z_t = j|Z_{t-1} = i)$ for $t \geq 1$. Let $P = \big(P(j|i)\big)$ be an ergodic transition matrix, and let $(p_1, \ldots, p_k)$ be the stationary distribution determined by P.

Lemma 7.7.6 *(Liu and Yang, 1996; Theorem 7) Let $\{Z_t, t \geq 0\}$ be a nonhomogeneous Markov chain with the initial distribution (7.108) and the transition matrix (7.109), and let g and $g_t, t \geq 1$, be functions defined on S. If the following conditions hold,*

$$\lim_{n\to\infty} \frac{1}{n}\sum_{t=1}^{n} |P_t(j|i) - P(j|i)| = 0, \ a.s., \quad \forall i, j \in S, \tag{7.110}$$

$$\lim_{n\to\infty} \frac{1}{n}\sum_{t=1}^{n} |g_t(i) - g(i)| = 0, \ a.s., \quad \forall i \in S, \tag{7.111}$$

then

$$\lim_{n\to\infty} \frac{1}{n}\sum_{t=1}^{n} g_t(Z_t) = \sum_{i=1}^{k} p_i g(i), \quad a.s.. \tag{7.112}$$

Let $(x_1, \theta_1), \ldots, (x_n, \theta_n)$ denote a set of samples generated by SAMC, where $\theta_t = (\theta_t^{(1)}, \ldots, \theta_t^{(m)})$, recalling that the sample space has been partitioned into m disjoint subregions $E_1, \ldots, E_m$. Without loss of generality, we assume that there are no empty subregions in the partition. Let $J(x_t)$ denote the index of the subregion that the sample x_t belongs to. Then $J(x_1), \ldots, J(x_n)$ forms a sample from a nonhomogeneous Markov chain defined on the finite state space $\{1, \ldots, m\}$. The Markov chain is nonhomogeneous as the target distribution $f_{\theta_t}(x)$ changes from iterations to iterations. Let

$$f_{\theta_*}(x) = \sum_{i=1}^{m} \frac{\pi_i \psi(x)}{\omega_i} I(x \in E_i), \tag{7.113}$$

where $\omega_i = \int_{E_i} \psi(x)dx$. It follows from (7.19) that $f_{\theta_t}(x) \to f_{\theta_*}(x)$ almost surely. Then, following from Lemma 7.7.3, we have

$$\lim_{t\to\infty} P_t(j|i) = P(j|i), \tag{7.114}$$

where the transition probability $P_t(j|i)$ is defined as

$$P_t(j|i) = \int_{E_i}\int_{E_j} \left[s_t(x, dy) + I(x \in dy)\left(1 - \int_{\mathcal{X}} s_t(x, dz)\right)\right] dx, \tag{7.115}$$

$s_t(x, dy) = q(x, dy)\min\{1, [f_{\theta_t}(y)q(y, x)]/[f_{\theta_t}(x)q(x, y)]\}$, and $P(j|i)$ can be defined similarly by replacing $f_{\theta_t}(\cdot)$ by $f_{\theta_*}(\cdot)$ in (7.115). Following from (7.114), (7.110) holds. It is obvious that π forms the stationary distribution of the Markov chain induced by the transition matrix P. Define the function

$$g_t(J(x_t)) = e^{\theta_t^{(J(x_t))}}, \quad t = 0, 1, 2, \ldots. \tag{7.116}$$

Following from (7.19), we have

$$\lim_{t\to\infty} g_t(i) = \omega_i/\pi_i, \quad \forall i \in S, \tag{7.117}$$

which implies (7.111) holds by defining $g(i) = \omega_i/\pi_i, i = 1\ldots, m$. Since both conditions (7.110) and (7.111) hold, we have, by Lemma 7.7.6, the following law of large numbers for the SAMC samples.

Proposition 7.7.1 *Assume the conditions* (A_1) *and* (B_2) *hold. For a set of samples generated by SAMC, we have*

$$\lim_{n\to\infty} \frac{1}{n}\sum_{t=1}^{n} e^{\theta_t^{(J(x_t))}} = \sum_{i=1}^{m} \omega_i, \quad a.s. \tag{7.118}$$

Let $\mathcal{A} \subset \mathcal{X}$ denote an arbitrary Borel set, and let $\mathcal{A}^c$ denote the complementary set of $\mathcal{A}$. Thus, $\tilde{E}_1 = E_1 \cap \mathcal{A}, \tilde{E}_2 = E_1 \cap \mathcal{A}^c, \ldots, \tilde{E}_{2m-1} = E_m \cap \mathcal{A}, \tilde{E}_{2m} = E_m \cap \mathcal{A}^c$ form a new partition of the sample space $\mathcal{X}$. In this paper, we call the new partition an induced partition by $\mathcal{A}$, and the subregion $\tilde{E}_i$ an induced subregion by $\mathcal{A}$. Let $(x_1, \theta_1), \ldots, (x_n, \theta_n)$ denote again a set of samples generated by SAMC with the partition $E_1, \ldots, E_m$, and let $\tilde{J}(x_t)$ denote the index of the induced subregion x_t belongs to. Then $\tilde{J}(x_1), \ldots, \tilde{J}(x_n)$ forms a sample from a nonhomogeneous Markov chain defined on the finite state space $\{1, \ldots, 2m\}$. The transition matrices of the Markov chain can be defined similarly to (7.115). The stationary distribution of the limiting Markov chain is then $(\tilde{p}_1, \ldots, \tilde{p}_{2m})$, where $\tilde{p}_{2i-1} = \pi_i P_f(\mathcal{A}|E_i)$ and $\tilde{p}_{2i} = \pi_i P_f(\mathcal{A}^c|E_i)$ for $i = 1, \ldots, m$, and $P_f(A|B) = \int_{A\cap B} f(x)dx / \int_B f(x)dx$. Define

$$\tilde{g}_t(\tilde{J}(x_t)) = e^{\theta_t^{(J(x_t))}} I(x_t \in \mathcal{A}), \quad t = 0, 1, 2, \ldots, \tag{7.119}$$

where $I(\cdot)$ is the indicator function. Following from (7.19), we have

$$\lim_{t\to\infty} \tilde{g}_t(2i-1) = \omega_i/\pi_i \quad \text{and} \quad \lim_{t\to\infty} \tilde{g}_t(2i) = 0, \quad i = 1, \ldots, m, \tag{7.120}$$

which implies (7.111) holds by defining $\tilde{g}(2i-1) = \omega_i/\pi_i$ and $\tilde{g}(2i) = 0, i = 1, \ldots, m$. With the similar arguments as for Proposition 7.7.1, we have Proposition 7.7.2:

Proposition 7.7.2 *Assume the conditions* (A_1) *and* (B_2) *hold. For a set of samples generated by SAMC, we have*

$$\lim_{n\to\infty} \frac{1}{n}\sum_{t=1}^{n} e^{\theta_t^{(J(x_t))}} I(x_t \in \mathcal{A}) = \sum_{i=1}^{m} \omega_i P_f(\mathcal{A}|E_i), \quad a.s. \tag{7.121}$$

Consider again the samples $(x_1, \theta_1), \ldots, (x_n, \theta_n)$ generated by SAMC with the partition $E_1, \ldots, E_m$. Let $y_1, \ldots, y_{n'}$ denote the distinct samples among

$x_1, \ldots, x_n$. Generate a random variable/vector Y such that

$$P(Y = y_i) = \frac{\sum_{t=1}^{n} e^{\theta_t^{(J(x_t))}} I(x_t = y_i)}{\sum_{t=1}^{n} e^{\theta_t^{(J(x_t))}}}, \quad i = 1, \ldots, n', \tag{7.122}$$

where $I(\cdot)$ is the indicator function. Since Θ has been restricted to a compact set in SAMC, $\theta_t^{(J(x_t))}$ is finite. The following theorem shows that Y is asymptotically distributed as $f(\cdot)$, which can be proved in a similar way to Theorem 6.2.1.

Theorem 7.7.3 *Assume the conditions* (A_1) *and* (B_2) *hold. For a set of samples generated by SAMC, the random variable/vector* Y *generated in (7.122) is asymptotically distributed as* $f(\cdot)$.

Theorem 7.7.3 implies that for an integrable function $\rho(x)$, the expectation $E_f \rho(x)$ can be estimated by

$$\widehat{E_f \rho(x)} = \frac{\sum_{t=1}^{n} e^{\theta_t^{(J(x_t))}} \rho(x_t)}{\sum_{t=1}^{n} e^{\theta_t^{(J(x_t))}}}. \tag{7.123}$$

As $n \to \infty$, $\widehat{E_f \rho(x)} \to E_f \rho(x)$ for the same reason that the usual importance sampling estimate converges (Geweke, 1989).

We note that SAMC falls into the class of dynamic weighting algorithms; that is, SAMC is (asymptotically) invariant with respect to the importance weights (IWIW) (Wong and Liang, 1997), which can be briefly described as follows. Let $g_t(x, w)$ be the joint distribution of the sample (x, w) drawn at iteration t, where $w = \exp(\theta_t^{(J(x_t))})$. The joint distribution $g_t(x, w)$ is said to be correctly weighted with respect to a distribution $f(x)$ if

$$\int w g_t(x, w) dw \propto f(x). \tag{7.124}$$

A transition rule is said to satisfy IWIW if it maintains the correctly weighted property for the joint distribution $g_t(x, w)$ whenever an initial joint distribution is correctly weighted.

As discussed in Chapter 6, IWIW is a much more general transition rule of MCMC than the detailed balance (or reversibility) condition required by conventional MCMC algorithms.

Theorem 7.7.4 *Assume the conditions* (A_1) *and* (B_2) *hold. SAMC asymptotically satisfies IWIW.*

Proof: Let $g_t(x, w_x)$ denote the joint distribution of the sample (x, w_x) generated by SAMC at iteration t, where $w_x = e^{\theta_t^{(J(x_t))}}$. If x can be regarded as a sample drawn from $f_{\theta_t}(x)$, then (x, w_x) has the joint distribution

$$g_t(x, w_x) = f_{\theta_t}(x) \delta(w_x = e^{\theta_t^{(J(x_t))}}), \tag{7.125}$$

where $\delta(\cdot)$ denotes a Dirac measure and corresponds to the conditional distribution of w_x given the sample x. It is easy to check that (7.125) is correctly weighted with respect to $f(x)$. The existence of such a sample is obvious, as it can, at least, be obtained after the convergence of θ_t.

Let (y, w_y) denote the sample generated at iteration $t+1$, and let $w'_y = e^{\theta_t^{(J(y))}}$. Then, $w_y = e^{\gamma_{t+1}(1-\pi_{J(y)})} w'_y$, and

$$
\begin{aligned}
&\int w_y g_{t+1}(y, w_y) dw_y \\
&= \int\int\int e^{\gamma_{t+1}(1-\pi_{J(y)})} w'_y g_t(x, w_x) P_{\theta_t}\left((x, w_x) \to (y, w'_y)\right) dx dw_x dw'_y \\
&= \int\int\int e^{\gamma_{t+1}(1-\pi_{J(y)})} w'_y \delta(w_x = e^{\theta_t^{(J(x))}}) f_{\theta_t}(x) P_{\theta_t}\left((x, w_x) \to (y, w'_y)\right) \\
&\qquad \times dx dw_x dw'_y \\
&= \int\int\int e^{\gamma_{t+1}(1-\pi_{J(y)})} w'_y \delta(w'_y = e^{\theta_t^{(J(y))}}) f_{\theta_t}(y) P_{\theta_t}\left((y, w'_y) \to (x, w_x)\right) \\
&\qquad \times dx dw_x dw'_y \\
&= \int e^{\gamma_{t+1}(1-\pi_{J(y)})} w'_y g_t(y, w'_y) dw'_y,
\end{aligned}
\tag{7.126}
$$

where $P_{\theta_t}(\cdot \to \cdot)$ denotes the MH transition kernel used in the t-th iteration of SAMC. Since $\gamma_t \to 0$, $\int w_y g_{t+1}(y, w_y) dw_y \propto f(y)$ holds asymptotically by the last line of (7.126). If $\pi_{J(y)}$ is independent of y, i.e., $\pi_1 = \cdots = \pi_m = 1/m$, then $\int w_y g_{t+1}(y, w_y) dw_y \propto f(y)$ holds exactly. ■

Theorem 7.7.4 suggests that setting $\boldsymbol{\pi}$ to be non-uniform over the subregions may lead to a slightly biased estimate of $E_f h(x)$ for a short run of SAMC for which the gain factor sequence has not yet decreased to be sufficiently small.

7.8 Trajectory Averaging: Toward the Optimal Convergence Rate

Consider the stochastic approximation algorithm:

$$\theta_{t+1} = \theta_t + \gamma_{t+1} H(\theta_t, X_{t+1}) + \gamma_{t+1}^{1+\tau} \eta(X_{t+1}), \tag{7.127}$$

where, as previously, γ_{t+1} denotes the gain factor, $\tau > 0$, $\eta(\cdot)$ is a bounded function, and X_{t+1} denotes a stochastic disturbance simulated from $f_{\theta_t}(x), x \in \mathcal{X} \subset \mathbb{R}^d$, using a MCMC sampler. This algorithm can be rewritten as an algorithm for search of zeros of a function $h(\theta)$,

$$\theta_{t+1} = \theta_t + \gamma_{t+1}[h(\theta_t) + \epsilon_{t+1}], \tag{7.128}$$

where $h(\theta_t) = \int_{\mathcal{X}} H(\theta_t, x) f_{\theta_t}(x) dx$ corresponds to the mean effect of $H(\theta_t, X_{t+1})$, and $\epsilon_{t+1} = H(\theta_t, X_{t+1}) - h(\theta_t) + \gamma_{t+1}^T \eta(X_{t+1})$ is the observation noise. It is well known that the optimal convergence rate of (7.128) can be achieved with $\gamma_t = -F^{-1}/t$, where $F = \partial h(\theta_*)/\partial\theta$, and θ_* denotes the zero point of $h(\theta)$. In this case, (7.128) is reduced to Newton's algorithm. Unfortunately, it is often impossible to use this algorithm, as the matrix F is generally unknown. Then, in a sequence of papers, Ruppert (1988), Polyak (1990), and Polyak and Juditsky (1992) show that the trajectory averaging estimator is asymptotically efficient; that is,

$$\bar{\theta}_n = \sum_{t=1}^{n} \theta_t / n \tag{7.129}$$

can converge in distribution to a normal random variable with mean θ_* and covariance matrix Σ, where Σ is the smallest possible covariance matrix in an appropriate sense. The trajectory averaging estimator allows the gain factor sequence $\{\gamma_t\}$ to be relatively large, decreasing slower than $O(1/t)$. As discussed by Polyak and Juditsky (1992), trajectory averaging is based on a paradoxical principle, a slow algorithm having less than optimal convergence rate must be averaged. Recently, the trajectory averaging technique has been further explored in literature (see e.g., Chen, 1993; Kushner and Yang, 1993, 1995; Dippon and Renz, 1997; Wang *et al.*, 1997; Tang *et al.*, 1999; Pelletier, 2000; and Kushner and Yin, 2003).

Liang (2009e) shows that the trajectory averaging estimator can also be applied to general stochastic approximation MCMC (SAMCMC) algorithms. In Liang (2009e), the bias term $\eta(x)$ in (7.127) is assumed to be 0, while some other conditions previously imposed on SAMC are relaxed. For example, the solution space Θ is relaxed from a compact set to $\mathbb{R}^d$ based on the varying truncation technique introduced by Chen (1993), and the uniform boundedness condition $\sup_{x \in \mathcal{X}} V(x) < \infty$ has been weakened to $\sup_{x \in \mathcal{X}_0} V(x) < \infty$ for a subset $\mathcal{X}_0 \subset \mathcal{X}$. In the following, we present a simplified version of the proof for the algorithm (7.127). In the simplified proof, Θ is still assumed to be compact, and the drift function is still assumed to be uniformly bounded. The reason is that these conditions simplify the implementation of SAMC, while can be naturally satisfied. For example, for SAMC and its variants, setting $V(x) = 1$ is always appropriate due to the bondedness of the function $H(\theta, X)$. Again, our aim here is to provide a theory to guide the applications of SAMC and its variants instead of providing a very general theory for stochastic approximation MCMC algorithms.

Although the algorithms considered in Chen (1993) and Kushner and Yin (2003) allow the observation noise to be state dependent, their results are not directly applicable to SAMCMC. Chen (1993) and Kushner and Yin (2003) assume directly that the observation noise has the decomposition $\epsilon_t = e_t + \nu_t$, where $\{e_t\}$ forms a martingale difference sequence and ν_t is a higher order term of $\gamma_t^{1/2}$. As shown in Lemma 7.8.2 (below), SAMCMC does not satisfy this decomposition. Instead, the observation noise $\{\epsilon_t\}$ of SAMCMC should be decomposed as

$$\epsilon_t = e_t + \nu_t + \varsigma_t, \tag{7.130}$$

where $\{e_t\}$ forms a martingale difference sequence, ν_t is a higher order term of $o(\gamma_t^{1/2})$, and ς_t is a tricky term with $\|\sum_{t=0}^{n}\gamma_t\varsigma_t\| = O(\gamma_n)$. Some cancellations occur in calculation of $\sum_{t=0}^{n}\gamma_t\varsigma_t$, while ς_t itself is of order $O(1)$. In other words, $\{\varsigma_t\}$ are terms that can be removed by averaging.

7.8.1 Trajectory Averaging for a SAMCMC Algorithm

Reconsider the stochastic approximation MCMC algorithm given in Section 7.7.1.1. To study the asymptotic efficiency of the trajectory averaging estimator for this algorithm, we assume the following conditions in additional to the drift condition (B_2) and the Lyapunov condition (B_3).

Conditions on step-sizes. (B_1') The gain factor sequence $\{\gamma_t\}$ is nonincreasing, positive, and satisfy the conditions:

$$\sum_{t=1}^{\infty}\gamma_t = \infty, \quad \lim_{t\to\infty}(t\gamma_t) = \infty, \quad \frac{\gamma_{t+1}-\gamma_t}{\gamma_t} = o(\gamma_{t+1}), \tag{7.131}$$

and for some $\zeta \in (0,1]$,

$$\sum_{t=1}^{\infty}\frac{\gamma_t^{(1+\zeta)/2}}{\sqrt{t}} < \infty. \tag{7.132}$$

For instance, we can set $\gamma_t = C_1/t^{\xi}$ for some constants $C_1 > 0$ and $\xi \in (\frac{1}{2}, 1)$. With this choice of γ_t, (7.132) is satisfied for any $\zeta > 1/\xi - 1$.

Stability condition on $h(\theta)$. (B_4') The mean field function $h(\theta)$ is measurable and locally bounded. There exist a stable matrix F (i.e., all eigenvalues of F are with negative real parts), $\delta > 0$, $\rho \in (\zeta, 1]$, and a constant c such that, for any $\theta_* \in \mathcal{L} = \{\theta : h(\theta) = 0\}$,

$$\|h(\theta) - F(\theta - \theta_*)\| \le c\|\theta - \theta_*\|^{1+\rho}, \quad \forall\theta \in \{\theta : \|\theta - \theta_*\| \le \delta\}.$$

This condition constrains the behavior of the mean field function $h(\theta)$ around θ_*. It makes the trajectory averaging estimator sensible both theoretically and practically. If $h(\theta)$ is differentiable, then the matrix F can be chosen to be the partial derivative of $h(\theta)$, that is, $\partial h(\theta)/\partial\theta$. Otherwise, certain approximations may be needed.

Theorem 7.8.1 concerns the convergence of the SAMCMC algorithm, whose proof follows the proof of Theorem 7.7.1 directly.

Theorem 7.8.1 *Assume the conditions* $(B_1'), (B_2)$ *and* (B_3) *hold,* $\tau \geq (1+\zeta)/2$ *with* ζ *being defined in* $(B_1'), \Theta$ *is compact, and* $\sup_{x \in \mathcal{X}} V(x) < \infty$. *Let the sequence* $\{\theta_t\}$ *be defined by (7.127). Then for all* $(x_0, \theta_0) \in \mathcal{X} \times \Theta$,

$$\theta_t \longrightarrow \theta_*, \quad a.s.$$

Theorem 7.8.2 concerns the asymptotic normality of the trajectory averaging estimator, whose proof is presented in Section 7.8.3.

Theorem 7.8.2 *Assume the conditions* $(B_1'), (B_2), (B_3)$ *and* (B_4') *hold,* $\tau \geq (1+\zeta)/2$ *with* ζ *being defined in* $(B_1'), \Theta$ *is compact, and* $\sup_{x \in \mathcal{X}} V(x) < \infty$. *Let the sequence* $\{\theta_t\}$ *be defined by (7.127). Then*

$$\sqrt{n}(\bar{\theta}_n - \theta_*) \longrightarrow N(\mathbf{0}, \Gamma)$$

where $\Gamma = F^{-1}Q(F^{-1})^T, F = \partial h(\theta_*)/\partial\theta$ *is negative definite,* $Q = \lim_{t\to\infty} E(e_t e_t^T)$, *and* e_t *is as defined in (7.130).*

As aforementioned, the asymptotic efficiency of the trajectory averaging estimator has been studied by quite a few authors. Following Tang *et al.* (1999), we set the following definition:

Definition 7.8.1 *Consider the stochastic approximation algorithm (7.128). Let* $\{Z_n\}$, *given as a function of* $\{\theta_n\}$, *be a sequence of estimators of* θ_*. *The algorithm* $\{Z_n\}$ *is said to be asymptotically efficient if*

$$\sqrt{n}(Z_n - \theta_*) \longrightarrow N\left(\mathbf{0}, F^{-1}\widetilde{Q}(F^{-1})^T\right), \tag{7.133}$$

where $F = \partial h(y^*)/\partial y$, *and* $\widetilde{Q}$ *is the asymptotic covariance matrix of* $(1/\sqrt{n})\sum_{t=1}^n \epsilon_t$.

As pointed out by Tang *et al.* (1999), $\widetilde{Q}$ is the smallest possible limit covariance matrix that an estimator based on the stochastic approximation algorithm (7.128) can achieve. If $\theta_t \to \theta_*$ and $\{\epsilon_t\}$ forms or asymptotically forms a martingale difference sequence, then we have $\widetilde{Q} = \lim_{t\to\infty} E(\epsilon_t\epsilon_t^T)$. This is also the case that the stochastic approximation MCMC algorithm belongs to. For some stochastic approximation algorithms, $\widetilde{Q}$ can also be expressed as $\widetilde{Q} = \lim_{t\to\infty} E(\epsilon_{t+1}\epsilon_{t+1}^T|\mathcal{F}_t)$ where $\mathcal{F}_t$ denotes a σ-filtration defined by $\{\theta_0, \theta_1, \ldots, \theta_t\}$. For the latter case, refer to Pelletier (2000) for the details. In the next theorem, we show that $\bar{\theta}_n$ is asymptotically efficient.

Theorem 7.8.3 *Assume the conditions* $(B_1'), (B_2), (B_3)$ *and* (B_4') *hold,* $\tau \geq (1+\zeta)/2$ *with* ζ *being defined in* $(B_1'), \Theta$ *is compact, and* $\sup_{x \in \mathcal{X}} V(x) < \infty$. *Let the sequence* $\{\theta_t\}$ *be defined by (7.127). Then* $\bar{\theta}_n$ *is asymptotically efficient.*

As implied by Theorem 7.8.3, the asymptotic efficiency of $\bar{\theta}_n$, which is measured by the asymptotic covariance matrix Γ, is independent of the choice of $\{\gamma_t\}$ as long as (B_1') is satisfied. The asymptotic efficiency of $\bar{\theta}_n$ can also be interpreted in terms of Fisher information theory. Refer to Pelletier (2000; Section 3) and the references therein for more discussions on this issue.

Trajectory averaging enables one to smooth the behavior of the algorithm, but meanwhile it slows down the numerical convergence because it takes longer for the algorithm to forget the first iterates. An alternative idea would be to consider moving window averaging algorithms (see, e.g., Kushner and Yang, 1993; Kushner and Yin, 2003). Extension of their results to general SAMCMC algorithms will be of great interest.

7.8.2 Trajectory Averaging for SAMC

To make the trajectory averaging estimator sensible for SAMC, the algorithm is modified as follows:

(a) *Sampling*. Simulate a sample x_{t+1} by a single MH update with the target distribution

$$f_{\theta_t}(x) \propto \sum_{i=1}^{m} \frac{\psi(x)}{e^{\theta_t^{(i)}}} I_{\{x \in E_i\}}. \tag{7.134}$$

(b) *Weight updating*. For $i = 1, \ldots, m$, set

$$\theta_{t+1}^{(i)} = \theta_t^{(i)} + \gamma_{t+1}\left(I_{\{x_{t+1} \in E_i\}} - \pi_i\right) - \gamma_{t+1}\left(I_{\{x_{t+1} \in E_m\}} - \pi_m\right), \tag{7.135}$$

provided that E_m is non-empty. Here E_m can be replaced by any other unempty subregion.

In terms of sampling, the modified algorithm is exactly the same with the original SAMC, as $f_{\theta_t}(x)$ is invariant with respect to a location shift of θ_t. The only effect by this modification is on the convergence point: For the modified SAMC, θ_t will converge to a single point θ_* as specified in Theorem 7.8.4, while, for the original SAMC, θ_t will converge to a set as specified in Theorem 7.4.1.

Theorem 7.8.4 *Assume the conditions (B_1') and (B_2) hold. Let $\{\theta_t\}$ be a sequence specified by (7.135). Then, as $t \to \infty$,*

$$\theta_t^{(i)} \to \theta_*^{(i)} = \begin{cases} \log\left(\frac{\int_{E_i}\psi(x)dx}{\int_{E_m}\psi(x)dx}\right) - \log\left(\frac{\pi_i+\nu}{\pi_m+\nu}\right), & \textit{if } E_i \neq \emptyset, \\ -\infty. & \textit{if } E_i = \emptyset, \end{cases} \tag{7.136}$$

provided that E_m *is unempty, where* $\nu = \sum_{j \in \{i: E_i = \emptyset\}} \pi_j/(m - m_0)$ *and* m_0 *is the number of empty subregions.*

Theorem 7.8.5 concerns the asymptotic normality and efficiency of $\bar{\theta}_n$.

Theorem 7.8.5 *Assume the conditions* (B_1') *and* (B_2) *hold. Let* $\{\theta_t\}$ *be a sequence specified by (7.135). Then* $\bar{\theta}_n$ *is asymptotically efficient; that is,*

$$\sqrt{n}(\bar{\theta}_n - \theta_*) \longrightarrow N(\mathbf{0}, \Gamma), \quad \text{as } n \to \infty,$$

where the covariance matrix $\Gamma = F^{-1}Q(F^{-1})^T, F = \partial h(\theta_*)/\partial \theta$ *is negative definite, and* $Q = \lim_{t \to \infty} E(e_t e_t^T)$.

Numerical Example. To illustrate the performance of trajectory averaging estimators, we reconsider Example 7.1. The distribution consists of 10 states with the unnormalized mass function $P(x)$ as given in Table 7.1. To simulate from this distribution using SAMC, the sample space was partitioned according to the mass function into five subregions: $E_1 = \{8\}, E_2 = \{2\}, E_3 = \{5, 6\}, E_4 = \{3, 9\}$ and $E_5 = \{1, 4, 7, 10\}$. Our goal is to estimate $\omega_i = \sum_{x \in E_i} P(x), i = 1, \ldots, 5$, which have the true values $(200, 100, 6, 4, 4)^T$.

In the simulation, we set $\psi(x) = P(x)$, set $\pi_i \propto 1/(1+i)$ for $i = 1, \ldots, 5$, set the transition proposal matrix as a stochastic matrix with each row being generated independently from the Dirichlet distribution $Dir(1, \ldots, 1)$, and tried the following gain factor sequences:

$$\gamma_t = \frac{T_0}{\max\{T_0, t^{\xi}\}}, \qquad T_0 = 10, \quad \xi \in \{0.7, 0.8, 0.9\}.$$

For each value of ξ, SAMC was run 100 times independently. Each run consisted of 5.01×10^5 iterations, for which the first 10^3 iterations were discarded for the burn-in process and the samples of θ_t generated in the remaining iterations were averaged.

Table 7.11 compares the two estimators $\widehat{\omega}$ and $\widetilde{\omega}$, which are computed based on $\bar{\theta}_t$ and θ_N, respectively. The θ_N refers to the estimate of θ obtained at the last iteration of the run. The comparison indicates that $\widehat{\omega}$ can be generally superior to $\widetilde{\omega}$ in terms of accuracy. It is hard for $\widetilde{\omega}$ to beat $\widehat{\omega}$ even with $\gamma_t = O(1/t)$. By showing the bias-variance decomposition, Table 7.11 demonstrates that the trajectory averaging estimator has a constant variance independent of the choice of the gain factor sequence (with $\xi < 1$). Trajectory averaging also improves the bias of the conventional SAMC estimator.

For some problems, the trajectory averaging estimator can directly benefit one's inference. A typical example is Bayesian model selection, for which the ratio ω_i/ω_j corresponds to the Bayesian factor of two models if one partitions the sample space according to the model index and imposes an uniform prior on the model space, as in Liang (2009a). Another example is Bayesian inference for spatial models with intractable normalizing constants,

Table 7.11 Comparison of trajectory averaging and conventional SAMC estimators.

	$\xi = 0.7$			$\xi = 0.8$			$\xi = 0.9$		
	bias	sd	rmse	bias	sd	rmse	bias	sd	rmse
	−0.81	0.52	0.96	−0.15	0.33	0.36	0.03	0.21	0.21
	0.71	0.49	0.87	0.14	0.31	0.34	−0.03	0.20	0.20
$\widetilde{\omega}$	0.04	0.03	0.05	0.01	0.01	0.02	0.00	0.01	0.01
	0.02	0.02	0.03	0.00	0.01	0.01	0.00	0.01	0.01
	0.03	0.02	0.04	0.01	0.01	0.02	0.01	0.01	0.01
	−0.24	0.09	0.26	−0.02	0.11	0.11	0.00	0.1	0.10
	0.19	0.09	0.21	0.00	0.10	0.10	−0.01	0.1	0.10
$\widehat{\omega}$	0.03	0.00	0.03	0.01	0.00	0.01	0.00	0.0	0.01
	0.00	0.00	0.01	0.00	0.00	0.01	0.00	0.0	0.00
	0.02	0.00	0.02	0.01	0.00	0.01	0.02	0.0	0.02

$\widehat{\omega}$: trajectory averaging estimates, where each row corresponds to one component of $\boldsymbol{\omega}$. $\widetilde{\omega}$: conventional SAMC estimates, where each row corresponds to one component of $\boldsymbol{\omega}$. The 'bias' and 'sd' are calculated based on 100 independent runs, and 'rmse' is calculated as the square root of 'bias^2+sd^2'.

for which Liang *et al.* (2007) demonstrate how SAMC can be applied to estimate the ratios of the normalizing constants of those models and how the estimates can then be used for inference of model parameters. Trajectory averaging technique will certainly improve the performance of SAMC on these problems.

7.8.3 Proof of Theorems 7.8.2 and 7.8.3.

Some lemmas. Lemma 7.8.1 is a restatement of Corollary 2.1.10 of Duflo (1997, pp. 46–47).

Lemma 7.8.1 *Let $\{S_{ni}, \mathcal{G}_{ni}, 1 \leq i \leq k_n, n \geq 1\}$ be a zero-mean, square-integrable martingale array with differences υ_{ni}, where $\mathcal{G}_{ni}$ denotes the σ-field. Suppose that the following assumptions apply:*

(i) The σ-fields are nested: $\mathcal{G}_{ni} \subseteq \mathcal{G}_{n+1,i}$ for $1 \leq i \leq k_n, n \geq 1$.

(ii) $\sum_{i=1}^{k_n} E(\upsilon_{ni}\upsilon_{ni}^T|\mathcal{G}_{n,i-1}) \to \Lambda$ in probability, where Λ is a positive definite matrix.

(iii) For any $\epsilon > 0, \sum_{i=1}^{k_n} E\left[\|\upsilon_{ni}\|^2 I_{(\|\upsilon_{ni}\| \geq \epsilon)}|\mathcal{G}_{n,i-1}\right] \to 0$ in probability.

Then $S_{nk_n} = \sum_{i=1}^{k_n} \upsilon_{ni} \to N(0, \Lambda)$ in distribution.

Lemma 7.8.2 *Assume the conditions* (B_1'), (B_2) *and* (B_3) *hold,* $\tau \geq (1+\zeta)/2$ *with* ζ *being defined in* (B_1'), Θ *is compact, and* $\sup_{x\in\mathcal{X}} V(x) < \infty$. *There exist* $\mathbb{R}^d$*-valued random processes* $\{e_k\}_{k\geq 1}$, $\{\nu_k\}_{k\geq 1}$, *and* $\{\varsigma_k\}_{k\geq 1}$ *defined on a probability space* $(\Omega, \mathcal{F}, \mathcal{P})$ *such that*

(i) $\epsilon_k = e_k + \nu_k + \varsigma_k$, *where* $\epsilon_k = H(\theta_{k-1}, X_k) - h(\theta_{k-1}) + \gamma_k^\tau \eta(X_k)$.

(ii) $\{e_k\}$ *is a martingale difference sequence, and* $\sum_{k=1}^n e_k/\sqrt{n} \longrightarrow N(0, Q)$ *in distribution, where* $Q = \lim_{k\to\infty} E(e_k e_k^T)$.

(iii) $\|\nu_k\| = O(\gamma_k^{(1+\zeta)/2})$, *where* ζ *is given in condition* (B_1').

(iv) $\|\sum_{k=0}^n \gamma_k \varsigma_k\| = O(\gamma_n)$.

Proof:

(i) Let $\epsilon_0 = \nu_0 = \varsigma_0 = 0$, and

$$\begin{aligned}
e_{k+1} &= u(\theta_k, x_{k+1}) - P_{\theta_k} u(\theta_k, x_k),\\
\nu_{k+1} &= P_{\theta_{k+1}} u(\theta_{k+1}, x_{k+1}) - P_{\theta_k} u(\theta_k, x_{k+1})\\
&\quad + \frac{\gamma_{k+2} - \gamma_{k+1}}{\gamma_{k+1}} P_{\theta_{k+1}} u(\theta_{k+1}, x_{k+1}) + \gamma_k^\tau \eta(X_k),\\
\tilde{\varsigma}_{k+1} &= \gamma_{k+1} P_{\theta_k} u(\theta_k, x_k),\\
\varsigma_{k+1} &= \frac{1}{\gamma_{k+1}} (\tilde{\varsigma}_{k+1} - \tilde{\varsigma}_{k+2}).
\end{aligned} \tag{7.137}$$

Following from the Poisson equation given in (C_2), it is easy to verify that (i) holds.

(ii) By (7.137), we have

$$E(e_{k+1}|\mathcal{F}_k) = E(u(\theta_k, x_{k+1})|\mathcal{F}_k) - P_{\theta_k} u(\theta_k, x_k) = 0,$$

where $\{\mathcal{F}_k\}_{k\geq 0}$ is a family of σ-algebras satisfying $\sigma\{\theta_0\} \subseteq \mathcal{F}_0$ and $\sigma\{\theta_1, e_1, \nu_1, \varsigma_1, \ldots, \theta_k, e_k, \nu_k, \varsigma_k\} \subseteq \mathcal{F}_k \subseteq \mathcal{F}_{k+1}$ for all $k \geq 0$. Hence, $\{e_k\}$ forms a martingale difference sequence.

Since Θ is compact, by (7.74) and the condition $\sup_{\mathcal{X}} V(x) < \infty$, $\|e_k\|$ is uniformly bounded with respect to k; that is, there exists a constant c such that

$$\|e_k\| < c, \quad \forall k \geq 0. \tag{7.138}$$

Hence, the martingale $s_n = \sum_{k=1}^n e_k$ is square integrable for all n.

By (7.137),

$$\begin{aligned}
E(e_{k+1}e_{k+1}^T|\mathcal{F}_k) =& E\left[u(\theta_k, x_{k+1})u(\theta_k, x_{k+1})^T|\mathcal{F}_k\right]\\
&- P_{\theta_k} u(\theta_k, x_k) P_{\theta_k} u(\theta_k, x_k)^T \triangleq l(\theta_k, x_k).
\end{aligned} \tag{7.139}$$

By (7.74), $\|l(\theta_k, x_k)\|$ is uniformly bounded with respect to k. This further implies that $\text{Var}(l(\theta_k, x_k))$ is uniformly bounded with respect to k, and $\sum_{j=1}^{k} \text{Var}\,(l(\theta_j, x_j))\,(\log j)^2/j^2 \rightarrow 0$ as $k \rightarrow 0$. Since $\{E(e_{k+1}e_{k+1}^T|\mathcal{F}_k) - E(e_{k+1}e_{k+1}^T)\}$forms a martingale difference sequence, the correlation coefficient $\text{Corr}\Big(l(\theta_i, x_i), l(\theta_j, x_j)\Big) = 0$ for all $i \neq j$. Following from the generalized law of large numbers (Serfling, 1970), we have, as $n \rightarrow \infty$,

$$\frac{1}{n}\sum_{k=1}^{n} l(\theta_k, x_k) \rightarrow \frac{1}{n}\sum_{k=1}^{n} El(\theta_k, x_k), \quad a.s. \tag{7.140}$$

Now we show that $El(\theta_k, x_k)$ also converges. By the continuum of Θ and the convergence of θ_k, $l(\theta_k, x)$ converges to $l(\theta_*, x)$ for any sample $x \in \mathcal{X}$. Since $l(\theta_k, x)$ is uniformly bounded with respect to k, it follows from Lebesgue's dominated convergence theorem that $El(\theta_k, x)$converges to $El(\theta_*, x)$. Putting together with (7.140), we have

$$\frac{1}{n}\sum_{k=1}^{n} l(\theta_k, x_k) \rightarrow El(\theta_*, x) = \lim_{k\rightarrow\infty} Ee_k e_k^T, \quad a.s. \tag{7.141}$$

Since $\|e_k\|$ is uniformly bounded with respect to k, the Lindeberg condition is satisfied, that is,

$$\sum_{i=1}^{n} E\left[\frac{\|e_i\|^2}{n} I_{(\frac{\|e_i\|}{\sqrt{n}} \geq \epsilon)} |\mathcal{F}_{i-1}\right] \rightarrow 0, \quad \text{as } n \rightarrow \infty.$$

Following from Lemma 7.8.1, we have $\sum_{i=1}^{n} e_i/\sqrt{n} \rightarrow N(0, Q)$ by identifying $e_i/\sqrt{n}$ to v_{ni}, n to k_n, and $\mathcal{F}_i$ to $\mathcal{G}_{ni}$.

(iii) By condition (B_1'), we have

$$\frac{\gamma_{k+2} - \gamma_{k+1}}{\gamma_{k+1}} = o(\gamma_{k+2}).$$

By (7.137) and (7.74), there exist constants c_1 and c_2 such that

$$\begin{aligned} \|\nu_{k+1}\| &\leq c_1 \left[\|\theta_{k+1} - \theta_k\|^{(1+\zeta)/2} + \gamma_{k+2}^{(1+\zeta)/2}\right] \\ &= c_1 \left[\|\gamma_{k+1} H(\theta_k, x_{k+1})\|^{(1+\zeta)/2} + \gamma_{k+2}^{(1+\zeta)/2}\right] \leq c_2 \gamma_{k+1}^{(1+\zeta)/2}, \end{aligned} \tag{7.142}$$

where ζ is given in (B_1'). Therefore, (iii) holds.

(iv) A straightforward calculation shows that

$$\sum_{i=0}^{k} \gamma_i \varsigma_i = \tilde{\varsigma}_{k+1} = \gamma_{k+1} P_{\theta_k} u(\theta_k, x_k),$$

By (7.74), (iv) holds.

This completes the proof of the lemma. ∎

To facilitate the theoretical analysis for the random process $\{\theta_k\}$, we define a reduced random process $\{\tilde{\theta}_k\}$ by

$$\tilde{\theta}_k = \theta_k + \tilde{\varsigma}_{k+1}, \quad k \geq 1. \tag{7.143}$$

For convenience, we also define

$$\tilde{\epsilon}_k = e_k + \nu_k, \quad k \geq 1. \tag{7.144}$$

It is easy to verify

$$\tilde{\theta}_{k+1} - \theta_* = (I + \gamma_{k+1}F)(\tilde{\theta}_k - \theta_*) + \gamma_{k+1}\left(h(\theta_k) - F(\tilde{\theta}_k - \theta_*)\right) + \gamma_{k+1}\tilde{\epsilon}_{k+1}, \tag{7.145}$$

which implies, for any $k_0 > 1$,

$$\begin{aligned}\tilde{\theta}_{k+1} - \theta_* = \Phi_{k,k_0}(\tilde{\theta}_{k_0} - \theta_*) + \sum_{j=k_0}^{k} \Phi_{k,j+1}\gamma_{j+1}\tilde{\epsilon}_{j+1} \\ + \sum_{j=k_0}^{k} \Phi_{k,j+1}\gamma_{j+1}\left(h(\theta_j) - F(\tilde{\theta}_j - \theta_*)\right),\end{aligned} \tag{7.146}$$

where $\Phi_{k,j} = \prod_{i=j}^{k}(I + \gamma_{i+1}F)$ if $k \geq j$, and $\Phi_{j,j+1} = I$, and I denotes the identity matrix.

For the δ specified in (B_4') and a deterministic integer k_0, define the stopping time $\mu = \min\{j : j > k_0, \|\theta_j - \theta_*\| \geq \delta\}$ and 0 if $\|\theta_{k_0} - \theta_*\| \geq \delta$. Define

$$A = \{i : k_0 \leq i < \mu\}, \tag{7.147}$$

and let $I_A(k)$ denote the indicator function; $I_A(k) = 1$ if $k \in A$ and 0 otherwise. Therefore, for all $k \geq k_0$,

$$\begin{aligned}&(\tilde{\theta}_{k+1} - \theta_*)I_A(k+1) \\ &= \Phi_{k,k_0}(\tilde{\theta}_{k_0} - \theta_*)I_A(k+1) + \left[\sum_{j=k_0}^{k} \Phi_{k,j+1}\gamma_{j+1}\tilde{\epsilon}_{j+1}I_A(j)\right] \\ &\quad \times I_A(k+1) + \left[\sum_{j=k_0}^{k} \Phi_{k,j+1}\gamma_{j+1}\left(h(\theta_j) - F(\tilde{\theta}_j - \theta_*)\right) I_A(j)\right] I_A(k+1).\end{aligned} \tag{7.148}$$

Lemma 7.8.3

(i) The following estimate takes place

$$\frac{\gamma_j}{\gamma_k} \leq \exp\left(o(1)\sum_{i=j}^{k}\gamma_i\right), \quad \forall k \geq j, \quad \forall j \geq 1, \tag{7.149}$$

where $o(1)$ denotes a magnitude that tends to zero as $j \to \infty$.

(ii) Let c be a positive constant, then there exists another constant c_1 such that

$$\sum_{i=1}^{k} \gamma_i^r \exp\left(-c \sum_{j=i+1}^{k} \gamma_j\right) \leq c_1, \qquad \forall k \geq 1, \quad \forall r \geq 1. \tag{7.150}$$

(iii) There exist constants $c_0 > 0$ and $c > 0$ such that

$$\|\Phi_{k,j}\| \leq c_0 \exp\left\{-c \sum_{i=j}^{k} \gamma_{i+1}\right\}, \qquad \forall k \geq j, \quad \forall j \geq 0. \tag{7.151}$$

(iv) Let $G_{k,j} = \sum_{i=j}^{k}(\gamma_j - \gamma_{i+1})\Phi_{i-1,j} + F^{-1}\Phi_{k,j}$. Then G_{kj} is uniformly bounded with respect to both k and j for $1 \leq j \leq k$, and

$$\frac{1}{k}\sum_{j=1}^{k} \|G_{k,j}\| \longrightarrow 0, \quad \text{as } k \to \infty. \tag{7.152}$$

Proof: Parts (i) and (iv) are a restatement of Lemma 3.4.1 of Chen (2002). The proof of part (ii) can be found in the proof of Lemma 3.3.2 of Chen (2002). The proof of part (iii) can be found in the proof of Lemma 3.1.1 of Chen (2002). ∎

Lemma 7.8.4 *Assume that Θ is compact, the conditions (B_1'), (B_2), (B_3) and (B_4') hold, $\tau \geq (1+\zeta)/2$ with ζ being defined in (B_1'), and $\sup_{x\in\mathcal{X}} V(x) < \infty$. Then*

$$\frac{1}{\gamma_{k+1}} E\|(\theta_{k+1} - \theta_*)I_A(k+1)\|^2$$

is uniformly bounded with respect to k, where the set A is as defined in (7.147).

Proof: By (7.143), we have

$$\begin{aligned}\frac{1}{\gamma_{k+1}}\|\theta_{k+1} - \theta_*\|^2 &= \frac{1}{\gamma_{k+1}}\|\tilde{\theta}_{k+1} - \theta_* - \varsigma_{k+2}\|^2 \\ &\leq \frac{2}{\gamma_{k+1}}\|\tilde{\theta}_{k+1} - \theta_*\|^2 + 2\frac{\gamma_{k+2}^2}{\gamma_{k+1}}\|P_{\theta_{k+1}}u(\theta_{k+1}, x_{k+1})\|^2.\end{aligned}$$

By (C_2), $\|P_{\theta_{k+1}}u(\theta_{k+1}, x_{k+1})\|^2$ is uniformly bounded with respect to k. Hence, to prove the lemma, it suffices to prove that $\frac{1}{\gamma_{k+1}}E\|(\tilde{\theta}_{k+1} - \theta_*)I_A(k+1)\|^2$ is uniformly bounded with respect to k.

By (7.143), (B_4') and (C_2), there exist constants c_1 and c_2 such that

$$\begin{aligned}\|h(\theta_j) - F(\tilde{\theta}_j - \theta_*)\|I_A(j) &= \|h(\theta_j) - F(\theta_j - \theta_*) - F\varsigma_{j+1}\|I_A(j) \\ &\leq \|h(\theta_j) - F(\theta_j - \theta_*)\|I_A(j) + c_2\gamma_{j+1} \\ &\leq c_1\|\theta_j - \theta_*\|^{1+\rho} + c_2\gamma_{j+1}.\end{aligned} \tag{7.153}$$

Similarly, there exists a constant $\tilde{\delta}$ such that

$$\begin{aligned}\|\tilde{\theta}_{k_0} - \theta_*\| I_A(k_0) &= \|\theta_{k_0} - \theta_* + \tilde{\varsigma}_{k_0+1}\| I_A(k_0) \\ &\leq \|\theta_{k_0} - \theta_*\| I_A(k_0) + \|\tilde{\varsigma}_{k_0+1}\| \leq \tilde{\delta}. \end{aligned} \tag{7.154}$$

By (7.148), (7.151) and (7.153), we have

$$\begin{aligned}
\frac{1}{\gamma_{k+1}} E\|(\tilde{\theta}_{k+1} - \theta_*) I_A(k+1)\|^2 &\leq \frac{5c_0\tilde{\delta}^2}{\gamma_{k+1}} \exp\left(-2c\sum_{i=k_0}^{k} \gamma_{i+1}\right) \\
&+ \frac{5c_0^2}{\gamma_{k+1}} \sum_{i=k_0}^{k}\sum_{j=k_0}^{k} \left[\exp\left(-c\sum_{s=j+1}^{k}\gamma_{s+1}\right)\gamma_{j+1}\exp\left(-c\sum_{s=i+1}^{k}\gamma_{s+1}\right)\right. \\
&\left.\times \gamma_{i+1}\|Ee_{i+1}e_{j+1}^T\|\right] \\
&+ \frac{5c_0^2}{\gamma_{k+1}} \sum_{i=k_0}^{k}\sum_{j=k_0}^{k} \left[\exp\left(-c\sum_{s=j+1}^{k}\gamma_{s+1}\right)\gamma_{j+1}\exp\left(-c\sum_{s=i+1}^{k}\gamma_{s+1}\right)\right. \\
&\left.\times \gamma_{i+1}\|E\nu_{i+1}\nu_{j+1}^T\|\right] \\
&+ \frac{5c_0^2c_2^2}{\gamma_{k+1}} \left[\sum_{j=k_0}^{k}\exp\left(-c\sum_{s=j+1}^{k}\gamma_{s+1}\right)\gamma_{j+1}^2\right]^2 \\
&+ \frac{5c_0^2c_1^2}{\gamma_{k+1}} E\left[\sum_{j=k_0}^{k}\exp\left(-c\sum_{s=j+1}^{k}\gamma_{s+1}\right)\gamma_{j+1}\|\theta_j - \theta_*\|^{1+\rho} I_A(j)\right]^2 \\
&\triangleq I_1 + I_2 + I_3 + I_4 + I_5
\end{aligned}$$

By (7.149), there exists a constant c_3 such that

$$\|I_1\| \leq \frac{5c_0c_3\tilde{\delta}^2}{\gamma_{k_0}} \exp\left(o(1)\sum_{i=k_0}^{k+1}\gamma_i\right)\exp\left(-2c\sum_{i=k_0}^{k}\gamma_{i+1}\right),$$

where $o(1) \to 0$ as $k_0 \to \infty$. This implies that $o(1) - 2c < 0$ if k_0 is large enough. Hence, I_1 is bounded if k_0 is large enough.

By (7.149) and (7.150), for large enough k_0, there exists a constant c_4 such that

$$\sum_{j=k_0}^{k} \frac{\gamma_{j+1}^2}{\gamma_{k+1}} \exp\left(-c\sum_{s=j+1}^{k}\gamma_{s+1}\right) \leq \sum_{j=k_0}^{k}\gamma_{j+1}\exp\left(-\frac{c}{2}\sum_{s=j+1}^{k}\gamma_{s+1}\right) \leq c_4. \tag{7.155}$$

Since $\{e_i\}$ forms a martingale difference sequence,

$$Ee_ie_j^T = E(E(e_i|\mathcal{F}_{i-1})e_j^T) = 0, \quad \forall i > j,$$

which implies that

$$\begin{aligned} I_2 &= \frac{5c_0^2}{\gamma_{k+1}} \sum_{i=k_0}^{k} \left[\gamma_{i+1}^2 \exp\left(-2c \sum_{s=j+1}^{k} \gamma_{s+1} \right) E\|e_{i+1}\|^2 \right] \\ &\leq 5c_0^2 \sup_i E\|e_{i+1}\|^2 \sum_{i=k_0}^{k} \left[\gamma_{i+1}^2 \exp\left(-2c \sum_{s=j+1}^{k} \gamma_{s+1} \right) \right]. \end{aligned}$$

By (7.138) and (7.150), I_2 is uniformly bounded with respect to k.

By (7.149) and Lemma 7.8.2, there exists a constant c_5 such that

$$\begin{aligned} I_3 &= 5c_0^2 \sum_{i=k_0}^{k} \sum_{j=k_0}^{k} \left[\exp\left(-c \sum_{s=j+1}^{k} \gamma_{s+1} \right) \frac{\gamma_{j+1}}{\sqrt{\gamma_{k+1}}} \exp\left(-c \sum_{s=i+1}^{k} \gamma_{s+1} \right) \right. \\ &\left. \times \frac{\gamma_{i+1}}{\sqrt{\gamma_{k+1}}} O(\gamma_{i+1}^{\frac{1+\varsigma}{2}}) O(\gamma_{j+1}^{\frac{1+\varsigma}{2}}) \right] \\ &\leq 5c_0^2 c_5 \sum_{i=k_0}^{k} \sum_{j=k_0}^{k} \left[\exp\left(-\frac{c}{2} \sum_{s=j+1}^{k} \gamma_{s+1} \right) {\gamma_{j+1}}^{\frac{1}{2}} \exp\left(-\frac{c}{2} \sum_{s=i+1}^{k} \gamma_{s+1} \right) \right. \\ &\left. \times {\gamma_{i+1}}^{\frac{1}{2}} \gamma_{i+1}^{\frac{1+\varsigma}{2}} \gamma_{j+1}^{\frac{1+\varsigma}{2}} \right] \\ &= 5c_0^2 c_5 \left\{ \sum_{j=k_0}^{k} \left[{\gamma_{j+1}}^{1+\frac{\varsigma}{2}} \exp\left(-\frac{c}{2} \sum_{s=j+1}^{k} \gamma_{s+1} \right) \right] \right\}^2 \end{aligned}$$

By (7.150), I_3 is uniformly bounded with respect to k.

By (7.150) and (7.155), there exists a constant c_6 such that

$$\begin{aligned} I_4 &= 5c_0^2 c_2^2 \\ &\left[\sum_{j=k_0}^{k} \frac{\gamma_{j+1}^2}{\gamma_{k+1}} \exp\left(-c \sum_{s=j+1}^{k} \gamma_{s+1} \right) \right] \left[\sum_{j=k_0}^{k} \gamma_{j+1}^2 \exp\left(-c \sum_{s=j+1}^{k} \gamma_{s+1} \right) \right] \\ &\leq 5c_0^2 c_2^2 c_6. \end{aligned}$$

Therefore, I_4 is uniformly bounded with respect to k.

The proof for the uniform boundedness of I_5 can be found in the proof of Lemma 3.4.3 of Chen (2002). ■

Lemma 7.8.5 *Assume that* Θ *is compact, the conditions* $(B_1'), (B_2), (B_3)$ *and* (B_4') *hold,* $\tau \geq (1+\zeta)/2$ *with* ζ *being defined in* (B_1')*, and* $\sup_{x\in\mathcal{X}} V(x) < \infty$. *Then, as* $k \to \infty$,

$$\frac{1}{\sqrt{k}}\sum_{i=1}^{k} \|h(\theta_i) - F(\tilde{\theta}_i - \theta_*)\| \longrightarrow 0 \quad a.s.$$

Proof: By (7.143), there exists a constant c such that

$$\frac{1}{\sqrt{k}}\sum_{i=1}^{k} \|h(\theta_i) - F(\tilde{\theta}_i - \theta_*)\| \leq \frac{1}{\sqrt{k}}\sum_{i=1}^{k} \|h(\theta_i) - F(\theta_i - \theta_*)\|$$
$$+\frac{c}{\sqrt{k}}\sum_{i=1}^{k}\gamma_{i+1} \stackrel{\triangle}{=} I_1 + I_2.$$

To prove the lemma, it suffices to prove that I_1 and I_2 both converge to zero as $k \to \infty$.

Condition (B_1') implies $\sum_{i=1}^{\infty}\frac{\gamma_i}{\sqrt{i}} < \infty$, which further implies $I_2 \to 0$ by Kronecker's lemma. The convergence $I_1 \to 0$ can be established as in Chen (2002, Lemma 3.4.4) using the condition (B_4') and Lemma 7.8.4. ∎

Proof of Theorem 7.8.2. By Theorem 7.8.4, θ_k converge to the zero point θ_* almost surely. Consequently, we have

$$\sqrt{k}(\bar{\theta}_k - \theta_*) = \frac{1}{\sqrt{k}}\sum_{i=1}^{k}(\theta_i - \theta_*) = \frac{1}{\sqrt{k}}\sum_{i=1}^{k}(\tilde{\theta}_i - \theta_*) - \frac{1}{\sqrt{k}}\sum_{i=1}^{k}\varsigma_{i+1}. \quad (7.156)$$

Condition (B_1') implies $\sum_{i=1}^{\infty}\frac{\gamma_i}{\sqrt{i}} < \infty$, which further implies $\frac{1}{\sqrt{k}}\sum_{i=1}^{k}\gamma_i \to 0$ by Kronecker's lemma. By (7.74), $\|P_{\theta_i}u(\theta_i, x_i)\|$ is uniformly bounded with respect to i, then

$$\frac{1}{\sqrt{k}}\sum_{i=1}^{k}\|\varsigma_{i+1}\| = \frac{1}{\sqrt{k}}\sum_{i=1}^{k}\gamma_{i+1}\|P_{\theta_i}u(\theta_i, x_i)\| \to 0. \quad (7.157)$$

By (7.146), (7.156) and (7.157), we have

$$\begin{aligned}
\sqrt{k}(\bar{\theta}_k - \theta_*) &= o(1) + \frac{1}{\sqrt{k}} \\
&\sum_{i=k_0}^{k}\Phi_{i-1,k_0}(\tilde{\theta}_{k_0} - \theta_*) + \frac{1}{\sqrt{k}}\sum_{i=k_0}^{k}\sum_{j=k_0}^{i-1}\Phi_{i-1,j+1}\gamma_{j+1}\tilde{\epsilon}_{j+1} \\
&+ \frac{1}{\sqrt{k}}\sum_{i=k_0}^{k}\sum_{j=k_0}^{i-1}\Phi_{i-1,j+1}\gamma_{j+1}\left(h(\theta_j) - F(\tilde{\theta}_j - \theta_*)\right) \\
&\stackrel{\triangle}{=} o(1) + I_1 + I_2 + I3.
\end{aligned} \quad (7.158)$$

By noticing that $\Phi_{k,j} = \Phi_{k-1,j} + \gamma_{k+1}F\Phi_{k-1,j}$, we have

$$\Phi_{k,j} = I + \sum_{i=j}^{k} \gamma_{i+1}F\Phi_{i-1,j}, \quad \text{and} \quad F^{-1}\Phi_{k,j} = F^{-1} + \sum_{i=j}^{k} \gamma_{i+1}\Phi_{i-1,j},$$

and thus

$$\gamma_j \sum_{i=j}^{k} \Phi_{i-1,j} = \sum_{i=j}^{k}(\gamma_j - \gamma_{i+1})\Phi_{i-1,j} + \sum_{i=j}^{k} \gamma_{i+1}\Phi_{i-1,j}.$$

By the definition of $G_{k,j}$ given in Lemma (7.8.3)(iv), we have

$$\gamma_j \sum_{i=j}^{k} \Phi_{i-1,j} = -F^{-1} + G_{k,j}, \tag{7.159}$$

which implies

$$I_1 = \frac{1}{\sqrt{k}\gamma_{k_0}}(-F^{-1} + G_{k,k_0})(\tilde{\theta}_{k_0} - \theta_*).$$

By Lemma 7.8.3, $G_{k,j}$ is bounded. Therefore, $I_1 \to 0$ as $k \to \infty$. The above arguments also imply that there exists a constant $c_0 > 0$ such that

$$\left\| \gamma_{j+1} \sum_{i=j+1}^{k} \Phi_{i-1,j+1} \right\| < c_0, \quad \forall k, \forall j < k. \tag{7.160}$$

By (7.160), we have

$$\begin{aligned}
\|I_3\| &= \frac{1}{\sqrt{k}} \left\| \sum_{j=k_0}^{k} \sum_{i=j+1}^{k} \Phi_{i-1,j+1}\gamma_{j+1} \left(h(\theta_j) - F(\tilde{\theta}_j - \theta_*\right) \right\| \\
&\leq \frac{c_0}{\sqrt{k}} \sum_{j=1}^{k} \|h(\theta_j) - F(\tilde{\theta}_j - \theta_*)\|.
\end{aligned}$$

It then follows from Lemma 7.8.5 that I_3 converges to zero almost surely as $k \to \infty$.

Now we consider I_2. By (7.159),

$$\begin{aligned}
I_2 &= -\frac{F^{-1}}{\sqrt{k}} \sum_{j=k_0}^{k} e_{j+1} + \frac{1}{\sqrt{k}} \sum_{j=k_0}^{k} G_{k,j+1}e_{j+1} + \frac{1}{\sqrt{k}} \sum_{j=k_0}^{k} (-F^{-1} + G_{k,j+1})\nu_{j+1} \\
&\triangleq J_1 + J_2 + J_3.
\end{aligned}$$

Since $\{e_j\}$ is a martingale difference sequence,

$$E(e_i^T G_{k,i}^T G_{k,j} e_j) = E[E(e_i|\mathcal{F}_{i-1})^T G_{k,i}^T G_{k,j} e_j] = 0, \quad \forall i > j,$$

which implies that

$$E\|J_2\|^2 = \frac{1}{k}\sum_{j=k_0}^{k} E\left(e_{j+1}^T G_{k,j+1}^T G_{k,j+1} e_{j+1}\right) \leq \frac{1}{k}\sum_{j=k_0}^{k} \|G_{k,j+1}\|^2 E\|e_{j+1}\|^2.$$

By (7.138), (7.152) and the uniform boundedness of G_{kj}, there exists a constant c_1 such that

$$E\|J_2\|^2 \leq \frac{c_1}{k}\sum_{j=k_0}^{k} \|G_{k,j+1}\| \to 0, \quad \text{as } k \to \infty. \tag{7.161}$$

Therefore, $J_2 \to 0$ in probability.

By Lemma 7.8.2 and condition (B_1'),

$$\sum_{j=1}^{\infty} \frac{1}{\sqrt{j}} E\|\nu_{j+1}\| = \sum_{j=1}^{\infty} \frac{1}{\sqrt{j}} O(\gamma_j^{(1+\varsigma)/2}) < \infty.$$

By Kronecker's lemma, we have

$$\frac{1}{\sqrt{k}}\sum_{j=k_0}^{k} E\|\nu_{j+1}\| \to 0, \quad \text{as } k \to \infty.$$

Since G_{kj} is uniformly bounded with respect to both k and j, there exists a constant c_2 such that

$$E\|J_3\| \leq \frac{c_2}{\sqrt{k}}\sum_{j=k_0}^{k} E\|\nu_{j+1}\| \to 0, \quad \text{as } k \to \infty.$$

Therefore, J_3 converges to zero in probability.

By Lemma 7.8.2, $J_1 \to N(0, S)$ in distribution. This, incorporating with the convergence results of I_1, I_3, J_2 and J_3, concludes the theorem.

Proof of Theorem 7.8.3. Since the order of ς_k is difficult to treat, we consider the following stochastic approximation MCMC algorithm

$$\tilde{\theta}_{k+1} = \tilde{\theta}_k + \gamma_{k+1}\left(h(\theta_k) + \tilde{\epsilon}_{k+1}\right), \tag{7.162}$$

where $\{\tilde{\theta}_k\}$ and $\{\tilde{\epsilon}_k\}$ are as defined in (7.143) and (7.144), respectively. Following from Lemma 7.8.2-(ii), $\{\tilde{\epsilon}_k\}$ forms a sequence of asymptotically unbiased estimator of 0.

Let $\bar{\tilde{\theta}}_n = \sum_{k=1}^{n} \tilde{\theta}_k/n$. To establish that $\bar{\tilde{\theta}}$ is an asymptotically efficient estimator of θ^*, we will first show (in step 1)

$$\sqrt{n}(\bar{\tilde{\theta}} - \theta^*) \to N(\mathbf{0}, \Gamma), \tag{7.163}$$

where $\Gamma = F^{-1}Q(F^{-1})^T, F = \partial h(\theta_*)/\partial\theta$, and $Q = \lim_{k\to\infty} E(e_k e_k^T)$; and then show (in step 2) that the asymptotic covariance matrix of $\sum_{k=1}^n \tilde{\epsilon}_k/\sqrt{n}$ is equal to Q.

Step 1. By (7.144), we have

$$\bar{\bar{\theta}}_n = \bar{\theta}_n + \frac{1}{n}\sum_{k=1}^n \tilde{\varsigma}_k. \tag{7.164}$$

By Lemma 7.7.1, $E\|P_{\theta_{k-1}}u(\theta_{k-1}, x_{k-1})\|$ is uniformly bounded for $k \geq 1$ and thus there exists a constant c such that

$$E\|\frac{1}{\sqrt{n}}\sum_{k=1}^n \tilde{\varsigma}_k\| = E\|\frac{1}{\sqrt{n}}\sum_{k=1}^n \gamma_k P_{\theta_{k-1}}u(\theta_{k-1}, x_{k-1})\| \leq \frac{c}{\sqrt{n}}\sum_{k=1}^n \gamma_k.$$

By Kronecker's lemma and $(B_1'), \frac{1}{\sqrt{n}}\sum_{k=1}^n \gamma_k \to 0$. Hence, $\frac{1}{\sqrt{n}}\sum_{k=1}^n \tilde{\varsigma}_k = o_p(1)$ and

$$\frac{1}{n}\sum_{k=1}^n \tilde{\varsigma}_k = o_p(n^{-1/2}), \tag{7.165}$$

where $o_p(\cdot)$ means

$$Y_k = o_p(Z_k) \text{ if and only if } Y_k/Z_k \to 0 \text{ in probability, as } k \to \infty.$$

Therefore,

$$\bar{\bar{\theta}}_n = \bar{\theta}_n + o_p(n^{-1/2}). \tag{7.166}$$

Following from Theorem 7.8.2 and Slutsky's theorem, (7.163) holds.

Step 2. Now we show the asymptotic covariance matrix of $\sum_{k=1}^n \tilde{\epsilon}_k/\sqrt{n}$ is equal to Q. Consider

$$\begin{aligned}
&E\left(\frac{1}{\sqrt{n}}\sum_{k=1}^n \tilde{\epsilon}_k\right)\left(\frac{1}{\sqrt{n}}\sum_{k=1}^n \tilde{\epsilon}_k\right)^T - \frac{1}{n}\left(\sum_{k=1}^n E(\tilde{\epsilon}_k)\right)\left(\sum_{k=1}^n E(\tilde{\epsilon}_k)\right)^T \\
&= \frac{1}{n}\sum_{k=1}^n E(\tilde{\epsilon}_k\tilde{\epsilon}_k^T) + \frac{1}{n}\sum\sum_{i\neq j} E(\tilde{\epsilon}_i\tilde{\epsilon}_j^T) - \frac{1}{n}\left[\sum_{k=1}^n E(\tilde{\epsilon}_k)\right]\left[\sum_{k=1}^n E(\tilde{\epsilon}_k)\right]^T \\
&= (I_1) + (I_2) + (I_3)
\end{aligned}$$

By Kronecker's lemma and (B_1'),

$$\frac{1}{\sqrt{n}}\sum_{k=1}^n \gamma_k^{(1+\zeta)/2} \to 0, \quad \text{as } n \to \infty, \tag{7.167}$$

where $\zeta \in (0,1)$ is defined in (B_1'). which, by Lemma 7.8.2-(iii), further implies

$$\frac{1}{\sqrt{n}} \sum_{k=1}^{n} \|\nu_k\| \to 0, \quad \text{as } n \to \infty. \tag{7.168}$$

By (7.144),

$$\begin{aligned}(I_1) &= \frac{1}{n}\sum_{k=1}^{n} E(e_k e_k^T) + \frac{2}{n}\sum_{k=1}^{n} E(e_k \nu_k^T) + \frac{1}{n}\sum_{k=1}^{n} E(\nu_k \nu_k^T) \\ &= (J_1) + (J_2) + (J_3).\end{aligned}$$

By Lemma 7.8.2, $\|\nu_k \nu_k^T\| = O(\gamma_k^{1+\zeta})$. Hence,

$$\frac{1}{n}\sum_{k=1}^{n} \|\nu_k \nu_k^T\| = \frac{1}{\sqrt{n}}\frac{1}{\sqrt{n}}\sum_{k=1}^{n} O(\gamma_k^{1+\zeta}),$$

which, by (7.167), implies $J_3 \to 0$ as $n \to \infty$.

Following from Lemma 7.7.1, $\{\|e_k\|\}$ is uniformly bounded with respect to k. Therefore, there exists a constant c such that

$$\frac{2}{n}\sum_{k=1}^{n} \|e_k \nu_k^T\| \le \frac{c}{n}\sum_{k=1}^{n} \|\nu_k\| \to 0,$$

where the limit follows (7.168). Hence, $J_2 \to 0$ as $n \to \infty$.

By (7.139), $E(e_{k+1}e_{k+1}^T) = El(\theta_k, x_k)$. Since $l(\theta, x)$ is continuous in θ, it follows from the convergence of θ_k that $l(\theta_k, x)$ converges to $l(\theta^*, x)$ a.s. for any $x \in \mathcal{X}$. Furthermore, following from Lemma 7.7.1 and Lebesgue's dominated convergence theorem, we conclude that $El(\theta_k, x_k)$ converges to $El(\theta^*, x)$, and thus

$$J_1 \to El(\theta^*, x) = \lim_{k\to\infty} E(e_k e_k^T) = Q.$$

Summarizing the convergence results of J_1, J_2 and J_3, we conclude that $(I_1) \to Q$ as $n \to \infty$.

By (7.144), for $i \neq j$, we have

$$\begin{aligned}E(\tilde{\epsilon}_i \tilde{\epsilon}_j^T) &= E\left\{(e_i + \nu_i)(e_j + \nu_j)^T\right\} = E\left(e_i e_j^T + \nu_i \nu_j^T + e_i \nu_j^T + \nu_i e_j^T\right) \\ &= E(\nu_i \nu_j^T),\end{aligned} \tag{7.169}$$

where the last equality follows from the result that $\{e_k\}$ is a martingale difference sequence [Lemma 7.8.2-(ii)]. By (7.142), there exists a constant c such that

$$\|\nu_i \nu_j^T\| \leq c\gamma_i^{\frac{1+\varsigma}{2}} \gamma_j^{\frac{1+\varsigma}{2}},$$

which implies

$$\left\| \frac{1}{n} \sum\sum_{i \neq j} \nu_i \nu_j^T \right\| \leq c \left[\frac{1}{\sqrt{n}} \sum_{i=1}^{n} \gamma_i^{\frac{1+\varsigma}{2}} \right] \left[\frac{1}{\sqrt{n}} \sum_{j=1}^{n} \gamma_j^{\frac{1+\varsigma}{2}} \right]. \tag{7.170}$$

By (7.167), we have

$$\frac{1}{n} \sum\sum_{i \neq j} E(\nu_i \nu_j^T) \to 0, \quad \text{as } n \to \infty. \tag{7.171}$$

In summary of (7.169) and (7.171),

$$(I_2) = \frac{1}{n} \sum\sum_{i \neq j} E(\tilde{\epsilon}_i \tilde{\epsilon}_j^T) \to 0, \quad \text{as } n \to \infty. \tag{7.172}$$

By Lemma 7.8.2 (i) and (ii), where it is shown that $\{e_k\}$ is a martingale difference sequence, we have

$$\begin{aligned}
(I_3) &= \frac{1}{n} \left[\sum_{k=1}^{n} E(e_k + \nu_k) \right] \left[\sum_{k=1}^{n} E(e_k + \nu_k) \right]^T \\
&= \left[\frac{1}{\sqrt{n}} \sum_{k=1}^{n} E(\nu_k) \right] \left[\frac{1}{\sqrt{n}} \sum_{k=1}^{n} E(\nu_k) \right]^T.
\end{aligned}$$

which implies $(I_3) \to 0$ by (7.168).

Summarizing the convergence results of $(I_1), (I_2)$ and (I_3), the asymptotic covariance matrix of $\sum_{k=1}^{n} \tilde{\epsilon}_k / \sqrt{n}$ is equal to Q. Combining with (7.163), we conclude that $\bar{\bar{\theta}}_k$ is an asymptotically efficient estimator of θ^*.

Since $\bar{\bar{\theta}}_k$ and $\bar{\theta}_k$ have the same asymptotic distribution $N(\mathbf{0}, \Gamma)$, $\bar{\theta}_k$ is also asymptotically efficient as an estimator of θ^* This concludes the proof of Theorem 7.8.3.

Proofs of Theorems 7.8.4 and 7.8.5. These theorems can be proved using Theorems 7.8.1 and 7.8.2 by verifying that the modified SAMC algorithm satisfies the conditions (B_2), (B_3) and (B_4'). The condition (B_2) can be verified as in Section 7.7.1.3. The condition (B_3) and (B_4') can be verified as follows.

- *Condition* (B_3). Since in the modified SAMC algorithm, $\theta_k^{(m)}$ is restricted to a constant, we redefine $\theta_k = (\theta_k^{(1)}, \ldots, \theta_k^{(m-1)})$ in this proof. To simplify notations, we also drop the subscript k, denoting θ_k by θ and x_k by x.

Since the invariant distribution of the MH kernel is $p_\theta(\boldsymbol{x})$, we have, for $i = 1, \ldots, m-1$,

$$\begin{aligned} E(I_{\{x \in E_i\}} - \pi_i) &= \int_{\mathcal{X}} (I_{\{x \in E_i\}} - \pi_i) p_\theta(x) dx \\ &= \frac{\int_{E_i} \psi(x)dx / e^{\theta^{(i)}}}{\sum_{j=1}^m [\int_{E_j} \psi(x)dx / e^{\theta^{(j)}}]} - \pi_i = \frac{S_i}{S} - \pi_i, \end{aligned} \tag{7.173}$$

where $S_i = \int_{E_i} \psi(x)dx / e^{\theta^{(i)}}$ and $S = \sum_{i=1}^{m-1} S_i + \int_{E_m} \psi(x)dx$. Thus,

$$h(\theta) = \int_{\mathcal{X}} H(\theta, x) p(dx) = \left(\frac{S_1}{S} - \pi_1, \ldots, \frac{S_{m-1}}{S} - \pi_{m-1} \right)^T.$$

It follows from (7.173) that $h(\theta)$ is a continuous function of θ. Let $\Lambda(\theta) = 1 - \frac{1}{2}\sum_{j=1}^{m-1}(\frac{S_j}{S} - \pi_j)^2$ and let $v(\theta) = -\log(\Lambda(\theta))$. As shown below, $v(\theta)$ has continuous partial derivatives of the first order. Since $0 \le \frac{1}{2}\sum_{j=1}^{m-1}(\frac{S_j}{S} - \pi_j)^2 < \frac{1}{2}[\sum_{j=1}^{m-1}(\frac{S_j}{S})^2 + \pi_j^2)] \le 1$ for all $\theta \in \Theta$, $v(\theta)$ takes values in the interval $[0, \infty)$.

Solving the system of equations formed by (7.173), we have the single solution $\theta_* = (\theta_*^{(1)}, \ldots, \theta_*^{(m-1)})^T$, where

$$\theta_*^{(i)} = c + \log(\int_{E_i} \psi(\boldsymbol{x}) d\boldsymbol{x}) - \log(\pi_i), i = 1, \ldots, m-1,$$

and $c = -\log(\int_{E_m} \psi(\boldsymbol{x}) d\boldsymbol{x}) + \log(\pi_m)$. It is obvious $v(\theta_*) = 0$.

To verify (B_3), we have the following calculations:

$$\begin{aligned} &\frac{\partial S}{\partial \theta^{(i)}} = \frac{\partial S_i}{\partial \theta^{(i)}} = -S_i, & &\frac{\partial S_i}{\partial \theta^{(j)}} = \frac{\partial S_j}{\partial \theta^{(i)}} = 0, \\ &\frac{\partial (\frac{S_i}{S})}{\partial \theta^{(i)}} = -\frac{S_i}{S}(1 - \frac{S_i}{S}), & &\frac{\partial (\frac{S_i}{S})}{\partial \theta^{(j)}} = \frac{\partial (\frac{S_j}{S})}{\partial \theta^{(j)}} = \frac{S_i S_j}{S^2}, \end{aligned} \tag{7.174}$$

for $i, j = 1, \ldots, m-1$ and $i \neq j$. Let $b = \sum_{j=1}^{m-1} S_j / S$, then we have

$$\begin{aligned} \frac{\partial v(\theta)}{\partial \theta^{(j)}} &= \frac{1}{2\Lambda(\theta)} \sum_{j=1}^{m-1} \frac{\partial (\frac{S_j}{S} - \pi_j)^2}{\partial \theta^{(j)}} \\ &= \frac{1}{\Lambda(\theta)} \left[\sum_{j \neq i} \left(\frac{S_j}{S} - \pi_j \right) \frac{S_i S_j}{S^2} - \left(\frac{S_i}{S} - \pi_i \right) \frac{S_i}{S} \left(1 - \frac{S_i}{S} \right) \right] \\ &= \frac{1}{\Lambda(\theta)} \left[\sum_{j=1}^{m-1} \left(\frac{S_j}{S} - \pi_j \right) \frac{S_i S_j}{S^2} - \left(\frac{S_i}{S} - \pi_i \right) \frac{S_i}{S} \right] \\ &= \frac{1}{\Lambda(\theta)} \left[b \mu_\xi \frac{S_i}{S} - \left(\frac{S_i}{S} - \pi_i \right) \frac{S_i}{S} \right], \end{aligned} \tag{7.175}$$

for $i = 1, \ldots, m-1$, where it is defined $\mu_\xi = \sum_{j=1}^{m-1}(\frac{S_j}{S} - \pi_j)\frac{S_j}{bS}$. Thus,

$$\begin{aligned}\langle \nabla v(\theta), h(\theta)\rangle &= \frac{1}{\Lambda(\theta)}\left[b^2\mu_\xi \sum_{i=1}^{m-1}\left(\frac{S_i}{S} - \pi_i\right)\frac{S_i}{bS} - b\sum_{i=1}^{m-1}\left(\frac{S_i}{S} - \pi_i\right)^2\frac{S_i}{bS}\right] \\ &= -\frac{1}{\Lambda(\theta)}\left[b\sum_{i=1}^{m}\left(\frac{S_i}{S} - \pi_i\right)^2\frac{S_i}{bS} - b^2\mu_\xi^2\right] \\ &= -\frac{1}{\Lambda(\theta)}\left(b\sigma_\xi^2 + b(1-b)\mu_\xi^2\right) \le 0,\end{aligned} \tag{7.176}$$

where σ_ξ^2 denotes the variance of the discrete distribution defined in Table 7.12.

If $\theta = \theta_*$, $\langle \nabla v(\theta), h(\theta)\rangle = 0$; otherwise, $\langle \nabla v(\theta), h(\theta)\rangle < 0$. Condition (B_3) is verified.

- *Condition* (B_4'). To verify this condition, we first show that $h(\theta)$ has bounded second derivatives. Continuing the calculation in (7.174), we have

$$\frac{\partial^2(\frac{S_i}{S})}{\partial(\theta^{(i)})^2} = \frac{S_i}{S}\left(1 - \frac{S_i}{S}\right)\left(1 - \frac{2S_i}{S}\right), \quad \frac{\partial^2(\frac{S_i}{S})}{\partial\theta^{(j)}\partial\theta^{(i)}} = -\frac{S_iS_j}{S^2}\left(1 - \frac{2S_i}{S}\right),$$

which implies that the second derivative of $h(\theta)$ is uniformly bounded by noting the inequality $0 < \frac{S_i}{S} < 1$.

Let $F = \partial h(\theta)/\partial\theta$. By (7.174), we have

$$F = \begin{pmatrix} -\frac{S_1}{S}(1 - \frac{S_1}{S}) & \frac{S_1S_2}{S^2} & \cdots & \frac{S_1S_{m-1}}{S^2} \\ \frac{S_2S_1}{S^2} & -\frac{S_2}{S}(1 - \frac{S_2}{S}) & \cdots & \frac{S_2S_{m-1}}{S^2} \\ \vdots & \ddots & \vdots & \vdots \\ \frac{S_{m-1}S_1}{S^2} & \cdots & \cdots & -\frac{S_{m-1}}{S}(1 - \frac{S_{m-1}}{S}) \end{pmatrix}.$$

Table 7.12 Discrete distribution of (ξ).

state (ξ)	$\frac{S_1}{S} - \pi_1$	$\cdots$	$\frac{S_{m-1}}{S} - \pi_{m-1}$
Prob.	$\frac{S_1}{bS}$	$\cdots$	$\frac{S_{m-1}}{bS}$

Table 7.13 Discrete distribution of Z.

state (Z)	z_1	$\cdots$	z_{m-1}
Prob.	$\frac{S_1}{bS}$	$\cdots$	$\frac{S_{m-1}}{bS}$

Thus, for any nonzero vector $\boldsymbol{z} = (z_1, \ldots, z_{m-1})^T$,

$$\begin{aligned}
\boldsymbol{z}^T F \boldsymbol{z} &= -\left[\sum_{i=1}^{m-1} z_i^2 \frac{S_i}{S} - \left(\sum_{i=1}^{m-1} z_i \frac{S_i}{S}\right)^2\right] \\
&= -b\left[\sum_{i=1}^{m-1} z_i^2 \frac{S_i}{bS} - \left(\sum_{i=1}^{m-1} z_i \frac{S_i}{bS}\right)^2\right] - b(1-b)\left(\sum_{i=1}^{m-1} z_i \frac{S_i}{bS}\right)^2 \\
&= -b\mathrm{Var}(Z) - b(1-b)\left(E(Z)\right)^2 < 0,
\end{aligned} \tag{7.177}$$

where $E(Z)$ and $\mathrm{Var}(Z)$ denote, respectively, the mean and variance of the discrete distribution defined by the Table 7.13.

This implies that the matrix F is negative definite and thus stable. Applying Taylor expansion to $h(\theta)$ at the point θ_*, we have

$$\|h(\theta) - F(\theta - \theta_*)\| \leq c\|\theta - \theta_*\|^{1+\rho},$$

for some constants $\rho \in (\zeta, 1]$ and $c > 0$, by noting that $h(\theta_*) = 0$ and that the second derivatives of $h(\theta)$ are uniformly bounded with respect to θ. Therefore, (B_4') is satisfied.

Exercises

7.1 Compare multicanonical Monte Carlo, the Wang-Landau algorithm, and SAMC for the 10-state problem described in Example 7.1.

7.2 Use SAMC to estimate the mean and covariance matrix of the following distribution:

$$\begin{aligned}
f(x) =& \frac{1}{3}N\left[\begin{pmatrix}-8\\-8\end{pmatrix}, \begin{pmatrix}1 & 0.9\\0.9 & 1\end{pmatrix}\right] + \frac{1}{3}N\left[\begin{pmatrix}6\\6\end{pmatrix}, \begin{pmatrix}1 & -0.9\\-0.9 & 1\end{pmatrix}\right] \\
&+ \frac{1}{3}N\left[\begin{pmatrix}0\\0\end{pmatrix}, \begin{pmatrix}1 & 0\\0 & 1\end{pmatrix}\right].
\end{aligned}$$

7.3 Implement SAMC for evaluating p-values of the two-sample t-test.

7.4 Implement the SAMC trajectory averaging estimator for the change-point problem described in Section 7.6.1, and compare it with the SSAMC and MSAMC estimators.

7.5 Implement CSAMC for the mixture Gaussian distribution specified in (7.56).

7.6 Consider the Ising model with the likelihood function

$$f(\boldsymbol{x}|\theta) = \exp\left(\theta \sum_{i \sim j} x_i x_j\right) / Z(\theta),$$

where $x_i = \pm 1$ and $i \sim j$ denotes the nearest neighbors on the lattice, and $Z(\theta)$ is an intractable normalizing constant. Estimate the function $Z(\theta), -1 \leq \theta \leq 1$, using CSAMC by viewing $Z(\theta)$ as a marginal density of $g(\boldsymbol{x}, \theta) \propto \exp(\theta \sum_{i \sim j} x_i x_j)$.

7.7 Implement ASAMC for the multimodal function $H(x)$ given in Example 7.2.

7.8 Implement AESAMC for the test functions listed in Appendix 7A.

7.9 Prove the convergence of the generalized $1/K$-ensemble sampling algorithm.

7.10 Prove the convergence of the generalized Wang-Landau algorithm.

7.11 Prove the convergence of the AESAMC algorithm.

7.12 Prove Theorem 7.7.3.

7.13 Prove the asymptotic normality of $\{\theta_t\}$ for the SAMCMC algorithm described in Section 7.7.1.1, assuming appropriate conditions.

Appendix 7A: Test Functions for Global Optimization

This appendix contains the description of 40 test functions used in Section 7.6.3.2. (Recompiled from Laguna and Martí, 2005; Hedar and Fukushima, 2006; and Hirsch *et al.*, 2006.)

1. *Branin function*

   Definition: $f(x) = \left(x_2 - (\frac{5}{4\pi^2})x_1^2 + \frac{5}{\pi}x_1 - 6\right)^2 + 10(1-\frac{1}{8\pi})\cos(x_1) + 10.$

   Search space: $-5 \leq x_1, x_2 \leq 15.$

   Global minima: $x^* = (-\pi, 12.275), (\pi, 2.275), (9.42478, 2.475)$; $f(x^*) = 0.$

2. *Bohachevsky function*

   Definition: $f(x) = x_1^2 + 2x_2^2 - 0.3\cos(3\pi x_1) - 0.4\cos(4\pi x_2) + 0.7.$

   Search space: $-50 \leq x_1, x_2 \leq 100.$

   Global minimum: $x^* = (0, 0)$; $f(x^*) = 0.$

3. *Easom function*

   Definition: $f(x) = -\cos(x_1)\cos(x_2)\exp(-(x_1 - \pi)^2 - (x_2 - \pi)^2).$

   Search space: $-100 \leq x_1, x_2 \leq 100.$

   Global minimum: $x^* = (\pi, \pi)$; $f(x^*) = -1.$

4. *Goldstein and Price function*

   Definition: $f(x) = [1 + (x_1 + x_2 + 1)^2(19 - 14x_1 + 13x_1^2 - 14x_2 + 6x_1x_2 + 3x_2^2)][(30 + (2x_1 - 3x_2)^2(18 - 32x_1 + 12x_1^2 - 48x_2 - 36x_1x_2 + 27x_2^2)].$

   Search space: $-2 \leq x_1, x_2 \leq 2.$

   Global minimum: $x^* = (0, -1)$; $f(x^*) = 3.$

5. *Shubert function*

   Definition: $f(x) = \left(\sum_{i=1}^5 i\cos((i+1)x_1 + i)\right)\left(\sum_{i=1}^5 i\cos((i+1)x_2 + i)\right).$

   Search space: $-10 \leq x_1, x_2 \leq 10.$

   Global minima: 18 global minima and $f(x^*) = -186.7309.$

6. *Beale function*

   Definition: $f(x) = (1.5 - x_1 + x_1x_2)^2 + (2.25 - x_1 + x_1x_2^2)^2 + (2.625 - x_1 + x_1x_2^3)^2.$

Search space: $-4.5 \leq x_1, x_2 \leq 4.5$.

Global minimum: $x^* = (3, 0.5)$ and $f(x^*) = 0$.

7. *Booth function*

   Definition: $f(x) = (x_1 + 2x_2 - 7)^2 + (2x_1 + x_2 - 5)^2$.

   Search space: $-10 \leq x_1, x_2 \leq 10$.

   Global minimum: $x^* = (1, 3)$ and $f(x^*) = 0$.

8. *Matyas function*

   Definition: $f(x) = 0.26(x_1^2 + x_2^2) - 0.48x_1x_2$.

   Search space: $-5 \leq x_1, x_2 \leq 10$.

   Global minimum: $x^* = (0, 0)$ and $f(x^*) = 0$.

9. *Hump function*

   Definition: $f(x) = 4x_1^2 - 2.1x_1^4 + \frac{1}{3}x_1^6 + x_1x_2 - 4x_2^2 + 4x_2^4$.

   Search space: $-5 \leq x_1, x_2 \leq 5$.

   Global minima: $x^* = (0.0898, -0.7126), (-0.0898, 0.7126)$ and $f(x^*) = 0$.

10. *Schwefel function SC_2*

    Definition: $f_n(x) = 418.9829n - \sum_{i=1}^{n}(x_i \sin\sqrt{|x_i|})$.

    Search space: $-500 \leq x_i \leq 500, i = 1, \ldots, n$.

    Global minimum: $x^* = (1, \ldots, 1)$ and $f_n(x^*) = 0$.

11. *Rosenbrock function R_2*

    Definition: $f_n(x) = \sum_{i=1}^{n-1}[100(x_i^2 - x_{i+1})^2 + (x_i - 1)^2]$.

    Search space: $-5 \leq x_i \leq 10, i = 1, 2, \ldots, n$.

    Global minimum: $x^* = (1, \ldots, 1)$ and $f(x^*) = 0$.

12. *Zakharov function Z_2*

    Definition: $f(x) = \sum_{i=1}^{n} x_i^2 + (\sum_{i=1}^{n} 0.5ix_i)^2 + (\sum_{i=1}^{n} 0.5ix_i)^4$.

    Search space: $-5 \leq x_i \leq 10, i = 1, 2, \ldots, n$.

    Global minimum: $x^* = (0, \ldots, 0)$ and $f(x^*) = 0$.

13. *De Joung function*

    Definition: $f(x) = x_1^2 + x_2^2 + x_3^2$.

Search space: $-2.56 \leq x_i \leq 5.12, i = 1, 2, 3$.

Global minimum: $x^* = (0, 0, 0)$ and $f(x^*) = 0$.

14. *Hartmann function* $H_{3,4}$

Definition: $f_{3,4}(x) = -\sum_{i=1}^{4} \alpha_i \exp(-\sum_{j=1}^{3} A_{ij}(x_j - P_{ij})^2), \alpha = (1, 1.2, 3, 3.2)$.

$$A = \begin{pmatrix} 3.0 & 10 & 30 \\ 0.1 & 10 & 35 \\ 3.0 & 10 & 30 \\ 0.1 & 10 & 35 \end{pmatrix}, \quad P = 10^{-4} \begin{pmatrix} 6890 & 1170 & 2673 \\ 4699 & 4387 & 7470 \\ 1091 & 8732 & 5547 \\ 381 & 5743 & 8828 \end{pmatrix}.$$

Search space: $0 \leq x_i \leq 1, i =, 1, 2, 3$.

Global minimum: $x^* = (0.114614, 0.555649, 0.852547)$ and $f(x^*) = -3.86278$.

15. *Colville function*

Definition: $f(x) = 100(x_1^2 - x_2)^2 + (x_1 - 1)^2 + (x_3 - 1)^2 + 90(x_3^2 - x_4)^2 + 10.1((x_2 - 1)^2 + (x_4 - 1)^2) + 19.8(x_2 - 1)(x_4 - 1)$.

Search space: $-10 \leq x_i \leq 10, i = 1, 2, 3, 4$.

Global minimum: $x^* = (1, 1, 1, 1)$ and $f(x^*) = 0$.

16. *Shekel function* $S_{4,5}$

Definition: $f_{4,m}(x) = -\sum_{j=1}^{m} [\sum_{i=1}^{4} (x_i - C_{ij})^2 + \beta_j]^{-1}, \beta = (0.1, 0.2, 0.2, 0.4, 0.4, 0.6, 0.3, 0.7, 0.5, 0.5)$

$$C = \begin{pmatrix} 4.0 & 1.0 & 8.0 & 6.0 & 3.0 & 2.0 & 5.0 & 8.0 & 6.0 & 7.0 \\ 4.0 & 1.0 & 8.0 & 6.0 & 7.0 & 9.0 & 5.0 & 1.0 & 2.0 & 3.6 \\ 4.0 & 1.0 & 8.0 & 6.0 & 3.0 & 2.0 & 3.0 & 8.0 & 6.0 & 7.0 \\ 4.0 & 1.0 & 8.0 & 6.0 & 7.0 & 9.0 & 3.0 & 1.0 & 2.0 & 3.6 \end{pmatrix}$$

Search space: $0 \leq x_i \leq 10, i = 1, \ldots, 4$.

Global minimum: $x^* = (4, \ldots, 4); f_{4,5}(x^*) = -10.1532, f_{4,7}(x^*) = -10.4029, f_{4,10}(x^*) = -10.5364$.

17. *Shekel function* $S_{4,7}$ (See function 16)

18. *Shekel function* $S_{4,10}$ (See function 16)

19. *Perm function* $P_{4,0.5}$

Definition: $f_{n,\beta}(x) = \sum_{k=1}^{n} [\sum_{i=1}^{n} (i^k + \beta)((x_i/i)^k - 1)]^2$.

Search space: $-n \leq x_i \leq n, i = 1, \ldots, n$.

Global minimum: $x^* = (1, 2, \ldots, n)$ and $f_{n,\beta}(x^*) = 0$.

20. *Perm function* $P^0_{4,0.5}$

Definition: $f^0_{n,\beta}(x) = \sum_{k=1}^n [\sum_{i=1}^n (i+\beta)(x_i^k - (1/i)^k)]^2$.

Search space: $-n \leq x_i \leq n, i = 1, \ldots, n$.

Global minimum: $x^* = (1, \frac{1}{2}, \ldots, \frac{1}{n})$ and $f^0_{n,\beta}(x^*) = 0$.

21. *Power sum function* $PS_{8,18,44,144}$

Definition: $f_{b_1,\ldots,b_n}(x) = \sum_{k=1}^n [(\sum_{i=1}^n x_i^k) - b_k]^2$.

Search space: $0 \leq x_i \leq n, i = 1, \ldots, n$.

Global minimum: $x^* = (1, 2, 2, 3)$ and $f_{b_1,\ldots,b_n}(x^*) = 0$.

22. *Hartmann function* $H_{6,4}$

Definition: $f_{6,4}(x) = -\sum_{i=1}^4 \alpha_i \exp(-\sum_{j=1}^6 B_{ij}(x_j - P_{ij})^2)$,

Table 7.14 Coefficients of $f_{6,4}(x)$.

i	B_{ij}						a_i
1	10	3	17	3.5	1.7	8	1
2	0.05	10	17	0.1	8	14	1.2
3	3	3.5	1.7	10	17	8	3
4	17	8	0.05	10	0.1	14	3.2

and

$$P = \begin{pmatrix} 0.1312 & 0.1696 & 0.5569 & 0.0124 & 0.8283 & 0.5886 \\ 0.2329 & 0.4135 & 0.8307 & 0.3736 & 0.1004 & 0.9991 \\ 0.2348 & 0.1451 & 0.3522 & 0.2883 & 0.3047 & 0.6650 \\ 0.4047 & 0.8828 & 0.8732 & 0.5743 & 0.1091 & 0.0381 \end{pmatrix}.$$

Search space: $0 \leq x_i \leq 1, i = 1, \ldots, 6$.

Global minimum: $x^* = (0.201690, 0.150011, 0.476874, 0.275332, 0.311652, 0.657300)$ and $f(x^*) = -3.32237$.

23. *Schwefel function* SC_6 (See function 10).

24. *Trid function* T_6

Definition: $f_n(x) = \sum_{i=1}^n (x_i - 1)^2 - \sum_{i=2}^n x_i x_{i-1}$.

Search space: $-n^2 \le x_i \le n^2, i = 1, \ldots, n$.

Global minima: (a) $n = 6, x_i^* = i(7-i), i = 1, \ldots, n, f_n(x^*) = -50$; (b) $n = 10, x_i^* = i(11-i), i = 1, \ldots, n, f_n(x^*) = -210$.

25. *Trid function* T_{10} (See function 24).

26. *Rastrigin* RT_{10}

    Definition: $f_n(x) = 10n + \sum_{i=1}^n (x_i^2 - 10\cos(2\pi x_i))$.

    Search space: $-2.56 \le x_i \le 5.12, i = 1, \ldots, n$.

    Global minimum: $x^* = (0, \ldots, 0$ and $f_n(x^*) = 0$.

27. *Griewank function* G_{10}

    Definition: $f_n(x) = \sum_{i=1}^n \frac{x_i^2}{4000} - \prod_{i=1}^n \cos(x_i/\sqrt{i}) + 1$.

    Search space: $-300 \le x_i \le 600, i = 1, \ldots, n$.

    Global minimum: $x^* = (0, \ldots, 0)$ and $f_n(x^*) = 0$.

28. *Sum squares function* SS_{10}

    Definition: $f_n(x) = \sum_{i=1}^n i x_i^2$.

    Search space: $-5 \le x_i \le 10, i = 1, \ldots, n$.

    Global minimum: $x^* = (0, \ldots, 0)$ and $f_n(x^*) = 0$.

29. *Rosenbrock function* R_{10} (See function 11).

30. *Zakharov function* Z_{10} (See function 12).

31. *Rastrigin function* RT_{20} (See function 26).

32. *Griewank function* G_{20} (See function 27).

33. *Sum squares function* SS_{20} (See function 28).

34. *Rosenbrock function* R_{20} (See function 11).

35. *Zakharov function* Z_{20} (See function 12).

36. *Powell function* PW_{24}

    Definition: $f_n(x) = \sum_{i=1}^{n/4} (x_{4i-3} + 10x_{4i-2})^2 + 5(x_{4i-1} - x_{4i})^2 + (x_{4i-2} - x_{4i-1})^4 + 10(x_{4i-3} - x_{4i})^4$.

    Search space: $-4 \le x_i \le 5, i = 1, \ldots, n$.

    Global minimum: $x^* = (3, -1, 0, 1, 3, \ldots, 3, -1, 0, 1)$ and $f_n(x^*) = 0$.

37. *Dixon and Price function* DP_{25}

Definition: $f_n(x) = (x_1 - 1)^2 + \sum_{i=2}^{n} i(2x_i^2 - x_{i-1})^2$.
Search space: $-10 \leq x_i \leq 10, i = 1, \ldots, n$.
Global minimum: $x_i^* = 2^{-\frac{2^i-2}{2^i}}$ and $f_n(x^*) = 0$.

38. *Levy function* L_{30}

Definition: $f_n(x) = \sin^2(\pi y_1) + \sum_{i=1}^{n-1} \left[(y_i - 1)^2(1 + 10\sin^2(\pi y_i + 1))\right] + (y_n - 1)^2(1 + 10\sin^2(2\pi y_n)), y_i = 1 + (x_i - 1)/4, i = 1, \ldots, n$.
Search space: $-10 \leq x_i \leq 10, i = 1, \ldots, n$.
Global minimum: $x^* = (1, \ldots, 1)$ and $f_n(x^*) = 0$.

39. *Sphere function* SR_{30}

Definition: $f_n(x) = \sum_{i=1}^{n} x_i^2$.
Search space: $-2.56 \leq x_i \leq 5.12, i = 1, \ldots, n$.
Global minimum: $x^* = (0, \ldots, 0)$ and $f_n(x^*) = 0$.

40. *Ackley function* AK_{30}

Definition: $f_n(x) = 20 + e - 20\exp\left\{-\frac{1}{5}\sqrt{\frac{1}{n}\sum_{i=1}^{n} x_i^2}\right\} - \exp\left\{-\frac{1}{n}\sum_{i=1}^{n}\cos(2\pi x_i)\right\}$. Search space: $-15 \leq x_i \leq 30, i = 1, \ldots, n$.
Global minimum: $x^* = (0, \ldots, 0)$ and $f_n(x^*) = 0$.

Chapter 8

Markov Chain Monte Carlo with Adaptive Proposals

The stochastic approximation Monte Carlo algorithms studied in Chapter 7 can be viewed as types of adaptive Markov chain Monte Carlo algorithms for which the invariant distribution varies from iteration to iteration, while keeping the samples correctly weighted with respect to the target distribution. In this chapter, we will study a different type of adaptive Markov chain Monte Carlo algorithm, for which the proposal distribution can be changed infinitely often during the course of simulation, while preserving stationarity of the target distribution.

It is known that the efficiency of the Metropolis-Hastings algorithm can be improved by careful tuning of the proposal distribution. On the optimal setting of proposal distributions, some theoretical results have been obtained in the literature. For example, under certain settings, Gelman *et al.* (1996) show that the optimal covariance matrix for a multivariate Gaussian random walk Metropolis algorithm is $(2.38^2/d)\Sigma_\pi$, where d is the dimension and Σ_π is the $d \times d$ covariance matrix of the target distribution $\pi(\cdot)$. However, Σ_π is generally unknown in advance. The MCMC algorithms we will study in this chapter aim to provide an automated tuning of proposal distributions such that certain optimal criteria, theoretical or intuitive, can be achieved dynamically.

To motivate the design of adaptive MCMC algorithms, we first consider an example, which was originally given in Andrieu and Moulines (2006). Consider the Markov transition probability matrix

$$P_\theta = \begin{pmatrix} 1-\theta & \theta \\ \theta & 1-\theta \end{pmatrix},$$

Advanced Markov Chain Monte Carlo Methods: Learning from Past Samples
Faming Liang, Chuanhai Liu and Raymond J. Carroll

on the state space $\mathcal{X} = \{1, 2\}$ for some fixed $\theta \in (0, 1)$. It is easy to see that for any θ, P_θ leaves $\pi = (1/2, 1/2)$ invariant; that is, $\pi P_\theta = \pi$. However, if we let $\theta : \mathcal{X} \to (0, 1)$ be a function of the current state X, that is, introducing self-adaptation of proposals, then the transition probability matrix becomes

$$\widetilde{P} = \begin{pmatrix} 1 - \theta(1) & \theta(1) \\ \theta(2) & 1 - \theta(2) \end{pmatrix},$$

which admits $(\theta(2)/(\theta(1) + \theta(2)), \theta(1)/(\theta(1) + \theta(2)))$ as its invariant distribution. This example implies that introduction of self-adaptation of proposals might not preserve the stationarity of the target distribution. Similar examples can be found in Gelfand and Sahu (1994) and Roberts and Rosenthal (2007).

To recover the target distribution π, one can either remove or diminish the dependence of P_θ on X with iterations. This has led to some adaptive MCMC algorithms developed in the literature. These can be divided roughly into the four categories listed below, and described in Sections 8.1–8.4.

- *Stochastic approximation-based adaptive algorithms*. The algorithms diminish the adaptation gradually with iterations.
- *Adaptive independent MH algorithms*. The algorithms work with proposals which are adaptive, but do not depend on the current state of the Markov chain. The diminishing adaptation condition may not necessarily hold for them.
- *Regeneration-based adaptive algorithms*. The algorithms are designed on a basic property of the Markov chain, whose future outputs become independent of the past after each regeneration point.
- *Population-based adaptive algorithms*. The algorithms work on an enlarged state space, which gives us much freedom to design adaptive proposals and to incorporate sophisticated computational techniques into MCMC simulations.

If the proposals are adapted properly, the above algorithms can converge faster than classical MCMC algorithms, since more information on the target distribution may be included in the proposals. Even so, in sampling from multimodal distributions, they may suffer the same local-trap problem as the classical MCMC algorithms, as the proposal adapting technique is not dedicated to accelerating the transitions between different modes.

8.1 Stochastic Approximation-Based Adaptive Algorithms

Haario *et al.* (2001) prescribe an adaptive Metropolis algorithm which learns to build an efficient proposal distribution on the fly; that is, the proposal distribution is updated at each iteration based on the past samples. Under certain settings, the proposal distribution will converge to the 'optimal' one. Andrieu and Robert (2001) observe that the algorithm of Haario *et al.* (2001) can be viewed as a stochastic approximation algorithm (Robbins and Monro, 1951). Under the framework of stochastic approximations, Atchadé and Rosenthal (2005) and Andrieu and Moulines (2006) prove the ergodicity of more general adaptive algorithms. Andrieu and Moulines (2006) also prove a central limit theorem result. The theory on the adaptive MCMC algorithms is further developed by Roberts and Rosenthal (2007) and Yang (2007). They present somewhat simpler conditions, which still ensure ergodicity for specific target distributions.

Below, we provide an overview of the theory of adaptive MCMC algorithms, and Section present the adaptive Metropolis algorithm and its variants developed under the framework of stochastic approximation.

8.1.1 Ergodicity and Weak Law of Large Numbers

Let $\pi(\cdot)$ be a fixed target distribution defined on a state space $\mathcal{X} \subset \mathbb{R}^d$ with σ-algebra $\mathcal{F}$. Let $\{P_\theta\}_{\theta\in\Theta}$ be a collection of Markov chain transition kernels on $\mathcal{X}$, each admitting $\pi(\cdot)$ as the stationary distribution, i.e., $(\pi P_\theta)(\cdot) = \pi(\cdot)$. Assuming that P_θ is irreducible and aperiodic, then P_θ is ergodic with respect to $\pi(\cdot)$; that is, $\lim_{n\to\infty} \|P_\theta^n(x,\cdot)-\pi(\cdot)\| = 0$ for all $x \in \mathcal{X}$, where $\|\mu(\cdot)-\nu(\cdot)\| = \sup_{A\in\mathcal{F}} \|\mu(A) - \nu(A)\|$ is the usual total variation distance. So, if θ is kept fixed, then the samples generated by P_θ will eventually converge to $\pi(\cdot)$ in distribution.

However, some transition kernel P_θ may lead to a far less efficient Markov chain than others, and it is hard to know in advance which transition kernel is preferable. To deal with this, adaptive MCMC algorithms allow the transition kernel to be changed at each iteration according to certain rules. Let Γ_n be a Θ-valued random variable, which specifies the transition kernel to be used at iteration n. Let X_n denote the state of the Markov chain at iteration n. Thus,

$$P_\theta(x, A) = P(X_{n+1} \in A | X_n = x, \Gamma_n = \theta, \mathcal{G}_n), \quad A \in \mathcal{F},$$

where $\mathcal{G}_n = \sigma(X_0, \ldots, X_n, \Gamma_0, \ldots, \Gamma_n)$ is a filtration generated by $\{X_i, \Gamma_i\}_{i\leq n}$. Define

$$A^n((x, \theta), B) = P(X_n \in B | X_0 = x, \Gamma_0 = \theta), \quad B \in \mathcal{F},$$

which denotes the conditional probabilities of X_n given the initial conditions $X_0 = x$ and $\Gamma_0 = \theta$, and

$$T(x, \theta, n) = \|A^n((x, \theta), \cdot) - \pi(\cdot)\| = \sup_{B\in\mathcal{F}} |A^n((x, \theta), B) - \pi(B)|,$$

which denotes the total variation distance between the distribution of X_n and the target distribution $\pi(\cdot)$.

To ensure the ergodicity of the adaptive Markov chain, that is, $\lim_{n\to\infty} T(x, \theta, n) = 0$ for all $x \in \mathcal{X}$ and $\theta \in \Theta$, Roberts and Rosenthal (2007) prescribe two conditions, namely the *bounded convergence* condition and the *diminishing adaptation* condition. Let

$$M_\epsilon(x, \theta) = \inf\{n \geq 1 : \|P_\theta^n(x, \cdot) - \pi(\cdot)\| \leq \epsilon\},$$

be the convergence time of the kernel P_θ when starting in state $x \in \mathcal{X}$. The *bounded convergence* condition is that for any $\epsilon > 0$, the stochastic process $\{M_\epsilon(X_n, \Gamma_n)\}$ is bounded in probability given the initial values $X_0 = x$ and $\Gamma_0 = \theta$. Let

$$D_n = \sup_{x \in \mathcal{X}} \|P_{\Gamma_{n+1}}(x, \cdot) - P_{\Gamma_n}(x, \cdot)\|.$$

The diminishing adaptation conditions states that $\lim_{n\to\infty} D_n = 0$, which can be achieved by modifying the parameters by smaller and smaller amounts, as in the adaptive Metropolis algorithm of Haario *et al.* (2001), or by doing the adaption with smaller and smaller probability, as in the adaptive evolutionary Monte Carlo algorithm of Ren *et al.* (2008). In summary, Roberts and Rosenthal (2007) prove the following theorem:

Theorem 8.1.1 *For an MCMC algorithm with adaptive proposals, if it satisfies the bounded convergence and diminishing adaptation conditions, then it is ergodic with respect to the stationary distribution* $\pi(\cdot)$.

Furthermore, they prove that the bounded convergence condition is satisfied whenever $\mathcal{X} \times \Theta$ is finite, or is compact in some topology in which either the transition kernels P_θ, or the MH proposals, have jointly continuous densities. Note that Theorem 8.1.1 does not require that Γ_n converges.

Since the quantity $M_\epsilon(x, \theta)$ is rather abstract, Roberts and Rosenthal (2007) give one condition, *simultaneous geometrical ergodicity*, which ensures the bounded convergence condition. A family $\{P_\theta\}_{\theta \in \Theta}$ of Markov chain transition kernels is said to be *simultaneously geometrically ergodic* if there exists $C \in \mathcal{F}$, a drift function $V : \mathcal{X} \to [1, \infty)$, $\delta > 0$, $\lambda < 1$ and $b < \infty$, such that $\sup_{x \in C} V(x) = v < \infty$ and the following conditions hold:

(i) *Minorization condition.* For each $\theta \in \Theta$, there exists a probability measure $\nu_\theta(\cdot)$ on C with $P_\theta(x, \cdot) \geq \delta \nu_\theta(\cdot)$ for all $x \in C$.

(ii) *Drift condition.* $P_\theta V \leq \lambda V + bI(x \in C)$, where $I(\cdot)$ is the indicator function.

This results in the following theorem:

Theorem 8.1.2 *For an MCMC algorithm with adaptive proposals, if it satisfies the diminishing adaptation condition and the family* $\{P_\theta\}_{\theta \in \Theta}$ *is simultaneously geometrically ergodic with* $E(V(X_0)) < \infty$, *then it is ergodic with respect to the stationary distribution* $\pi(\cdot)$.

In addition to the ergodicity of the Markov chain, in practice, one is often interested in the weak law of large numbers (WLLN), that is, whether or not the sample path average $(1/n)\sum_{i=1}^{n} h(X_i)$ will converge to the mean $E_\pi h(x) = \int h(x)\pi(x)dx$ for some function $h(x)$. For adaptive MCMC algorithms, this is to some extent even more important than the ergodicity. Under slightly stronger conditions, the *simultaneous uniform ergodicity* and *diminishing adaptation* conditions, Roberts and Rosenthal (2007) show that $(1/n)\sum_{i=1}^{n} h(X_i)$ will converge to $E_\pi h(x)$, provided that $h(x)$ is bounded. A family $\{P_\theta\}_{\theta\in\Theta}$ of Markov chain transition kernels is said to be *simultaneously uniformly ergodic* if $\sup_{(x,\theta)\in\mathcal{X}\times\Theta} M_\epsilon(x,\theta) < \infty$. In summary, the WLLN for adaptive MCMC algorithms can be stated as:

Theorem 8.1.3 *For an MCMC algorithm with adaptive proposals, if it satisfies the simultaneous uniform ergodicity and diminishing adaptation conditions, then for any starting values $x \in \mathcal{X}$ and $\theta \in \Theta$,*

$$\frac{1}{n}\sum_{i=1}^{n} h(X_i) \to E_\pi h(x),$$

provided that $h : \mathcal{X} \to \mathbb{R}$ is a bounded measurable function.

In Theorem 8.1.3, the function $h(x)$ is restricted to be bounded. This is different from the WLLN established for conventional MCMC algorithms (see, e.g., Tierney, 1994). For MCMC algorithms with adaptive proposals, some constraints on the boundedness of $h(x)$ may be necessary. A counterexample is constructed in Yang (2007), that the WLLN may not hold for unbounded measurable functions even if the ergodicity holds. In Andrieu and Moulines (2006), more restrictive conditions are imposed on the transition kernels, for example, to establish WLLN, $h(x)$ is still required to be bounded by the drift function in a certain form. Further research on this issue is of great interest from both theoretical and practical perspectives.

8.1.2 Adaptive Metropolis Algorithms

Adaptive Metropolis Algorithm. Let $\pi(x)$ denote the target distribution. Consider a Gaussian random walk MH algorithm, for which the proposal distribution is $q(x,y) = N(y; x, \Sigma)$, and $N(y; x, \Sigma)$ denotes the density of a multivariate Gaussian with mean x and covariance matrix Σ. It is known that either too small or too large a covariance matrix will lead to a highly correlated Markov chain. Under certain settings, Gelman *et al.* (1996) show that the 'optimal' covariance matrix for the Gaussian random walk MH algorithm is $(2.38^2/d)\Sigma_\pi$, where d is the dimension of x and Σ_π is the true covariance matrix of the target distribution $\pi(\cdot)$. Haario *et al.* (2001) propose learning Σ_π 'on the fly'; that is, estimating Σ_π from the empirical distribution of the available Markov chain outputs, and thus adapting the estimate of Σ while the algorithm runs. Let $\{\gamma_k\}$ denote a gain factor sequence. As for the standard stochastic approximation algorithm, it is required to satisfy the following conditions:

$$\sum_{k=1}^{\infty} \gamma_k = \infty, \qquad \sum_{k=1}^{\infty} \gamma_k^{1+\delta} < \infty,$$

for some $\delta \in (0, 1]$. As mentioned in Chapter 7, the former condition ensures that any point of Θ can eventually be reached, and the latter condition ensures that the noise introduced by new observations is contained and does not prevent convergence. Haario *et al.* (2001) suggest the choice $\gamma_k = O(1/k)$; with the above notations, the algorithm of Haario *et al.* (2001) can be summarized as follows:

Adaptive Metropolis Algorithm I

1. Initialize X_0, μ_0 and Σ_0.

2. At iteration $k + 1$, given X_k, μ_k and Σ_k

   (a) Generate X_{k+1} via the MH kernel $P_{\theta_k}(X_k, \cdot)$, where $\theta_k = (\mu_k, \Sigma_k)$.

   (b) Update

$$\begin{aligned}\mu_{k+1} &= \mu_k + \gamma_{k+1}(X_{k+1} - \mu_k),\\ \Sigma_{k+1} &= \Sigma_k + \gamma_{k+1}\left[(X_{k+1} - \mu_k)(X_{k+1} - \mu_k)^T - \Sigma_k\right].\end{aligned}$$

Haario *et al.* (2001) showed that under certain conditions, such as $\mathcal{X}$ is compact and ϵI_d is added to each empirical covariance matrix at each iteration (which implies Θ to be compact), this algorithm is ergodic and the WLLN holds for bounded functionals. These results can also be implied from Theorems 8.1.1 and 8.1.3 presented in Section 8.1.1.

The adaptive Metropolis algorithm has been generalized in various ways by C. Andrieu and his coauthors. In what follows we give several variants of the algorithm, which have been presented in Andrieu and Thoms (2008).

An Improved Adaptive Metropolis Algorithm. Let $\lambda\Sigma_k$ denote the covariance matrix used by the adaptive Metropolis algorithm at iteration k, where λ is preset at $2.38^2/d$ as suggested by Gelman *et al.* (1996). As pointed out by Andrieu and Thoms (2008), if $\lambda\Sigma_k$ is either too large in some directions or too small in all directions, the algorithm will have either a very small or a very large acceptance probability, rendering a slow learning of Σ_π due to limited exploration of the sample space $\mathcal{X}$.

To alleviate this difficulty, Andrieu and Thoms (2008) propose simultaneously adapting the parameter λ and the covariance matrix Σ in order to coerce the acceptance probability to a preset and sensible value, such as 0.234. Roberts and Rosenthal (2001) show that, for large d, the optimal acceptance rate of the random walk Metropolis algorithm is 0.234 when the components of $\pi(\cdot)$ are approximately uncorrelated but heterogeneously scaled. Assuming that for any fixed covariance matrix Σ the corresponding expected acceptance rate is a nonincreasing function of λ, the following recursion can be used for learning λ according to the Robbins-Monro algorithm (Robbins and Monro, 1951):

$$\log(\lambda_{k+1}) = \log(\lambda_k) + \gamma_{k+1}[\alpha(X_k, X^*) - \alpha^*],$$

where X^* denotes the proposed value, and α^* denotes the targeted acceptance rate, for example, 0.234.

In summary, the modified adaptive Metropolis algorithm can be described as follows:

Adaptive Metropolis Algorithm II (Steps)

1. Initialize X_0, λ_0, μ_0 and Σ_0.

2. At iteration $k+1$, given X_k, λ_k, μ_k and Σ_k

   (a) Draw X^* from the proposal distribution $N(X_k, \lambda_k\Sigma_k)$, set $X_{k+1} = X^*$ with probability $\alpha(X_k, X^*)$, specified by the MH rule and set $X_{k+1} = X_k$ with the remaining probability.

   (b) Update

$$\begin{aligned}
\log(\lambda_{k+1}) &= \log(\lambda_k) + \gamma_{k+1}[\alpha(X_k, X^*) - \alpha^*], \\
\mu_{k+1} &= \mu_k + \gamma_{k+1}(X_{k+1} - \mu_k), \\
\Sigma_{k+1} &= \Sigma_k + \gamma_{k+1}\left[(X_{k+1} - \mu_k)(X_{k+1} - \mu_k)^T - \Sigma_k\right].
\end{aligned}$$

The adaption of λ can be very useful in the early stage of the simulation, although it is likely not needed in the long run.

Adaptive Metropolis-within-Gibbs Algorithm. In algorithms I and II, all components are updated simultaneously, this might not be efficient when the dimension d is high. The reason is that the scaling of $\lambda\Sigma$ may be correct in some directions, but incorrect in others. Motivated by this, Andrieu and Thoms (2008) propose the following adaptive Metropolis algorithm, which can be viewed as an adaptive version of the Metropolis-with-Gibbs sampler with random scanning.

Adaptive Metropolis Algorithm III (Steps)

1. Initialize X_0, μ_0, Σ_0, and $\lambda_0^{(1)}, \ldots, \lambda_0^{(d)}$.

2. At iteration $k+1$, given X_k, μ_k, Σ_k, and $\lambda_k^{(1)}, \ldots, \lambda_k^{(d)}$.

   (a) Choose a component $i \sim Uniform\{1, \ldots, d\}$.

   (b) Draw Z from $N(0, \lambda_k^{(i)}[\Sigma_k]_{i,i})$, where $[\Sigma_k]_{i,i}$ denotes the ith diagonal element of Σ_k. Set $X_{k+1} = X_k + Ze_i$ with probability $\alpha(X_k, X_k + Ze_i)$, and set $X_{k+1} = X_k$ with the remaining probability, where e_i is a vector with zeros everywhere but 1 for its i-th component.

   (c) Update

$$\begin{aligned}\log(\lambda_{k+1}^{(i)}) &= \log(\lambda_k^{(i)}) + \gamma_{k+1}[\alpha(X_k, X_k + Ze_i) - \alpha^*],\\ \mu_{k+1} &= \mu_k + \gamma_{k+1}(X_{k+1} - \mu_k),\\ \Sigma_{k+1} &= \Sigma_k + \gamma_{k+1}\left[(X_{k+1} - \mu_k)(X_{k+1} - \mu_k)^T - \Sigma_k\right].\end{aligned}$$

   and set $\lambda_{k+1}^{(j)} = \lambda_k^{(j)}$ for $j \neq i$.

For this algorithm, α^* is usually set to a high value, for example, 0.44, in which there is only a single component updated at each iteration. Note that it is redundant to use both the scaling $\lambda^{(i)}$ and $[\Sigma]_{i,i}$ in this algorithm. In practice, one may choose to combine both quantities into a single scaling factor. One potential advantage of algorithm III is that it gathers information about $\pi(\cdot)$, which may be used by more sophisticated and more global updates during simulations.

8.2 Adaptive Independent Metropolis-Hastings Algorithms

Holden *et al.* (2009) describe an independent MH algorithm, for which the proposal is adaptive with (part of) past samples, but avoids the requirement of diminishing adaptation.

Let $q_t(z|\boldsymbol{y}_{t-1})$ denote the proposal used at iteration t, where $\boldsymbol{y}_{t-1}$ denotes the set of past samples used in forming the proposal. Since the basic requirement for the independent MH algorithm is that its proposal is independent of the current state x_t, $\boldsymbol{y}_{t-1}$ can not include x_t as an element. Suppose that z has been generated from the proposal $q_t(z|\boldsymbol{y}_{t-1})$. If it is accepted, then set $x_{t+1} = z$ and append $\boldsymbol{y}_{t-1}$ with x_t. Otherwise, set $x_{t+1} = x_t$ and append $\boldsymbol{y}_{t-1}$ with z. The difference between the traditional independent MH algorithm and the adaptive independent MH algorithm is only that the proposal function $q_t(\cdot)$ may depend on a history vector, which can include all samples that $\pi(x)$ has been evaluated except for the current state of the Markov chain. The algorithm can be summarized as follows:

Adaptive Independent MH Algorithm (Steps)

1. Set $\boldsymbol{y}_0 = \emptyset$, and generate an initial sample x_0 in $\mathcal{X}$.

2. For $t = 1, \ldots, n$:

   (a) Generate a state z from the proposal $q_t(z|\boldsymbol{y}_{t-1})$, and calculate the acceptance probability

   $$\alpha(z, x_t, \boldsymbol{y}_{t-1}) = \min\left\{1, \frac{\pi(z)q_t(x_t|\boldsymbol{y}_{t-1})}{\pi(x_t)q_t(z|\boldsymbol{y}_{t-1})}\right\}.$$

   (b) If it is accepted, set $x_{t+1} = z$ and $\boldsymbol{y}_t = \boldsymbol{y}_{t-1} \cup \{x_t\}$. Otherwise, set $x_{t+1} = x_t$ and $\boldsymbol{y}_t = \boldsymbol{y}_{t-1} \cup \{z\}$.

Holden *et al.* (2009) show the following theorem for the algorithm, which implies that the chain never leaves the stationary distribution $\pi(x)$ once it is reached.

Theorem 8.2.1 *The target distribution $\pi(x)$ is invariant for the adaptive independent MH algorithm; that is, $p_t(x_t|\boldsymbol{y}_{t-1}) = \pi(x_t)$ implies $p_t(x_{t+1}|\boldsymbol{y}_t) = \pi(x_{t+1})$, where $p_t(\cdot|\cdot)$ denotes the distribution of x_t conditional on the past samples.*

Proof: Assume $p_t(x_t|\boldsymbol{y}_{t-1}) = \pi(x_t)$ holds. Let $f_t(\boldsymbol{y}_t)$ denote the joint distribution of $\boldsymbol{y}_t$, and let w be the state appended to the history at iteration $t+1$. Then,

$$\begin{aligned}
p_{t+1}(x_{t+1}|\boldsymbol{y}_t)f_t(\boldsymbol{y}_t) &= f_{t-1}(\boldsymbol{y}_{t-1})\Big\{\pi(w)q_t(x_{t+1}|\boldsymbol{y}_{t-1})\alpha(x_{t+1}, w, \boldsymbol{y}_{t-1}) \\
&\quad + \pi(x_{t+1})q_t(w|\boldsymbol{y}_{t-1})[1 - \alpha(w, x_{t+1}, \boldsymbol{y}_{t-1})]\Big\} \\
&= f_{t-1}(\boldsymbol{y}_{t-1})\Big\{\pi(x_{t+1})q_t(w|\boldsymbol{y}_{t-1}) + \pi(w)q_t(x_{t+1}|\boldsymbol{y}_{t-1})\alpha(x_{t+1}, w, \boldsymbol{y}_{t-1}) \\
&\quad - \pi(x_{t+1})q_t(w|\boldsymbol{y}_{t-1})\alpha(w, x_{t+1}, \boldsymbol{y}_{t-1})\Big\} \\
&= \pi(x_{t+1})q_t(w|\boldsymbol{y}_{t-1})f_{t-1}(\boldsymbol{y}_{t-1}),
\end{aligned}$$

which implies $p_{t+1}(x_{t+1}|\boldsymbol{y}_t) = \pi(x_{t+1})$. ■

The proof tells us that if $q_t(\cdot)$ depends on x_t, then w has to be integrated out to get $p_{t+1}(x_{t+1}|\boldsymbol{y}_t) = \pi(x_{t+1})$. This is also the reason why $\boldsymbol{y}_{t-1}$ can not be extended to include x_t.

In addition, Holden *et al.* (2009) show that the adaptive independent MH algorithm can converge geometrically under a strong Doeblin condition: for each proposal distribution, there exists a function $\beta_t(\boldsymbol{y}_{t-1}) \in [0, 1]$ such that

$$q_t(z|\boldsymbol{y}_{t-1}) \geq \beta_t(\boldsymbol{y}_{t-1})\pi(z), \quad \text{for all } (z, x, \boldsymbol{y}_{t-1}) \in \mathcal{X}^{t+1} \text{ and all } t > 1, \quad (8.1)$$

which essentially requires that all the proposal distributions have uniformly heavier tails than the target distribution.

Theorem 8.2.2 *Assume the adaptive independent MH algorithm satisfies (8.1). Then*

$$\|p_t - \pi\| \leq 2E\left(\prod_{i=1}^{t}(1 - \beta_i(\boldsymbol{y}_{i-1}))\right),$$

where $\|\cdot\|$ *denotes the total variation norm.*

Clearly, if the product $\prod_{i=1}^{t}(1-\beta_i(\boldsymbol{y}_{i-1}))$ goes to zero when $t \to \infty$, then the algorithm converges. Furthermore, if there exists a constant β such that $\beta_i(\boldsymbol{y}_{i-1}) \geq \beta > 0$, then the algorithm converges geometrically.

The adaptive independent MH algorithm gives one much freedom to choose the proposal, although designing a proposal which satisfies (8.1) may not be easy. A frequently used strategy is to fit $\boldsymbol{y}_t$ by a mixture normal distribution. This can be done recursively as follows. Given a multivariate normal mixture

$$g_\theta(y) = \sum_{i=1}^{k} \alpha_i \phi(y; \mu_i, \Sigma_i), \tag{8.2}$$

where $\phi(\cdot)$ denotes the multivariate normal density, and $\theta = (\alpha_1, \mu_1, \Sigma_1, \ldots, \alpha_k, \mu_k, \Sigma_k)$ denotes the vector of unknown parameters of the distribution. Suppose that a sequence of samples $\{y_t\}$ have been drawn from $g_\theta(y)$. Then θ can be estimated by a recursive procedure, described by Titterington (1984) as follows:

$$\begin{aligned}
\widehat{\mu}_i^{(t+1)} &= \widehat{\mu}_i^{(t)} + \frac{w_i^{(t)}}{t\widehat{\alpha}_i^{(t)}}(y_{t+1} - \widehat{\mu}_i^{(t)}),\\
\widehat{\Sigma}_i^{(t+1)} &= \widehat{\Sigma}_i^{(t)} + \frac{w_i^{(t)}}{t\widehat{\alpha}_i^{(t)}}\left[(y_{t+1} - \widehat{\mu}_i^{(t)})(y_{t+1} - \widehat{\mu}_i^{(t)})^T - \widehat{\Sigma}_i^{(t)}\right],\\
\widehat{\alpha}_i^{(t+1)} &= \widehat{\alpha}_i^{(t)} + \frac{1}{j}(w_i^{(t)} - \widehat{\alpha}_i^{(t)}),
\end{aligned} \tag{8.3}$$

where $w_i^{(t)} \propto \widehat{\alpha}_i^{(t)} \phi(y_{t+1}; \widehat{\mu}_i^{(t)}, \widehat{\Sigma}_i^{(t)})$ with $\sum_{i=1}^{k} w_i^{(t)} = 1$. Although this estimator is not consistent, it can still work well for proposal distributions. Alternatively, the parameters can be estimated by an on-line EM algorithm (Andrieu and Moulines, 2006). Giordani and Kohn (2009) also gave a fast method for estimation of mixtures of normals.

Given (8.2) and (8.3), the proposal can be specified as

$$q_t(z|\boldsymbol{y}_{t-1}) = \eta_t q_0(z) + (1 - \eta_t) g_{\theta_{t-1}}(z), \tag{8.4}$$

where $q_0(z)$ is a heavy tail density, $g_{\theta_{t-1}}$ is estimated in (8.3) based on the history vector $\boldsymbol{y}_{t-1}$, and η_t is a sequence starting with 1 and decreasing to $\epsilon \geq 0$. As the proposal distribution gets closer to the target distribution, the convergence of the adaptive independent MH algorithm is expected to be accelerated. A local step, for which the proposal depends on the current state, can also be inserted into the run of the algorithm, while not disturbing the stationarity of the target distribution. The limitation is that information gained from doing the local steps cannot be used in constructing the independent proposals.

8.3 Regeneration-Based Adaptive Algorithms

Loosely speaking, the regeneration time of a Markov chain is a time at which its future becomes independent of the past (see, e.g., Mykland *et al.*, 1995). Based on this concept, Gilks *et al.* (1998) describe a framework for Markov chain adaptation, which allows the proposal to be modified infinitely often, but preserves the stationarity of the target distribution, and maintains consistency of the sample path averages. At each regeneration time, the proposal can be modified based on all past samples up to that time. Although every ergodic Markov chain is regenerative (see, e.g., Meyn and Tweedie, 1993), identification of the regeneration times is generally difficult, unless the chain takes values in a fine state space. For most chains, no recurrent proper atom exists, and it is not always easy to use the splitting method of Nummelin (1978) to identify the regeneration times.

To address this difficulty, Brockwell and Kadane (2005) propose a method of identifying regeneration times, which relies on simulating a Markov chain on an augmented state space. Sahu and Zhigljavsky (2003) propose à self-regenerative version of the adaptive MH based on an auxiliary chain with some other stationary distribution. We describe below how the proposal can be adapted for a Markov chain at regeneration times, using Brockwell and Kadane's method.

8.3.1 Identification of Regeneration Times

Let $\pi(x) = \psi(x)/C$ denote the target distribution with the support set $\mathcal{X}$ and an unknown normalizing constant C. The main idea of Brockwell and Kadane (2005) is to enlarge the state space from $\mathcal{X}$ to

$$\mathcal{X}^* = \mathcal{X} \cup \{\alpha\},$$

where α is a new state called the artificial atom. Then, define on $\mathcal{X}^*$ a new distribution:

$$\pi^*(x) = \begin{cases} \psi(x), & x \in \mathcal{X}, \\ q, & x = \alpha, \end{cases} \tag{8.5}$$

which assigns mass $p = q/(C + q)$ to the new state α and mass $1 - p$ to the original distribution $\pi(x)$.

Let $\{Y_t\}$ be a Markov chain which admits $\pi^*(x)$ as the stationary distribution. Let $\tau(j)$, $j = 1, 2, \ldots$, denote the jth time that the chain visits state α, with $\tau(0) = 0$, so

$$\tau(j) = \min\{k > \tau(j-1) : Y_k = \alpha\}, \quad j = 1, 2, 3, \ldots.$$

Define the tours $Y^1, Y^2, \ldots$ to be the segments of the chain between the visiting times of state α; that is,

$$Y^j = \{Y_t : \tau(j-1) < t \leq \tau(j)\}, \quad j = 1, 2, \ldots.$$

By observing that the state α occurs exactly once at the end of each tour Y^j, we construct a chain $\{Z_t\}$ by string together the tours Y^j whose length is longer than one, after removing the last element from each one. Let T_j denote the time at which the $(j+1)$th tour of $\{Z_t\}$ begins, for $j = 0, 1, 2, \ldots$, with the convention that $T_0 = 0$. Brockwell and Kadane (2005) prove the following theorem, which states that $\{Z_t\}$ is an ergodic Markov chain with stationary distribution $\pi(x)$:

Theorem 8.3.1 *Suppose that $\{Y_t\}$ is an ergodic Markov chain defined on $\mathcal{X}^*$ with $Y_0 = \alpha$ and stationary distribution $\pi_p^*(x)$ given in (8.5). Let $\{Z_t\}$ be constructed as described above. Then $\{Z_t\}$ is ergodic with respect to $\pi(x)$, and the time T_j, $j = 0, 1, 2, \ldots$ are regeneration times of the chain.*

We next describe a simulation method for constructing $\{Y_t\}$ using a hybrid kernel method. Let P_θ denote a Markov transition kernel, for example, the MH kernel, which admits $\pi(x)$ as its invariant distribution. The notation P_θ indicates that the transition kernel can depend on a parameter θ and that it is adaptable. The hybrid kernel method consists of two steps:

1. *Component Sampling.* If the current state is α, keep it unchanged. Otherwise, update the state using a π-invariant kernel P_θ.

2. *Component Mixing.* If the current state is α, draw a new state from a re-entry proposal distribution $\varphi(\cdot)$ defined on $\mathcal{X}$. Otherwise, propose the new state to be α. Denote the current state by V and denote the proposed state by W. According to the MH rule, the state W can be accepted with probability $\min\{1, r\}$, where

$$r = \begin{cases} \dfrac{\psi(W)}{q\varphi(W)}, & V = \alpha, \\[2ex] \dfrac{q\varphi(V)}{\psi(V)}, & V \in \mathcal{X}. \end{cases}$$

It is easy to see that, as long as the re-entry proposal density $\varphi(\cdot)$ has the same support as $\pi(\cdot)$, the chain generated by the hybrid kernel method admits $\pi^*(x)$ as its invariant distribution. To have a reasonable acceptance rate at the component mixing step, it is desirable to choose $\varphi(\cdot)$ to be reasonably close to $\pi(\cdot)$, and to choose q to be roughly of the order of magnitude of $\varphi(x)/\psi(x)$ with x being some point in a high-density region of $\pi(\cdot)$.

■ **Example 8.1 Markov Chain Regeneration**

Consider the unnormalized density $\psi(x) = \exp(-x^2/2)$ (Brockwell and Kadane, 2005). Let P_θ be a MH kernel constructed with a Gaussian random walk proposal $N(Y_t, \theta)$. Let $\varphi(x) = \exp(-x^2/5)/\sqrt{5\pi}$ be the re-entry proposal density, and set $q = 1$. Starting with $Y_0 = \alpha$, then Y_{t+1} can be generated conditional on Y_t in the following procedure:

- If $Y_t = \alpha$, then set $V = \alpha$. Otherwise, generate Z via the kernel $P_\theta(Y_t, \cdot)$, and set $V = Z$.
- If $V \in \mathbb{R}$, then set

$$Y_{t+1} = \begin{cases} \alpha, & \text{with probability } \min\{1, \varphi(V)/\psi(V)\}, \\ V, & \text{otherwise.} \end{cases}$$

Otherwise, if $V = \alpha$, draw $W \sim \varphi$, and set

$$Y_{t+1} = \begin{cases} W, & \text{with probability } \min\{1, \pi(W)/\varphi(W)\}, \\ \alpha, & \text{otherwise.} \end{cases}$$

8.3.2 Proposal Adaptation at Regeneration Times

Brockwell and Kadane (2005) suggest adapting the proposal at regeneration times with a mixture normal distribution, starting with an arbitrary proposal distribution, and as the simulation goes on, transforming it progressively into a mixture normal distribution learned from past samples. Let the mixture normal distribution be estimated in (8.3). Similar to (8.4), the proposal distribution used in the component sampling step can be adapted with simulations as follows:

$$f_{\theta_m}(y|x_t) = \eta_m g_0(y|x_t) + (1 - \eta_m) g_{\theta_m}(y), \quad m = 0, 1, 2, \ldots, \tag{8.6}$$

where $g_0(\cdot)$ denote an arbitrary proposal distribution, m denotes the number of tours, and $\{\eta_m\}$ is a sequence starting with $\eta_0 = 1$ and decreasing to $\epsilon \geq 0$. The MH kernel P_{θ_m} can be constructed accordingly. As pointed out by Brockwell and Kadane (2005), choosing $\epsilon = 0$ allows the kernel to be completely replaced by the independence sampler, and is potentially dangerous if η_m becomes very close to 0 before g_{θ_m} provides a reasonable approximation to the target distribution.

8.4 Population-Based Adaptive Algorithms

8.4.1 ADS, EMC, NKC and More

The idea of adapting proposals with a population of iid samples can be easily traced back to the adaptive direction sampler (ADS) (Gilks *et al.*, 1994), which has been described in Section 5.1. To set notation, we let $\pi(x)$ denote the target distribution. In the ADS, the target distribution is re-defined as $\pi(x_1, \ldots, x_n) = \prod_{i=1}^n \pi(x_i)$, a n-dimensional product density on $\mathcal{X}^n$. The joint state $(x_1, \ldots, x_n)$ is called a population of samples. At each iteration, a sample is selected at random to undergo an update along a direction toward another sample which is selected at random from the remaining set of the population. Apparently, the ADS is adaptive in the sense that the proposal used at each iteration depends on the past samples, although within a fixed horizon.

Later, ADS was generalized by Liang and Wong (2000, 2001a) and Warnes (2001). Liang and Wong (2000, 2001a) attached a different temperature to each chain (sample) and included some genetic operators to have the population updated in a global manner. They called the generalized algorithm the evolutionary Monte Carlo (EMC), see Section 5.5. In Warnes' generalization, the so-called normal kernel coupler (NKC), each sample is updated with a mixture Gaussian proposal, which is obtained by fitting a Gaussian kernel to the current population.

Note that NKC is theoretically different from EMC and ADS. In EMC and ADS, each sample is updated with a proposal that depends only on the remaining samples of the population, and thus they fall into the class of Metropolis-within-Gibbs algorithms. As shown, for instance, by Besag *et al.* 1995, such a Markov chain admits the joint distribution $\pi(x_1, \ldots, x_n)$ as its stationary distribution. This result can also be implied from a generalized Gibbs sampler based on conditional moves along the traces of groups of transformations in the joint state space $\mathcal{X}^n$ by Liu and Sabatti (2000), whereas, in NKC, each sample is updated with a proposal that depends on all samples of the current population. Hence, the ergodicity of NKC needs to be proved from first principles.

Under the framework of the Metropolis-with-Gibbs sampler, Cai *et al.* (2008) propose a new generalization of ADS, where each chain is updated with a proposal of a mixture of triangular and trapezoidal densities. The proposal is constructed based on the remaining samples of the current population.

8.4.2 Adaptive EMC

The development of the adaptive EMC algorithm (Ren *et al.*, 2008) is motivated by sensor placement applications in engineering, which requires optimizing certain complicated black-box objective functions over a given region. Adaptive EMC combines EMC with a tree-based search procedure, the classification and regression trees (CART) proposed by Breiman *et al.* (1984). In return, it significantly improves the convergence of EMC.

To set notation, let $H(x)$ denote the objective function to be minimized over a compact space $\mathcal{X} \subset \mathbb{R}^d$, and let

$$\pi_i(x) = \frac{1}{Z_i} \exp\{-H(x)/t_i\}, \quad i = 1, \ldots, n,$$

denote the sequence of distributions to be simulated in adaptive EMC, where n represents the population size, and $t_1 > t_2 > \cdots > t_n$ forms a temperature ladder to be used in the simulation. As in EMC, assume the samples in the population are mutually independent. Then the distribution of the population is given by

$$\pi(\boldsymbol{x}) = \prod_{i=1}^{n} \pi_i(x).$$

In AEMC, we first run a number of iterations of EMC and then use CART to learn a proposal distribution from the samples produced by EMC. Denote by $\mathcal{Y}_k$ the set of samples we have retained after iteration k. From $\mathcal{Y}_k$, we define high performance samples to be those with relatively small $H(x)$ values. The high performance samples are the representatives of the promising search regions. We denote by $H_k^{(h)}$ the h percentile of the $H(x)$ values in $\mathcal{Y}_k$. Then, the set of high performance samples at iteration k is defined by $B_k = \{x : x \in \mathcal{Y}_k, H(x) \le H_k^{(h)}\}$. As a result, the samples in $\mathcal{Y}_k$ are grouped into two classes, the high performance samples in B_k and the others. Treating these samples as a training dataset, we then fit a CART model to a two-class classification problem. Using the prediction from the resulting CART model, we can partition the sample space into rectangular regions, some of which have small $H(x)$ values and are therefore deemed as the promising regions, while other regions as non promising regions.

The promising regions produced by CART are represented as $a_j^{(k)} \le x_{ij} \le b_j^{(k)}$ for $i = 1, \ldots, n$ and $j = 1, \ldots, d$, where x_{ij} denotes the j-th component of x_i. Since $\mathcal{X}$ is compact, there is a lower bound l_j and an upper bound u_j in the j-th dimension of the sample space. Clearly we have $l_j \le a_j^{(k)} \le b_j^{(k)} \le u_j$. CART may produce multiple promising regions. We denote by m_k the number of regions. Then, the collection of promising regions can be specified by

$$a_{js}^{(k)} \le x_{ij} \le b_{js}^{(k)}, \quad j = 1, \ldots, d, i = 1, \ldots, n, s = 1, \ldots, m_k.$$

As the algorithm goes on, we continuously update $\mathcal{Y}_k$, and hence $a_{js}^{(k)}$ and $b_{js}^{(k)}$.

After the promising regions have been identified, we construct the proposal density by getting a sample from the promising regions with probability α, and from elsewhere with probability $1-\alpha$, respectively. We recommend using a relatively large α, say $\alpha = 0.9$. Since there may be multiple promising regions identified by CART, we denote the proposal density associated with each region by $q_{ks}(x)$, $s = 1, \ldots, m_k$. In what follows we describe a Metropolis-within-Gibbs procedure (Müller, 1991) to generate new samples.

For $i = 1, \ldots, n$, denote the population after the k-th iteration by $\boldsymbol{x}^{(k+1,i-1)} = (x_1^{(k+1)}, \ldots, x_{i-1}^{(k+1)}, x_i^{(k)}, \ldots, x_n^{(k)})$, of which the first $i - 1$

samples have been updated, and the Metropolis-within-Gibbs procedure is about to generate the i-th new sample. Note that $\boldsymbol{x}^{(k+1,0)} = (x_1^{(k)}, \ldots, x_n^{(k)})$.

- Draw S randomly from the set $\{1, \ldots, m_k\}$. Generate a sample x_i' from the proposal density $q_{kS}(\cdot)$.

$$q_{kS}(x_i') = \prod_{j=1}^{d} \left(\beta \frac{I(a_{jS}^{(k)} \le x_{ij}' \le b_{jS}^{(k)})}{b_{jS}^{(k)} - a_{jS}^{(k)}} + (1-\beta) \frac{I(x_{ij}' < a_{jS}^{(k)} \text{ or } x_{ij}' > b_{jS}^{(k)})}{(u_j - l_j) - (b_{jS}^{(k)} - a_{jS}^{(k)})} \right),$$

  where $I(\cdot)$ is the indicator function. Here β is the probability of sampling uniformly within the range specified by the CART rules on each dimension. Since all dimensions are mutually independent, we set $\beta = \alpha^{1/d}$.

- Construct a new population $\boldsymbol{x}^{(k+1,i)}$ by replacing $x_i^{(k)}$ with x_i', and accept the new population with probability min(1,r), where

$$\begin{aligned} r &= \frac{\pi(\boldsymbol{x}^{(k+1,i)})}{\pi(\boldsymbol{x}^{(k+1,i-1)})} \frac{T(\boldsymbol{x}^{(k+1,i-1)}|\boldsymbol{x}^{(k+1,i)})}{T(\boldsymbol{x}^{(k+1,i)}|\boldsymbol{x}^{(k+1,i-1)})} \\ &= \exp\left\{ -\frac{H(x_i') - H(x_i^{(k)})}{t_i} \right\} \frac{T(\boldsymbol{x}^{(k+1,i-1)}|\boldsymbol{x}^{(k+1,i)})}{T(\boldsymbol{x}^{(k+1,i)}|\boldsymbol{x}^{(k+1,i-1)})}, \end{aligned} \tag{8.7}$$

  If the proposal is rejected, $\boldsymbol{x}^{(k+1,i)} = \boldsymbol{x}^{(k+1,i-1)}$.

Since only a single sample is changed in each Metropolis-within-Gibbs step, the proposal probability in (8.7) can be expressed as

$$T(x_i^{(k)} \to x_i' | \boldsymbol{x}_{[-i]}^{(k)}) = \sum_{s=1}^{m^{(k)}} \frac{1}{m_k} q_{ks}(x_i').$$

where $\boldsymbol{x}_{[-i]}^{(k)} = (x_1^{(k+1)}, \ldots, x_{i-1}^{(k+1)}, x_{i+1}^{(k)}, \ldots, x_n^{(k)})$.

Now we are ready to present a summary of the AEMC algorithm, which consists of two modes: the EMC mode and the data-mining mode.

Adaptive EMC Algorithm

1. Set $k = 0$. Start with an initial population $\boldsymbol{x}^{(0)}$ by uniformly sampling n samples over $\mathcal{X}$ and a temperature ladder $t_1 > \ldots > t_n$.

2. EMC mode: run EMC until a switching condition is met. Apply mutation, crossover, and exchange operators (described in Section 5.5) to the population $\boldsymbol{x}^{(k)}$ and accept the updated population according to the Metropolis-Hastings rule. Set $k = k + 1$.

3. Run the data-mining mode until a switching condition is met.
   (a) With probability P_k, use the CART method to update the promising regions, that is, update the values of $a_{js}^{(k+1)}$ and $b_{js}^{(k+1)}$.
   (b) With probability $1 - P_k$, set $a_{js}^{(k+1)} = a_{js}^{(k)}$ and $b_{js}^{(k+1)} = b_{js}^{(k)}$, and then Generate n new samples using the Metropolis-within-Gibbs procedure described above. Set $k = k + 1$.
4. Alternate between modes 3(a) and 3(b) until a stopping rule is met. The algorithm could terminate when the computational budget (the number of iterations) is consumed or when the change in the best $H(x)$ value does not exceed a given threshold for several consecutive iterations.

As implied by Theorem 8.1.1 and Theorem 8.1.3, adaptive EMC is ergodic and the weak law of large numbers holds for bounded functionals as long as $\lim_{k\to\infty} P_k = 0$, which ensures the diminishing adaptation condition to be satisfied. Since $\mathcal{X}$ is compact, Θ, the solution space of a_{js}'s and b_{js}'s, is also compact. Hence, Theorems 8.1.1 and 8.1.3 follow as long as the diminishing adaptation condition holds.

Example 8.2 Mixture Gaussian Distribution

Compare the Adaptive EMC, EMC and MH algorithms with a 5-D mixture Gaussian distribution

$$\pi(x) = \frac{1}{3}N_5(\mathbf{0}, I_5) + \frac{2}{3}N_5(\mathbf{5}, I_5), \quad x \in [-10, 10]^5,$$

where $\mathbf{0} = (0, 0, 0, 0, 0)^T$ and $\mathbf{5} = (5, 5, 5, 5, 5)^T$; see Liang and Wong (2001a).

In our simulations, each algorithm was run until 10^5 samples were produced and all numerical results were averages of 10 independent runs. The MH algorithm was applied with a uniform proposal density $U[x-2, x+2]^5$. The acceptance rate was 0.22. The MH algorithm cannot escape from the mode in which it started. To compare EMC and adaptive EMC, we look at only the samples at the first dimension, as all dimensions are mutually independent. Specifically, we divide the interval $[-10, 10]$ into 40 intervals with a resolution of 0.5, and calculate the true and estimated probability mass respectively in each of the 40 intervals.

In AEMC, we set $h = 25\%$ so that samples from both modes can be obtained. If h is too small, AEMC will focus only on the peaks of the function and thus only samples around the mode $\mathbf{5}$ can be obtained, because the probability of the mode $\mathbf{5}$ is twice as large as that of the mode $\mathbf{0}$. The other parameters, such as the mutation and crossover rates, were set as in Liang and Wong (2001a). Figure 8.1 shows the L^2 distance (between the estimated mass vector and its true value) versus the number of samples for the three methods in comparison. The results show that adaptive EMC converges significantly faster than EMC and the MH algorithm.

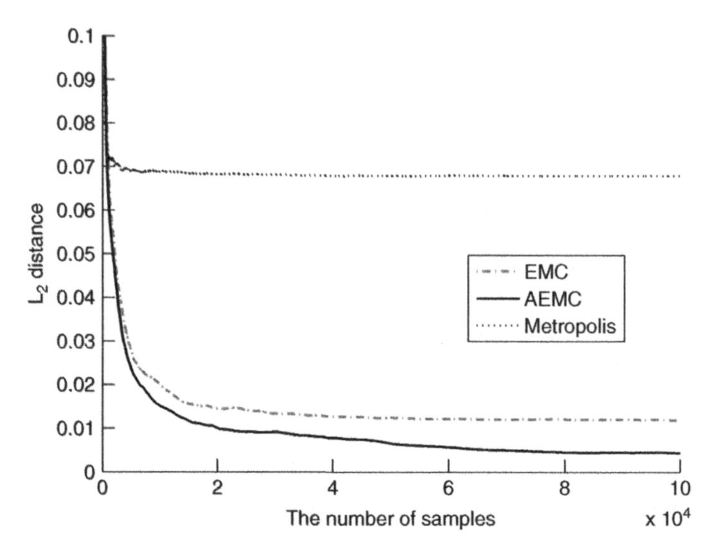

Figure 8.1 Comparison of convergence rates of different algorithms: adaptive EMC (AEMC), EMC and the MH algorithm (Ren *et al.*, 2008).

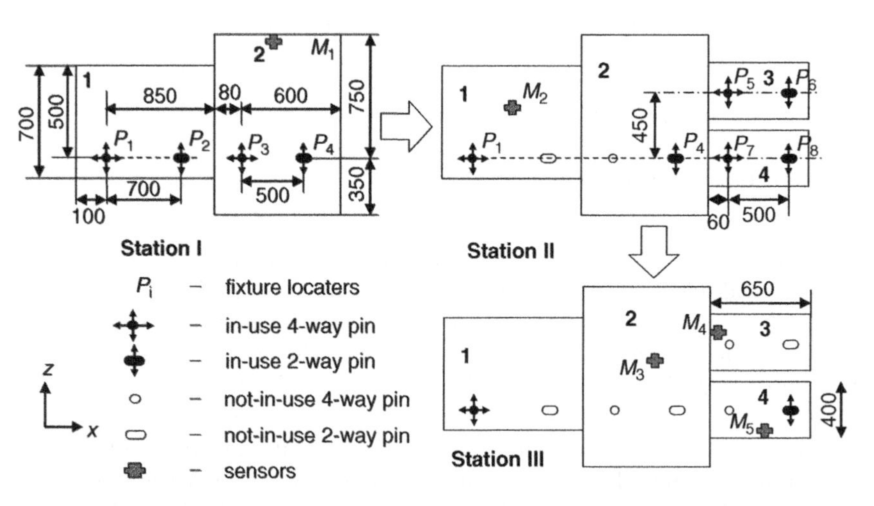

Figure 8.2 Illustrative example: a multi-station assembly process. The process proceeds as follows: (i) at the station I, part 1 and part 2 are assembled; (ii) at the station II, the subassembly consisting of part 1 and part 2 receives part 3 and part 4; and (iii) at the station III, no assembly operation is performed but the final assembly is inspected. The 4-way pins constrain the part motion in both the x- and the z-axes, and the 2-way pins constrain the part motion in the z-axis (Ren *et al.*, 2008).

8.4.3 Application to Sensor Placement Problems

The sensor placement problem is a constrained optimization problem, in which one wants to determine the number and locations of multiple sensors so that certain design criteria can be optimized within a given budget. Sensor placement issues have been encountered in various applications, such as manufacturing quality control, transportation management, and security surveillance. Depending on applications, the design criteria to be optimized include, among others, sensitivity, detection probability, and coverage.

Ren *et al.* (2008) consider the problem of finding an optimal sensor placement strategy in a three-station two-dimensional assembly process. The problem is illustrated by Figure 8.2, where coordinate sensors are distributed throughout the assembly process to monitor the dimensional quality of the final assembly and/or of the intermediate subassemblies, and $M_1 - M_5$ are 5 coordinate sensors that are currently in place on the stations; that is simply one instance of, out of hundreds of thousands of other possible, sensor placements. The goal of having these coordinate sensors is to estimate the dimensional deviation at the fixture locators on different stations, labeled as P_i, $i = 1, .., 8$. Researchers have established physical models connecting the sensor measurements to the deviations associated with the fixture locators (see, e.g., Mandroli *et al.*, 2006). Similar to optimal designs, people choose to optimize an alphabetic optimality criterion, such as D-optimality or E-optimality, of the information matrix determined by the sensor placement. Ren *et al.* (2008) used the E-optimality as a measure of the sensor system sensitivity.

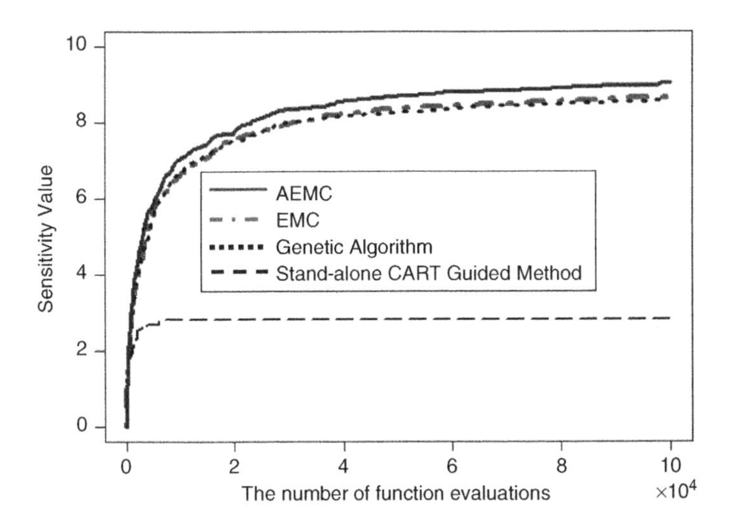

Figure 8.3 Performance comparison of adaptive EMC, EMC, genetic algorithm, and the stand-alone CART guided method for the 20-sensor placement problem (Ren *et al.*, 2008).

Ren *et al.* (2008) compare adaptive EMC, EMC, genetic algorithm (Holland, 1992), and the stand-alone CART guided method for this problem. For two cases of 9 and 20 sensors, each algorithm was run for 10^5 function evaluations. Figure 8.3 shows the result for 20 sensors, which is similar to the result for 9 sensors. It can bee seen from Figure 8.3 that adaptive EMC outperforms all other algorithms, EMC and the genetic algorithm perform similarly, and the stand-alone CART guided method performs much worse than the others. This indicates that incorporation of CART into EMC improves its convergence. More importantly, this application provides a framework on how to incorporate an optimization procedure into MCMC simulations through an adaptive process.

Exercises

8.1 (Roberts and Rosenthal, 2007). Let $\mathcal{X} = \{1, 2, 3, 4\}$, with $\pi(1) = \pi(2) = \pi(3) = 2/7$ and $\pi(4) = 1/7$. Let $P_1(1,2) = P_1(3,1) = P_1(4,3) = 1$ and $P_1(2,3) = P_1(2,4) = 1/2$. Similarly, let $P_2(2,1) = P_2(3,2) = P_2(4,3) = 1$ and $P_2(1,3) = P_2(1,4) = 1/2$. Verify that each of P_1 and P_2 are irreducible, aperiodic, and has stationary distribution $\pi(\cdot)$, but an adaptive algorithm that alternates p_1 and P_2 fails to be irreducible.

8.2 (Haario *et al.*, 2001). Compare the adaptive Metropolis algorithm and the conventional Metropolis algorithm on the following distribution (banana-shaped distribution)

$$\pi_B(x_1, x_2, \ldots, x_d) \propto \exp\left\{-\frac{1}{200}x_1^2 - \frac{1}{2}(x_2 + Bx_1^2 - 100B)^2 - \frac{1}{2}(x_3^2 + x_4^2 + \cdots + x_d)^2\right\},$$

where, for instance, set $d = 20$ and $B = 0.1$.

8.3 The Metropolis adjusted Langevin algorithm (Roberts and Tweedie, 1996) is a useful Monte Carlo algorithm when the gradient of the log-target density function is available. Describe an adaptive Langevin algorithm. Roberts and Rosenthal (1998) show that, for large d, the optimal acceptance rate for the Metropolis adjusted Langevin algorithm is 0.574.

8.4 Let $\pi(x)$ be a density function

$$\pi(x) \propto 4\min\left\{\left(1 + \frac{2}{3}\right)^\alpha, \left(\frac{4}{3} - x\right)^\alpha\right\} + \min\{(x + \frac{1}{3})^\alpha, (\frac{5}{3} - x)^\alpha\}, \quad x \in (0, 1),$$

where $\alpha = 30$. Simulate $\pi(x)$ using the adaptive Metropolis-Hastings algorithm.

8.5 (Yang, 2007). Consider an adaptive MCMC algorithm with diminishing adaptation, such that there is $C \in \mathcal{F}$, $V : \mathcal{X} \to [1, \infty)$ such that $\int V(x)\pi(x) < \infty$, $\delta > 0$, and $b < \infty$, with $\sup_{x \in C} V(x) < \infty$, and

- **(i)** For each $\theta \in \Theta$, there exists a probability measure $\nu_\theta(\cdot)$ on C with $P_\theta(x, \cdot) \geq \delta\nu_\theta(\cdot)$ for all $x \in C$.
- **(ii)** $P_\theta V \leq V - 1 + bI_C$ for each $\theta \in \Theta$.
- **(iii)** The set $\Delta = \{\theta \in \Theta : P_\theta \leq V - 1 + bI_C\}$ is compact with respect to the metric $d(\theta_1, \theta_2) = \sup_{x \in \mathcal{X}} \|P_{\theta_1}(x, \cdot) - P_{\theta_2}(x, \cdot)\|$.
- **(iv)** The sequence $\{V(X_n)\}_{n \geq 0}$ is bounded in probability.
  - **(a)** (Ergodicity) Show the adaptive MCMC algorithm is ergodic with respect to $\pi(\cdot)$.
  - **(b)** (WLLN) Let $h : \mathcal{X} \to \mathbb{R}$ be a bounded measurable function. Show

$$\frac{1}{n}\sum_{i=1}^{n} h(X_i) \to \int h(x)\pi(x)dx, \quad \text{in probability.}$$

8.6 Discuss how to incorporate the conjugate gradient algorithm into a MH simulation through the diminishing adaptation technique.

8.7 Prove Theorem 8.1.1.

8.8 Prove Theorem 8.1.2.

8.9 Prove Theorem 8.1.3.

8.10 Prove Theorem 8.2.1.

References

Aerts, M., Augustyns, I., and Janssen, P. (1997) Smoothing sparse multinomial data using local polynomial fitting. *Journal of Nonparametric Statistics*, **8** 127–147.

Albert, J.H. and Chib, S. (1993) Bayesian analysis of binary and polychotomous response data. **88** (422), 669–679.

Altekar, G., Dwarkadas, S., Huelsenbeck, J.P., *et al.* (2004) Parallel Metropolis coupled Markov chain Monte Carlo for Bayesian phylogenetic inference. *Bioinformatics*, **20** (3), 407–415.

Amato, S., Apolloni, B., Caporali, G., *et al.* (1991) Simulated annealing approach in back-propagation. *Neurocomputing*, **3** (5–6), 207–220.

Andrews, D.F. and Herzberg, A.M. (1985) *Data,* Springer, New York.

Andrieu, C. and Moulines, É. (2006) On the ergodicity properties of some adaptive MCMC algorithms. *Annals of Applied Probability*, **16** (3), 1462–1505.

Andrieu, C., Moulines, É., and Priouret, P. (2005) Stability of stochastic approximation under verifiable conditions. *SIAM Journal of Control and Optimization*, **44** (1), 283–312.

Andrieu, C. and Robert, C.P. (2001) Controlled MCMC for optimal sampling. Technical Report 0125, Cahiers de Mathématiques du Ceremade, Université Paris-Dauphine.

Andrieu, C. and Roberts, G.O. (2009) The pseudo-marginal approach for efficient Monte Carlo computations. *Annals of Statistics*, **37** (2), 697–725.

Andrieu, C. and Thoms, J. (2008) A tutorial on adaptive MCMC. *Statistics and Computing*, **18** (4), 343–373.

Anquist, L. and Hössjer, O. (2004) Using importance sampling to improve simulation in linkage analysis. *Statistical Applications in Genetics and Molecular Biology*, **3** (1), article 5.

Atchadé Y.F. and Rosenthal, J.S. (2005) On adaptive Markov chain Monte Carlo algorithms. Bernoulli, **11** (5), 815–828.

Balram, N. and Moura, J.M.F. (1993) Noncausal Gauss Markov random fields: parameter structure and estimation. *IEEE Transactions on Information Theory*, **39** (4), 1333–1355.

Barker, A.A. (1965) Monte Carlo calculations of the radial distribution functions for a proton-electron plasma. *Australian Journal of Physics*, **18**, 119–133.

Barry, D. and Hartigan, J.A. (1993) A Bayesian analysis for change point problems. *Journal of the American Statistical Association*, **88**, 309–319.

Bartz, K., Blitzstein, J., and Liu, J. (2008) Monte Carlo maximum likelihood for exponential random graph models: from snowballs to umbrella densities. Technical Report, Department of Statistics, Harvard University.

Bastolla, U., Frauenkron, H., Gerstner, E., *et al.* (1998) Testing a new Monte Carlo algorithm for protein folding. *Proteins*, **32** (1), 52–66.

Baum, E.B. and Lang, K.J. (1991) Constructing hidden units using examples and queries, in *Advances in Neural Information Processing Systems, 3,* Morgan Kaufmann, San Mateo, pp. 904–910.

Bayes, T. (1764) An essay towards solving a problem in the doctrine of chances. *Philosophical Transactions,* **53**. The Royal Society, London. Reprinted 1958, *Biometrika*, **45** (3–4), 293–315.

Beaumont, M.A. (2003) Estimation of population growth or decline in genetically monitored populations. *Genetics*, **164**, 1139–1160.

Bélisle, C.J.P., Romeijn, H.E., and Smith, R.L. (1993) Hit–and–run algorithms for generating multivariate distributions. *Mathematics of Operations Research*, **18**, 255–266.

Benveniste, A., Métivier, M., and Priouret, P. (1990) *Adaptive Algorithms and Stochastic Approximations,* Springer-Verlag, New York.

Berger, J.O. (1985) *Statistical Decision Theory and Bayesian Analysis,* 2nd edn, Springer-Verlag, New York.

Berger, J.O. (1993) The present and future of Bayesian multivariate analysis, in *Multivariate Analysis: Future Directions* (ed. C.R. Rao), North-Holland, Amsterdam, pp. 25–53.

Bernardo, J.M. (1979) Reference posterior distributions for Bayesian inference. *Journal of the Royal Statistical Society, Series B*, **41** (2), 113–147.

Bernardo, J.M. and Smith, A.F.M. (1994) *Bayesian Theory,* John Wiley & Sons Ltd, Chichester.

Bernstein, B.E., Kamal, M., Lindblad-Toh, K., *et al.* (2005) Genomic maps and comparative analysis of histone modifications in human and mouse. *Cell*, **120** (2), 169–181.

Bertone, P., Stolc, V., Royce, T.E., *et al.* (2004) Global identification of human transcribed sequences with genome tiling arrays. *Science*, **306** (5705), 2242–2246.

Besag, J.E. (1974) Spatial interaction and the statistical analysis of lattice systems. *Journal of the Royal Statistical Society, Series B*, **36**, 192–326.

Besag, J.E. (1994) Comment on 'representations of knowledge in complex systems' by Grenander and Miller. *Journal of the Royal Statistical Society, Series B*, **56**, 591–592.

Besag, J.E., Green, P.J., Higdon, D., and Mengersen, K. (1995) Bayesian computation and stochastic systems, *Statistical Science*, **10** (1), 3–41.

Besag, J. and Kooperberg, C. (1995) On conditional and intrinsic autoregressions. *Biometrika*, **82**, 733–746.

Besag, J.E., and Moran, P.A.P. (1975) On the estimation and testing of spatial interaction in Gaussian lattice processes. *Biometrika*, **62**, 555–562.

Berg, B.A., Meirovitch, H., Novotny, M.A., *et al.* (eds) (1999) Monte Carlo and structure optimization methods for biology, chemistry and physics. Electronic Proceedings, `http://wwww.scri.fsu.edu/MCatSCRI/proceedings`, Florida State University.

Berg, B.A., and Neuhaus, T. (1991) Multicanonical algorithms for first order phase transitions. *Physics Letters B*, **267** (2), 249–253.

Berg, B.A., and Neuhaus, T. (1992) Multicanonical ensemble: a new approach to simulate first-order phase transitions. *Physics Review Letters*, **68** (1), 9–12.

Bickel, P.J. and Doksum, K.A. (2000) *Mathematical Statistics: Basic Ideas and Selected Topics,* vol. **1**, 2nd edn, Prentice Hall, Englewood Cliffs.

Billingsley, P. (1986) *Probability and Measure,* 2nd edn, John Wiley & Sons, Inc., New York.

Bolstad, B.M., Irizarry, R.A., Astrand, M., *et al.* (2003) A comparison of normalization methods for high density ologonucleotide array data based on variance and bias. *Bioinformatics*, **19** (2), 185–193.

Boneh, A. and Golan, A. (1979) Constraints' redundancy and feasible region boundedness by random feasible point generator (RFPG). *Third European Congress on Operations Research*, EURO III, Amsterdam.

Bonomi, E. and Lutton, J.L. (1984) The n-city traveling salesman problem: statistical mechanics and the Metropolis algorithm. *SIAM Review*, **26**, 551–568.

Booth, J.G. and Hobert, J.P. (1999) Maximizing generalized linear mixed model likelihoods with an automated Monte Carlo EM algorithm. *Journal of the Royal Statistical Society, Series B*, **61** (1), 265–285.

Box, G.E.P. and Jenkins, G.M. (1970) *Time Series Analysis, Forecast and Control,* Holden Day, San Francisco.

Box, G.E.P. and Tiao, G.C. (1973) *Bayesian Inference in Statistical Analysis,* John Wiley & Sons, Inc., New York.

Breiman, L., Friedman, J., Stone, C.J., and Olshen, R.A. (1984) *Classification and Regression Trees,* Chapman & Hall, London.

Brockwell, A.E. and Kadane, J.B. (2005) Identification of regeneration times in MCMC simulation, with application to adaptive schemes. *Journal of the Royal Statistical Society, Series B*, **14** (2), 436–458.

Brooks, S.P. and Gelman, A. (1998) General methods for monitoring convergence of iterative simulations. *Journal of Computational and Graphical Statistics*, **7** (4), 434–455.

Brooks, S.P. Giudici, P., and Roberts, G. (2003) Efficient construction of reversible jump Markov chain Monte Carlo proposal distributions. *Journal of the Royal Statistical Society, Series B*, **65**, 3–55.

Broyden, C.G. (1970) The convergence of a class of double-rank minimization algorithms. *IMA Journal of Applied Mathematics*, **6** (3), 76–90.

Burman, P. (1987) Smoothing sparse contingency tables. *Sankhyā*, **49**, series A, 24–36.

Cai, B., Meyer, R., and Perron, F. (2008) Metropolis-Hastings algorithms with adaptive proposals. *Statistics and Computing*, **18** (4), 421–433.

Cappé O., Guillin, A., Marin, J.M., *et al.* (2004) Population Monte Carlo. *Journal of Computational and Graphical Statistics*, **13** (4), 907–929.

Carroll, J.S., Liu, X.S., Brodsky, A.S., *et al.* (2005) Chromosome-wide mapping of estrogen receptor binding reveals long-range regulation requiring the forkhead protein FoxA1. *Cell*, **122** (1), 33–43.

Cawley, S., Bekiranov, S., Ng, H.H., *et al.* (2004) Unbiased mapping of transcription factor binding sites along human chromosomes 21 and 22 points to widespread regulation of noncoding RNAs. *Cell*, **116** (4), 499–509.

Chan, K.S. (1989) A note on the geometric ergodicity of a Markov chain. *Advances in Applied Probability*, **21**, 702–704.

Chan, H.S. and Dill, K.A. (1993) Energy landscapes and the collapse dynamics of homopolymers. *Journal of Chemical Physics*, **99**, 2116–2127.

Chatterjee, S., Carrera, C., and Lynch, L.A. (1996) Genetic algorithms and traveling salesman problems. *European Journal of Operational Research*, **93** (3), 490–510.

Chen, D., Riedl, T., Washbrook, E., *et al.* (2000) Activation of estrogen receptor α by S118 phosphorylation involves a ligand-dependent interaction with TFIIH and participation of CDK7. *Molecular Cell*, **6** (1), 127–137.

Chen, H.F. (1993) Asymptotic efficient stochastic approximation. *Stochastics and Stochastics Reports*, **45**, 1–16.

Chen, H.F. (2002) *Stochastic Approximation and Its Applications*. Kluwer Academic Publishers, Dordrecht.

Chen, M.-H. (1994) Importance-weighted marginal Bayesian posterior density estimation. *Journal of the American Statistical Association*, **89**, 818–824.

Chen, M.-H., and Schmeiser, B.W. (1993) Performance of the Gibbs, hit-and-run, and Metropolis samplers. *Journal of Computational and Graphical Statistics*, **2**, 251–272.

Chen, M.-H. and Schmeiser, B.W. (1996) General hit-and-run Monte Carlo sampling for evaluating multidimensional integrals. *Operations Research Letters*, **19** (4), 161–169.

Chen, M.-H. and Schmeiser, B.W. (1998) Towards black-box sampling. *Journal of Computational and Graphical Statistics*, **7** (1), 1–22.

Chen, M.-H. and Shao, Q.-M. (1997) On Monte Carlo methods for estimating ratios of normalizing constants. *Annals of Statistics*, **25** (4), 1563–1594.

Chen, M.-H., Shao, Q.-M., and Ibrahim, J.G. (2000) *Monte Carlo Methods in Bayesian Computation,* Springer, New York.

Cheng, R.C.H. and Feast, G.M. (1979) Some simple gamma variate generators. *Journal of the Royal Statistical Society, Series C* (Applied Statistics), **28** (3), 290–295.

Cheng, R.C.H. and Feast, G.M. (1980) Gamma variate generators with increased shape parameter range. *Communications of the ACM*, **23** (7), 389–394.

Cheon, S. and Liang, F. (2008) Phylogenetic tree reconstruction using sequential stochastic approximation Monte Carlo. *Biosystems*, **91** (1), 94–107.

Cheon, S. and Liang, F. (2009) Bayesian phylogeny analysis via stochastic approximation Monte Carlo. *Molecular Phylogenetics and Evolution*, **53** (2), 394–403.

Chickering, D.M. (1996) Learning Bayesian networks is NP–complete, in *Learning from Data: Artificial Intelligence and Statistics* (eds D. Fisher and H.J. Lenz) Springer-Verlag, New York, pp. 121–130.

Childs, A.M., Patterson, R.B., and MacKay, D.J. (2001) Exact sampling from nonattractive distributions using summary states. *Physical Review E*, **63**, 036113.

Cios, K.J., Wedding, D.K., and Liu, N. (1997) CLIP3: Cover learning using integer programming. *Kybernetes*, **26** (5), 513–536.

Cong, J., Kong, T., Xu, D., *et al.* (1999) Relaxed simulated tempering for VLSI floorplan designs, in *Proceedings of the Asia and South Pacific Design Automation Conference*, Hong Kong, pp. 13–16.

Cooper, G.F. and Yoo. C. (1999) Causal discovery from a mixture of experimental and observational data, in *Proceedings of the Fifteenth Conference on Uncertainty in Artificial Intelligence*, Morgan Kaufmann, CA, pp. 116–125.

Cowles, M.K. and Carlin, B.P. (1996) Markov chain Monte Carlo convergence diagnostics: a comparative review. *Journal of the Americal Statistical Association*, **91**, 883–904.

Creutz, M. (1980) Monte Carlo study of quantized SU (2) gauge theory. *Physical Review D*, **21**, 2308–2315.

Crippen, G.M. (1991) Prediction of protein folding from amino acid sequence over discrete conformation spaces. *Biochemistry*, **30** (17), 4232–4237.

Damien, P., Wakefield, J.C., and Walker, S.G. (1999) Gibbs sampling for Bayesian nonconjugate and hierarchical models by using auxiliary variables. *Journal of the Royal Statistical Society, Series B*, **61** (2), 331–344.

Dawid, A.P. (1985) Calibration-based empirical probability. *Annals of Statistics*, **13** (4), 1251–1285.

de Campos, L.M., and Huete, J.F. (2000) A new approach for learning belief networks using independence criteria. *International Journal of Approximating Reasoning*, **24**, 11–37.

de Candia, A.D. and Coniglio, A. (2002) Spin and density overlaps in the frustrated Ising lattice gas. *Physical Review E*, **65**, 016132.

de Meo, M. and Oh, S.K. (1992) Wolff algorithm and anisotropic continuous-spin models: an application to the spin-van der Waals model. *Physical Review B*, **46**, 257–260.

Dempster, A.P. (1966) New methods for reasoning towards posterior distributions based on sample data. *Annals of Mathematical Statistics*, **37**, 355–374.

Dempster, A.P., Laird, N.M., and Rubin, D.B. (1977) Maximum likelihood from incomplete data via the EM algorithm. *Journal of the Royal Statistical Society, Series B*, **39**, 1–38.

Dempster, A.P. (2008) The Dempster-Shafer calculus for statisticians. *International Journal of Approximating Reasoning*, **48**, 365–377.

Denison, D.G.T., Mallick, B.K., and Smith, A.F.M. (1998) Automatic Bayesian curve fitting. *Journal of the Royal Statistical Society, Series B*, **60**, 333–350.

Devroye L. (1986) *Non-Uniform Random Variate Generation,* Springer-Verlag, New York.

Dippon, J. and Renz, J. (1997) Weighted means in stochastic approximation of minima. *SIAM Journal on Control and Optimization*, **35**, 1811–1827.

Djurić, P.M., Huang, Y., and Ghirmai, T. (2002) Perfect sampling: a review and applications to signal processing. *IEEE Transactions on Signal Processing*, **50**, 345–356.

Dong, J. and Simonoff, J.S. (1994) The construction and properties of boundary kernels for smoothing sparse multinomials. *Journal of Computational and Graphical Statistics*, **3**, 57–66.

Dong, S. and Kraemer, E. (2004) Calculation, visualization, and manipulation of MASTs (maximum agreement subtrees). In, *Proceedings of the 2004 IEEE Computational Systems Bioinformatics Conference,* IEEE Computer Society, Washington, DC, pp. 405–414.

Duflo, M. (1997) *Random Iterative Models,* Springer, Berlin.

Durbin, R., Eddy, S., Krogh, A., and Mitchison, G. (1998) *Biological Sequence Analysis: Probabilistic Models of Proteins and Nucleic Acids,* Cambridge University Press, Cambridge.

Dryden, I., Ippoliti, L., and Romagnoli, L. (2002) Adjusted maximum likelihood and pseudolikelihood estimation for noisy Gaussian Markov random fields. *Journal of Computational and Graphical Statistics*, **11**, 370–388.

Edwards, R.G. and Sokal, A.D. (1988) Generalization of the Fortuin-Kasteleyn-Swendsen-Wang representation and Monte Carlo algorithm. *Physical Review D*, **38**, 2009–2012.

Efron, B. (2004) Large-scale simultaneous hypothesis testing: the choice of a null hypothesis. *Journal of the American Statistical Association*, **99**, 96–104.

Efron, B. and Tibshirani, R.J. (1993) *An Introduction to the Bootstrap,* Chapman & Hall, London.

Ellis, B. and Wong, W.H. (2008) Learning causal Bayesian network structures from experimental data. *Journal of the American Statistical Association*, **103** (482), 778–789.

Emahazion, T., Feuk, L., Jobs, M., *et al.* (2001) SNP association studies in Alzheimer's disease highlight problems for complex disease analysis. *Trends in Genetics*, **17** (7), 407–413.

Fahlman, S.E. and Lebiere, C. (1990) The cascade-correlation learning architecture, in *Advances in Neural Information Processing Systems 2* (ed. D.S. Touretzky), Morgan Kaufmann, San Mateo, pp. 524–532.

Fan, J., Heckman, N.E., and Wand, M.P. (1995) Local polynomial kernel regression for generalized linear models and quasi-likelihood functions. *Journal of the American Statistical Association*, **90**, 141–150.

Farach, M., Przytycka, T., and Thorup, M. (1995) On the agreement of many trees. *Information Processing Letters* **55**, 297–301.

Felsenstein, J. (1981) Evolutionary trees from DNA sequences: a maximum likelihood approach. *Journal of Molecular Evolution*, **17**, 368–376.

Felsenstein, J. (1993) *PHYLIP* (phylogenetic inference package), version 3.5. University of Washington, Seattle.

Feng, X., Buell, D.A., Rose, J.R., *et al.* (2003) Parallel algorithms for Bayesian phylogenetic inference. *Journal of Parallel and Distributed Computing*, **63** (7–8), 707–718.

Fernández, G., Ley, E., and Steel, M.F.J. (2001) Benchmark priors for Bayesian model averaging. *Journal of Econometrics*, **100** (2), 381–427.

Fill, J.A. (1998) An interruptible algorithm for perfect sampling via Markov chains. *Annals of Applied Probability*, **8**, 131–162.

Fisher, R.A. (1973) *Statistical Methods for Scientific Induction,* 3rd edn, Hafner Publishing Company, New York.

Fitch, W.M. (1971) Toward defining the course of evolution: minimal change for a specific tree topology. *Systematic Zoology*, **20**, 406–416.

Fletcher, R. (1970) A new approach to variable metric algorithms. *The Computer Journal*, **13** (3), 317–322.

Folland, G.B. (1990) Remainder estimates in Taylor's theorem. *American Mathematical Monthly*, **97**, 233–235.

Francois, O., Ancelet, S., and Guillot, G. (2006) Bayesian clustering using hidden Markov random fields in spatial population genetics. *Genetics*, **174** (2), 805–816.

Frank, O. and Strauss, D. (1986) Markov graphs. *Journal of the American Statistical Association*, **81**, 832–842.

Friedman, J.H., and Silverman, B.W. (1989) Flexible parsimonious smoothing and additive modeling. *Technometrics*, **31** (1), 3–21.

Friedman, N., Linial, M., Naachman, I., *et al.* (2000) Using Bayesian network to analyze expression data. *Journal of Computational Biology*, **7** (3–4), 601–620.

Friedman, N. and Koller, D. (2003) Being Bayesian about network structure: a Bayesian approach to structure discovery in Bayesian networks. *Machine Learning*, **50**, 95–125.

Gabr, M.M. and Subba Rao, T. (1981) The estimation and prediction of subset bilinear time series models with applications. *Journal of Time Series Analysis*, **2**, 155–171.

Galtier, N., Gascuel, O., and Jean-Marie, A. (2005) Markov models in molecular evolution, in, *Statistical Methods in Molecular Evolution,* (ed. R. Nielsen), Springer, New York, pp. 3–24.

Gelfand, A.E. and Sahu, S.K. (1994) On Markov chain Monte Carlo acceleration. *Journal of Computational and Graphical Statistics*, **3**, 261–276.

Gelfand, A.E., Sahu, S., and Carlin, B. (1995) Efficient parameterization for normal linear mixed effects models. *Biometrika*, **82**, 479–488.

Gelfand, A.E. and Smith, A.F.M. (1990) Sampling based approaches to calculating marginal densities. *Journal of the American Statistical Association*, **85**, 398–409.

Gelfand, A.E., Smith, A.F.M., and Lee, T.M. (1992) Bayesian analysis of constrained parameter and truncated data problems using Gibbs sampling. *Journal of the American Statistical Association*, **85**, 398–405.

Gelman, A., Carlin, J.B., Stern, H.S., and Rubin, D.B. (2004) *Bayesian Data Analysis,* 2nd edn, Chapman & Hall, London.

Gelman, A., King, G., and Liu, C. (1998) Not asked or not answered: multiple imputation for multiple surveys. *Journal of the American Statistical Association*, **93**, 846–874.

Gelman, A. and Meng, X.L. (1998) Simulating normalizing constants: from importance sampling to bridge sampling to path sampling. *Statistical Science*, **13**, 163–185.

Gelman A. and Speed T.P. (1993) Characterizing a joint probability distribution by conditionals. *Journal of the Royal Statistical Society, Series B*, **55**, 185–188.

Gelman, A., Roberts, R.O., and Gilks, W.R. (1996) Efficient Metropolis jumping rules, in *Bayesian Statistics 5* (eds J.M. Bernardo, J.O. Berger, A.P. Dawid and A.F.M. Smith) Oxford University Press, New York, pp. 599–607.

Gelman, A. and Rubin, D.B. (1992) Inference from iterative simulation using multiple sequences. *Statistical Science*, **7**, 457–511.

Geman, S. and Geman, D. (1984) Stochastic relaxation, Gibbs distributions and the Bayesian restoration of images. *IEEE Transactions on Pattern Analysis and Machine Intelligence*, **6**, 721–741.

Geweke, J. (1989) Bayesian inference in econometric models using Monte Carlo integration. *Econometrica*, **57**, 1317–1339.

Geyer, C.J. (1991) Markov chain Monte Carlo maximum likelihood, in *Computing Science and Statistics: Proceedings of the 23rd Symposium on the Interface,* (ed. E.M. Keramigas), pp. 153–163.

Geyer, C.J. (1992) Practical Monte Carlo Markov chain. *Statistical Science*, **7**, 473–511.

Geyer, C.J. and Thompson, E.A. (1992) Constrained Monte Carlo maximum likelihood for dependent data. *Journal of the Royal Statistical Society, Series B*, **54**, 657–699.

Geyer, C.J. and Thompson, E.A. (1995) Annealing Markov chain Monte Carlo with applications to ancestral inference. *Journal of the American Statistical Association*, **90**, 909–920.

Gilks, W.R. (1992) *Derivative-free adaptive rejection sampling for Gibbs sampling. Bayesian Statistics 4,* (eds. J. Bernardo, J. Berger, A.P. Dawid, and A.F.M. Smith), Oxford University Press, Oxford.

Gilks, W.R., Best, N.G., and Tan, K.K.C. (1995) Adaptive rejection Metropolis sampling. *Applied Statistics*, **44**, 455–472.

Gilks, W.R., Roberts, G.O., and George, E.I. (1994) Adaptive Direction Sampling, *The Statistician*, **43**, 179–189.

Gilks, W.R., Roberts, G.O., and Sahu, S.K. (1998) Adaptive Markov chain Monte Carlo through regeneration. *Journal of the American Statistical Association*, **93**, 1045–1054.

Gilks, W.R. and Wild, P. (1992) Adaptive rejection sampling for Gibbs sampling. *Applied Statistics*, **41**, 337–348.

Giordani, P. and Kohn, R. (2009) Adaptive independent Metropolis-Hastings by fast estimation of mixtures of normals. *Journal of Computational and Graphical Statistics*; in press.

Giudici, P. and Green, P. (1999) Decomposable graphical Gaussian model determination. *Biometrika*, **86**, 785–801.

Glover, F. and Laguna, M. (1997) *Tabu Search,* Kluwer Academic Publishers, Norwell, MA.

Goddard, W., Kubicka, E., Kubicki, G., *et al.* (1994) The agreement metric for labeled binary trees. *Mathematical Biosciences*, **123**, 215–226.

Goldberg, D.E. (1989) *Goldberg Algorithms in Search, Optimization, and Machine Learning,* Addison Wesley.

Golden, B., and Skiscim, C. (1986) Using simulated annealing to solve routing and location problems. *Naval Research Logistic Quarterly*, **33**, 261–279.

Goldfarb, D. (1970) A family of variable metric methods derived by variational means. *Mathematics of Computation*, **24**, 23–26.

Good, P.I. (2005) *Permutation, Parametric, and Bootstrap Tests of Hypotheses,* 3rd edn, Springer, New York.

Goswami, G. and Liu, J.S. (2007) On learning strategies for evolutionary Monte Carlo. *Statistics and Computing*, **17**, 23–38.

Gottardo, R., Li, W., Johnson, W.E., *et al.* (2008) A flexible and powerful Bayesian hierarchical model for ChIP–chip experiments. *Biometrics*, **64**, 468–478.

Grassberger, P. (1997) Pruned-enriched Rosenbluth method: simulation of θ polymers of chain length up to 1,000,000. *Physical Review E*, 3682–3693.

Green, P.J. (1995) Reversible jump Markov chain Monte Carlo computation and Bayesian model determination. *Biometrika*, **82**, 711–732.

Green, P.J. and Mira, A. (2001) Delayed rejection in reversible jump Metropolis-Hastings. *Biometrika*, **88**, 1035–1053.

Green, P.J. and Richardson, S. (2002) Hidden Markov Models and disease mapping. *Journal of the American Statistical Association*, **97**, 1055–1070.

Grenander, U. and Miller, M.I. (1994) Representations of knowledge in complex systems. *Journal of the Royal Statistical Society, Series B*, **56**, 549–603.

Gu, C. (1993) Smoothing spline density estimation: A dimensionless automatic algorithm. *Journal of the American Statistical Association*, **88**, 495–504.

Gu, C. and Qiu, C. (1993) Smoothing spline density estimation: theory. *Annals of Statistics*, **21**, 217–234.

Haario, H., Saksmaan, E., and Tamminen, J. (2001) An adaptive Metropolis Algorithm. *Bernoulli*, **7**, 223–242.

Hall, P. and Heyde, C.C. (1980) *Martingale Limit Theory and its Application,* Academic Press, New York.

Hall, P. and Titterington, D.M. (1987) On smoothing sparse multinomial data. *Australian Journal of Statistics*, **29**, 19–37.

Hall, P., Lahiri, S.N., and Truong, Y.K. (1995) On bandwidth choice for density estimation with dependent data. *Annals of Statistics*, **23**, 2241–2263.

Hamada, M. and Wu, C.F.J. (1995) Analysis of censored data from fractionated experiment: a Bayesian approach, *Journal of the American Statistical Association*, **90**, 467–477.

Handcock, M. (2003) Statistical models for social networks: degeneracy and inference, in *Dynamic Social Network Modeling and Analysis* (eds R. Breiger, K. Carley, and P. Pattison), National Academies Press, Washington, DC, pp. 229–240.

Hansmann, U.H.E., and Okamoto, Y. (1997) Numerical comparisons of three recently proposed algorithms in the protein folding problems. *Journal of Computational Chemistry*, **18**, 920–933.

Hart, J.D. and Vieu, P. (1990) Data-driven bandwidth choice for density estimation based on dependent data. *The Annals of Statistics*, **18**, 873–890.

Hasegawa, M., Kishino, H., and Yano, T. (1985) Dating of the human-ape splitting by a molecular clock of mitochondrial DNA. *Journal of Molecular Evolution*, **22**, 160–174.

Hastie, T.H. and Tibshirani, R.J. (1990) *Generalized Additive Models,* Chapman & Hall, London.

Hastings, W.K. (1970) Monte Carlo sampling methods using Markov chain and their applications. *Biometrika*, **57**, 97–109.

Heckerman, D., Geiger, D., and Chickering, D.M. (1995) Learning Bayesian networks: the combination of knowledge and statistical data. *Machine Learning*, **20**, 197–243.

Hedar, A.R. and Fukushima, M. (2006) Tabu search directed by direct search methods for nonlinear global optimization. *European Journal of Operational Research*, **170**, 329–349.

Hernandez-Lerma, O., and Lasserre, J.B. (2001) Further criteria for positive Harris recurrence of Markov chains. *Proceedings of the American Mathematical Society*, **129**, 1521–1524.

Herskovits, E. and Cooper, G.F. (1990) Kutató: an entropy-driven system for the construction of probabilistic expert systems from datasets, in *Proceedings of the Sixth Conference on Uncertainty in Artificial Intelligence* (ed. P. Bonissone), Cambridge, pp. 54–62.

Hesselbo, B. and Stinchcombe, R.B. (1995) Monte Carlo simulation and global optimization without parameters. *Physical Review Letters*, **74**, 2151–2155.

Higdon, D.M. (1998) Auxiliary variable methods for Markov chain Monte Carlo with applications. *Journal of the American Statistical Association*, **93**, 585–595.

Hills, S.E. and Smith, A.F.M. (1992) Parameterization issues in Bayesian inference, in *Bayesian Statistics 4*, (eds. J.M. Bernardo, J.O. Berger, A.P. Dawid, and A.F.M. Smith) Oxford University Press, Oxford, pp. 641–649.

Hills, T., O'Connor, M., and Remus, W. (1996) Neural network models for time series forecasts. *Management Science*, **42**, 1082–1092.

Hirsch, M.J., Meneses, C.N., Pardalos, P.M., *et al.* (2007) Global optimization by continuous GRASP. *Optimization Letters*, **1**, 201–212.

Hirsch, M.J., Pardalos, P.M., and Resende, M.G.C. (2006) Speeding up continuous GRASP. Submitted to the *European Journal of Operational Research.*

Holden, L., Hauge, R., and Holden, M. (2009) Adaptive independent Metropolis-Hastings. *Annals of Applied Probability*, **19**, 395–413.

Holland, J.H. (1992) *Adaptation in Natural and Artificial Systems,* University of Michigan Press, Ann Arbor, MI.

Holley, R.A., Kusuoka, S., and Stroock, D.W. (1989) Asymptotics of the spectral gap with applications to the theory of simulated annealing. *Journal of Functional Analysis*, **83**, 333–347.

Huang, F. and Ogata, Y. (1999) Improvements of the maximum pseudo-likelihood estimators in various spatial statistical models. *Journal of Computational and Graphical Statistics*, **8**, 510–530.

Huang, F. and Ogata, Y. (2002) Generalized pseudo-likelihood estimates for Markov random fields on lattice. *Annals of the Institute of Statistical Mathematics*, **54**, 1–18.

Huber, W., Toedling, J., Steinmetz, L.M. (2006) Transcript mapping with high-density oligonucleotide tiling arrays. *Bioinformatics*, **22** (16), 1963–1970.

Huelsenbeck, J.P. and Ronquist, F. (2001) MrBayes: Bayesian inference of phylogenetic trees. *Bioinformatics*, **17** (8), 754–755.

Hukushima K. and Nemoto, K. (1996) Exchange Monte Carlo method and application to spin glass simulations. *Journal of the Physical Society of Japan*, **65** (6), 1604–1608.

Hull, J. and White, A. (1987) The pricing of options on assets with stochastic volatility. *Journal of Finance*, **42** (2), 281–300.

Humburg, P., Bulger, D., Stone, G. (2008) Parameter estimation for robust HMM analysis of ChIP-chip data. *BMC Bioinformatics*, **9**, 343.

Hunter, D. (2007) Curved exponential family models for social network. *Social Networks*, **29** (2), 216–230.

Hunter, D. and Handcock, M. (2006) Inference in curved exponential family models for network. *Journal of Computational and Graphical Statistics*, **15**, 565–583.

Hunter, D.R., Goodreau, S.M., and Handcock, M.S. (2008) Goodness of fit of social network models. *Journal of the American Statistical Association*, **103**, 248–258.

Hurn, M., Husby, O. and Rue, H. (2003) A tutorial on image analysis. *Lecture Notes in Statistics*, **173**, 87–141.

Jarner, S. and Hansen, E. (2000) Geometric ergodicity of Metropolis algorithms. *Stochastic Processes and their Applications*, **85**, 341–361.

Jasra, A., Stephens, D., and Holmes, C. (2007) Population-based reversible jump Markov chain Monte Carlo. *Biometrika*, **94**, 787–807.

Jasra, A., Stephens, D., and Holmes, C. (2007b) On population-based simulation for static inference. *Statistics and Computing*, **17** (3), 263–279.

Jeffreys, H. (1961) *Theory of Probability,* 3rd edn, Oxford University Press, Oxford.

Jin, I.H. and Liang, F. (2009) Bayesian analysis for exponential random graph models using the double Metropolis-Hastings sampler. Technical Report, Department of Statistics, Texas A&M University.

Jobson, J.D. (1992) *Applied Multivariate Data Analysis* (vol. II): *Categorical and Multivariate Methods*. Springer-Verlag, New York.

Jukes, T. and Cantor, C. (1969) Evolution of protein molecules, in *Mammalian Protein Metabolism,* (ed. H.N. Munro), Academic Press, New York.

Kang, S. (1991) An investigation of the use of feedforward neural networks for forecasting. Ph.D. Dissertation, Kent State University, Kent, Ohio.

Kass, R.E. and Raftery, A.E. (1995) Bayes factors. *Journal of the American Statistical Association*, **90**, 773–795.

Kass, R.E. and Wasserman, L. (1996) The selection of prior distributions by formal rules. *Journal of the American Statistical Association*, **91**, 1343–1370.

Keles, S. (2007) Mixture modeling for genome-wide localization of transcription factors. *Biometrics*, **63**, 10–21.

Keles, S., Van Der Laan, M.J., Dudoit, S., *et al.* (2006) Multiple testing methods for ChIP-chip high density oligonucleotide array data. *Journal of Computational Biology*, **13** (3), 579–613.

Kidd, K.K. and Sgaramella-Zonta, L.A. (1971) Phylogenetic analysis: concepts and methods. *American Journal of Human Genetics*, **23** (3), 235–252.

Kimmel, G. and Shamir, R. (2006) A fast method for computing high-significance disease association in large population-based studies. *American Journal of Human Genetics*, **79**, 481–492.

Kimura, M. (1980) A simple method for estimating evolutionary rate in a finite population due to mutational production of neutral and nearly neutral base substitution through comparative studies of nucleotide sequences. *Journal of Molecular Biology*, **16**, 111–120.

Kinderman, A.J. and Monahan, J.F. (1977) Computer generation of random variables using the ratio of uniform deviates. *ACM Transactions on Mathematical Software*, **3** (3), 257–260.

Kirkpatrick, S., Gelatt, C.D., and Vecchi, M.P. (1983) Optimization by simulated annealing. *Science*, **220**, 671–680.

Kohn, R. and Ansley, C. (1987) A new algorithm for spline smoothing based on smoothing a stochastic process. *SIAM Journal on Scientific and Statistical Computing*, **8**, 33–48.

Koo, J.Y. (1997) Spline estimation of discontinuous regression functions. *Journal of Computational and Graphical Statistics*, **6**, 266–284.

Kooperberg, C. (1998) Bivariate density estimation with an application to survival analysis. *Journal of Computational and Graphical Statistics*, **7**, 322–341.

Kooperberg, C. and Stone, C.J. (1991) A Study of logspline density estimation. *Computational Statistics and Data Analysis*, **12**, 327–348.

Kou, S.C., Zhou, Q., and Wong, W.H. (2006) Equi-energy sampler with applications to statistical inference and statistical mechanics. *Annals of Statistics*, **32**, 1581–1619.

Kurgan, L.A., Cios, K.J., Tadeusiewicz, R., *et al.* (2001) Knowledge discovery approach to automated cardiac SPECT diagnosis. *Artificial Intelligence in Medicine*, **23** (2), 149–169.

Kushner, H.J. and Yang, J. (1993) Stochastic approximation with averaging of the iterates: optimal asymptotic rate of convergence for general processes. *SIAM Journal on Control and Optimization*, **31**, 1045–1062.

Kushner, H.J. and Yang, J. (1995) Stochastic approximation with averaging and feedback: rapidly convergent 'on-line' algorithms. *IEEE Transactions on Automatic Control*, **40**, 24–34.

Kushner, H.J. and Yin, G.G. (2003) *Stochastic Approximation and Recursive Algorithms and Applications,* 2nd edn, Springer, New York.

Laguna, M. and Martí, R. (2005) Experimental testing of advanced scatter search designs for global optimization of multimodal functions. *Journal of Global Optimization*, **33**, 235–255.

Lam, W. and Bacchus, F. (1994) Learning Bayesian belief networks: an approach based on the MDL principle. *Computational Intelligence*, **10**, 269–293.

Lang, K.J. and Witbrock, M.J. (1989) Learning to tell two spirals apart, in *Proceedings of the 1988 Connectionist Models* (eds D. Touretzky, G. Hinton, and T. Sejnowski), Morgan Kaufmann, San Mateo, pp. 52–59.

Lange, K., Little, R.J.A., and Taylor, J.M.G. (1989) Robust statistical inference using the t distribution. *Journal of the American Statistical Association*, **84**, 881–896.

Larget, B. and Simon, D. (1999) Markov chain Monte Carlo algorithms for the Bayesian analysis of phylogenetic trees. *Molecular Biology and Evolution*, **16**, 750–759.

Lau, K.F. and Dill, K.A. (1990) Theory for protein mutability and biogenesis. *Proceedings of the National Academy of Sciences USA*, **87** (2), 638–642.

Lawrence, E., Bingham, D., Liu, C., *et al.* (2008) Bayesian inference for multivariate ordinal data using parameter expansion, *Technometrics*, **50**, 182–191.

Lawrence, E, Liu, C., Vander Wiel, S., *et al.* (2009) A new method for multinomial inference using Dempster-Shafer theory. Preprint.

Levine, R.A. and Casella, G. (2008) Comment: on random scan Gibbs samplers. *Statistical Science*, **23** (2), 192–195.

Lewandowski, A. and Liu, C. (2008). The sample Metropolis-Hastings algorithm. Technical Report, Department of Statistics, Purdue University.

Lewandowski, A., Liu, C., and Vander Wiel, S. (2010) Parameter expansion and efficient inference. Technical Report, Department of Statistics, Purdue University.

Li, K.H. (1988) Imputation using Markov chains, *Journal of Statistical Computing and Simulation*, **30**, 57–79.

Li, S., Pearl, D., and Doss, H. (2000) Phylogenetic tree construction using Markov chain Monte Carlo. *Journal of the American Statistical Association*, **95**, 493–508.

Li, W., Meyer, C.A., and Liu, X.S. (2005) A hidden Markov model for analyzing ChIP-chip experiments on genome tiling arrays and its application to p53 binding sequences. *Bioinformatics*, **21** (supplement), 274–282.

Li, Y., Protopopescu, V.A., and Gorin, A. (2004) Accelerated simulated tempering, *Physics Letters A*, **328** (4), 274–283.

Liang, F. (2002a) Some connections between Bayesian and non-Bayesian methods for regression model selection. *Statistics & Probability Letters*, **57**, 53–63.

Liang, F. (2002b) Dynamically weighted importance sampling in Monte Carlo Computation. *Journal of the American Statistical Association*, **97**, 807–821.

Liang, F. (2003) Use of sequential structure in simulation from high-dimensional systems. *Physical Review E*, **67**, 056101.

Liang, F. (2004) Generalized $1/k$-ensemble algorithm. *Physical Review E*, **69**, 66701–66707.

Liang, F. (2005a) Bayesian neural networks for non-linear time series forecasting. *Statistics and Computing*, **15**, 13–29.

Liang, F. (2005b) Generalized Wang-Landau algorithm for Monte Carlo computation. *Journal of the American Statistical Association*, **100**, 1311–1327.

Liang, F. (2006) A theory on flat histogram Monte Carlo algorithms. *Journal of Statistical Physics*, **122**, 511–529.

Liang, F. (2007a) Continuous contour Monte Carlo for marginal density estimation with an application to a spatial statistical model. *Journal of Computational and Graphical Statistics*, **16** (3), 608–632.

Liang, F. (2007b) Annealing stochastic approximation Monte Carlo for neural network training. *Machine Learning*, **68**, 201–233.

Liang, F. (2008a) Clustering gene expression profiles using mixture model ensemble averaging approach. *JP Journal of Biostatistics*, **2** (1), 57–80.

Liang, F. (2008b) Stochastic approximation Monte Carlo for MLP learning, in *Encyclopedia of Artificial Intelligence,* (eds. J.R.R. Dopico, J.D. de la Calle, and A.P. Sierra), pp. 1482–1489.

Liang, F. (2009a) Improving SAMC using smoothing methods: theory and applications to Bayesian model selection problems. *Annals of Statistics*, **37**, 2626–2654.

Liang, F. (2009b) On the use of SAMC for Monte Carlo integration. *Statistics & Probability Letters*, **79**, 581–587.

Liang, F. (2009c) A double Metropolis-Hastings sampler for spatial models with intractable normalizing constants. *Journal of Statistical Computing and Simulation*, in press.

Liang, F. (2009d) Evolutionary stochastic approximation Monte Carlo for global optimization. *Statistics and Computing*, revised edn.

Liang, F. (2009e) Trajectory averaging for stochastic approximation MCMC algorithms. *Annals of Statistics*, in press.

Liang, F. and Cheon, S. (2009) Monte Carlo dynamically weighted importance sampling for spatial models with intractable normalizing constants. *Journal of Physics* conference series, **197**, 012004.

Liang, F. and Jin, I.K. (2010) A Monte Carlo Metropolis-Hastings algorithm for sampling from distributions with intractable normalizing constants. Technical Report, Department of Statistics, Texas A&M University.

Liang, F. and Kuk, Y.C.A. (2004) A finite population estimation study with Bayesian neural networks. *Survey Methodology*, **30**, 219–234.

Liang, F. and Liu, C. (2005) Efficient MCMC estimation of discrete distributions. *Computational Statistics and Data Analysis*, **49**, 1039–1052.

Liang, F., Liu, C., and Carroll, R.J. (2007) Stochastic approximation in Monte Carlo computation. *Journal of the American Statistical Association*, **102**, 305–320.

Liang, F., Truong, Y.K., and Wong, W.H. (2001) Automatic Bayesian model averaging for linear regression and applications in Bayesian curve fitting. *Statistica Sinica*, **11**, 1005–1029.

Liang, F. and Wong, W.H. (1999) Dynamic weighting in simulations of spin systems. *Physics Letters A*, **252**, 257–262.

Liang, F. and Wong, W.H. (2000) Evolutionary Monte Carlo: application to *Cp* model sampling and change point problem. *Statistica Sinica*, **10**, 317–342.

Liang, F. and Wong, W.H. (2001a) Real parameter evolutionary Monte Carlo with applications in Bayesian mixture models. *Journal of the American Statistical Association*, **96**, 653–666.

Liang, F. and Wong, W.H. (2001b) Evolutionary Monte Carlo for protein folding simulations, *Journal of Chemical Physics*, **115**, 3374–3380.

Liang, F. and Zhang, J. (2008) Estimating FDR under general dependence using stochastic approximation. *Biometrika*, **95**, 961–977.

Liang, F. and Zhang, J. (2009) Learning Bayesian networks for discrete data. *Computational Statistics and Data Analysis*, **53**, 865–876.

Lin, S. and Kernighan, B.W. (1973) An effective heuristic algorithm for the traveling salesman problem. *Operations Research*, **21**, 498–516.

Little, R.J.A. and Rubin, D.B. (1987) *Statistical Analysis with Missing Data,* John Wiley, Inc., New York.

Liu, C. (1993) Bartlett's decomposition of the posterior distribution of the covariance for normal monotone ignorable missing data, *Journal of Multivariate Analysis*, **46** (2), 198–206.

Liu, C. (1995) Monotone data augmentation using the multivariate t distribution, *Journal of Multivariate Analysis*, **53** (1), 139–158.

Liu, C. (1996) Bayesian robust multivariate linear regression with incomplete data, *Journal of the American Statistical Association*, **91**, 1219–1227.

Liu, C. (1997) ML estimation of the multivariate t distribution and the EM algorithms, *Journal of Multivariate Analysis*, **63** (2), 296–312.

Liu, C. (2003) Alternating subspace-spanning resampling to accelerate Markov chain Monte Carlo simulation, *Journal of the American Statistical Association*, **98**, 110–117.

Liu, C. (2004) Robit regression: a simple robust alternative to logistic regression and probit regression, in *Missing Data and Bayesian Methods in Practice* (eds. A. Gelman and X. Meng), Chapman & Hall, London.

Liu, C. and Rubin, D.B. (1994) The ECME algorithm: a simple extension of EM and ECM with faster monotone convergence, *Biometrika*, **81**, 633–648.

Liu, C. and Rubin, D.B. (1995) ML estimation of the multivariate t distribution, *Statistica Sinica*, **5**, 19–39.

Liu, C. and Rubin, D.B. (1996). Markov-normal analysis of iterative simulations before their convergence, *Econometrics*, **75**, 69–78.

Liu, C. and Rubin, D.B. (1998) Ellipsoidally symmetric extensions of the general location model for mixed categorical and continuous data, *Biometrika*, **85**, 673–688.

Liu, C. and Rubin, D.B. (2002) Model-based analysis to improve the performance of iterative simulations, *Statistica Sinica*, **12**, 751–767.

Liu, C. Rubin, D.B., and Wu, Y. (1998) Parameter expansion for EM acceleration – the PX-EM algorithm, *Biometrika* **85**, 655–770.

Liu, C. and Sun, D.X. (2000) Analysis of interval-censored data from fractional experiments using covariance adjustment, *Technometrics*, **42**, 353–365.

Liu, J.S. (1994) The collapsed Gibbs sampler with applications to a gene regulation problem. *Journal of the American Statistical Association*, **89**, 958–966.

Liu, J.S. (1996) Peskun's theorem and a modified discrete-state Gibbs sampler. *Biometrika*, **83**, 681–682.

Liu, J.S. (2001) *Monte Carlo Strategies in Scientific Computing,* Springer, New York.

Liu, J.S., and Chen, R. (1998) Sequential Monte Carlo methods for dynamic systems. *Journal of the American Statistical Association*, **93**, 1032–1044.

Liu, J.S., Chen, R., and Wong, W.H. (1998) Rejection control and sequential importance sampling. *Journal of the American Statistical Association*, **93**, 1022–1031.

Liu, J.S., Liang, F., and Wong, W.H. (2000) The use of multiple-try method and local optimization in Metropolis sampling. *Journal of the American Statistical Association*, **94**, 121–134.

Liu, J.S., Liang, F., and Wong, W.H. (2001) A theory for dynamic weighting in Monte Carlo. *Journal of the American Statistical Association*, **96**, 561–573.

Liu, J.S. and Sabatti, C. (2000) Generalized Gibbs sampler and multigrid Monte Carlo for Bayesian computation. *Biometrika*, **87**, 353–369.

Liu, J.S., Wong, W.H., and Kong, A. (1994) Covariance structure of the Gibbs sampler with applications to the comparisons of estimators and augmentation schemes. *Biometrika*, **81**, 27–40.

Liu, J.S., and Wu, Y.N. (1999) Parameter expansion for data augmentation. *Journal of the American Statistical Association*, **94**, 1264–1274.

Liu, W. and Yang, W. (1996) An extension of Shannon-McMillan theorem and some limit properties for nonhomogeneous Markov chains. *Stochastic Processes and their Applications*, **61**, 129–145.

Loader, C. (1999) *Local Regression and Likelihood,* Springer, New York.

Loftsgaarden, P.O. and Quesenberry, C.P. (1965) A nonparametric estimate of a multivariate probability density function. *Annals of Mathematical Statistics*, **28**, 1049–1051.

MacEachern, S.N., Clyde, M., and Liu, J.S. (1999) Sequential importance sampling for non-parametric Bayes models: the next generation. *Canadian Journal of Statistics*, **27**, 251–267.

Maddison, D.R. (1991) The discovery and importance of multiple islands of most parsimonious trees. *Systematic Zoology*, **40**, 315–328.

Madigan, D. and Raftery, E. (1994) Model selection and accounting for model uncertainty in graphical models using Occam's window. *Journal of the American Statistical Association*, **89**, 1535–1546.

Madigan, D. and York, J. (1995) Bayesian graphical models for discrete data. *International Statistical Review*, **63**, 215–232.

Mallows, C.L. (1973) Some comments on *Cp*. *Technometrics*, **15**, 661–676.

Mandroli, S.S., Shrivastava, A.K., and Ding, Y. (2006) A survey of inspection strategy and sensor distribution studies in discrete-part manufacturing processes. *IIE Transactions*, **38**, 309–328.

Marinari, E. and Parisi, G. (1992) Simulated tempering: a new Monte Carlo scheme. *Europhysics Letters*, **19** (6), 451–458.

Marske, D. (1967) Biomedical oxygen demand data interpretation using sums of squares surface. Unpublished Master's thesis, University of Wisconsin.

Martin, R., Zhang, J., and Liu, C. (2009) Situation-specific inference using Dempster-Shafer theory. Preprint.

Martinez, J.N., Liang, F., Zhou, L., *et al.* (2009) Longitudinal functional principal component modeling via stochastic approximation Monte Carlo. *Canadian Journal of Statistics*; in press.

Matsumoto, M. and Nishimura, T. (1998) Mersenne twister: a 623-dimensionally equidistributed uniform pseudo-random number generator. *ACM Transactions on Modeling and Computer Simulation*, **8** 3.

Matthews, P. (1993) A slowly mixing Markov chain and its implication for Gibbs sampling. *Statistics and Probability Letters*, **17**, 231–236.

Mau, B. (1996) Bayesian phylogenetic inference via Markov chain Monte Carlo methods. Ph.D. thesis. University of Wisconsin-Madison, Department of Statistics.

Mau, B. and Newton, M.A. (1997) Phylogenetic inference for binary data on dendrograms using Markov chain Monte Carlo. *Journal of Computational and Graphical Statistics*, **6**, 122–131.

Mau, B., Newton, M.A., and Larget, B. (1999) Bayesian phylogenetic inference via Markov chain Monte Carlo. *Biometrics*, **55**, 1–12.

McCulloch, C.E., Searle, S.R., and Neuhaus, J.M. (2008) *Generalized, Linear, and Mixed Models,* 2nd edn, John Wiley & Sons, Hoboken.

Meng, X.L. and Rubin, D.B. (1993) Maximum likelihood estimation via the ECM algorithm: a general framework. *Biometrika*, **80**, 267–278.

Meng, X.L. and van Dyk, D.A. (1997) The EM algorithm – an old folk song sung to a fast new tune. *Journal of the Royal Statistical Society, Series B*, **59**, 511–567.

Meng, X.L. and van Dyk, D. (1999) Seeking efficient data augmentation schemes via conditional and marginal augmentation, *Biometrika*, **86**, 301–320.

Meng, X.L. and Wong, W.H. (1996) Simulating ratios of normalizing constants via a simple identity: a theoretical exploration. *Statistica Sinica*, **6**, 831–860.

Mengersen, K.L., Robert, C.P., and Guihenneuc-Jouyaux, C. (1999) MCMC convergence diagnostics: a 'reviewww', in *Bayesian Statistics 6* (eds J. Berger, J. Bernardo, A.P. Dawid and A.F.M. Smith), 415–440. Oxford Sciences Publications, Oxford.

Mengerson, K.L. and Tweedie, R.L. (1996) Rates of convergence of the Hastings and Metropolis algorithms. *Annals of Statistics*, **24**, 101–121.

Metropolis N., Rosenbluth A.W., Rosenbluth M.N., *et al.* (1953) Equation of state calculations by fast computing machines. *Journal of Chemical Physics*, **21**, 1087–1091.

Metropolis, N. and Ulam, S. (1949) The Monte Carlo method. *Journal of the American Statistical Association*, **44**, 335–341.

Meyn, S.P. and Tweedie, R.L. (1993) *Markov chains and Stochastic Stability,* Springer, New York.

Michalewicz, Z. and Nazhiyath, G. (1995) Genocop III: A co-evolutionary algorithm for numerical optimization problems with nonlinear constraints, in *Proceedings of the Second IEEE ICEC*, Perth, Australia.

Mira, A. and Tierney, L. (2002) Efficiency and convergence properties of slice samplers. *Scandinavian Journal of Statistics*, **29**, 1–12.

Mo, Q. and Liang, F. (2009) Bayesian modeling of ChIP–chip data through a high-order Ising model. *Biometrics*, in press.

Møller, J., Pettitt, A.N., Reeves, R., and Berthelsen, K.K. (2006) An efficient Markov chain Monte Carlo method for distributions with intractable normalizing constants. *Biometrika*, **93**, 451–458.

Mudholkar, G.S. and George, E.O. (1978) A remark on the shape of the logistic distribution. *Biometrika*, **65**, 667–668.

Müller, H.G. and Stadtmüller, U. (1987) Variable bandwidth kernel estimators of regression curves. *Annals of Statistics*, **15**, 182–201.

Müller, P. (1991) A generic approach to posterior integration and Gibbs sampling. Technical Report, Purdue University, West Lafayette IN.

Müller, P. (1993) Alternatives to the Gibbs sampling scheme. Technical Report, Institute of Statistics and Decision Sciences, Duke University.

Munch, K., Gardner, P.P., Arctander, P., *et al.* (2006) A hidden Markov model approach for determining expression from genomic tiling microarrays. *BMC Bioinformatics*, **7**, 239.

Murdoch, D.J. and Green, P.J. (1998) Exact sampling from a continuous state space. *Scandinavian Journal of Statistics*, **25**, 483–502.

Murray, I., Ghahramani, Z., and MacKay, D.J.C. (2006) MCMC for doubly-intractable distributions, in *Proceedings of the 22nd Annual Conference on Uncertainty in Artificial Intelligence (UAI).*

Mykland, P., Tierney, L., and Yu, B. (1995) Regeneration in Markov chain samplers. *Journal of the American Statistical Association*, **90**, 233–241.

Nahar, S., Sahni, S., and Shragowitz, E. (1989) Simulated annealing and combinational optimization. *International Journal of Computer Aided VLSI Design*, **1**, 1–23.

Neal, R.M. (1996) *Bayesian Learning for Neural Networks.* Springer, New York.

Neal, R.M. (2001) Annealed importance sampling. *Statistics and Computing*, **11**, 125–139.

Neal, R.M. (2003) Slice sampling. *Annals of Statistics*, **31**, 705–767.

Neirotti, J.P., Freeman, D.L., and Doll, J.D. (2000) Approach to ergodicity in Monte Carlo simulations. *Physical Review E*, **62**, 7445–7461.

Nevel'son, M.B. and Has'minskiĭ, R.Z. (1973) *Stochastic Approximation and Recursive Estimation,* the American Mathematical Society, Rhode Island.

Newton, M.A., Mau, B., and Larget, B. (1999) Markov chain Monte Carlo for the Bayesian analysis of evolutionary trees from aligned molecular sequences, in *Statistics in Molecular Biology and Genetics* (ed. F. Seillier-Moseiwitch) IMS Lecture Notes Monograph Series, vol. **3**, pp. 143–162.

Niedermayer, F. (1988) General Cluster Updating Method for Monte Carlo Simulations. *Physical Review Letters*, **61**, 2026–2029.

Nummelin, E. (1978). A splitting technique for Harris recurrent Markov chains. *Zeitschrift für Wahrscheinlichkeitstheorie und Vervandte Gebiete*, **43**, 309–318.

Nummelin, E. (1984) *General Irreducible Markov Chains and Non-Negative Operators,* Cambridge University Press, Cambridge.

Onsager, L. (1949) Statistical hydrodynamics. *Il Nuovo Cimento* (supplement), **6**, 279–287.

Owen, C.B. and Abunawass, A.M. (1993) Applications of simulated annealing to the back-propagation model improves convergence. *SPIE Proceedings*, *1966*, 269–276.

Park, Y.R., Murray, T.J., and Chen, C. (1996) Predicting sunspots using a layered perceptron neural network. *IEEE Transactions on Neural Networks*, **7**, 501–505.

Patton, A.L., Punch III, W.F. and Goodman, E.D. (1995) A standard GA approach to native protein conformation prediction, in *Proceedings of the Sixth International Conference on Genetic Algorithms*, pp. 574–581.

Pearl, J. (1988) *Probabilistic Reasoning in Intelligent Systems: Networks of Plausible Inference*, Morgan Kaufmann, San Mateo.

Pearson, T.A. and Manolio, T.A. (2008) How to interpret a genome-wide association study. *Journal of the American Medical Association*, **299**, 1335–1344.

Pelletier, M. (2000) Asymptotic almost sure efficiency of averaged stochastic algorithms. *SIAM Journal of Control and Optimization*, **39**, 49–72.

Pérez, C.J., Martí, J., Rojano, C., *et al.* (2008) Efficient generation of random vectors by using the ratio-of-uniforms method with ellipsoidal envelopes. Statistics and Computing **18**, 209–217.

Penny, W.D. and Roberts, S.J. (2000) Bayesian methods for autoregressive models. *Proceedings of Neural Networks for Signal Processing*, Sydney.

Perrone, M.P. (1993) Improving regression estimation: averaging methods for variance reduction with extension, to general convex measure optimization. Ph.D. thesis, Brown University, Rhode Island.

Peskun, P.H. (1973) Optimal Monte-Carlo sampling using Markov chains. *Biometrika*, **60**, 607–612.

Pinheiro, J.C., Liu, C., and Wu, Y. (2001) Efficient algorithms for robust estimation in linear mixed-effects models using the multivariate t-distribution, *Journal of Computational and Graphical Statistics*, **10**, 249–276.

Plummer, M., Best, N., Cowles, K., *et al.* (2006) CODA: output analysis and diagnostics for Markov Chain Monte Carlo simulations. `http://www.cran-R-project.org`.

Polyak, B.T. (1990) New stochastic approximation type procedures. (In Russian.) *Avtomat. i Telemekh*, **7**, 98–107.

Polyak, B. and Juditsky, A.B. (1992) Acceleration of stochastic approximation by averaging. *SIAM Journal of Control and Optimization*, **30**, 838–855.

Preisler, H.K. (1993) Modelling spatial patterns of trees attacked by barkbettles. *Applied Statistics*, **42**, 501–514.

Press, W.H., Teukolsky, S.A., Vetterling, W.T., and Flannery, B.P. (1992) *Numerical Recipes in C,* 2nd edn, Cambridge University Press, Cambridge.

Propp, J.G. and Wilson, D.B. (1996) Exact sampling with coupled Markov chains and applications to statistical mechanics. *Random Structures and Algorithms*, **9**, 223–252.

Qi, Y., Rolfe, A., MacIsaac, K.D., *et al.* (2006) High-resolution computational models of genome binding events. *Nature Biotechnology*, **24** (8), 963–970.

Rabiner, L.R. (1989) A tutorial on hidden Markov models and selected applications in speech recognition. *Proceedings of the IEEE*, **77**, 257–286.

Raftery, A.E., Madigan, D., and Hoeting, J.A. (1997) Bayesian model averaging for linear regression models. *Journal of the American Statistical Association*, **92**, 179–191.

Rannala, B. and Yang, Z. (1996) Probability distribution of molecular evolutionary trees: a new method of phylogenetic inference. *Journal of Molecular Evolution*, **43**, 304–311.

Reinelt, G. (1994) *The Traveling Salesman: Computational Solutions for TSP Applications,* Springer, New York.

Reiss, D.J., Facciotti, M.T., and Baliga, N.S. (2008) Model-based deconvolution of genome-wide DNA binding. *Bioinformatics*, **24** (3), 396–403.

Ren, Y., Ding, Y., and Liang, F. (2008) Adaptive evolutionary Monte Carlo algorithm for optimization with applications to sensor placement problems. *Statistics and Computing*, **18**, 375–390.

Resnick, M.D., Bearman, P.S., Blum, R.W., *et al.* (1997) Protecting adolescents from harm. Findings from the National Longitudinal Study on Adolescent Health. *Journal of the American Medical Association*, **278**, 823–832.

Revuz, D. (1975) *Markov Chains*. North-Holland, Amsterdam.

Rice, J.A. (2007) *Mathematical Statistics and Data Analysis,* 3rd edn, Brooks/Cole, Florence, KY.

Richardson, S. and Green, P. (1997) On Bayesian analysis of mixtures with an unknown number of components. *Journal of the Royal Statistical Society, Series B*, **59**, 731–792.

Riggan, W.B., Creason, J.P., Nelson, W.C., *et al.* (1987) *U.S. Cancer Mortality Rates and Trends, 1950–1979: Maps* (vol. IV). Government Printing Office, Washington, DC.

Ripley, B.D. (1987) *Stochastic Simulation,* John Wiley & Sons, Inc., New York.

Ritter, C. and Tanner, M.A. (1992) Facilitating the Gibbs sampler: the Gibbs stopper and the griddy-Gibbs sampler. *Journal of the American Statistical Association*, **87**, 861–868.

Robbins, H. and Monro, S. (1951) A stochastic approximation method. *Annals of Mathematical Statistics*, **22**, 400–407.

Robert, C.P. and Casella, G. (2004) *Monte Carlo Statistical Methods* (2nd edition) Springer, New York.

Roberts, G.O. (1996) Markov chain concepts related to sampling algorithms, in *Markov Chain Monte Carlo in Practice,* (eds. W.R. Gilks, S. Richardson, and D.J. Spiegelhalter), Chapman & Hall, London, pp. 45–57.

Roberts, G.O. and Gilks, W.R. (1994) Convergence of adaptive direction sampling. *Journal of Multivariate Analysis*, **49**, 287–298.

Roberts, G.O. and Rosenthal, J.S. (1998) Optimal scaling of discrete approximations to Langevin diffusions. *Journal of the Royal Statistical Society, Series B*, **60**, 255–268.

Roberts, G.O. and Rosenthal, J.S. (1999) Convergence of slice sampler Markov chains. *Journal of the Royal Statistical Society, Series B*, **61**, 643–660.

Roberts, G.O. and Rosenthal, J.S. (2001) Optimal scaling for various Metropolis-Hastings algorithms. *Statistical Science*, **16**, 351–367.

Roberts, G.O. and Rosenthal, J.S. (2007) Coupling and ergodicity of adaptive Markov chain Monte Carlo algorithms. *Journal of Applied Probability*, **44**, 458–475.

Roberts, G.O. and Sahu, S.K. (1997) Updating schemes, correlation structure, blocking and parameterization for the Gibbs sampler. *Journal of the Royal Statistical Society, Series B*, **59**, 291–317.

Roberts, G.O. and Stramer, O. (2002) Langevin diffusions and Metropolis-Hastings algorithms. *Methodology and Computing in Applied Probability*, **4**, 337–357.

Roberts, G.O. and Tweedie, R.L. (1996) Geometric convergence and central limit theorems for multidimensional Hastings and Metropolis algorithms. *Biometrika*, **83**, 95–110.

Robins, G., Snijers, T., Wang, P., *et al.* (2007) Recent development in exponential random graph models for social networks. *Social Networks*, **29**, 192–215.

Rosenthal, J.S. (1995) Minorization conditions and convergence rate for Markov chain Monte Carlo. *Journal of the American Statistical Association*, **90**, 558–566.

Ross, S. (1998) *A First Course in Probability,* 5th edn, Prentice-Hall, Inc., Saddle River, NJ.

Rossier, Y., Troyon, M., and Liebling, T.M. (1986) Probabilistic exchange algorithms and Euclidean traveling salesman problems. *OR Spektrum*, **8** (3), 151–164.

Rubin, D.B. (1984) Bayesianly justifiable and relevant frequency calculations for the applied statistician. *Annals of Statistics*, **12**, 1151–1172.

Rubin, D.B. and Schafer, J.L. (1990) Efficiently creating multiple imputations for incomplete multivariate normal data. *ASA 1990 Proceedings of the Statistical Computing Section*, 83–88.

Rubinstein, R.Y. (1981) *Simulation and the Monte Carlo Method*. John Wiley & Sons, Inc., New York.

Rue, H. and Held, L. (2005), *Gaussian Markov Random Fields: Theory and Applications*. Chapman & Hall, London.

Rumelhart, D.E., Hinton, G.E., and Williams, R.J. (1986) Learning internal representations by back-propagating errors, in *Parallel Distributed Processing: Explorations in the Microstructure of Cognition,* vol. 1 (ed., D.E. Ruppert, and J.L. MacLelland. MIT Press, pp. 318–362.

Ruppert, D. (1988) Efficient estimation from a slowly convergent Robbins-Monro procedure. Technical Report 781, School of Operations Research and Industrial Engineering, Cornell University.

Ryu, D., Liang, F. and Mallick, B.K. (2009) A dynamically weighted particle filter with radial basis function networks to sea surface temperature modeling. Technical Report, Department of Statistics, Texas A&M University.

Rzhetsky, A. and Nei, M. (1992) A simple method for estimating and testing minimum evolution trees. *Molecular Biology and Evolution*, **9**, 945–967.

Sahu, S.K. and Zhigljavsky, A.A. (2003) Self-regenerative Markov chain Monte Carlo with adaptation. *Bernoulli*, **9**, 395–422.

Saitou, N. and Nei, M. (1987) The neighbor-joining method: a new method for reconstructing phylogenetic trees. *Molecular Biology and Evolution*, **4**, 406–425.

Sali, A., Shahknovich, E., and Karplus, M. (1994) How does a protein fold? *Nature*, **369**, 248–251.

Salter, L.A. and Pearl, D.K. (2001) Stochastic search strategy for estimation of maximum likelihood phylogenetic trees. *Systematic Biology*, **50**, 7–17.

Savage, L.J. (1967a) Difficulties in the theory of personal probability. *Philosophy of Science*, **34**, 305–310. Reprinted in translation in *Mathématiques et Sciences Humaines*, **21**, 1968, 5–9.

Savage, L.J. (1967b) Implication of personal probability for induction. *Journal of Philosophy*, **64**, 593–607.

Schafer, J.L. (1997) *Analysis of Incomplete Multivariate Data*. Chapman & Hall, London.

Schmitt, L.M. (2001) Theory of genetic algorithms. *Theoretical Computer Science*, **259**, 1–61.

Shanno, D.F. (1970) Conditioning of quasi-Newton methods for function minimization. Mathematics of Computation, **24**, 647–656.

Serfling, R.J. (1970) Convergence properties of *Sn* under moment restrictions. *Annals of Mathematical Statistics*, **41**, 1235–1248.

Shafer, G. (1982) Lindley's paradox (with discussion). *Journal of the American Statistical Association*, **77**, 325–351.

Sherman, M., Apanasovich, T.V., and Carroll, R.J. (2006) On estimation in binary autologistic spatial models. *Journal of Statistical Computation and Simulation*, **76**, 167–179.

Shi, J., Siegmund, D. and Yakir, B. (2007) Importance sampling for estimating p-values in linkage analysis. *Journal of the American Statistical Association*, **102**, 929–937.

Simonoff, J.S. (1998) Three sides of smoothing: categorical data smoothing, nonparametric regression, and density estimation. *International Statistical Review*, **66**, 137–156.

Smith, A.F.M., and Roberts, G.O. (1993) Bayesian computation via the Gibbs sampler and related Markov chain Monte Carlo methods. *Journal of the Royal Statistical Society, Series B*, **55**, 3–23.

Smith, R.L. (1980) A Monte Carlo procedure for the random generation of feasible solutions to mathematical programming problems. *Bulletin of the TIMS/ORSA Joint National Meeting*, Washington, DC, p. 101.

Smith, R.L. (1984) Efficient Monte Carlo procedures for generating points uniformly distributed over bounded regions. *Operations Research*, **32**, 1297–1308.

Snijders, T.A.B. (2002) Markov chain Monte Carlo estimation of exponential random graph models. *Journal of Social Structure*, **3** (2), 64.

Snijders, T.A.B., Pattison, P.E., Robins, G.J., *et al.* (2006) New specifications for exponential random graph models. *Sociological Methodology*, 99–153.

Spirtes, P., Glymour, C., and Scheines, R. (1993) *Causation, Prediction and Search*, Springer-Verlag, New York.

Stadlober, E. (1989). Ratio of uniforms as a convenient method for sampling from classical discrete distributions. *Proceedings of the 21st conference on Winter simulation*, Washington, DC, pp. 484–489.

Stephens, M. (2000) Bayesian analysis of mixture models with an unknown number of components: an alternative to reversible jump methods. *Annals of Statistics*, **28**, 40–74.

Stramer, O. and Tweedie, R.L. (1999a) Langevin-type models I: diffusions with given stationary distributions and their discretizations. *Methodology and Computing in Applied Probability*, **1**, 283–306.

Stramer, O. and Tweedie, R.L. (1999b) Langevin-type models II: self-targeting candidates for MCMC algorithms. *Methodology and Computing in Applied Probability*, **1**, 307–328.

Strauss, D. and Ikeda, M. (1990) Pseudolikelihood estimation for social networks. *Journal of the American Statistical Association*, **85**, 204–212.

Swendsen, R.H. and Wang, J.S. (1987) Nonuniversal critical dynamics in Monte Carlo simulations. *Physical Review Letters*, **58**, 86–88.

Tadić, V. (1997) On the convergence of stochastic iterative algorithms and their applications to machine learning. A short version of this paper was published in the *Proceedings of the 36th Conference on Decision and Control*, San Diego, pp. 2281–2286.

Taguchi, G. (1987) *System of Experimental Design*. Unipub/Kraus International Publications, White Plains, NY.

Tan, C.M. (2008) *Simulated Annealing*, In-Tech. `http://www.intechweb.org`.

Tanizaki, H. (2008) A simple gamma random number generator for arbitrary shape parameters. *Economics Bulletin*, **3**, 1–10.

Tang, Q.Y., L'Ecuyer, P., and Chen, H.F. (1999) Asymptotic efficiency of perturbation-analysis-based stochastic approximation with averaging. *SIAM Journal of Control and Optimization*, **37**, 1822–1847.

Tanner, M.A. and Wong, W.H. (1987) The calculation of posterior distributions by data augmentation. *Journal of the American Statistical Association*, **82**, 528–550.

Taroni, F., Bozza, S., and Biedermann, A. (2006) Two items of evidence, no putative source: an inference problem in forensic intelligence. *Journal of Forensic Sciences*, **51**, 1350–1361.

Terrell, G.R. and Scott, D.W. (1992) Variable kernel density estimation. *Annals of Statistics*, **20**, 1236–1265.

Tibshirani, R. and Hastie, T. (1987) Local likelihood estimation. *Journal of the American Statistical Association*, **82**, 559–568.

Tierney, L. (1994) Markov chains for exploring posterior distributions (with discussion) *Annals of Statistics*, **22**, 1701–1762.

Titterington, D.M. (1984) Recursive parameter estimation using incomplete data. *Journal of the Royal Statistical Society, Series B*, **46**, 257–267.

Tong, H. (1990) *Non-Linear Time Series: A Dynamical System Approach,* Oxford University Press, Oxford.

Tong, H. and Lim, K.S. (1980) Threshold autoregression, limit cycles and cyclical data. *Journal of the Royal Statistical Society, Series B*, **42**, 245–292.

Torrie, G.M. and Valleau, J.P. (1977) Nonphysical Sampling Distribution in Monte Carlo Free-energy Estimation: Umbrella Sampling. Journal of Computational Physics, **23**, 187–199.

Udry, J.R. and Bearman, P.S. (1998) New methods for new research on adolescent sexual behavior, in *New Perspectives on Adolescent Risk Behavior* (ed. R. Jessor), Cambridge University Press, New York, pp. 241–269.

Unger, R. and Moult, J. (1993) Genetic algorithms for protein folding simulations. *Journal of Molecular Biology*, **231**, 75–81.

van Dyk, D.A. and Meng, X.L. (2001) The art of data augmentation. *Journal of Computational and Graphical Statistics*, **10**, 1–111.

Venables, W.N. and Ripley, B.D. (1999) *Modern Applied Statistics with S-PLUS*, 3rd edn, Springer, New York.

Verdinelli, I. and Wasserman, L. (1995) Computing Bayes factors using a generalization of the Savage-Dickey density ratio. *Journal of the American Statistical Association*, **90**, 614–618.

Wakefield, J.C., Gelfand, A.E., and Smith, A.F.M. (1991) Efficient generation of random variates via the ratio-of-uniforms method. Stat. Comput., **1**, 129–133.

Waldmeirer, M. (1961) *The Sunspot Activity in the Years 1610–1960*. Schulthess, Zürich.

Wallace, C.S. and Korb, K.B. (1999) Learning linear causal models by MML sampling, in *Causal Models and Intelligent Data Management* (ed. A. Gammerman) Springer-Verlag, Heidelberg.

Wand, M.P. and Jones, M.C. (1995) *Kernel Smoothing,* Chapman & Hall, London.

Wang, F. and Landau, D.P. (2001) Efficient, multiple-range random walk algorithm to calculate the density of states. *Physical Review Letters*, **86**, 2050–2053.

Wang, I.J., Chong, E.K.P., and Kulkarni, S.R. (1997) Weighted averaging and stochastic approximation. *Mathematics of control, Signals, and Systems*, **10**, 41–60.

Wang, Y. (1995) Jump and sharp cusp detection by wavelets. *Biometrika*, **82**, 385–397.

Warnes, G.R. (2001) The normal kernel coupler: an adaptive Markov chain Monte Carlo method for efficiently sampling from multimodal distributions. Technical Report No. 39, Department of Statistics, University of Washington.

Wei, G. and Tanner, M. (1990), A Monte Carlo Implementation of the EM Algorithm and the Poor Man's Data Augmentation Algorithm. *Journal of the American Statistical Association*, **85**, 699–704.

Weigend, A.S., Huberman, B.A., and Rumelhart, D.E. (1990) Predicting the future: a connectionist approach. International Journal of Neural Systems, **1**, 193–209.

Weisberg, S. (1985) *Applied Linear Regression,* John Wiley & Sons, Inc., New York.

Welch, B.L. and Peers, H.W. (1963) On formulae for confidence points based on integrals of weighted likelihoods. *Journal of the Royal Statistical Society, Series B*, **25**, 318–329.

Wermuth, N. and Lauritzen, S. (1983) Graphical and recursive models for contingency tables. *Biometrika*, **72**, 537–552.

Wolff, U. (1989) Collective Monte Carlo updating for spin systems. *Physical Review Letters*, **62**, 361–364.

Wong, D.F., Leong, H.W., and Liu, C.L. (1988) *Simulated Annealing for VLSI Design*. Kluwer Academic, Boston.

Wong, W.H. and Liang, F. (1997) Dynamic weighting in Monte Carlo and optimization. *Proceedings of the National Academy of Sciences USA*, **94**, 14220–14224.

Wright, A.H. (1991) Genetic algorithms for real parameter optimization, in *Foundations of Genetic Algorithms* (ed. G.J.E. Rawlins), Morgan Kaufmann, San Mateo, CA:, pp. 205–218.

Wu, H. and Huffer, F.W. (1997) Modeling the distribution of plant species using the autologistic regression model. *Ecological Statistics*, **4**, 49–64.

Wu, M., Liang, F., and Tian, Y. (2009) Bayesian modeling of ChIP-chip data using latent variables. *BMC Bioinformatics*, **10**, 352.

Yan, Q. and de Pablo, J.J. (2000) Hyperparallel tempering Monte Carlo simulation of polymeric systems. *Journal of Chemical Physics*, **113**, 1276–1282.

Yan, Q. and de Pablo, J.J. (2003) Fast calculation of the density of states of a fluid by Monte Carlo simulations. *Physical Review Letters*, **90**, 035701.

Yang, C. (2007) Recurrent and ergodicity properties of adaptive MCMC. Technical Report, Department of Mathematics, University of Toronto.

Yang, C.N. (1952) The spontaneous magnetization of a 2-dimensional Ising model. *Physical Review Letters*, **85**, 808–815.

Yu, B. (1993) Density Estimation in the L^∞ Norm for Dependent Data, with Applications to the Gibbs Sampler. *Annals of Statistics*, **21**, 711–735.

Yu, K., and Liang, F. (2009) Efficient p-value evaluation for resampling-based tests. Technical Report, Division of Cancer Epidemiology and Genetics, NCI.

Yu, Y. and Meng, X.L. (2008) Espousing classical statistics with modern computation: sufficiency, ancillarity and an interweaving generation of MCMC. Technical Report, Department of Statistics, University of California, Irvine, CA.

Zhang, J., Li, J., and Liu, C. (2009) Robust factor analysis. Technical Report, Department of Statistics, Purdue University.

Zhang, X., Clarenz, O., Cokus, S., *et al.* (2007) Whole-genome analysis of histone H3 lysine 27 trimethylation in arabidopsis. *PLoS Biology*, **5** (5), 129.

Zhang, X., Yazakij, J., Sundaresan, A., *et al.* (2006) Genome-wide high-resolution mapping and functional analysis of DNA methylation in arabidopsis. *Cell*, **126**, 1189–1201.

Zheng, M., Barrera, L.O., Ren, B., *et al.* (2007) ChIP-chip: data, model, and analysis. *Biometrics*, **63** (3), 787–796.

Index

Advanced Markov Chain Monte Carlo Methods: Learning from Past Samples
Faming Liang, Chuanhai Liu and Raymond J. Carroll

Printed and bound by CPI Group (UK) Ltd, Croydon, CR0 4YY

16/04/2025

14658497-0001